Climate Change and Neglected Underutilized Crops

Mukhtar Ahmed
Editor

Climate Change and Neglected Underutilized Crops

Foundations, Cereals, and Emerging Legumes
Volume 1

Springer

Editor
Mukhtar Ahmed
Department of Agronomy
Pir Mehr Ali Shah Arid Agriculture
University
Rawalpindi, Pakistan

ISBN 978-3-032-21004-3 ISBN 978-3-032-21005-0 (eBook)
https://doi.org/10.1007/978-3-032-21005-0

This Springer imprint is published by the registered company Springer Nature Switzerland AG
The registered company address is: Gewerbestrasse 11, 6330 Cham, Switzerland

This volume is dedicated to farming communities worldwide, whose resilience sustains global food systems, and to researchers, practitioners, and policymakers working to advance sustainable and climate-resilient agriculture.

It is also dedicated to future generations, in the hope that the knowledge shared here contributes to more resilient, equitable, and food-secure systems.

Foreword

Climate change is reshaping agriculture and food systems at an unprecedented pace. Rising temperatures, changing rainfall patterns, increasing climate extremes, and declining soil health threaten the stability, productivity, and nutritional quality of global food supplies. These challenges are most acute for vulnerable communities that rely on smallholder and resource-constrained farming systems. Addressing these pressures requires innovative, evidence-based strategies that integrate resilience, sustainability, and food security.

Climate Change and Neglected Underutilized Crops: Foundations, Cereals, and Emerging Legumes Volume 1 provides a timely and comprehensive examination of the role of neglected and underutilized crops (NUCs) in responding to these challenges. By highlighting cereals, pseudocereals, millets, and emerging legumes, this volume demonstrates how these crops can contribute to climate adaptation, diversification of cropping systems, improved nutrition, and ecosystem services. The integration of scientific research, traditional knowledge, and policy perspectives ensures that the book is both theoretically robust and practically relevant.

The editors and contributing authors have brought together interdisciplinary expertise from agronomy, soil science, ecology, climate science, and socio-economics, producing a volume that serves multiple audiences. Researchers and students will find rigorous data, analyses, and methodological approaches. Practitioners and extension professionals will benefit from practical insights and strategies for implementing NUC-based systems. Policymakers and development agencies will gain evidence-based guidance for promoting resilient and sustainable food systems.

This volume is a significant contribution to the literature on climate-resilient agriculture. By focusing on underutilized crops that have historically received limited attention, it challenges conventional paradigms and opens new pathways for sustainable food production. It is my hope that this work will inspire continued research, inform policy, and support the adoption of innovative agricultural practices that strengthen resilience and ensure food security in the face of climate change.

Islamabad, Pakistan
January 2026

Dr. M. Aslam Gill
Ex-Commissioner Ministry of National Food, Security and Research Pakistan

Preface

Climate Change and Neglected Underutilized Crops: Foundations, Cereals, and Emerging Legumes Volume 1 examines the role of neglected and underutilized crops (NUCs) in addressing the challenges posed by climate change to agricultural production and food systems. Increasing temperatures, altered precipitation regimes, more frequent extreme events, and declining soil health are placing significant pressure on conventional cropping systems, with particularly severe consequences for smallholder and resource-constrained farming communities.

NUCs, including minor cereals, pseudocereals, millets, and food legumes, are characterized by their adaptability to marginal environments, low input requirements, and high nutritional value. Although these crops have long contributed to local food systems, they have received limited attention in mainstream agricultural research, policy frameworks, and value chains. This volume considers NUCs as an important component of diversified and climate-resilient agricultural systems, with potential contributions to food and nutrition security, ecosystem services, and sustainable resource use.

The book is organized into three interrelated parts. Part I provides the conceptual and contextual foundations, addressing the role of NUCs in global food security, biodiversity conservation, environmental protection, and the water–energy–food nexus under changing climatic conditions. Part II focuses on cereal and pseudocereal crops, including quinoa, amaranth, teff, buckwheat, triticale, and a range of millets, examining their adaptation to climate stress, resource-use efficiency, and contributions to soil health and sustainable production systems. Part III addresses emerging food legume crops, highlighting their roles in biological nitrogen fixation, soil fertility improvement, water-use efficiency, and integration into diversified cropping systems.

Across its chapters, the volume integrates agronomic, ecological, and systems-based perspectives, drawing on empirical studies, case analyses, and recent advances in climate-smart and agroecological approaches. Attention is given to both adaptation and mitigation dimensions, as well as to the technological, institutional, and policy contexts that influence the adoption and scaling of NUCs.

This edited volume brings together contributions from an international group of researchers representing a range of disciplines. It is intended for researchers, graduate students, extension professionals, development practitioners, and policymakers seeking a structured and evidence-based understanding of the potential of neglected and underutilized crops in climate-resilient food systems. By situating NUCs within contemporary debates on climate change and agricultural sustainability, this volume aims to contribute to informed research, practice, and policy development.

Rawalpindi, Pakistan
January 2026

Mukhtar Ahmed

Acknowledgments

The Editor sincerely acknowledges the contributions of all chapter authors whose scholarly work forms the foundation of this volume. Their expertise, commitment, and interdisciplinary perspectives have been essential in advancing knowledge on neglected and underutilized crops and their role in climate resilience, sustainable agriculture, and food security.

The Editor is grateful to the reviewers for their thoughtful evaluations and constructive suggestions, which greatly improved the scientific rigor, clarity, and coherence of the chapters. Appreciation is also extended to the contributing institutions and research organizations for providing an enabling academic and professional environment that supported the development of this book.

Special thanks are due to colleagues and collaborators who offered guidance, discussion, and encouragement during the preparation of this volume. Finally, the Editor expresses sincere gratitude to family members for their patience, understanding, and continued support throughout the completion of this work.

January 2026 Mukhtar Ahmed, Ph.D.

Contents

Contributors

Zameer Abbas Department of Agronomy, Pir Mehr Ali Shah Arid Agriculture University, Rawalpindi, Pakistan

Muhammad Abid UNESCO Chair On Knowledge Systems for IWRM Director Interdisciplinary Research, Islamabad, Pakistan;
Department of Mechanical Engineering, COMSATS University Islamabad, Islamabad, Pakistan;
Department of Mechanical Engineering, Faculty of Engineering, Islamic University of Madinah, Madinah, Saudi Arabia

Munir Ahmad Department of Plant Breeding and Genetics, Pir Mehr Ali Shah Arid Agriculture University, Rawalpindi, Pakistan

Shakeel Ahmad Faculty of Agricultural Sciences and Technology, Institute of Agronomy, Bahauddin Zakariya University, Multan, Pakistan

Mukhtar Ahmed Institute of Geo-Information and Earth Observation, Pir Mehr Ali Shah Arid Agriculture University, Rawalpindi, Pakistan;
Department of Agronomy, Pir Mehr Ali Shah Arid Agriculture University, Rawalpindi, Pakistan;
Swedish University of Agricultural Sciences, Umea, Sweden;
Department of Biological Systems Engineering, Washington State University, Pullman, USA

Murtaza Ali Department of Agronomy, Pir Mehr Ali Shah Arid Agriculture University, Rawalpindi, Pakistan

Adnan Arshad State-Key Laboratory Herbage Improvement and Grassland Agroecosystem, Key Laboratory of Grassland Livestock Industry Innovation, Ministry of Agriculture and Rural Affairs, Engineering Research Center of Grassland Industry, Ministry of Education, College of Pastoral Agriculture Science and Technology, Lanzhou University, Lanzhou, China;
Department of Climate-smart Agriculture, National Cotton R&D Center, Xinjiang Academy of Agricultural Sciences, Ürümqi, China

Muhammad Asif Department of Agronomy, Pir Mehr Ali Shah Arid Agriculture University, Rawalpindi, Pakistan

Umar Farooq Department of Agronomy, Pir Mehr Ali Shah Arid Agriculture University, Rawalpindi, Pakistan;
Institute of Geo-Information and Earth Observation, Pir Mehr Ali Shah Arid Agriculture University, Rawalpindi, Pakistan

Zohaa Fatima Department of Agronomy, Pir Mehr Ali Shah Arid Agriculture University, Rawalpindi, Pakistan

Wajahat Hussain Institute of Geo-Information and Earth Observation, Pir Mehr Ali Shah Arid Agriculture University, Rawalpindi, Pakistan;
Department of Agronomy, Pir Mehr Ali Shah Arid Agriculture University, Rawalpindi, Pakistan

Abdullah Jan Department of Agronomy, Pir Mehr Ali Shah Arid Agriculture University, Rawalpindi, Pakistan

Mustafa Kan Department of Agricultural Economics, Agricultural Faculty, Kirsehir Ahi Evran University, Kırşehir, Türkiye

Ahmed M. S. Kheir Julius Kühn Institute (JKI)—Federal Research Centre for Cultivated Plants, Institute for Strategies and Technology Assessment, Kleinmachnow, Germany;
Soils, Water and Environment Research Institute, Agricultural Research Center, Giza, Egypt

Amjad Malik Kenanow Bachelor of Education Program, Faculty of Education, University College of the North (UCN), Thompson, MB, Canada

Shahid Riaz Malik Crop Sciences Institute, National Agricultural Research Center, Pakistan Agricultural Research Council, Islamabad, Pakistan

Laiba Noor Department of Agronomy, Pir Mehr Ali Shah Arid Agriculture University, Rawalpindi, Pakistan

Muqaddas Noor Department of Agronomy, Pir Mehr Ali Shah Arid Agriculture University, Rawalpindi, Pakistan

P. V. Vara Prasad Department of Agronomy, Kansas State University, Manhattan, KS, USA

Muhammad Imtiaz Rashid Center of Excellence in Environmental Studies, Department of Environmental Sciences, King Abdulaziz University, Jeddah, Saudi Arabia

Aashir Sameen Department of Agronomy, Pir Mehr Ali Shah Arid Agriculture University, Rawalpindi, Pakistan

Farid Asif Shaheen Department of Entomology, Pir Mehr Ali Shah Arid Agriculture University, Rawalpindi, Pakistan

P. V. Vara Prasad Department of Agronomy, Kansas State University, Manhattan, KS, USA

Abbreviations

ABA	Abscisic Acid
AI/ML	Artificial Intelligence/Machine Learning
AI	Artificial Intelligence
AMF	Arbuscular Mycorrhizal Fungi
ANOVA	Analysis of Variance
APSIM	Agricultural Production Systems sIMulator
APX	Ascorbate Peroxidase
ATP	Adenosine Triphosphate
BBCH	Biologische Bundesanstalt, Bundessortenamt und Chemische Industrie (growth stage scale)
BCR	Benefit–Cost Ratio
BD	Bulk Density
BGB	Brown Ground Bean
BIOM	Biomass Organic Matter Pool (APSIM)
BM	Aboveground Biomass
Bombus spp.	Bumblebee Species
bZIP	Basic Leucine Zipper (transcription factor family)
C4/C_4	C4 Photosynthetic Pathway
CA	Conservation Agriculture
CAT	Catalase
CET	Cyclic Electron Transfer
CFU	Colony-Forming Units
CGDD	Cumulative Growing Degree Days
CI	Confidence Interval
CO_2	Carbon Dioxide
CO_2-eq ha^{-1}	Carbon Dioxide Equivalent per Hectare
CRISPR/Cas9	Clustered Regularly Interspaced Short Palindromic Repeats/Cas9
CROPGRO	Crop Growth Simulation Model
CSA	Central Statistical Agency (Ethiopia)

CV	Coefficient of Variation
CWSI	Crop Water Stress Index
DAS	Days After Sowing
DCGB	Deep-Cream Ground Bean
DM	Dry Matter
DNA	Deoxyribonucleic Acid
DNDC	Denitrification–Decomposition Model
DREB	Dehydration Responsive Element Binding
DSSAT	Decision Support System for Agrotechnology Transfer
EC	Electrical Conductivity
EE	Energy Efficiency
ETc	Seasonal Crop Evapotranspiration
EVI	Enhanced Vegetation Index
FAO	Food and Agriculture Organization of the United Nations
FAOSTAT	Food and Agriculture Organization Corporate Statistical Database
FOM	Fresh Organic Matter Pool
FRW	Fresh Weight
FYM	Farmyard Manure
GDD5	Growing Degree Days (base temperature 5 °C)
GDD	Growing Degree Days
GHG	Greenhouse Gas
GIS	Geographic Information System
GPAW	Gas Plasma-Activated Water
GR	Glutathione Reductase
GY	Grain Yield
H_2O_2	Hydrogen Peroxide
HSW	Hundred Seed Weight
Hymenoptera	Order of Bees, Wasps, and Ants
ILLPA	Agrarian Experimental Station ILLPA (Puno, Peru)
INFO	InfoCrop: A generic crop model
INIA	Instituto Nacional de Innovación Agraria
IoT	Internet of Things
IPCC	Intergovernmental Panel on Climate Change
IPM	Integrated Pest Management
ISSR	Inter-Simple Sequence Repeat
Kc	Crop Coefficient
KY	Yield Response Factor to Water Stress
LAI	Leaf Area Index
LCA	Life Cycle Assessment
LER	Land Equivalent Ratio
LGBG	Light Brown Ground Bean
LOESS	Locally Estimated Scatterplot Smoothing
LST	Land Surface Temperature
MAS	Marker-Assisted Selection

MDA	Malondialdehyde
MENA	Middle East and North Africa
MJ	Mega Joule
MODIS	Moderate Resolution Imaging Spectroradiometer
NDVI	Normalized Difference Vegetation Index
NDWI	Normalized Difference Water Index
N	Nitrogen
NPK	Nitrogen–Phosphorus–Potassium
NPQ	Non-Photochemical Quenching
NUCs	Neglected Underutilized Crops
NUE	Nitrogen Use Efficiency
O_2-	Superoxide Radical
PCA	Principal Component Analysis
PGPR	Plant Growth-Promoting Rhizobacteria
PLH	Plant Height
PNL	Panicle Length
POD	Peroxidase
PRI	Photochemical Reflectance Index
PSII	Photosystem II
QTL	Quantitative Trait Loci
RCBD	Randomized Complete Block Design
RCP	Representative Concentration Pathway
RDF	Recommended Dose of Fertilizer
RGBG	Reddish-Brown Ground Bean
RMSD	Root Mean Square Deviation
ROS	Reactive Oxygen Species
RUSLE	Revised Universal Soil Loss Equation
RWC	Relative Water Content
SAVI	Soil Adjusted Vegetation Index
SDY	Seed Yield
SEM	Standard Error of Mean
SNP	Single Nucleotide Polymorphism
SOC	Soil Organic Carbon
SOM	Soil Organic Matter
SPAC	Soil–Plant–Atmosphere Continuum
SPI	Standardized Precipitation Index
SSR	Simple Sequence Repeat
STY	Straw Yield
TF	Transcription Factor
TIE	Total Input Energy
TLC	Tiller Count
TOE	Total Output Energy
UAV	Unmanned Aerial Vehicle

UN	United Nations
VHI	Vegetation Health Index
WDI	Water Deficit Index
WRKY	WRKY Transcription Factor Family
WUE	Water Use Efficiency

Part I
Foundations and Context

Chapter 1
Role of Neglected and Underutilized Crops in Global Food Security, Biodiversity, and Climate Resilience

Mukhtar Ahmed, Ahmed M. S. Kheir, Shahid Riaz Malik, and P. V. Vara Prasad

Abstract This chapter analyses how neglected and underutilized crops (NUCs) have the potential to convert food systems to a changing climate, ecological, and nutrition environment. The study is motivated by the fact that the global agricultural sector has continued to be vulnerable because of overdependence on few staple crops that has contributed to environmental degradation, food staleness and increased risks of production with the advent of climate change. The chapter is based on a mixed-method approach to qualitative analysis and involves research synthesizing over 120 peer-reviewed articles, 32 cases of countries and 15 global policies to evaluate the adaptability nature evidences, socioeconomic advantages, and obstacles of NUCs scales up. The evaluation indicates that solutions such as NUCs have the potential to enhance the resilience of smallholders (25–40%) due to enhanced drought and heat tolerance, and low soil fertility, adding dietary diversity (30–60% greater vitamin and nutrient content than primary staples). The case studies in Africa, Asia and Latin America demonstrate yield stability gains of 20–35% when the case was under stress and incomes enhanced by 10–28% when the value chains were established. Nevertheless, its use remains not popular due to systemic barriers, such as poor seed systems,

M. Ahmed (✉)
Department of Agronomy, Pir Mehr Ali Shah Arid Agriculture University, Rawalpindi, Pakistan
e-mail: ahmadmukhtar@uaar.edu.pk

Swedish University of Agricultural Sciences, Umea, Sweden

Department of Biological Systems Engineering, Washington State University, Pullman, WA, USA

A. M. S. Kheir
Julius Kühn Institute (JKI)—Federal Research Centre for Cultivated Plants, Institute for Strategies and Technology Assessment, Kleinmachnow, Germany

Soils, Water and Environment Research Institute, Agricultural Research Center, Giza, Egypt

S. R. Malik
Crop Sciences Institute, National Agricultural Research Center, Pakistan Agricultural Research Council, Islamabad, Pakistan

P. V. Vara Prasad
Department of Agronomy, Kansas State University, Manhattan, KS, USA

M. Ahmed (ed.), *Climate Change and Neglected Underutilized Crops*,
https://doi.org/10.1007/978-3-032-21005-0_1

lack of research funding (a small part, 3–5% of crop research), and lack of brand awareness. The chapter finds that mainstreaming NUCs would need a concerted effort to address research, policy, and markets, and identifies genomics-assisted breeding, integration of climate-smarter agriculture, public–private collaboration and facilitating the global structure both through Food and Agriculture Organization (FAO), International Fund for Agricultural Development (IFAD), Consultative Group on International Agricultural Research (CGIAR), and the UN Food Systems Summit as priorities. The efforts to reinforce these pathways will prove highly beneficial in achieving Sustainable Development Goal (SDG) 2 by 2030, zero hunger due to the effective and healthy food systems.

Keywords Neglected and underutilized crops · Food security · Drought and heat tolerance · Dietary diversity

1.1 Introduction

The combination of climate change, demographic pressure, geopolitical instability, and economic shocks are causing an increased stress on global food systems. They all are threats to availability, accessibility, and stability of food supplies (Das et al., 2025). Even though there has been a significant advancement in crop breeding, biotechnology, irrigation, mechanization, and integration of markets, the global community remains far away in achieving Sustainable Development Goal 2 (Zero Hunger) (Zewdu et al., 2025). Approximately 690–783 million people were faced with undernourishment in 2023, and no remarkable improvement was observed in the years that followed (FAO, 2024). The vulnerabilities have risen especially in the low and middle-income countries where economic resilience is low and food system is vulnerable to external shocks (Bai et al., 2025; Das et al., 2025). The chapter is an analysis of the numerous pressures affecting global food systems by 2023–2025 using three fundamental concepts, which are (i) structural vulnerability, (ii) the spillovers of global crises, and (iii) the growing popularity of the idea of Neglected and Underutilized Crops (NUCs) as an alternative to climate crises. The argument that it puts forward is that the inclusion of NUCs in mainstream food systems is not only desirable, but necessary to overcome the present and future risks of food security.

There is also the extreme unprecedented stress in the global food systems (Mehrabi et al., 2022). The prevailing and persistent climatic change, surging population, degradation of resources and geopolitical insecurities continue to redefine the world agriculture in a way that endangers the productivity, nutritional security as well as stability of the farming societies (Yuan et al., 2024). Regardless of the advanced scientific breakthroughs in crop breeding and agronomic controls, global agriculture continues to be extremely reliant on three staple crops that include wheat, rice, and maize. These crops provide over 60% of the worldwide energy consumption (Raza et al., 2025).

This weak agricultural foundation makes them more susceptible to systems especially when the climatic shocks become severer and the environmental resources are dwindling. Population growth, climate change, and limited dietary diversity pose growing threats to food security. This predicament is exacerbated by a small number of staple crops with limited genetic diversity, which constrains their adaptability to diseases, pests, and environmental changes. Orphan crops, with greater biodiversity, nutritional value, and local adaptability, could contribute to overcoming these challenges (Fig. 1.1). Crop species that have been neglected and underused (NUCs) or orphan crops provide a good and untapped alternative approach to agricultural system diversification (Huang et al., 2025; Purba & Krishnaswamy, 2025). These consist of crops, including the local cereals and pulses, as well as root, tuber, and leafy crops, which have been sustaining rural lifestyles and the traditional diets over centuries. Yet, contemporary agricultural policies, commercialization demands and research interests have mostly prevented them to be replaced by a small range of high-yield staples. The new interest in NUCs across the globe indicates the expanded awareness of the ecological, nutritional, and cultural significance and their capabilities to enhance the resilience in the future climate conditions (Shafiq et al., 2025).

1.2 Global Food Security Challenges (2023–2025)

The period between 2023 and 2025 has seen significant decrease in global food security, which has positioned the world well behind the schedule of Sustainable Development Goal 2 (Zero Hunger) being achieved by 2030 (Sporchia et al., 2024). Millions of people are now exposed to hunger and malnutrition due to the convergence of health crises, the instability of geopolitical situations, economic shocks, and structural inequalities (Odoms-Young et al., 2024). These changes represent fundamental system weaknesses in world food systems that are becoming incapable of responding to multi-dimensional threats.

1.2.1 Remnants of Socioeconomic Post-Pandemic Pathways

Although the world entered the post-pandemic stage, foods security was still shaken by the socioeconomic impacts of COVID-19. The agricultural production was disrupted by lockdowns, restrictions on mobility, and income losses of the millions of people in the 2020–2022 season, which impacts and weakens the labor market and reduces the income of people who rely on informal labor, mainly low- and middle-income countries (Barchielli et al., 2022). In Asia and Africa, the long-term income losses and the lack of savings by many households decreased the purchasing power of healthy food even after the markets were opened (Günal et al., 2025).

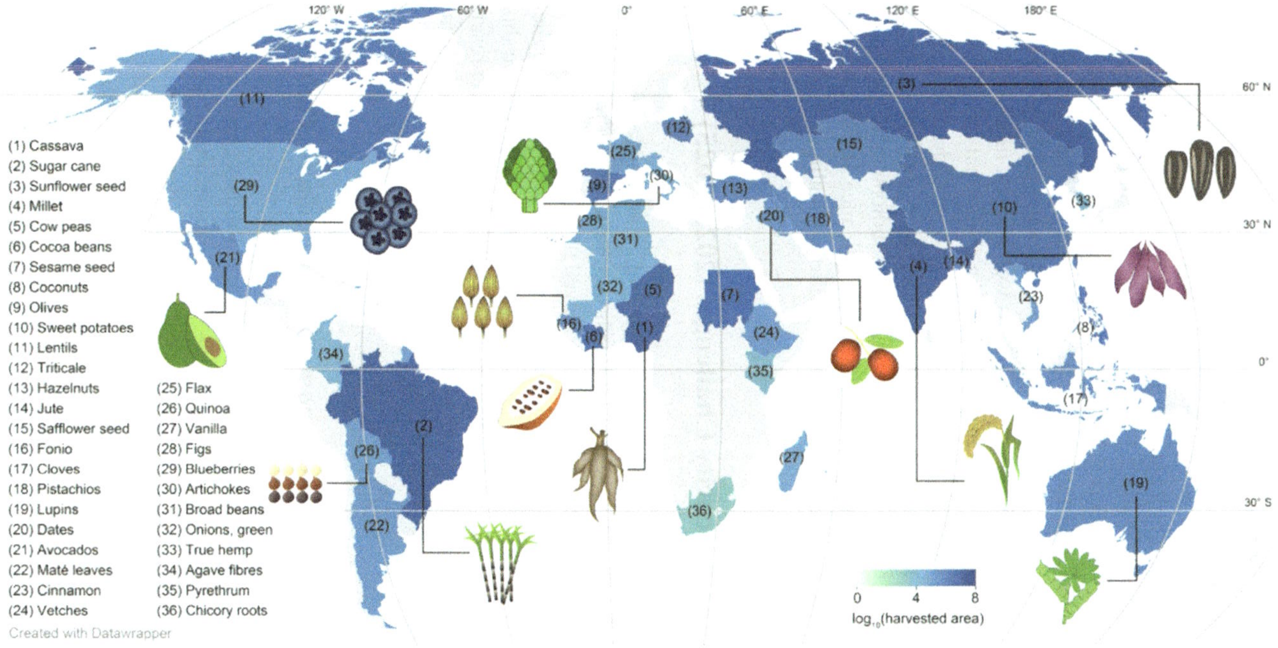

Fig. 1.1 Geographic distribution of cultivation centers for neglected and underused (NUCs) or orphan crops

Furthermore, interruptions in the transportation of goods around the world caused by the pandemic also led to the supply chain bottlenecks that increased food transportation expenses. In spite of the fact that the global trade was resumed, but slowly, structural instability, including a diminished transport capacity, labor shortage in ports, and higher freight prices, still affected food availability and prices well into 2023–2024 (Li et al., 2023).

1.2.2 Geopolitics and Shocks in the Commodity Market

One of the most important systemic shocks in the food systems of the world is the ongoing Russia-Ukraine conflict. Russia and Ukraine have a historical volume of approximately 30% of the world wheat exports, 20% of the maize exports, and 75% of the sunflower oil exports (Mottaleb et al., 2022). The war badly interfered with the exports of Black Sea grains, declining fertilizer exports, and increasing the cost of energy, all of which also led to the skyrocketing food prices internationally. This has had implication effects especially with respect to fertilizers (Urak et al., 2024). The major producers of potash and nitrogen fertilizers are Russia and Belarus and exports were cut off as well as sanctions, which led to the doubling of global fertilizer prices in most developing areas in 2022–2024 (Vos et al., 2025). The real outcomes of these shortages were lower crop yields and higher production costs to the smallholder farmers in the African continent, South Asia, and the Latin America, further deteriorating food system resilience (Saccone & Vallino, 2025).

1.2.3 Global Inflation and Rising Input Costs

There has been a concomitant increase in the global inflation rates which has been caused by the price of energy, currency fluctuations, and the costs of inputs that have increased food insecurity. According to the International Monetary Fund, in 2022, the world recorded an inflation rate of 8.7% and it still stood at elevated levels in 2023, although in many developing economies, food inflation was progressing much faster than headline inflation (Das et al., 2025; Omotayo et al., 2025). This has been experienced by the households in sub-Saharan Africa and south Asia, where food takes more than half of the household spending (Anwar et al., 2024). High inflation decreases the wages and purchasing power, which impact negatively low-income families and the rural population more than a rural population does. Millions of people could not afford to eat beef, vegetables, and other healthy foods even in areas whose food supply was sufficient, which raised the chances of undernutrition, micronutrient deficiencies and diet-related diseases (Algieri et al., 2025; Anwar et al., 2024; Samal et al., 2022).

1.2.4 Expanding Income Disparities

The post-pandemic economic recovery exhibits regional disparities, with affluent countries demonstrating a more rapid rebound attributed to robust safety nets and resilient economic structures. Conversely, low-income countries encounter significant debt burdens, restricted fiscal capacity, and a protracted economic recovery (Ngono et al., 2025). Increasing inequality has two immediate implications for food security:

- Decreased capacity to manage price fluctuations and
- Prolonged susceptibility to hunger and malnutrition, particularly affecting children, women, and marginalized populations.

The United Nations Development Programme have forecasted that increasing inequalities will continue to be a major factor contributing to global food insecurity until 2025, unless substantial redistributive policies, targeted subsidies, and social protection initiatives are enacted (UNDP, 2020).

1.2.5 Climate Shocks as a Cross Cutting Threat

Climate extremes were a top cross-cutting issue and food insecurity during the period of 2023–2025 (Wang et al., 2025) resulted to socioeconomic/geopolitical pressure and market strain. As it is stated by the World Meteorological Organization, 2023 was the hottest year in the history of the planet, and global mean temperatures reached to 1.45 °C above pre-industrial levels (Sandford et al., 2024). These changes in the temperatures also created high heatwave, long periods of droughts and unpredictability in the rainfall distribution in key agricultural areas. Extreme heat in South Asia affected the yield of wheat and rice, with Pakistan and India having a shorter grain-filling and increased evapotranspiration (Chandio et al., 2023; Fan et al., 2022; Heino et al., 2023; Khan et al., 2021). East Africa had to deal with both multi-year drought and the extreme rain and flooding events, which had a severe impact on the seasonal planting (Odongo et al., 2025). In North America, U.S. Midwest drought decreased the yield of maize, whereas wildfires and heatwaves decreased the quality of crops in some parts of Canada. According to the IPCC (2023), global food supply shocks due to such climate shocks increase price volatility and diminish stability because it narrows planting and shrinks water supplies, as well as increases the speed of pests and disease outbreak. The coincidence of these weather occurrences with geopolitical instabilities especially the Russia-Ukraine war increased the pressure on global grain markets (Saccone & Vallino, 2025).

1.2.6 Production Volatility Induced by Climate

Combination of climate shocks and inflation pressure aggravated volatility of production. Countries that were already pestered with the increasing input prices faced further pressure as droughts led to low irrigation water supply or floods destroyed important agricultural infrastructures (Tchonkouang et al., 2024). The losses of crops in Southeast Asia, East Africa, and some parts of Latin America caused by climate increased dependency on imports, making the areas more susceptible to changes in the global market (Liu et al., 2025). According to Bouteska et al. (2024), the impact of extreme climate events now accounts for the slowing growth in agricultural productivity. It has been reported that 30% of yearly yield variability in susceptible nations is attributable to climate variability. This movement highlights the fact that the global food systems have become more vulnerable (Ali et al., 2017; Janjua et al., 2021; Knight et al., 2024).

1.3 Regional Climate Hotspots and Escalating Food Insecurity

Since past 2 years (2023–2024) climate-related crises exacerbated food insecurity in the areas where already socioeconomic and geopolitical pressure, putting millions of people in the realm of famine (Saccone & Vallino, 2025). The East African cycles of droughts and floods wiped out crop lands, pasture land, and undermined livestock systems, and led susceptible households into negative coping mechanisms. South Asia have experienced severe heat waves and unnatural precipitation which impacted the production of cereals and the accessibility of food in the households (Fan et al., 2022). In the Sahel, the pastoral and agricultural life was thrown into turmoil by spreading desertification, unpredictable rain and extended momentary impacts (Eboreime et al., 2025; Etienne et al., 2025). In Latin America, especially Central America and Andes region, there was a high El Nino event which intensified drought and cut down maize, sugarcane and bean crops and put subsistence farmers at risk (Pérez et al., 2024). The shocks were accompanied by drastic food price increases all over the globe, lack of humanitarian access, and increased poverty, moving millions of people to Integrated Food security phase classification (IPC) Phase 3 and beyond (Frankenberger et al., 2011). The IPC warnings indicate that communities at Phase 3 are already experiencing critical food deficient circulation and disastrous coping strategies-such as skipping meals or selling vital resources. Furthermore, communities in Phase 4 have entered into emergency levels of malnutrition and some hotspots have already entered into Phase 5 conditions such as famine as visible in parts of Afghanistan, Yemen, Somalia, Ethiopia and South Sudan (Maxwell et al., 2020). As acute hunger is rising at a rate that surpasses the humanitarian resources, agencies have to juggle between impossible decisions to serve more people with less food, or focus on the most at-risk households. Such compounding climate and economic

Table 1.1 Regional food insecurity hotspots and key drivers (2023–2025) (Al Sharjabi et al., 2024; Dossa & Miassi, 2025; Otekunrin, 2025)

Region	Primary drivers of food insecurity (2023–2025)	Climate-related shocks	Estimated population in IPC phase 3+ (millions)
East Africa	Conflict, economic shocks, high food prices	Multi-year drought, floods (2023–2024)	55–60
South Asia	Inflation, low income recovery, import reliance	Extreme heatwaves, erratic monsoon	35–40
Sahel (West Africa)	Conflict, displacement, market disruptions	High temperatures, rainfall variability	40–45
Latin America (Central America and Andes)	Economic stagnation, migration pressure	El Niño drought, crop failures	15–18
Middle east and north Africa	Currency crises, import dependence	Water scarcity, heat extremes	30–32
Southern Africa	Low productivity, economic instability	Cyclones, drought episodes	20–25

pressures highlight the need to act early, invest adequately and have food systems that are climate resistant to avoid mass starvation (Maxwell et al., 2011) (Table 1.1).

1.4 Climate Change Projections for 2030 and Agricultural Vulnerability

According to the IPCC AR6, there is a medium to high probability that by the year before 2030 world average annual temperatures may exceed +1.5 °C (IPCC, 2023). The agricultural output will be greatly affected by extreme weather conditions, irregular rainfalls, and increased evapotranspiration (Syed et al., 2022). Staple crops like wheat may lose 4–6% yield per °C warming, while maize yields are projected to decline substantially across Sub-Saharan Africa and South Asia by 2030 (Heikonen et al., 2025) (Tables 1.2 and 1.3).

1.5 Structural Weaknesses in the Global Food System

The food system in the world has various structural challenges that disrupt its resilience and sustainability. To start with, the world diets are still overly reliant on three primary cereals; wheat, rice, and maize, which contribute over 50% of the

Table 1.2 Global food insecurity and malnutrition indicators (2023–2024)

Indicator	Value	Trend	Notes
People facing moderate to severe food insecurity	~2.4 billion	Increasing	Highest in Sub-Saharan Africa and Western Asia
Chronically undernourished population (projected for 2030)	590–620 million	Rising	Off-track for SDG 2
Global stunting prevalence	148 million children <5	Stable	Concentrated in low-income regions
Global wasting prevalence	45 million children <5	Unchanged	Sensitive to conflict and climate shocks
Regions most affected by food insecurity	SSA, Western Asia, conflict zones	Persistent hotspots	Linked to climate + conflict

Table 1.3 Climate projections to 2030 and their agricultural implications

Climate variable	Projection (2030)	Implications for agriculture
Rise in global mean temperature	0.8–1.3 °C	Heat stress during flowering/grain fill
Increase in extreme heat days	10–20%	Reduces yields of wheat, maize, rice
Rainfall variability	Higher-intensity storms and longer dry spells	Soil erosion + drought risk
Wet season changes	Shorter, erratic	Reduced sowing windows
Frequency of droughts	Increasing	Affects rainfed agriculture severely
Shifts in agroecological zones	Expansion of heat-prone drylands	Staple crops decline; NUCs preferred

global calories, posing a threatening level of concentration in case of weather and market shocks (Said et al., 2025). This dependence is further aggravated by weaknesses within globalized supply chains, in which the rise of geopolitical tensions, pandemics, or trade restrictions may cause a quick shift in food supply, as the recent example of the cascading effects of the Russia Ukraine conflict on global grain and fertilizer supplies has seen (López et al., 2025). The agricultural productivity is further limited by environmental degradation, such as soil erosion, the decreasing water resources, and greenhouse gas emissions, and climate change raises the number of extreme weather events that directly lower yields, particularly in climate-sensitive tropical and subtropical areas (Kabato et al., 2025). Monoculture systems have also contributed to increasing the agro-biodiversity loss because of their high penetration in reducing genetic diversity, diminishing ecological resilience, and increasing vulnerability to pests, diseases, and climatic stressors (Kaur et al., 2024). These nutritional flaws manifest in global nutrition outcomes: the global calorie adequacy is not sufficient due to the lack of micronutrients, which is the result of simple diets that are based on starchy food and not the nutrient-dense variety of foods (Meenakshi & Quisumbing, 2025). Together, they show that the world food system is ecologically and, crucially, nutritionally unbalanced, thus in dire need of diversification and resilience-inculcating measures of all kinds.

1.6 Why Diversification is Critical in Climate-Unstable Scenarios

The concept of crop diversification has become increasingly important as one of the key strategies to stabilize the food systems that are exposed to rapidly increasing climate hazards. The dependence on a small number of major cereals such as wheat, rice and maize introduce systemic vulnerability whereby, they are very sensitive to heat stress, irregular rainfall and spreading pests and diseases. Dispersal of the biophysical and economic risk, lessening the likelihood of a complete loss of crop and enhancing adaptive capacity, both at the farm and national levels, are all benefits of diversifying the production landscapes. Diversified systems help in overcoming climatic socio-economic volatility conditions by sustaining ecosystem services like retention of soil moisture, pollination, and recycling of nutrients and by offering a range of harvesting periods diminishing climate fluctuations. There is evidence to indicate that diversified farms tend to perform better in extreme years than monocultures because their crop characteristics are complementary and they are not affected by failure at the same time (Mohamed-Shaffril et al., 2024).

1.7 Role of NUS/NUCs in Ecological Resilience

The Neglected and Underutilized Species (NUS/NUCs) have been identified as the most appropriate in formation of ecological resilience since most of them are naturally suited to the marginal or stress-prone environment. Millet, quinoa, amaranth, teff and buckwheat are examples of crops that are drought, salty, soils and high temperature tolerant than the major cereals (Duodu & Banerjee, 2025; Joshi, 2023; Kaya et al., 2015; Pudasaini & Raizada, 2025; Woo et al., 2016). Their extensive root systems, reduced water needs, and susceptibility to grow under the lowest levels of external inputs qualify them to be excellent constituents of climate-autonomous agroecosystems. The inclusion of these species also benefits the landscape by increasing diversity, promoting useful insects and decreasing the pressure of pathogens and pests due to non-host effects. Besides, NUS promote agroecosystems stability through enhancement of soil structure, carbon uptake, and year-round ground cover, which are crucial characteristics that enhance resilience during climate change (Purba & Krishnaswamy, 2025).

1.8 Benefits of NUS to Nutrition Security

NUS/NUCs are nutrient rich crops which are a possible solution to hidden hunger, micronutrient deficiencies and monotony in the diet. The concentrations of proteins, essential amino acids, iron, zinc, dietary fiber, antioxidants, etc., in many NUS are much higher in comparison with the content of major staple cereals. As an example, finger millet has up to three times the amount of calcium as milk or quinoa has a complete amino-acid profile that is better than that of wheat or rice. NUS can enhance the diversity of food and decrease reliance on fortified foods by diversifying diets and increasing the availability of culturally important, traditional foods. They can be particularly relevant in the context of the low-income and rural population, the dietary diversity of which is still insufficient, and the threat of malnutrition is enhanced by climate shocks. The genetic abundance that is particular to NUCs is significant in ensuring the maintenance of biodiversity (Luthar et al., 2021). This increases their ability to withstand pests, diseases, and environmental stress, and thus such crops become more resistant to the unpredictable weather conditions. This means that by introducing NUCs in the current cropping systems, small holder farmers will get an added source of income. A large number of them are sold on the value-added products like gluten-free flours, which have a higher price in the markets (Talucder et al., 2024).

In addition, some abandoned leguminous crops have the capacity to fix nitrogen in the atmosphere and thus enhance the fertility of the soil and decrease the use of synthetic fertilizations (Talucder et al., 2024). The effect of the plant rotation of traditional crops with NUCs or its increment is also the disruption of pest and disease patterns, which reduces the utilization of chemical pesticides. Besides the agronomic value, there are numerous undervalued crops that have historical and cultural significance to the local communities. The U-turn of them in the agricultural systems thus helps in the maintenance of traditional knowledge and cultural background. Figure 1.2 compares the nutritional value of NUCs to rice and wheat. NUCs have an average of significantly larger quantities of fat (235%), calcium (222%), zinc (30%), thiamin (vitamin B1) (103%), riboflavin (vitamin B 2) (310%), and dietary fiber (775 mg) in comparison to rice and wheat (Chivenge et al., 2015). Nevertheless, they have a mean caloric value which is about 32% lower than these leading cereals.

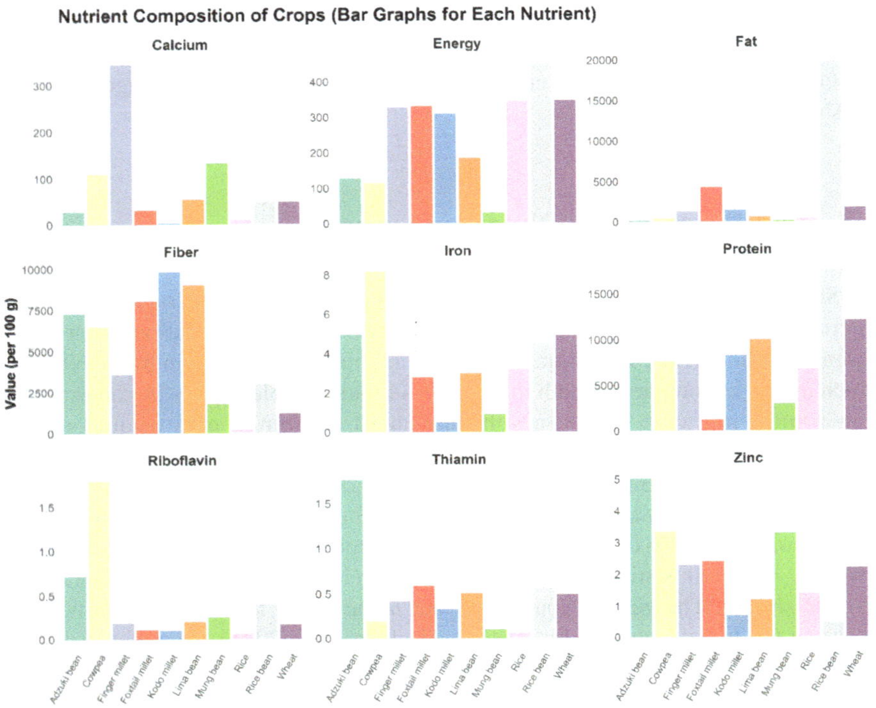

Fig. 1.2 Nutritional profile of NUCs and major cereal crops

1.9 Economic and Livelihood Benefits to Smallholders

NUS/NUCs can be used to mitigate financial risk faced by smallholders as they can provide new livelihoods. A large number of NUS demand fewer inputs which include fertilizers, pesticides, and irrigation and thus the cost of production is low and the yields are optimal in marginal environments (Talucder et al., 2024). Crop diversification also allows the farmers to enter into new niche markets such as gluten-free products, health foods, functional foods, and traditional-value markets. Other NUS have good prospects of value addition (e.g., millet flour, amaranth snacks, products based on quinoa), with additional sources of income other than raw grain sales. Moreover, diversification of crops enhances market risks, that is, when the prices of major cereals are unstable, value chains based on NUS provide alternative sources of income that help improve the economic security of rural families (Purba & Krishnaswamy, 2025) (Table 1.4).

Table 1.4 Different parameters showing comparison between NUCs and major cereal crops (Shafiq et al., 2025)

Parameter	NUS/NUCs (e.g., millets, quinoa, amaranth)	Major cereals (wheat, rice, maize)
Drought tolerance	High (many grow with 250–400 mm rainfall)	Moderate to low
Heat tolerance	40–45 °C for several NUS (e.g., millets)	30–35 °C threshold for yield decline
Water requirement	Low (30–50% less than rice or maize)	High for rice; moderate for wheat/maize
Yield stability under stress	High—adapted to marginal lands	Low to moderate—yield collapses in extreme years
Soil requirements	Thrive in poor, sandy, rocky soils	Require fertile soils for optimal yields
Protein quality	High (complete amino acids in quinoa)	Incomplete amino-acid profile
Micronutrients	Rich in Fe, Ca, Zn, fiber, antioxidants	Generally lower micronutrient density
Input requirements	Low (minimal fertilizers/pesticides)	Moderate to high inputs for high yield
Market type	Emerging niche and health markets	Large, established global markets
Carbon footprint	Lower—adapted to low-input systems	Higher due to fertilizer and irrigation dependence

1.10 Why Underutilized and Neglected Crops Are Important Today

Due to their exceptional ability to withstand the complex stresses resulting from climate change, NUCs have attracted renewed attention on a global scale (Talucder et al., 2024). These crops have a variety of adaptive characteristics that enable them to flourish in the exact conditions where major cereals are increasingly failing, such as remarkable drought and heat tolerance, low water requirements, and the capacity to flourish in nutrient-poor or degraded soils (Chivenge et al., 2015; Mittler et al., 2025). They are of a great genetic diversity that will provide a ready pool of genetic material to be used in further plant breeding projects particularly in improving the nutritional value and stress resistance of mainstream crops. Most NUCs, including millets, amaranth, teff, cowpea, Bambara groundnut, taro, and yam, have better micronutrient density than commonly consumed staples, including wheat and rice, and are important in overcoming the hidden hunger and increasing dietary diversity (Purba & Krishnaswamy, 2025). Table 1.5 displays the important resilience features, nutritional benefits, and other advantages of specific NUCs (Li et al., 2020). Combined

Table 1.5 Key traits and advantages of selected NUCs

Crops	Climate resilience traits	Key nutritional benefits	Additional advantages
Pearl millet	Drought and heat tolerant	High iron, zinc	Short duration; ideal for drylands
Finger millet	Thrives in low-fertility soils	Calcium-rich	Long shelf life
Teff	Tolerant to heat and poor soils	High-quality protein	Versatile grain
Amaranth (grain and leaves)	Heat and drought resistant	Antioxidants, essential amino acids	Dual-use crop (leafy + grain)
Cowpea	Drought tolerant	Protein, iron, zinc	Nitrogen-fixing
Bambara groundnut	Survives extreme drought	Balanced protein	Known as a "complete food" crop
Taro / Yam	Tolerant to rainfall variability	Energy + micronutrients	Staple for tropical regions

altogether, these characteristics position NUCs as the leading group in nutrition-sensitive and climate-smart agriculture that can make the food system more resilient and reduce the range of income sources.

1.11 NUCs' Potential to Improve Nutrition and Food Safety

The NUCs have a role to play in the fight against hidden hunger, a form of malnutrition that is manifested by lack of vital micronutrients, due to the fact that the micronutrient concentrations of NUCs are considerably higher than those of most major cereals. Most NUCs contain superior levels of iron, calcium, zinc, magnesium, essential amino acids and bioactive compounds. As an example, leafy NUCs are a good source of vitamin A and folate, amaranth is a good source of high-quality protein and antioxidants and finger millet is an exceptionally good source of calcium (Shafiq et al., 2025). These nutritional advantages help in the diversification of diets and reductions of reliance on staples which are rich in energy yet deficient in nutrients. Placing NUCs in food systems can help community's micronutrient-enrich their diets, as well as construct more balanced diets to prevent vulnerable groups, such as women, children, and the poor.

1.11.1 Climate Resilience (Heat Tolerance: Salinity: Drought) Characteristics

The resilience mechanisms that NUCs possess are extensive and enable it to thrive during the periods of climatic extremes which will grow even worse within the upcoming few decades (Yesuph et al., 2025). Due to their lengthy root systems, use of short life cycle, excellence in the physiological adaptation to usage of water, some species exhibit extraordinary drought resilience (Onyeneke et al., 2025). Others are adapted to coastal zones and saline prone soils affected by sea level rise or poor irrigation methods as they can withstand salinity. The other difference is heat tolerance where crops such as millets and teff can be used in the high temperatures when those such as wheat and maize are yielding very low (Singh et al., 2025). Due to these climate-smart features, NUCs play a crucial role in sustaining agricultural productivity in semi-arid and arid regions and also in climatically-stressed regions (Yadav and Upasana, 2026). Different indices as given below could be used to evaluate stress tolerance of NUCs and calculation have been shown in Fig. 1.3.

$$Heat\ Tolerance\ Index(HTI) = \frac{Y_{HS} \times Y_{NS}}{\left(\overline{Y_{NS}}\right)^2}$$

$$Drought\ Tolerance\ Index(DTI) = \frac{Y_{DS} \times Y_{NS}}{\left(\overline{Y_{NS}}\right)^2}$$

$$Salinity\ Tolerance\ Index(DTI) = \frac{Y_{SS}}{Y_{NS}} \times 100$$

$$Stress\ Susceptibity\ Index(SSI) = \frac{1 - \left(\frac{Y_s}{Y_p}\right)}{1 - \left(\frac{\overline{Y_s}}{\overline{Y_p}}\right)}$$

$$Tolernace\ Index\ (TOL) = Y_p - Y_s$$

$$Mean\ productivity(MP) = \frac{Y_p + Y_s}{2}$$

$$Geometric\ Mean\ productivity(GMP) = \sqrt{Y_p \times Y_s}$$

where

Y_{HS} = Yield under heat stress.
Y_{DS} = Yield under drought stress.
Y_{SS} = Yield under salinity stress.
Y_{NS} = Yield under normal conditions.
$\overline{Y_{NS}}$ = Mean yield of all genotypes under normal conditions.

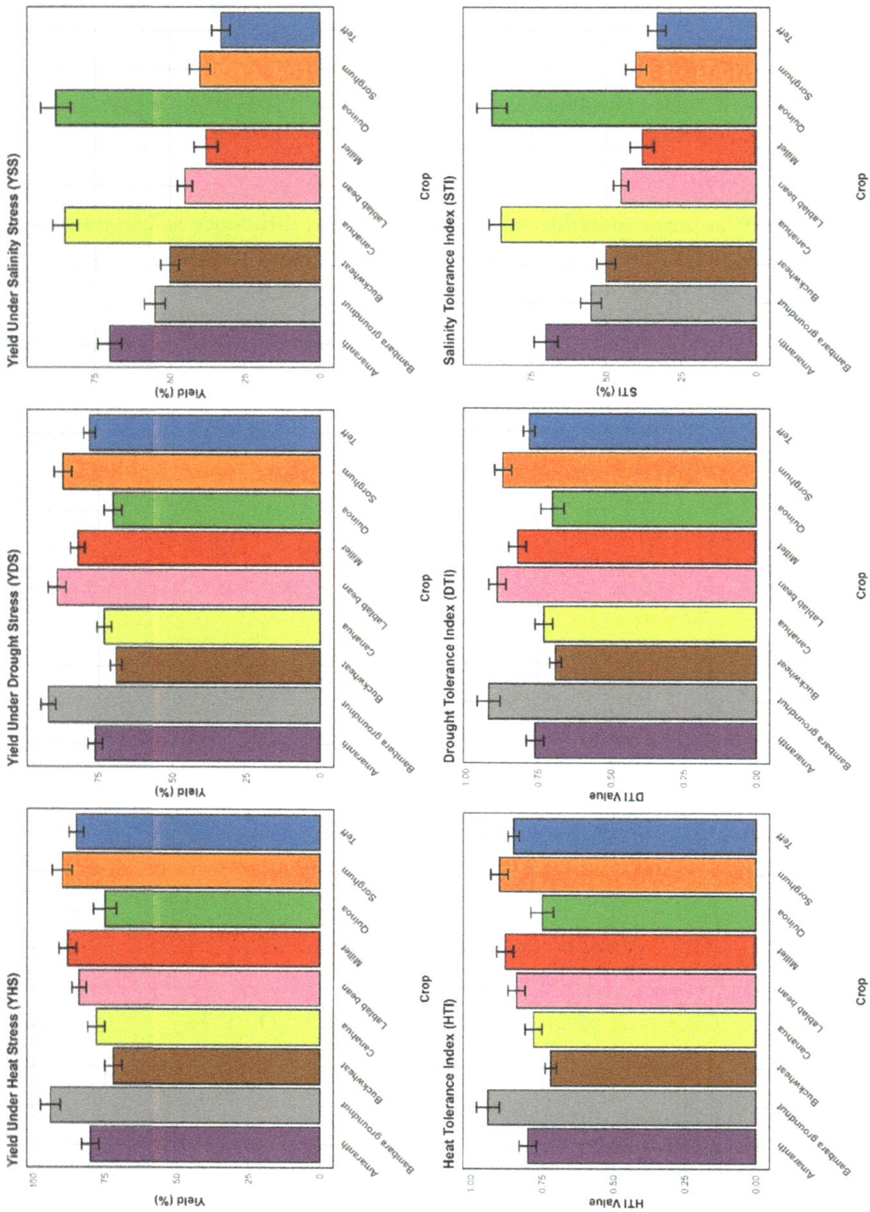

Fig. 1.3 Stress tolerance indices of different NUCs

1.11.2 *Suitability for Marginal Lands and Low Input Requirements*

Most NUCs are able to grow in poor, degraded or marginal soils with minimal external input since the species have adapted in the challenging environment. Especially in the rainfed systems, they have low requirements of artificial fertilizers, pesticides, and irrigation, leading to low production costs, and can be afforded by farmers with limited resources. Crops, such as cowpea and Bambara groundnut, repair the atmosphere with nitrogen and improve the quality of soil and ecosystems sustainability. Due to the fact that they are versatile, NUCs can enhance the productive capacity of marginal landscapes as they serve as reliable sources of food in areas where major cereals cannot be produced in a profitable manner (Mmbando, 2025).

1.11.3 *NUCs in Traditional and Indigenous Food Systems*

NUCs have become a part of the agriculture, cuisine and culture of many indigenous communities in Africa, Asia, and Latin America (Shafiq et al., 2025). People have cultivated these crops over centuries as staple food, medicine, ritualistic, and dietary elements that are able to adjust to climatic conditions. The fact that they are integrated into ancient agroecosystems promotes community-level resilience and encourages culinary traditions and enhances local biodiversity (Purba & Krishnaswamy, 2025). Besides enhancing nutrition and environmental sustainability, NUCs revival in indigenous food systems enhances the economic growth of local communities, food sovereignty, and cultural identity. The traditional knowledge on NUC growing and utilization would be an important aspect in future food security policies as world food systems seek alternatives to delicate monocultures (Omotayo et al., 2025) (Fig. 1.4).

1.12 NUC-SDG Nexus: Theoretical Bases

The overlooked and under-exploited crops (NUC) are one of the critical strategic aspects, which continue to be systematically ignored in the agri-food systems of low- and middle-income countries (LMICs) (Devkota et al., 2025; Singh et al., 2025). NUC are a multidimensional way of sustainable development and the realization of the Sustainable Development Goals (SDGs), especially SDG 2 (Zero Hunger) in Pakistan and South Asia, and at the same time create social, economic and environmental co-benefits. Due to their high nutritional quality, great genetic diversity, endurance to climatic stresses, as well as low external inputs needed relative to the dominant cereal-based systems, NUC are highly adapted to marginal environments, rainfed agriculture and resource-based smallholder farming systems. Notably, the applicability of NUC to the SDGs must not be viewed in isolation results, but they

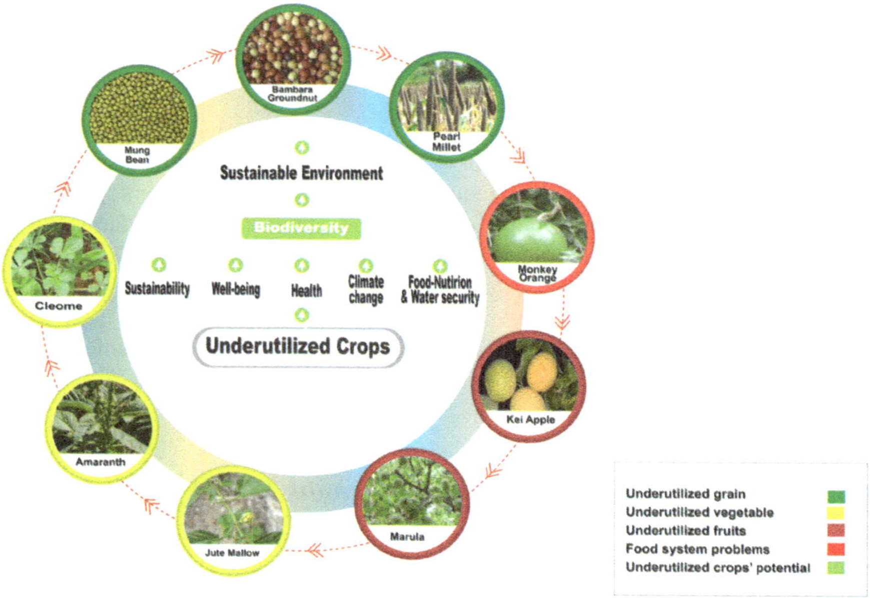

Fig. 1.4 Diagrammatic representation of NUCs potential to change food system. *Source* Omotayo et al. (2025)

have systematic impacts, working via a nexus of networks of results. This is a nexus between NUC and SDG that is expressed by better nutrition and food security, better livelihoods by smallholders, better climate resilience and preserving agrobiodiversity. These connections, in the context of Pakistan, are observable in agroecological region, with NUC, including millets and moth bean in Sindh, underutilized pulses in Punjab, and sorghum, millets, and native fruits in Baluchistan, all showing how NUC act as leverage points to support the achievement of multiple SDGs in the agroecological systems maintained largely by climatically vulnerable and resource-constrained food systems (Fig. 1.5).

1.13 Value Chain Development and Market Integration

1.13.1 Present NUC Market Restrictions

Despite their agronomic and nutritional advantages, NUCs continued to be a marginal component in mainstream markets. Low consumer awareness, a lack of market demand, broken supply chains, and disparate product quality standards all contribute to their limited commercialization. Since NUCs are still viewed as poor man crops or subsistence foods, their popularity in cities is low in most regions. Furthermore,

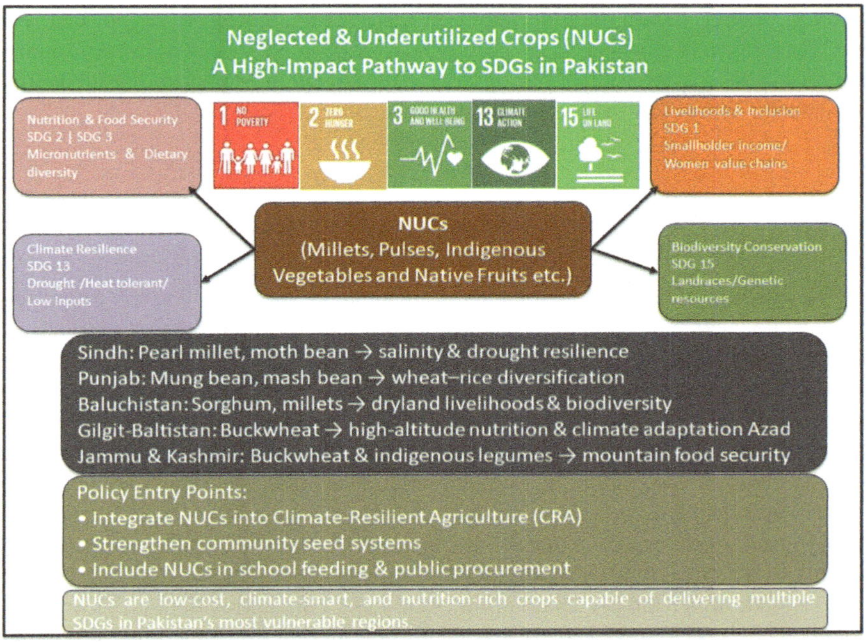

Fig. 1.5 NUC–SDG nexus: nutrition, livelihoods, and climate resilience in Pakistan

farmers frequently lack access to better seed varieties, appropriate post-harvest technologies, and organized marketing channels. Additionally, post-harvest losses diminish their existence and profitability in formal retail markets, especially for perishable NUCs like tubers and leafy vegetables. Because of this, the majority of NUCs are still restricted to unofficial village-based trade networks, which hinders economic gains and scale-up (Purba & Krishnaswamy, 2025).

1.13.2 Commercialization Possibilities

NUCs now have unprecedented commercialization opportunities due to growing interest in nutrient-dense, sustainable, and climate-smart foods. Health-conscious consumers are increasingly drawn to foods high in antioxidants, plant proteins, fiber, and micronutrients qualities shared by many NUCs (Freymond et al., 2023). The growing demand for gluten-free products creates a strong market niche for crops like millets, teff, and amaranth. NUCs can also be used to make high-value processed products like breakfast cereals, plant-based beverages, flours, snacks, and nutraceutical ingredients. The expanding natural and organic food sector further enhances commercial prospects. Rural economies can benefit from premium pricing and value addition through niche branding based on geographic identity (e.g., "African teff" and "Himalayan buckwheat") and sustainable sourcing. NUCs can become profitable commodities on a large scale through commercialization, transforming them from subsistence crops (Santunione & Montevecchi, 2025).

1.13.3 The Function of Agri-tech and Digital Agriculture

Digital agriculture has the potential to eliminate many structural barriers that hinder NUC value chains. Mobile-based extension tools can boost cultivation productivity and efficiency by giving farmers access to real-time agronomic advice, market prices, weather alerts, and pest diagnosis. By removing the need for middlemen and ensuring more fair prices, digital marketplaces and e-commerce platforms can connect buyers and smallholders directly. Technologies like blockchain and traceability aid in the establishment of food safety and quality assurance standards, which are necessary to enter official retail and export markets. Additionally, digital tools facilitate group marketing through farmer cooperatives, support demand forecasting for processors and retailers, and aggregate distributed production. Agri-tech startups that concentrate on data-driven supply-chain optimization and climate-resilient crops are positioning NUCs as future-ready value chain commodities in emerging economies (Vărzaru, 2025).

1.13.4 Institutional Support, Investment, and Policy Requirements

Strong institutional support throughout the whole value chain, strategic investment, and enabling policy changes are all necessary to fully realize the market potential of NUCs. National agricultural strategies that improve extension services, advance crop-breeding programs, bolster seed systems, and encourage climate-resilient crops should give priority to NUCs. Establishing processing facilities, storage infrastructure, and quality certification systems will require public–private partnerships. Additionally, policies like minimum support prices (MSPs), crop insurance, accessible credit programs, and targeted procurement initiatives can encourage farmers to diversify beyond major cereals. Research and development (R&D) expenditures are essential for generating interest in the agrifood sector, especially in post-harvest technologies, nutritional profiling, and value-added product development. Through their support of cross-border trade initiatives, market intelligence, and policy advocacy, regional and international organizations also play a critical role in mainstreaming NUCs. In order to ensure sustainable value chain growth and facilitate NUCs' integration into formal markets, institutions must work together (Barbosa, 2024).

1.14 Case Studies

There has also been a resurgence of millets in India and Africa mostly due to the mounting demands of climate change, water scarcity and dwindling soil fertility. India has lost interest in millets because of the policy priorities focusing on rice and wheat, but recent efforts, such as the declaring of 2023 as the International Year of Millets, helped to revive production. States like Karnataka and Odisha have made millet planting a part of nutritional schemes, cooperative organizations of farmers and development of value chains, which have established these robust cereals as climate-resistant and less vulnerable to heat stress and low water resources (Harish et al., 2024). Similarly, there are other initiatives being done in Africa such as in Niger, Nigeria, Mali, Ethiopia and Uganda, where better hybrid varieties, community seed banks, and farmers training schemes have also reinforced production. Millets maintain unpredictable yields even with unpredictable rainfalls, which affirm their role in offering food and nutrition security in the climate vulnerable areas (Kheya et al., 2023).

Quinoa is a plant that has grown quickly in the Andean region within its native highland areas in Bolivia, Peru, and Ecuador to a new terrain and markets all over the world. Quinoa with its high nutritional value along with the resistance to drought, frost, salinity, and low quality soils has become a main crop in attempt to increase climate resilience. Organic certification and export markets in Bolvia, on their part, have helped prec autonomous cooperatives in the Altiplanos to boost incomes, whereas Peru has managed to get quinoa cultivation to the coastal lowlands by using salinity tolerant ecotypes and improved irrigation techniques. Ecuadorian education systems have incorporated instances of quinoa in school lunch to secure the poverty health of its population. Despite the economic opportunities associated with global demand, it has created sustainability challenges associated with land-use alterations, soil erosion in delicate ecosystems, and fluctuations in the market, and thus there is the need to have balanced and community-based approaches towards the management approaches (Jaikishun et al., 2019).

African legumes are becoming a highly significant climate-resistant crop in Asia with bambara groundnut being one of such nutrient-rich foods. Its adaptation to be grown in marginal soils, withstand drought, and supply high quality protein makes it very valuable in the rainfed agricultural systems. In Malaysia and Thailand research centers are putting its efforts in breeding programs to stabilize yield, adaptation to photoperiod as well as seed quality and in India pilot projects in Karnataka and Odisha are implementing Bambara groundnut into the diversified cropping systems to increase dietary and income diversity. The growth of small scale value chain of processed products is also being experienced in Indonesia. These efforts are being encouraged by regional and international research networks that are producing new information on agronomy, water-use efficiency, and adjustment in conditions of Asian climatic environment (Abberton et al., 2022).

Little-used legumes such as cowpea, pigeon pea, tepary bean, hyacinth bean among the drought resistant species, are becoming increasingly significant in arid and semi-arid areas throughout Africa, South Asia and Latin America (Vilakazi et al., 2025). These legumes can easily cope with harsh environment where temperatures are high and have low rainfall because of the deep root systems, water-efficient absorption, and have good capacity to fix nitrogen. Where cowpea and pigeon pea are present, African Sahel and East Africa, the two beans keep the house running when there is a period of drought and this helps occur when there is an advancement of better varieties of peas that withhold pests like Striga. Pigeon pea and moth bean are still part of mixed dryland cropping systems in South Asia especially in India and Pakistan which enables the farmers to sustain productivity amidst climatic uncertainties. The tepary bean which historically was grown by Indigenous people in Latin America has now received a new resurgence due to its exceptional ability to survive in extreme hot regions as well as its quick turn around. Altogether, these legumes can make the agriculture climate-resistant by improving soil health, decreasing the use of fertilizers, and providing the local diets with high nutritional value (Akchaya et al., 2025).

1.15 Challenges Hindering Mainstreaming of NUCs

Specifically, the mainstreaming of neglected and underutilized crops (NUCs) still encounters a number of challenges that are interrelated and limit the entry of agricultural technologies into modern agricultural systems. The second limitation is that there are no specific research and crop breeding programs since in the past NUCs have had little scientific import compared to the staple crops like rice, wheat and maize. This has led to poor genetic enhancement, incompetence of their physiology and agronomy and lack of enhanced varieties that meet the needs of different agro ecological conditions. These limitations are further aggravated by the weaknesses of seed systems because there are most reliable seed supply chains of NUC which are quite fragmented, informal, or nonexistent. Local exchange or self-saving of seed results in quality being inconsistent and farmers not having access to high yielding or stress resistant varieties. Accompanied by limitation of production, consumer awareness of NUCs as well as the market demand is still low at most points. The perception of these crops is that it is food of poor man or the foodstuff of tradition which has little culinary value hence it may not sieved into the urban markets, commercial value chain and food processing sector. Lastly, institutional and policy climate has been apathetic towards NUCs in the past and this has been largely because of priorities in agriculture investment which is directed to the major cereals. Expansive policy incentives, poor extension aid and insufficient integration of NUCs to national food security plans have all acted against their general adoption. The way out of these limitations is to combine efforts, research institutions, the seed sector, governments, and market actors to realize the potential of NUCs in climate-resilient and nutrition-sensitive agriculture (Purba & Krishnaswamy, 2025).

1.16 Strategic Interventions for Scaling Up NUCs

The process of scaling up neglected and underutilized crops (NUCs) entails a set of strategic interventions integrated in a holistic way to deal with scientific, institutional, and market-type gaps. More research needs to be invested in breeding, genomics, and the characterization of traits to unlock the potential in NUCs and come up with better plants with higher yields, nutrient makeup, and resistance to low soil fertility, drought, heat, pests, and chickens. The use of these crops as part of climate-smart agricultural systems increases their applicability further because NUCs inherently fit in adaptation and mitigation purposes due to their efficiency in water consumption, ability to endure in adverse climatic conditions, and reduced reliance on artificial inputs (Shafiq et al., 2025). In addition to farm enhancements, it is important to strengthen the rural value chains in an attempt to produce sustainable demand and economic incentives on their part. This comprises after harvest treatments, processing methods, warehouses, connections in the market, and the advertisement of the many-sided food items manufactured using NUCs (Liu et al., 2023; Sun et al., 2025). The role of the public- private partnerships is also significant since they mobilize investment, develop the system of seed and grow processing enterprises and bridge the small holder farmers with the local and international market. Organizations operating at the international tier (FAO, IFAD and the CGIAR) as well as international systems (the UN Food Systems Summit) are starting to appreciate the role of NUCs in solving issues of food security, nutrition, and climate resiliency (FAO, 2016). Their involvement in research partnerships, policymaking, funding programs, and capacity building activities can make the mainstreaming of NUCs a much faster and more efficient process and makes them a mandatory element of resilient and sustainable food systems.

1.17 Conclusion

NUCs have the ability to transform food system especially in climate-stressed zones, ecologically devitalized areas, with nutritional insecurity. The combination of their natural endurance, adaptation to marginal settings and nutrient dense diets makes them key resources towards the construction of diversified and climate resolute agricultural landscapes. It is crucial to rethink both national and international food systems in order to realize this potential, going beyond crop-level interventions. In order to transform NUCs from marginalized crops to mainstream components of sustainable agricultural production, research gaps must be filled, seed systems must be strengthened, value chains must be upgraded, and supportive policies must be established. By helping to achieve zero hunger, improving farmer livelihoods, and strengthening resilience to climate shocks, incorporating NUCs into farming systems could greatly advance SDG 2 as the world moves closer to 2030. The world community can pursue sustainable, equitable, and climate-adaptive agriculture by transforming NUCs from neglected species into key pillars of a food-system revolution through strategic investment, innovation, and inclusive governance.

References

Abberton, M., Paliwal, R., Faloye, B., Marimagne, T., Moriam, A., & Oyatomi, O. (2022). Indigenous African orphan legumes: Potential for food and nutrition security in SSA. *Frontiers in Sustainable Food Systems, 6*, 708124. https://doi.org/10.3389/fsufs.2022.708124

Akchaya, K., Parasuraman, P., Pandian, K., Vijayakumar, S., Thirukumaran, K., Mustaffa, M. R., et al. (2025). Boosting resource use efficiency, soil fertility, food security, ecosystem services, and climate resilience with legume intercropping: A review. *Frontiers in Sustainable Food Systems, 9*, 1527256. https://doi.org/10.3389/fsufs.2025.1527256

Al Sharjabi, S. J., Al Jawaldeh, A., Hassan, O. E. H., & Dureab, F. (2024). Understanding the food and nutrition insecurity drivers in some emergency-affected countries in the eastern mediterranean region from 2020 to 2024. *Nutrients, 16*(22), 3853.

Algieri, B., Kornher, L., & von Braun, J. (2025). The changing drivers of inflation—the case of food: Macroeconomics, speculation, climate change and war. *Structural Change and Economic Dynamics, 75*, 782–800. https://doi.org/10.1016/j.strueco.2025.10.006

Ali, S., Liu, Y., Ishaq, M., Shah, T., Abdullah, S., Ilyas, A., & Din, I. U. (2017). Climate change and its impact on the yield of major food crops: Evidence from Pakistan. *Foods, 6*(6), 39. https://doi.org/10.3390/foods6060039

Anwar, M., Shair, W., & Hussain, S. (2024). Effect of coping strategies on household food insecurity in Pakistan amid global economic crisis. *International Social Science Journal, 74*(254), 1397–1421. https://doi.org/10.1111/issj.12520

Bai, H., Xiao, D., Liu, D. L., Tao, F., Liu, F., & Tang, J. (2025). Impacts of future climate change and management practices to yield, eco-efficiency and global warming potential for rice–wheat rotation system. *Journal of the Food Science and Agriculture, 105*, 6138–6150. https://doi.org/10.1002/jsfa.14322

Barbosa, M. W. (2024). Government support mechanisms for sustainable agriculture: A systematic literature review and future research agenda. *Sustainability, 16*(5), 2185.

Barchielli, B., Cricenti, C., Gallè, F., Sabella, E. A., Liguori, F., Da Molin, G., et al. (2022). Climate changes, natural resources depletion, COVID-19 pandemic, and Russian-Ukrainian war: What is the impact on habits change and mental health? *International Journal of Environmental Research and Public Health, 19*(19), 11929.

Bouteska, A., Sharif, T., Bhuiyan, F., & Abedin, M. Z. (2024). Impacts of the changing climate on agricultural productivity and food security: Evidence from Ethiopia. *Journal of Cleaner Production, 449*, 141793. https://doi.org/10.1016/j.jclepro.2024.141793

Chandio, A. A., Dash, D. P., Nathaniel, S. P., Sargani, G. R., & Jiang, Y. (2023). Mitigation pathways towards climate change: Modelling the impact of climatological factors on wheat production in top six regions of China. *Ecological Modelling, 481*, 110381. https://doi.org/10.1016/j.ecolmodel.2023.110381

Chivenge, P., Mabhaudhi, T., Modi, A. T., & Mafongoya, P. (2015). The potential role of neglected and underutilised crop species as future crops under water scarce conditions in Sub-Saharan Africa. *International Journal of Environmental Research and Public Health, 12*(6), 5685–5711.

Das, A., Kumar, S., Kasala, K., Nedumaran, S., Paithankar, P., Kumar, A., et al. (2025). Effects of climate change on food security and nutrition in India: A systematic review. *Current Research in Environmental Sustainability, 9*, 100286. https://doi.org/10.1016/j.crsust.2025.100286

Devkota, K. P., Devkota, M., Mabhaudhi, T., Nangia, V., Attaher, S., Boroto, R. J., et al. (2025). A blueprint for building resilience and food security in MENA and SSA drylands: Diversifying agriculture with neglected and underutilized species. *Food Energy Security, 14*(1), e70046. https://doi.org/10.1002/fes3.70046

Dossa, K. F., & Miassi, Y. E. (2025). Nexus between food insecurity, poverty, and climate change: A cross-regional multifactorial analysis. *Frontiers in Environmental Economics, 4*, 1689025. https://doi.org/10.3389/frevc.2025.1689025

Duodu, K. G., & Banerjee, R. (2025). Chapter 18—Sorghum and millets: Quality management systems. In J. R. N. Taylor, S. R. Bean, & K. G. Duodu (Eds.), *Sorghum and millets* (3rd ed., pp. 573–596). Elsevier. https://doi.org/10.1016/B978-0-443-23954-0.00004-5

Eboreime, E., Anjorin, O., Obi-Jeff, C., Ojo, T. M., & Hertelendy, A. (2025). From drought to displacement: Assessing the impacts of climate change on conflict and forced migration in West Africa's Sahel region. *The Journal of Climate Change and Health, 23*, 100448. https://doi.org/10.1016/j.joclim.2025.100448

Etienne, N., Joseph, N., & Martin, M. (2025). A blessing or a curse? Climate change and political economy effects on the agricultural sector: Empirical evidence from Sub-Saharan Africa. *International Journal of Climate Change Strategies and Management, 17*(1), 707–742. https://doi.org/10.1108/IJCCSM-11-2024-0199

Fan, X., Zhu, D., Sun, X., Wang, J., Wang, M., Wang, S., & Watson, A. E. (2022). Impacts of extreme temperature and precipitation on crops during the growing season in South Asia. *Remote Sensing, 14*(23), 6093.

FAO. (2016). Public–private partnerships for agribusiness development. In M. Rankin, E. G. Nogales, P. Santacoloma, N. Mhlanga, & C. Rizzo (Eds.), *A review of international experiences*.

FAO. (2024). *FAOSTAT database*. Food and Agriculture Organization of the United Nations. http://faostat.fao.org/. Accessed 15 Feburary 2024.

Frankenberger, T. R., Verduijn, R. J., & Nations, A. O. (2011). *Integrated food security phase classification (IPC)* (pp. 1–90).

Freymond, M., van Zutphen-Küffer, K. G., & Kraemer, K. (2023). Innovative research for nutrition- and climate-smart food systems in low- and middle-income countries. *Nutrients, 15*(13), 3020.

Günal, A. M., Cantürk, S., Yılmaz, S., Boz, C., & Karabay, D. (2025). Examining the interconnections among income, food prices, food insecurity, and health expenditure: A multicausality approach. *BMC Public Health, 25*(1), 2778. https://doi.org/10.1186/s12889-025-24153-6

Harish, M. S., Bhuker, A., & Chauhan, B. S. (2024). Millet production, challenges, and opportunities in the Asia-pacific region: A comprehensive review. *Frontiers in Sustainable Food Systems, 8*, 1386469. https://doi.org/10.3389/fsufs.2024.1386469

Heikonen, S., Heino, M., Jalava, M., Siebert, S., Viviroli, D., & Kummu, M. (2025). Climate change threatens crop diversity at low latitudes. *Nature Food, 6*(4), 331–342. https://doi.org/10.1038/s43016-025-01135-w

Heino, M., Kinnunen, P., Anderson, W., Ray, D. K., Puma, M. J., Varis, O., et al. (2023). Increased probability of hot and dry weather extremes during the growing season threatens global crop yields. *Scientific Reports, 13*(1), 3583. https://doi.org/10.1038/s41598-023-29378-2

Huang, X., Su, D., & Xu, C. (2025). Revitalizing orphan crops to combat food insecurity. *Nature Communications, 16*(1), 10596. https://doi.org/10.1038/s41467-025-66020-3

IPCC. (2023). Summary for policymakers. In H. Lee, & J. Romero (Eds.), *Climate change 2023: Synthesis report. A report of the intergovernmental panel on climate change. Contribution of working groups I, II and III to the sixth assessment report of the intergovernmental panel on climate change [Core Writing Team].*

Jaikishun, S., Li, W., Yang, Z., & Song, S. (2019). Quinoa: In perspective of global challenges. *Agronomy, 9*(4), 176.

Janjua, A. A., Aslam, M., Sultana, N., & Batool, Z. (2021). Identification of climate induced optimal rice yield and vulnerable districts rankings of the Punjab, Pakistan. *Scientific Reports, 11*(1), 23393. https://doi.org/10.1038/s41598-021-02691-4

Joshi, B. K. (2023). Chapter 7—Buckwheat (Fagopyrum esculentum Moench and *F. tataricum* Gaertn.). In M. Farooq, & K. H. M. Siddique (Eds.), *Neglected and underutilized crops* (pp. 151–200). Academic Press. https://doi.org/10.1016/B978-0-323-90537-4.00016-8

Kabato, W., Hailegnaw, N., Mutum, L., & Molnar, Z. (2025). Managing soil health for climate resilience and crop productivity in a changing environment. *Science of the Total Environment, 1000*, 180460. https://doi.org/10.1016/j.scitotenv.2025.180460

Kaur, S., Bedi, M., Singh, S., Kour, N., Bhatti, S. S., Bhatia, A., et al. (2024). Chapter 8—monoculture of crops: A challenge in attaining food security. In A. Sharma, M. Kumar, & P. Sharma (Eds.), *Advances in food security and sustainability* (Vol. 9, pp. 197–213). Elsevier. https://doi.org/10.1016/bs.af2s.2024.07.008

Kaya, Ç. I., Yazar, A., & Sezen, S. M. (2015). SALTMED model performance on simulation of soil moisture and crop yield for quinoa irrigated using different irrigation systems, irrigation strategies and water qualities in Turkey. *Agriculture and Agricultural Science Procedia, 4*, 108–118. https://doi.org/10.1016/j.aaspro.2015.03.013

Khan, A., Ahmad, M., Ahmed, M., & Iftikhar-Hussain, M. (2021). Rising atmospheric temperature impact on wheat and thermotolerance strategies. *Plants, 10*(1), 43.

Kheya, S. A., Talukder, S. K., Datta, P., Yeasmin, S., Rashid, M. H., Hasan, A. K., et al. (2023). Millets: The future crops for the tropics—status, challenges and future prospects. *Heliyon, 9*(11), e22123. https://doi.org/10.1016/j.heliyon.2023.e22123

Knight, C., Khouakhi, A., & Waine, T. W. (2024). The impact of weather patterns on inter-annual crop yield variability. *Science of the Total Environment, 955*, 177181. https://doi.org/10.1016/j.scitotenv.2024.177181

Li, X., Yadav, R., & Siddique, K. H. M. (2020). Neglected and underutilized crop species: The key to improving dietary diversity and fighting hunger and malnutrition in Asia and the Pacific. *Frontiers in Nutrition, 7*, 593711. https://doi.org/10.3389/fnut.2020.593711

Li, Z., Gu, W., & Meng, Q. (2023). The impact of COVID-19 on logistics and coping strategies: A literature review. *Regional Science Policy and Practice, 15*(8), 1768–1795. https://doi.org/10.1111/rsp3.12665

Liu, J., Wu, J., Jiang, D., Chen, S., Hao, M., Ding, F., et al. (2025). Research on the impact of climate change on food security in Africa. *Scientific Reports, 15*(1), 31251. https://doi.org/10.1038/s41598-025-14560-5

Liu, L., Ross, H., & Ariyawardana, A. (2023). Building rural resilience through agri-food value chains and community interactions: A vegetable case study in Wuhan, China. *Journal of Rural Studies, 101*, 103047. https://doi.org/10.1016/j.jrurstud.2023.103047

López, C., Morales-Contreras, M. F., Langella, I. M., & Alonso-Monge, J. (2025). Modeling supply chain disruptions due to geopolitical reasons: A systematic literature review. *Transportation Research Part e: Logistics and Transportation Review, 202*, 104290. https://doi.org/10.1016/j.tre.2025.104290

Luthar, Z., Golob, A., Germ, M., Vombergar, B., & Kreft, I. (2021). Tartary buckwheat in human nutrition. *Plants, 10*(4), 700.

Maxwell, D., Khalif, A., Hailey, P., & Checchi, F. (2020). Viewpoint: Determining famine: Multi-dimensional analysis for the twenty-first century. *Food Policy, 92*, 101832. https://doi.org/10.1016/j.foodpol.2020.101832

Maxwell, D., Young, H., Jaspars, S., Frize, J., & Burns, J. (2011). Targeting and distribution in complex emergencies: Participatory management of humanitarian food assistance. *Food Policy, 36*(4), 535–543. https://doi.org/10.1016/j.foodpol.2011.03.010

Meenakshi, J. V., & Quisumbing, A. (2025). Diet quality and micronutrient intakes in nutritional value chains: A synthesis and suggestions for further research. *Food Policy, 130*, 102789. https://doi.org/10.1016/j.foodpol.2024.102789

Mehrabi, Z., Delzeit, R., Ignaciuk, A., Levers, C., Braich, G., Bajaj, K., et al. (2022). Research priorities for global food security under extreme events. *One Earth, 5*(7), 756–766. https://doi.org/10.1016/j.oneear.2022.06.008

Mittler, R., Karlova, R., Bassham, D. C., & Lawson, T. (2025). Crops under stress: Can we mitigate the impacts of climate change on agriculture and launch the 'resilience revolution'? *Philosophical Transactions of the Royal Society b: Biological Sciences, 380*(1927), 228. https://doi.org/10.1098/rstb.2024.0228

Mmbando, G. S. (2025). The current challenges and solutions on utilizing the potential of underutilized and neglected crops for Africa's food security. *Discover Plants, 2*(1), 163. https://doi.org/10.1007/s44372-025-00261-w

Mohamed-Shaffril, H. A., Abu-Samah, A., Samsuddin, S. F., Ahmad, N., Tangang, F., Ahmad-Sidique, S. F., et al. (2024). Diversification of agriculture practices as a response to climate change impacts among farmers in low-income countries: A systematic literature review. *Climate Services, 35*, 100508. https://doi.org/10.1016/j.cliser.2024.100508

Mottaleb, K. A., Kruseman, G., & Snapp, S. (2022). Potential impacts of Ukraine-Russia armed conflict on global wheat food security: A quantitative exploration. *Global Food Security, 35*, 100659. https://doi.org/10.1016/j.gfs.2022.100659

Ngono, A. M., Ongo-Nkoa, B. E., Depret, M.-H., & Asafor, C. E. (2025). Economic resilience in Central Africa in the face of COVID-19: Corruption, a hindrance or a key factor? *BMC Health Services Research, 25*(1), 1204. https://doi.org/10.1186/s12913-025-13163-0

Odoms-Young, A., Brown, A. G. M., Agurs-Collins, T., & Glanz, K. (2024). Food insecurity, neighborhood food environment, and health disparities: State of the science, research gaps and opportunities. *American Journal of Clinical Nutrition, 119*(3), 850–861. https://doi.org/10.1016/j.ajcnut.2023.12.019

Odongo, R. A., Schrieks, T., Streefkerk, I., de Moel, H., Busker, T., Haer, T., et al. (2025). Drought impacts and community adaptation: Perspectives on the 2020–2023 drought in East Africa. *International Journal of Disaster Risk Reduction, 119*, 105309. https://doi.org/10.1016/j.ijdrr.2025.105309

Omotayo, A. O., Omotoso, A. B., & Asong, J. A. (2025). Leveraging Africa's underutilized crops to combat climate change, water scarcity, and food insecurity in South Africa. *Scientific Reports, 15*(1), 19404. https://doi.org/10.1038/s41598-025-03853-4

Onyeneke, R. U., Chidiebere-Mark, N. M., Ukoha, I. I., Onyeneke, C. J., & Onoh, P. A. (2025). Assessing the impact of climatic factors on underutilized crops: Novel evidence from Nigeria. *Theoretical and Applied Climatology, 156*(9), 474. https://doi.org/10.1007/s00704-025-05709-3

Otekunrin, O. A. (2025). A critical assessment of the interplay of conflict, hunger, poverty, and food insecurity in Africa. *Food and Humanity, 4*, 100544. https://doi.org/10.1016/j.foohum.2025.100544

Pérez, C. F., Vicario, S. A., van Cauwenbergh, N., Garcia, M., & Werner, M. (2024). Disentangling the socio-natural dynamics of drought and water scarcity in Colombia's Tropical Andes. *Journal of Hydrology: Regional Studies, 56*, 102068. https://doi.org/10.1016/j.ejrh.2024.102068

Pudasaini, R., & Raizada, M. N. (2025). Potential of short-duration millets as a post-winter wheat crop in Ontario, Canada. *Agronomy Journal, 117*(1), e21740. https://doi.org/10.1002/agj2.21740

Purba, N. H., & Krishnaswamy, K. (2025). Exploring the potentials of neglected underutilized crops (NUCs): An integrative review for developing a sustainable food system model. *NPJ Science of Food, 9*(1), 199. https://doi.org/10.1038/s41538-025-00554-0

Raza, A., Khare, T., Zhang, X., Rahman, M. M., Hussain, M., Gill, S. S., et al. (2025). Novel strategies for designing climate-smart crops to ensure sustainable agriculture and future food security. *Journal of Sustainable Agriculture and Environment, 4*(2), e70048. https://doi.org/10.1002/sae2.70048

Saccone, D., & Vallino, E. (2025). Global food security in a turbulent world: Reviewing the impacts of the pandemic, the war and climate change. *Agricultural and Food Economics, 13*(1), 47. https://doi.org/10.1186/s40100-025-00388-0

Said, Z., Vigneshwaran, P., Shaik, S., Rauf, A., & Ahmad, Z. (2025). Climate and carbon policy pathways for sustainable food systems. *Environmental and Sustainability Indicators, 27*, 100730. https://doi.org/10.1016/j.indic.2025.100730

Samal, A., Ummalla, M., & Goyari, P. (2022). The impact of macroeconomic factors on food price inflation: An evidence from India. *Future Business Journal, 8*(1), 15. https://doi.org/10.1186/s43093-022-00127-7

Sandford, C., Dunn, R., Titchner, H., Kendon, M., Rayner, N., Morice, C., et al. (2024). Global and regional climate in 2023. *Weather, 79*(12), 400–412. https://doi.org/10.1002/wea.7636

Santunione, G., & Montevecchi, G. (2025). Superfoods: Exploring sustainability perspectives between nutrient synthesizers and accumulators. *Frontiers in Food Science and Technology, 5*, 1507933. https://doi.org/10.3389/frfst.2025.1507933

Shafiq, S., Zia-Ul-Haq, M., Naqvi, S. A. R., Sarfaraz, W., Ali, H., Islam, M. M., et al. (2025). Integrating neglected and underutilized crops (NUCs) in South Asian cropping systems and diets: Challenges and prospects. *Regional Sustainability, 6*(4), 100242. https://doi.org/10.1016/j.regsus.2025.100242

Singh, V., Vijay, A., & Choudhary, P. (2025). Climate change and food production: Small millets cultivation ensures global food supply and nutritional security. *Plant, Cell and Environment*. https://doi.org/10.1111/pce.70120

Sporchia, F., Antonelli, M., Aguilar-Martínez, A., Bach-Faig, A., Caro, D., Davis, K. F., et al. (2024). Zero hunger: Future challenges and the way forward towards the achievement of sustainable development goal 2. *Sustainable Earth Reviews, 7*(1), 10. https://doi.org/10.1186/s42055-024-00078-7

Sun, J., Zhou, M., & Wang, S. (2025). Localized practices of rural tourism makers from a resilience perspective: A comparative study in China. *Journal of Rural Studies, 119*, 103722. https://doi.org/10.1016/j.jrurstud.2025.103722

Syed, A., Raza, T., Bhatti, T. T., & Eash, N. S. (2022). Climate impacts on the agricultural sector of Pakistan: Risks and solutions. *Environmental Challenges, 6*, 100433. https://doi.org/10.1016/j.envc.2021.100433

Talucder, M. S. A., Ruba, U. B., & Robi, M. A. S. (2024). Potentiality of neglected and underutilized species (NUS) as a future resilient food: A systematic review. *Journal of Agriculture and Food Research, 16*, 101116. https://doi.org/10.1016/j.jafr.2024.101116

Tchonkouang, R. D., Onyeaka, H., & Nkoutchou, H. (2024). Assessing the vulnerability of food supply chains to climate change-induced disruptions. *Science of the Total Environment, 920*, 171047. https://doi.org/10.1016/j.scitotenv.2024.171047

UNDP. (2020). *Human development report 2020-the next frontier: Human development and the anthropocene*. https://hdr.undp.org/content/human-development-report-2020

Urak, F., Bilgic, A., Florkowski, W. J., & Bozma, G. (2024). Confluence of COVID-19 and the Russia-Ukraine conflict: Effects on agricultural commodity prices and food security. *Borsa Istanbul Review, 24*(3), 506–519. https://doi.org/10.1016/j.bir.2024.02.008

Vărzaru, A. A. (2025). Digital revolution in agriculture: Using predictive models to enhance agricultural performance through digital technology. *Agriculture, 15*(3), 258.

Vilakazi, B., Mafongoya, P. L., Odindo, A. O., & Phophi, M. M. (2025). The role of neglected grain legumes in food and nutrition security and human health. *Sustainability, 17*(1), 350.

Vos, R., Glauber, J., Hebebrand, C., & Rice, B. (2025). Global shocks to fertilizer markets: Impacts on prices, demand and farm profitability. *Food Policy, 133*, 102790. https://doi.org/10.1016/j.foodpol.2024.102790

Wang, S., Mustafa, P. G., Mwalupaso, G. E., & Li, Z. (2025). Nexus of climate change adaptation and household wealth in climate risk hotspots: Insights from rural farm households of Pakistan. *Climate Risk Management, 48*, 100704. https://doi.org/10.1016/j.crm.2025.100704

Woo, S. H., Roy, S. K., Kwon, S. J., Cho, S. W., Sarker, K., Lee, M. S., et al. (2016). Chapter three—concepts, prospects, and potentiality in buckwheat (*Fagopyrum esculentum* Moench): A research perspective. In M. Zhou, I. Kreft, S.-H. Woo, N. Chrungoo, & G. Wieslander (Eds.), *Molecular breeding and nutritional aspects of buckwheat* (pp. 21–49). Academic Press. https://doi.org/10.1016/B978-0-12-803692-1.00003-1

Yadav, L., & Upasana, J. (2026). Chapter 3—heritage crops and forgotten foods: Rediscovering ancient grains and varieties. In T. Sarkar, & S. Smaoui (Eds.), *Health, nutrition and sustainability* (pp. 43–56). Academic Press. https://doi.org/10.1016/B978-0-443-32920-3.00002-1

Yesuph, D. S., Dalka, D. D., & Baye, M. T. (2025). Underutilized and neglected crop species and their role in enhancing household food security amid climate change, Wolaita Zone, Ethiopia. *Heliyon, 11*(3), e42345. https://doi.org/10.1016/j.heliyon.2025.e42345

Yuan, X., Li, S., Chen, J., Yu, H., Yang, T., Wang, C., et al. (2024). Impacts of global climate change on agricultural production: A comprehensive review. *Agronomy, 14*(7), 1360.

Zewdu, D., Krishnan, C. M., Raj, P. P. N., Arlikatti, S., & McAleavy, T. (2025). Climate-smart innovation practices and sustainable rural livelihoods: A systematic literature review. *Technology in Society, 82*, 102914. https://doi.org/10.1016/j.techsoc.2025.102914

Chapter 2
Challenges and Opportunities in the Production of Neglected and Underutilized Crops Under Climate Change

Mukhtar Ahmed, Munir Ahmad, Ahmed M. S. Kheir, Mustafa Kan, and P. V. Vara Prasad

Abstract The problem of climate change is a major threat to global food security as it increases the magnitude of abiotic stress, disrupts crop production, and determines the fragility of food systems favoring a few large crops. The crops that are neglected and underutilized (NUCs) have a great potential to increase food and nutritional security, climate resilience, and sustainable agricultural development, especially in marginal and resource-constrained settings. This chapter is a synthesis of the current understanding of the role of NUCs in climate-resilient food systems, through the critical analysis of the biological, agronomic, environmental, socio-economic, institutional, and policy-related factors that affect their production and use. The marginalization of NUCs is associated with the limited genetic enhancement, ineffective seed production systems, non-standardized agronomic protocols, poor value chains, low product

---

M. Ahmed (✉)
Department of Agronomy, Pir Mehr Ali Shah Arid Agriculture University, Rawalpindi, Pakistan
e-mail: ahmadmukhtar@uaar.edu.pk

Swedish University of Agricultural Sciences, Umea, Sweden

Department of Biological Systems Engineering, Washington State University, Pullman, WA, USA

M. Ahmad
Department of Plant Breeding and Genetics, Pir Mehr Ali Shah Arid Agriculture University, Rawalpindi, Pakistan

A. M. S. Kheir
Julius Kühn Institute (JKI)—Federal Research Centre for Cultivated Plants, Institute for Strategies and Technology Assessment, Kleinmachnow, Germany

Soils, Water and Environment Research Institute, Agricultural Research Center, Giza, Egypt

M. Kan
Department of Agricultural Economics, Agricultural Faculty, Kirsehir Ahi Evran University, Kırşehir, Türkiye

P. V. Vara Prasad
Department of Agronomy, Kansas State University, Manhattan, KS, USA

M. Ahmed (ed.), *Climate Change and Neglected Underutilized Crops*,
https://doi.org/10.1007/978-3-032-21005-0_2

quality, low demand, and insufficient policy. The chapter takes a systems viewpoint in examining the issues and opportunities, which come with mainstreaming NUCs in varying climatic conditions. It uses systematic analysis systems such as SWOT analysis and constraint-opportunity evaluations to find out crucial leverage points on how to enhance productivity, adoption, and integration of NUCs to current farming systems. To a large extent, special attention is given to regional variability and context-related limitations, and the production capabilities and problems of NUCs in Pakistan are discussed in detail. Moreover, the chapter presents the potential that is offered by new tools and technologies, including genomics and other omics methods, crop and climate modeling, remote sensing, geographic information systems, artificial intelligence, and digital agriculture, to overcome research gaps, expedite genetic returns, enhance climate-risk assessment, and reinforce decision support systems. All in all, the chapter reinforces the need to pursue research coordination, which allows policies, institutional support, and market development to realize the full potential of NUCs to establish resilient, nutrition-sensitive, and sustainable food systems in the impact of climate change.

Keywords Neglected and underutilized crops · Climate change · Food security · Climate resilience · Sustainable agriculture · Food systems

2.1 Introduction

Natural catastrophes, economic crises, pandemics, and others have produced extreme shocks that have repeatedly revealed weak spots in food systems, many of which have long-term impacts (Kc et al., 2024). The COVID-19 pandemic alone drove 83–132 more people into hunger and malnutrition by 2020 (FAO, 2020). These crises demonstrate the importance of learning how food systems react to stressors and to learn the lessons of shocks in the past to make systems better resilient to future shocks. The essence of food system resilience (FSR) has been a problematic idea since the introduction of the Zero Hunger (SDG 2) objective by the United Nations in 2015. To achieve FSR, it is necessary to have the capacity to quickly evaluate the food systems, track reactions of households to countries and determine their resilience to unexpected shocks. Conceptualization and systematic evaluation of FSR will thus become necessary in the future to inform the formulation of effective policies and interventions that will ensure the protection of food security in a more uncertain international environment.

Resilience is a concept originally created in ecology, but which has been applied to food systems to explain what defines their ability to survive shocks without affecting their long-term operations (Béné et al., 2019). Food system resilience (FSR) refers to the ability of a food system and food system units (at different levels) to generate adequate, suitable, and available food to everyone, during different and even unpredictable disruptions (Tendall et al., 2015). FSR focuses on learning and improvement as opposed to status quo and is based on three themes: capacities, temporal variation, and spatial variation. Capacities consist of absorbing, recovering, adaptable as well as transformation capacity, which has a biophysical, socio-economic as well as diversification dimension. Temporal variation covers both the gradual changes (climatic change) and abrupt changes whereas spatial variation involves multi-level governance and scale-specific resilience indicators (Kc et al., 2024).

The global food system is a complicated system of systems, which includes related economic, social, ecological, health, technological, and financial subsystems. It is gradually emerging as a social-ecological system, in which multi-scalar and interdependent elements work both over time and space (Allen & Prosperi, 2016) (Fig. 2.1). Food systems are global, regional, national, and local, with local ones being extremely diverse and situation-specific. Internal forces such as productivity increase and external forces such as population increase, conflict, urbanization, or political upheaval cause those changes, some of which, including trade or infrastructure are both. There are cascading effects and feedbacks, either in real-time or delayed, making them more complex than linear dynamics and not predictable, a reality that was emphasized by COVID-19 and recent food price crises (Dinku et al., 2023).

The outcomes of food systems are food and nutrition security, environmentally friendly, and social equity. To achieve resilience, it is necessary to go beyond siloed solutions, recognize complexity of systems and deal with interdependent drivers, vulnerabilities and shocks to protect the long-term food security. The food system in the modern world generates more than it has ever generated, but unseen costs are to the health, the environment, and even social inclusion, which are significant dangers in the long term (Fanzo et al., 2021). Productivity, low price, and availability policies have had a threefold increase in the yield of cereal harvest since 1961 (FAO, 2024) and decreased the price of staple commodities, contributing to the increase of the population, as well as a decrease in famines (Davis et al., 2024). Nevertheless, this attention has deferred expenses, which contributes to endemic diet-related diseases, environmental pollution a strain that has been termed as the paradox of productivity. The solution to this paradox can be found in the NUCs. Production diversification through NUCs can improve nutrition, environmental sustainability, and social inclusion by producing other food items other than high-yield staples, and add to a more robust and balanced food system (Sultanbawa & Sivakumar, 2022; Tamburini et al., 2020).

As economies grow and develop, and countries become open to the global economy, they tend to transform their traditional, rural food systems into industrialized, modernized ones. The result of this development is the rise of agricultural productivity, concentration of the production process, reduction in the costs of transactions, and involvement in international trade. More individuals are able to afford a variety of diets and more people tend to migrate to other non-farm or more remunerative jobs in urban areas (Reardon & Timmer, 2007). Nevertheless, these changes are trade-offs: productivity and affordability are not necessarily translated into better nutrition, environmental sustainability, and social inclusion. The gaps can be filled through NUCs. NUCs improve the quality of diets, minimize negative effects on the

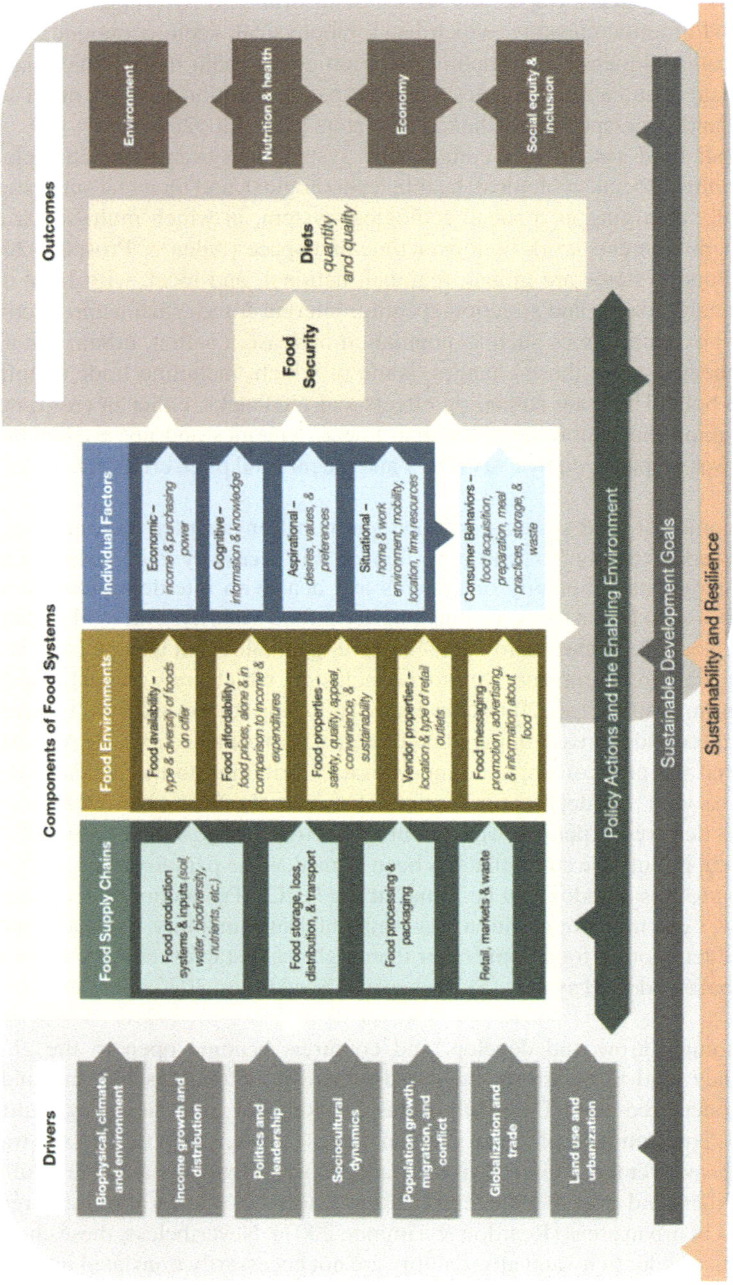

Fig. 2.1 The food system

environment, and promote resiliency, which is a kind of a way to a more sustainable and inclusive food system (Deng et al., 2025).

Even with the highest food production rates, the access to healthy diets is still lacking- the inability to afford a healthy diet is experienced by more than three billion people. As hunger and poverty are returning, so are obesity and non-communicable diseases that are caused by diet. Food system activities have violated five of nine planetary boundaries, such as biodiversity, freshwater consumption, acreage growth, GHG emissions, and nutrient application, and augment extreme weather events diminish ecological resilience (IPCC, 2023). The existing food systems produce both negative and positive externalities, including health and environmental harm and positive effects such as pollination and carbon sequestration, and climate resilience. The combination of trade-offs that can be realized by mitigating these disadvantages and unmet needs in diversified food systems through the introduction of NUCs to mitigate their consequences, as well as considering negative externalities and increasing beneficial externalities. This chapter has been written with the objectives that includes (i) To examine the challenges facing the production of neglected and underutilized crops under climate change, (ii) To evaluate the role of NUCs in addressing the paradox of productivity and enhancing nutrition, environmental sustainability, and social inclusion, (iii) To analyze the systemic drivers, trade-offs, and externalities affecting global and local food systems and (iv) To propose strategies for integrating NUCs into resilient and diversified food systems to improve food security and climate adaptation.

2.2 Challenges in the Production of NUCs

Although the NUCs have been identified to have certain nutritional potential, climate resilience, and could play a role in transforming the food systems into more sustainable ones, they are still sidelined in the mainstream agricultural practices (Fig. 2.2). The few cases of their introduction and commercialization are attributed to various, compound factors in the biological, agronomic, social-economic, institutional, and technological spheres. The causes of these problems lie in historical unimportance, policy favoring of the major staples, and structural changes of food systems around the world. These social, economic, environmental, agronomic and institutional challenges, not only restrict conservation and utilization of neglected and underutilized crops, but also hinder the achievement of various SDGs, especially SDGs 2, 12, 13 and 15 which highlights the importance of interventions between policy and research (Table 2.1)(Purba & Krishnaswamy, 2025).

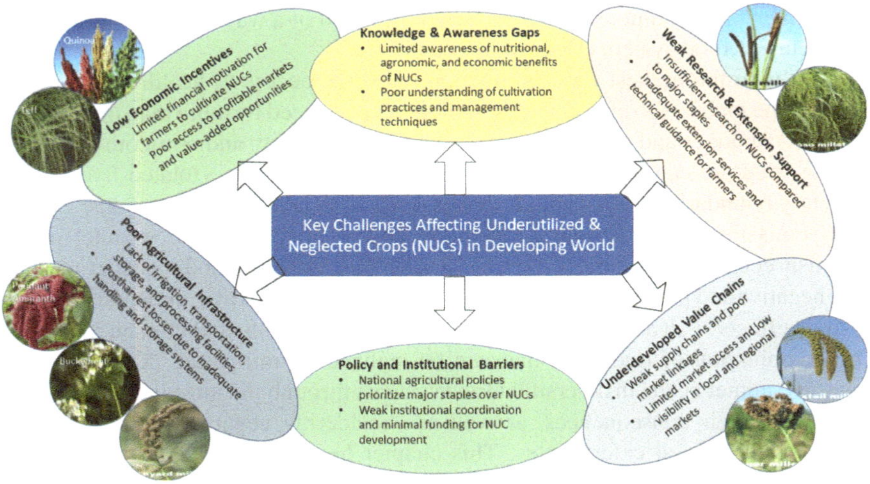

Fig. 2.2 Key challenges affecting NUCs

2.2.1 Biological and Genetic Challenges

The majority of NUCs do not have well-defined germplasm, reference genomes, and the molecular breeding tools. They are usually farmed as unimproved landraces that have low yields potential, phenotypic variability and asynchronous maturity. Lack of organized breeding programs and good seed systems has restricted genetic gain, and the process of breeding stress-tolerant as well as high-performing varieties (Tuggle et al., 2024).

2.2.2 Agro-environmental and Economic Problems

NUCs are mostly cultivated in marginal conditions where there is poor soils, scarce water, and high levels of climate extremities like drought, heat, salinity, and new pest pressures. Absence of a set of standardized agronomic packages regarding planting density, nutrient management, water use, and pests lead to low consistency in productivity and increased risk in production.

2.2.3 Market Challenges and Socio-economic Challenges

Limited processing and storage facilities, weak value chains and low market integration can play a significant role in making NUCs less economically attractive. The

Table 2.1 Challenge to NUCs by category and their links to the sustainable development goals (SDGs) (Purba & Krishnaswamy, 2025)

Constraint category	Key issues	Relevant SDGs
Social	Replacement of traditional crops with modern varieties; dietary shifts due to urbanization; erosion of indigenous knowledge; perception of native foods as low-status; rural–urban migration; unsustainable use of natural resources	*SDG 2* (zero hunger), *SDG 3* (good health and well-being), *SDG 4* (quality education), *SDG 11* (sustainable cities and communities), *SDG 12* (responsible consumption and production)
Economic	Land-use changes favoring commercial crops; weak market infrastructure; low economic value and poor competitiveness of NUCs; lack of market niches; absence of financial incentives for farmers	*SDG 1* (no poverty), *SDG 2* (zero hunger), *SDG 8* (decent work and economic growth), *SDG 9* (industry, innovation and infrastructure), *SDG 12* (responsible consumption and production)
Environmental	Genetic erosion due to pests, diseases, droughts, fires, overgrazing, deforestation, and land degradation; climate change impacts; pollution; ecosystem degradation	*SDG 13* (climate action), *SDG 15* (life on land), *SDG 6* (clean water and sanitation), *SDG 12* (responsible consumption and production)
Agronomic	Inadequate seed supply systems; limited availability of quality propagation material; lack of trained human resources; excessive use of fertilizers and agrochemicals	*SDG 2* (zero hunger), *SDG 4* (quality education), *SDG 12* (responsible consumption and production), *SDG 15* (life on land)
Political / institutional	Low prioritization of NUCs in national agendas; insufficient funding for conservation and research; lack of infrastructure and energy for in situ conservation; absence of breeding and characterization data; weak or fragmented policies, programs, and legal frameworks	*SDG 2* (zero hunger), *SDG 16* (peace, justice and strong institutions), *SDG 17* (partnerships for the goals), *SDG 9* (industry, innovation and infrastructure)

level of consumer awareness is low and the continued attitudes of NUCs as food of the poor or famine food further de-demand. Formal markets within the NUCs are disadvantaged by competition with subsidized and highly marketed staple crops.

2.2.4 Policy, Institutional and Research Problems

NUCs have been neglected in national agricultural policies, food security measures as well as government investment programs. Funding on research and development is low, extension services are incompetent in NUC and the institutional coordination among seed systems, markets and nutrition programs is low. This policy does not help break the cycle of low visibility and the lack of adoption.

2.2.5 Technological Challenges

Poorly designed mechanization to suit small seeded crops and structurally different crops make work harder and less efficient. The availability of limited data does not allow NUCs to be included into crop models, climate risk models, and digital agriculture systems. There is a poor level of post-harvest technologies and value-addition innovations especially in the midstream processing sector.

2.3 Identification of NUC Challenges Using SWOT Analysis

In order to evaluate these challenges systematically and determine strategic entry points where the intervention could be done, a SWOT analysis framework was used. In this strategy, the internal and external forces affecting NUCs would be categorized as strengths, weaknesses, opportunities and threats thus converting observed problems into actionable knowledge (Fig. 2.3) (Purba & Krishnaswamy, 2025).

2.3.1 Weaknesses (Internal Constraints)

The Internal Constraints (Weaknesses) are strengths, resources, and capabilities that fail to enhance the organization's strategy, mission, and vision. The above biological, agronomic, and technological issues are captured as main weaknesses in the SWOT model. These are fewer genetic enhancement, no uniform production methods, low outputs, a poor processing and value-addition facility, and inadequate mechanization and digitalization. All these weaknesses lower productivity, competitiveness, and scalability of NUCs.

2.3.2 Threats (External Pressures)

Tremendous external threats are socio-economic and policy related challenges. The role of NUCs in food systems is compromised by negative consumer attitudes, ostracism in the market, the dominance of key staple crops, urbanization, and cultural homogenization. Their sustainability and conservation are also jeopardized by policy biasness, low investment by the people, and susceptibility to climatic extremities.

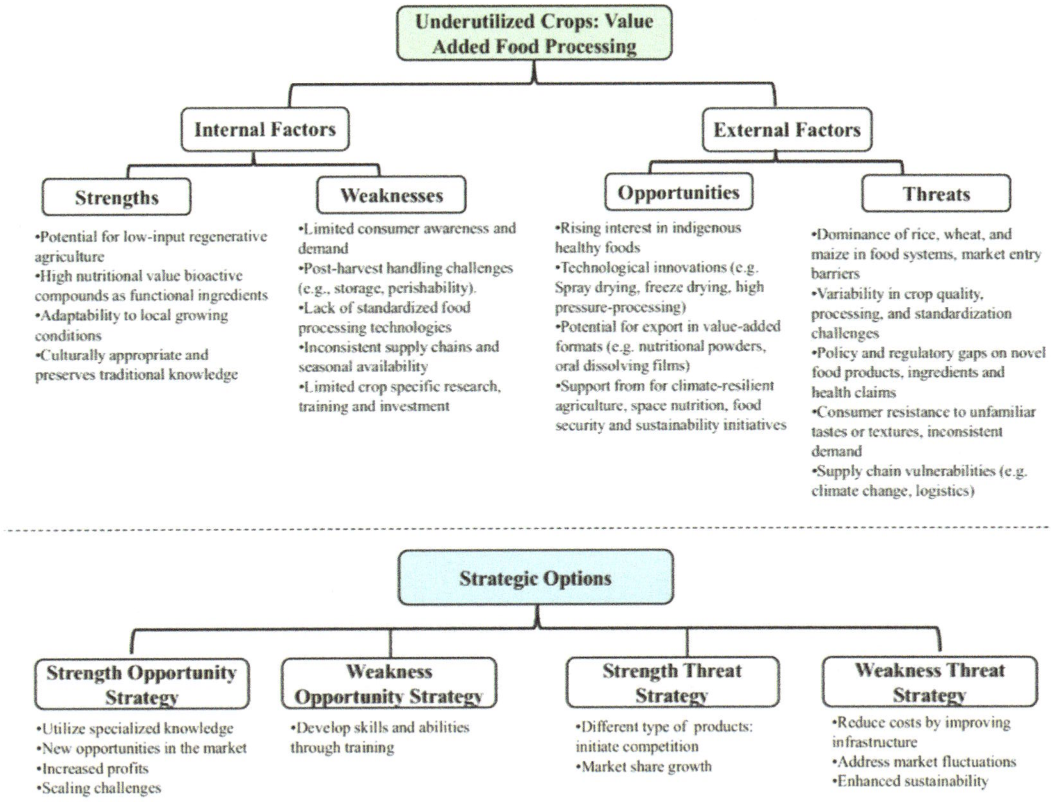

Fig. 2.3 SWOT analysis of NUCs. *Source* Purba and Krishnaswamy (2025)

2.3.3 Strengths (Inherent Advantages)

In spite of such limitations, NUCs have some major strengths, which offset most of the identified challenges. They are nutritional density, resistance to climate change, marginal environment adaptation, low-input needs, and connection to regional food cultures and traditional knowledge systems. Such attributes offer a solid basis of procuring food and nutrition security in the climate change.

2.3.4 Opportunities (Pathways for Overcoming Challenges)

New opportunities match well with the solutions to the given challenges. The rising consumerism in healthy and sustainable diets, the development of niche and functional food markets, more NUCs in agroecological processes, and the international efforts of promoting SDG 2 open a promising perspective of reintegrating NUCs into conventional agriculture. The further opportunities to scale NUCs are also provided by the evolution of plant breeding, the use of digital agriculture, and all-inclusive policy frameworks.

2.3.5 Synthesis

Purba and Krishnaswamy (2025) emphasized that the weakness of NUCs lies not in its nature but in the system by initially identifying the multidimensional nature of the challenges associated with neglected and underused crops, and then putting them into focus via a SWOT analysis. The SWOT model explains how the current weaknesses and threats are overcome by the inherent strengths and emergent opportunities to offer a systematic foundation on how to integrate NUCs in mainstreaming of resilient, nutrition-sensitive and sustainable food systems figure.

Figure 2.4 provides a combination of inputs and outputs and impacts of mainstreaming NUCs into resilient food systems to achieve well-specified final results. The framework starts with various inputs such as awareness creation on multilingual media, public engagement program, capacity building, stakeholder inclusion, targeted fundraising, contextual planning triggered by climate and nutrition, research gap identification, priority setting, educational integration, and powerful monitoring and evaluation mechanisms. The inputs are channeled to the joint efforts like training and workshops, dissemination through social platforms and visible materials, development of strategies to address NUCs specific to a region, development of pre- and post-harvest technologies, integration of NUCs in food systems and biofortification efforts, creation of resilient linear and circular food supply chains, and good knowledge and information management. The overall effects of these processes can be seen in final results with environmental, social, and economic impacts: environmentally, these processes will result in better conservation of biodiversity and improved soil health with positive effects on human well-being; socially, in strengthened empowerment of women and increased inclusion of marginalized communities; and, economically, in the development of new value-added products and diversified livelihood opportunities and employment, which will render NUCs a strategic solution to resilient, inclusive, and sustainable agri-food systems.

2.4 Regional Constraints and Opportunities for Nutrient-Rich, Climate-Resilient Neglected and Underutilized Crops in Pakistan

The constraints opportunities matrix demonstrates the fact that there is a strong regional heterogeneity in the issues and opportunity of the NUCs in the diverse agro-ecological areas of Pakistan (Table 2.2). Even though prevalent limitations, including poor seed regimes, inadequate market integration, loss of native knowledge and low prioritization of policies are present throughout the country, their manifestation and intensity differ significantly across provinces. Meanwhile, every region offers specific opportunities, which are directly connected to its agro-climatic conditions and crop diversity drought- and salinity-tolerant NUCs in Sindh and Balochistan, high-value and climate-resilient grains in Gilgit-Baltistan and Khyber Pakhtunkhwa and nutrition sensitive diversification choices in Punjab and Azar Jammu and Kashmir. The identification of these region-specific opportunities must explicitly take into account crop species that have high potential nutritional and climate resiliency. An example is that traditional millets like pearl millet (*Pensisetum glaucum*) and foxtail millet and amaranth and cowpea in Punjab have better micronutrient profiles including calcium, iron, zinc, dietary fiber, and essential amino acids and show high tolerance to heat and water stress. Sorghum (*Sorghum bicolor*), pearl millet, and quinoa are good plants that can be used in the dryland, salty, and marginal soil conditions, with nutritional value and tolerance to abiotic stress. On the same note, barley, buckwheat, and wild leafy vegetables in KP, GB, and AJK are better adapted to cooler and mountainous conditions and are more likely to provide more protein, fiber, and micronutrients than most traditional staple cereals. Altogether, these province-specific NUCs illustrate how efforts to enhance the localized constraints by means of policy support, adaptive research and extension, market development, and community-based conservation can collectively help increase the level of climate resilience, enhance nutrition security as well as sustainable livelihoods in Pakistan.

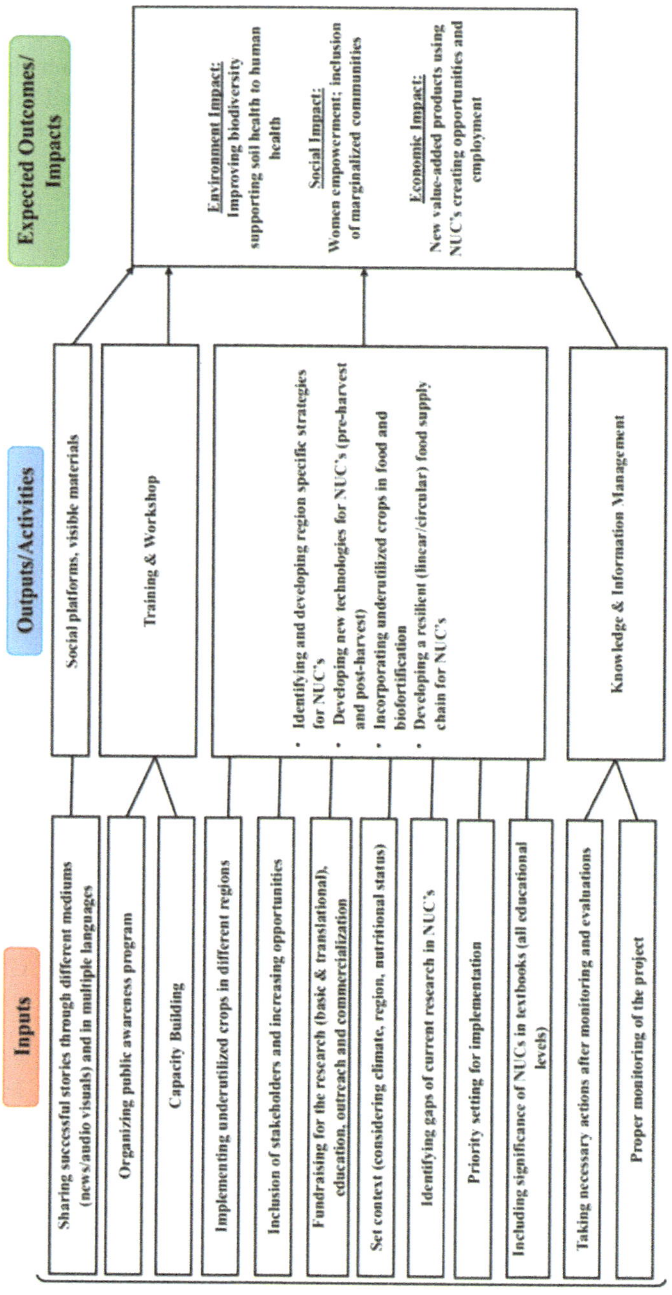

Fig. 2.4 Conceptual logic framework model incorporating neglected underutilized crops (NUCs) for sustainable food system. *Source* Purba and Krishnaswamy (2025)

Table 2.2 Constraints–opportunities matrix for NUCs in Pakistan (by province)

Province	Key NUCs	Constraint categories	Priority opportunities
Punjab	Millets (e.g., bajra/pearl millet), taro (arvi), amaranth, cowpea	*Social*: urban preference for rice/wheat and low status of traditional foods *Economic*: weak markets; lack of incentives *Agronomic*: poor seed systems	• *Awareness and demand creation*: nutrition campaigns, school feeding programs using millets and amaranth • *Market development*: strengthen rural markets, value addition (snacks/flours) • *Seed systems*: community seed banks and quality seed production hubs
Sindh	Sorghum, fonio/black millet, taro, jackbean	*Environmental*: salinity, drought and climate stress *Economic*: low competitiveness *Political*: limited policy focus	• *Climate-resilient varieties*: breeding for drought/salinity tolerance • *Policy integration*: include NUCs in provincial agriculture plans • *Private sector linkages*: support agro-processors and market niches
Khyber Pakhtunkhwa (KP)	Buckwheat, barley (local), chestnut, wild leafy vegetables	*Social*: loss of indigenous knowledge *Agronomic*: limited extension and training *Institutional*: weak research funding	• *Documentation and KB*: ethnobotanical surveys, digital knowledge hubs • *Capacity building*: extension training on NUC agronomy • *Research support*: local breeding and performance trials
Balochistan	Quinoa (adapted genotypes), teff, lentils (local landraces), wild fruits	*Environmental*: harsh climate and degradation *Agronomic*: seed/propagation gaps *Economic*: market access challenges	• *Drought-resilient practices*: water-efficient cropping and agroforestry • *Seed programs*: decentralized seed hubs and mobile labs • *Market linkage support*: value chains for hardy crops

(continued)

Table 2.2 (continued)

Province	Key NUCs	Constraint categories	Priority opportunities
Gilgit-Baltistan (GB)	Buckwheat, barley, hemp seeds, apricot kernels	*Social*: outmigration and knowledge loss *Economic*: weak infrastructure and trade links *Environmental*: fragile ecosystems	• *Community-based stewardship*: local resource protection hubs • *Value-added products*: specialty foods (high value) for tourism • *Connectivity and cold chains*: improved rural logistics
Azad Jammu and Kashmir (AJK)	Buckwheat, wild greens, pulses (local), oats	*Political*: low policy prioritization *Agronomic*: insufficient extension support *Market*: limited commercialization	• *Provincial policy inclusion*: integrate NUC agenda in AJK plans • *Farmer field schools*: local agronomy capacity development • *Market platforms*: farmers markets, cooperatives

2.5 Opportunities for Enhancing NUC Production for Food Sustainability

Crops that are NUCs have great potential in improving food security, nutrition, and agricultural sustainability especially in the face of climate change. Their broader use needs to be a holistic effort that included research on genetic diversity, better propagation and breeding, agronomic research, better soil health, market development, and value addition. Other activities that are equally important include public awareness campaigns, policy support, documentation of traditional knowledge, farmer engagement and capacity building along the value chain. These proposed initiatives are summarized in Table 2.3, and the strategies to research, commercialize, community involvement, and create climate-resilient and nutritious diets, focusing on mainstreaming NUCs to climate-resilient food systems.

The actions span research, breeding, agronomy, markets, policy, community engagement, and value addition, providing a holistic framework for sustainable food systems.

2.6 Integration of Modern Tools and Technologies

Although the NUCs have been identified to have the potential to improve food security, nutrition, and climate resilience, their application at a global scale has been limited by lack of genetic characterization, breeding programs, lack of agronomic information, and effective integration into the contemporary value chains. Current

Table 2.3 Key initiatives for promoting neglected and underutilized crops (NUCs) to enhance food security, nutrition, and climate-resilient agricultural systems

Areas of action	Proposed initiatives
Strengthening genetic diversity research	• Characterization and evaluation of NUC germplasm • Development of technically sound recommendations for sustainable use of wild and collected species • Surveys and long-term monitoring to assess genetic erosion • Analysis of land-use patterns to optimize conservation and utilization of NUC genetic resources
Advancing propagation and breeding methods	• Identification and optimization of appropriate breeding and propagation technologies • Enabling modern breeding tools (marker-assisted selection, genomics) • Reintroduction of lost or underrepresented alleles to enhance resilience and productivity
Enhancing agronomic research	• Studies on growth dynamics, yield stability, nutritional composition, harvesting practices, and pest and disease resistance • Evaluation of adaptability across agro-ecological zones
Soil health improvement and biodiversity conservation	• Use of NUCs as cover crops (e.g., teff, amaranth, cowpea) to improve soil structure, reduce erosion, enhance nitrogen fixation, and promote on-farm biodiversity
Improving market competitiveness and scalability	• Strengthening linkages between growers, traders, and processors • Identification of new cultivation regions and niche environments • Assessment of eco-geographical distribution and environmental adaptability
Public awareness, knowledge generation, and documentation	• Ethnobotanical surveys and documentation of traditional uses • Compilation of data on genetic resources and indigenous knowledge • Development of educational and awareness materials for diverse audiences
Policy, legal, and regulatory support	• Review of laws and regulations affecting NUC utilization • Organization of policy dialogues, seminars, and stakeholder consultations • Addressing market distortions caused by subsidies and policies favoring a narrow range of crops

(continued)

Table 2.3 (continued)

Areas of action	Proposed initiatives
Public awareness and behavior change campaigns	• Implementation of targeted strategies to promote NUC adoption • Impact assessments of value-addition strategies • Evaluation of conservation status and implementation of appropriate in situ and ex situ measures
Preservation of traditional knowledge systems	• Documentation and digitization of indigenous knowledge • Strengthening local knowledge systems • Increased funding for research on underutilized crops
Farmer and user participation	• Active engagement of farmers in promotion and innovation processes • Dissemination of farmer success stories and best practices
Sustainable food processing and value addition	• Reduction of post-harvest losses and food waste • Research on physical, chemical, and nutritional properties of NUCs • Development of low-carbon processing technologies and equipment • Product development to enhance nutritional value and consumer acceptability
Commercialization and market development	• Research on pricing, packaging, and product presentation • Strategies for diversified and inclusive markets • Integration into modern retail systems, including supermarkets
Alternative income generation and livelihood diversification	• Evaluation of diversified farming systems using NUCs • Assessment of crop compatibility and intercropping options • Promotion of youth, early adopters, and new entrants into NUC farming
Evidence dissemination and science–policy interface	• Sharing regional success stories in local languages • Use of multimedia platforms (text, audio-visual) • Application of Social and Behavior Change Communication (SBCC) approaches

(continued)

Table 2.3 (continued)

Areas of action	Proposed initiatives
Capacity building across the value chain	• Training programs, field demonstrations, and hands-on workshops • Strengthening farmer organizations and cooperatives • Development of practical tools for capacity enhancement
Community engagement and partnerships	• Collaboration with local communities and farmer organizations • Establishment of community-based conservation and protection hubs
Multi-stakeholder cooperation	• Strengthening collaboration at local, national, and international levels • Ensuring active participation and empowerment of women • Development of inclusive multi-stakeholder platforms
Development of climate-resilient and nutritious diets	• Promotion of region-specific NUC-based foods • Integration of NUCs into traditional diets • Research on nutritional complementarities with staple foods
Creating systemic growth opportunities	• Use of NUCs in food fortification programs • Integration into school feeding programs and school gardens • Promotion through agro-tourism initiatives • Inclusion in public procurement and national food programs

modern tools and technologies are giving greater opportunities than ever before to break these systemic barriers and mainstream NUCs into sustainable food systems. Omics sciences, advanced breeding technology, and climate projections, digital agriculture, and market systems based on data can present scalable and cost-effective options to increase the productivity, resiliency, and use of NUCs and other situations of diversified agro-ecology.

2.6.1 Omics Technologies in NUC Improvement

Genomics has emerged as a pillar towards the realization of the genetic potential of NUCs. The use of genome sequencing such as reference genomes, pan genomes allows characterization of genetic diversity, detecting adaptive traits, and storing valuable alleles, which are commonly underrepresented in major crops. Genome-wide association studies (GWAS) and marker-assisted breeding (MAS) support the study of the relationship between loci and yield stability, nutritional quality, and resistance to abiotic and biotic stresses, which is a global issue because many of the breeding histories of NUCs are limited.

Transcriptomics and proteomics fill the gaps in genomic studies since they clarify gene expression levels and protein dynamics with environmental pressures like drought, heat, and salinity, which are major limitations to NUC productivity across the world. The tools allow stress responsive genes and protein markers related to better nutritional qualities to be identified to formulate targeted crop improvement strategies.

Metabolomics and phenomics are important in providing essential information on biochemical structure and phenotype functionality of NUCs. High-throughput phenotyping platforms aid in the quick determination of multifaceted characteristics (drought resistance, biomass rise, and phenological adaptability) in varied situations with the assistance of metabolomic profiling that assists in recognizing bioactive molecules and nutritional hotspots. These strategies are especially important when it comes to NUCs that grow in the marginal and heterogeneous environment.

Epigenomics is a new frontier in NUC study, which provides information on hereditary responses to stress. The concept of epigenetic regulation can help to determine how NUCs adapt to repeated environmental stresses and what epigenetic signatures of climate resilience can be identified, which would be important in the long-term adaptation of NUCs to the shifting climatic conditions.

2.6.2 Modern Breeding Tools for NUCs

Breeding technologies of the modern breeding may tremendously increase genetic gains in NUCs and minimize the time and cost used in traditional breeding. Genome editing with CRISPR/Cas-9 provides precise and effective and efficient editing of nutritional quality, stress tolerance and yield related genes, yet without extensive linkage drag. The rapid generation cycle and breeding speed allows better varieties to be developed faster to overcome the slow pace of breeding in many NUCs. Genomic selection (GS) offers an effective solution to enhancing complex traits in NUCs including situations where there is lack of data. GS allows prediction and prior accurate selection of a limited set of phenotype and genotype data through the use of predictive models, thereby increasing the breeding efficacy. Parallel to that, pre-breeding with wild relatives and landraces is necessary to enhance the genetic diversity of NUCs and reintroduction of lost or unused alleles with resistance to environmental stressors.

2.6.3 Climate Modeling and Predictive Tools

Climate change and variability are a major threat to global NUC production especially in rainfed and marginal production. Applications of crop simulation models (like APSIM, DSSAT, and AquaCrop) can be modified to use in the NUCs to predict the growth, yield, and water use under various climatic conditions. These models are useful in assessing the management practices, examining climate risk, and determining future climate-smart zone that NUC can expand to.

Machine learning tools and artificial intelligence (AI) broaden predictive capacity by facilitating the prediction of yields, early detection of diseases outbreaks and stresses, and recommendations on data usage to select and implement varieties and to manage them. These tools, when combined with climate, soil, and crop data, help support the use of precision agriculture based on the multi-environmental setting of NUCs production.

2.6.4 Remote Sensing, GIS, and Digital Agriculture

Remote sensing technologies can be applied by the large-scale monitoring of crop health, stress, biomass and phenology to give timely and spatially explicit information that can be used to make decisions. Satelite based and UAV based information is currently being applied on evaluating drought effects, soil moisture movement, and vegetation coverage to help in risk management and adaptation to changes in NUC-based systems.

Suitable mapping in terms of GIS, based on ecological niche modeling and multi-criteria decision analysis (MCDA) can be used to identify areas of priority in terms of NUC cultivation and upscaling. These aids can be used to ensure that crop preferences are as per the environmental suitability, availability of resources, and the socio-economic conditions to curb land-use and scalability limitations.

Digital agricultural technology, such as smartphone based advisory services, IoT-based sensor and UAV-based phenotyping, improve farmers, researchers and extension services accessibility to real-time information. The technologies are especially applicable to systems with a large number of smallholders where NUCs are usually cultivated.

2.6.5 Strengthening NUC Value Chains Through Technology

Technological advances are also very important in addressing market and value chain limitations related to NUCs. Blockchain technologies and digital traceability enhance transparency, quality assurance and consumer trust, which enables premium and niche markets. Small scale processing innovations minimize after harvesting losses, increase shelf life and provide value addition with minimal environmental impacts.

Moreover, the market analytics based on AI can be used to forecast prices, analyze demand, and market intelligence, allowing producers and entrepreneurs to make sound decisions and react to the market dynamics. These technological interventions, together, are useful in changing NUCs into competitive, resilient, and economically viable participants of the global food systems.

The use of the modern tools and technologies to integrate NUCs directly aligns with several Sustainable Development Goals. The application of genomics, transcriptomics, proteomics, and metabolomics help to create nutrient-rich and climate-resistant crops, developing SDG 2 (zero hunger) and SDG 3 (good health and well-being). New breeding methods, such as genome editing, rapid breeding, and genomic selection, enhance the rapidity of the crop enhancement, and the use of genetic resources, and they lead to SDG 9 (industry, innovation and infrastructure) and SDG 12 (responsible consumption and production). Climate modeling, artificial intelligence and predicting tools enhance adaptive capacity and resilience to environmental stress and are in line with SDG 13 (climate action) and SDG 15 (life on land). Remote sensing, GIS and digital agriculture improve monitoring, management and efficient utilization of resources and help to achieve SDG 6 (clean water and sanitation) and SDG 11 (sustainable cities and communities). Lastly, value chain innovations, supported by technology such as digital traceability, AI-based market analytics, small-scale processing, etc., allow equal market access and sustainable livelihoods, which support SDG 8 (decent work and economic growth) and SDG 17 (partnerships for the goals). Combined, these concerted efforts indicate that the global limitation of NUCs should be overcome with the help of technological innovation and the further development of resilient, nutrition-sensitive, and sustainable food systems on the global scale. Figure presents connections between each technology to the constraints it directly solves and shows its role in supporting several SDGs, which would form a clear picture of where global NUC development and research should focus (Fig. 2.5).

2.7 Synthesis: Roadmap for Scaling NUCs

The multidimensional approach involving scientific innovation, policy support, market development, and networks should be used to scale NUCs into mainstream food systems. The key element of this roadmap is synergetic application of the modern technologies, such as omics-based breeding, crop modeling, and remote sensing. Through a combination of genomic knowledge and predictive models, scientists will be able to discover the most appropriate varieties to grow in certain areas, manage them most efficiently, as well as predict the effects of changes in climate variability. Subsequent digital agricultural tools and remote sensing are used to monitor the health of crops, potential yield, and stressors, whereas high-throughput phenotyping enhances the rapid selection of traits. This combined technology will make NUCs resistant to the climate and nutritionally sound to harness their maximum potential in food and nutrition security (Fig. 2.6).

Technology is however not adequate without policy reforms that are proactive in ensuring that NUC research, production and market integration is encouraged.

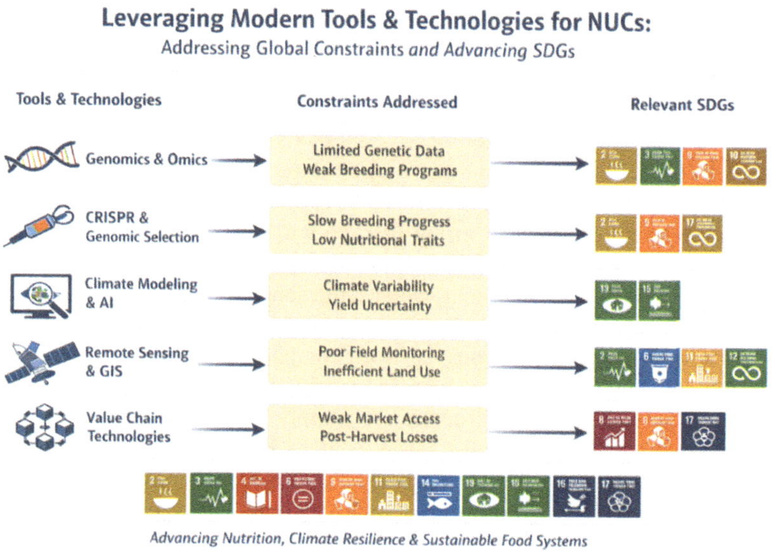

Fig. 2.5 Integration of modern tools and technologies for NUCs: addressing global constraints and advancing SDGs

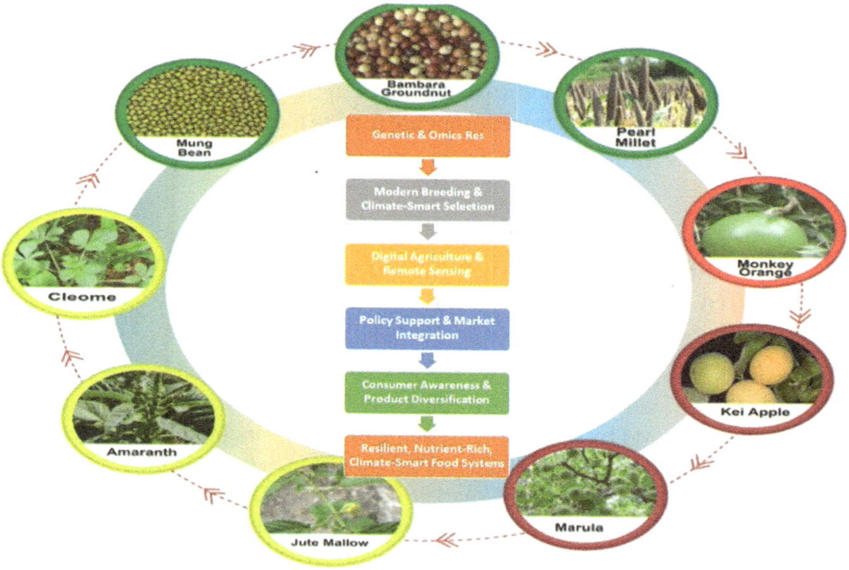

Fig. 2.6 Roadmap for scaling NUCs into mainstream food systems

Governments need to ensure that they understand the strategic importance of underutilized crops by offering them national agricultural priority, subsidies and incentives to adopt these crops by small farmers and by encouraging research and extension services. The policy frameworks that promote agroecological systems, safeguard genetic resources, and incentivize effective methods of agricultural intensification can also help NUCs to be thriving across the various agro-climatic environments. On the same note, developing quality, traceability and value addition standards will build confidence and acceptance in the market and these crops will no longer be considered niche products, but rather consumed as standard food.

Multi-stakeholder collaboration, as well as, the concept of public–private partnerships (PPP) are also very important in breaking historical bottlenecks in the deployment of NUCs. Research institutions, agritech companies, farmer cooperatives, and NGOs can combine their resources, knowledge and speed up the process of creating better varieties, processing infrastructure and supply chains. Enhancing global and regional research networks, such as CGIAR centers, FAO, and regional agricultural research centres, is a way of ensuring that activities are coordinated in breeding, climate adaptation and best practice spreading. Such networks have the potential to promote transfer of knowledge, capacity building and adoption of technology among the small holder farmers to help them exploit the full potential of the NUCs.

To sum up, scaling NUCs roadmap focuses on the integration in terms of technology, policy, research, and markets. Through a range of overlapping measures to improve genetics, climate-smart production, value chain, and policy enabling, NUCs can make the shift to use under-exploited species as resilient, nutritious, and economical parts of global food systems. Such a combined strategy, in addition to improving food security and nutrition, can make agricultural systems more resilient to climate change, so that underused crops are used to meaningfully meet Sustainable Development Goal 2 (zero hunger).

2.8 Conclusion: Cultivating a Food-Secure, Sustainable Future with NUCs

The transformative potential of the NUCs is to increase world food security, nutrition, climate resilience and sustainable livelihoods. These crops are constrained across Pakistan and the world by multidimensional constraints, including biological, agronomic, environmental, economic, and institutional, which result in the restrictiveness of the productivity, utilization, and market integration. Regional studies emphasize the dissimilarity of these challenges: whereas in certain regions there are species with drought and salinity tolerance, in others, other crops are good value cereals, including millets, buckwheat, barley, and wild leafy vegetables. Not only do such crops have better nutritional value, such as protein, fiber, and micronutrients, but are also heat-resistant, drought-resistant and able to grow on poor soils, they offer context-specific answers to climate-smart and nutrition-sensitive agriculture.

Although NUCs have ecological and nutritional benefit, they are not fully utilized because their quality and processing facilities are not consistent as well as the absence of a standard and not well defined regulatory environment regarding novel foods. The food systems are still overly reliant on a small number of staple crops globally that generate ecological instability, loss of biodiversity, and low resilience. The process of the introduction of NUCs into modern food systems will involve coordinated interventions based on modern technologies, including genomics, transcriptomics, metabolomics, genome editing, speed breeding, climate modeling, AI, remote sensing, GIS, and digital agriculture. These technologies enhance the rapidity of genetic enhancements, the finding of stress-tolerant and nutritious kinds, agronomy of positioning, and value chain fortification via market intelligence, digital provenance, and mini-processing new technologies.

The sustainable food system in the future is based on the acceptance of biodiversity and variety in agriculture, beyond the technological and policy solutions. NUCs are an array of ecological wealth and resilience, and can convert the monoculture landscapes to a highly colorful, climate-sensitive, and nutrition-sensitive ecosystems. When we place importance on these forgotten crops, we will be able to produce just, regenerative and resilient food systems- where the nature of biodiversity is honored, soils are cared about, and human nutrition is improved. This change is not just the meal-to-meal level, but the beginning of multi-generational sustainability, food sovereignty, and a better understanding of the interrelationship of all life, soils to humans.

To sum up, mainstreaming NUCs needs a two-fold strategy with scientific and technological innovation to achieve the desired level of performance through current limitations, and a philosophical and ethical dedication to biodiversity and ecological care. With the incorporation of current tools, location-specific studies, market-building, policy-enhancing, and community-enhancing, NUCs will be able to change the marginalized and underutilized crops into the major pillars of sustainable, climate-resistant, and nutritional food systems around the globe. It is high time to shift the focus away to the known and towards the unknown and unlock the potential of the crops that are underutilized to develop a food-secure, fair, and sustainable future.

References

Allen, T., & Prosperi, P. (2016). Modeling sustainable food systems. *Environmental Management, 57*(5), 956–975. https://doi.org/10.1007/s00267-016-0664-8

Béné, C., Oosterveer, P., Lamotte, L., Brouwer, I. D., de Haan, S., Prager, S. D., et al. (2019). When food systems meet sustainability: Current narratives and implications for actions. *World Development, 113*, 116–130. https://doi.org/10.1016/j.worlddev.2018.08.011

Davis, B., de la O'Campos, A. P., Farrae, M., & Winters, P. (2024). Whither the agricultural productivity-led model? Reconsidering resilient and inclusive rural transformation in the context of agrifood systems. *Global Food Security, 43*, 100812. https://doi.org/10.1016/j.gfs.2024.100812

Deng, Z., Hu, Y., Wang, X., Li, C., Wang, J., He, P., et al. (2025). Transitioning to healthy and sustainable diets has higher environmental and affordability trade-offs for emerging and developing economies. *Nature Communications, 16*(1), 3948. https://doi.org/10.1038/s41467-025-59275-3

Dinku, A. M., Mekonnen, T. C., & Adilu, G. S. (2023). Urban food systems: Factors associated with food insecurity in the urban settings evidence from Dessie and Combolcha cities, north-central Ethiopia. *Heliyon, 9*(3), e14482. https://doi.org/10.1016/j.heliyon.2023.e14482

Fanzo, J., Bellows, A. L., Spiker, M. L., Thorne-Lyman, A. L., & Bloem, M. W. (2021). The importance of food systems and the environment for nutrition. *American Journal of Clinical Nutrition, 113*(1), 7–16. https://doi.org/10.1093/ajcn/nqaa313

FAO. (2020). *The state of food security and nutrition in the world 2020. Transforming food systems for affordable healthy diets.*

FAO. (2024). *FAOSTAT database.* Food and Agriculture Organization of the United Nations. http://faostat.fao.org/. Accessed 15 Feburary 2024.

IPCC. (2023). Summary for policymakers. In H. Lee, & J. Romero (Eds.), *Climate change 2023: Synthesis report. A report of the intergovernmental panel on climate change. Contribution of working groups I, II and III to the sixth assessment report of the intergovernmental panel on climate change* [Core Writing Team].

Kc, U., Campbell-Ross, H., Godde, C., Friedman, R., Lim-Camacho, L., & Crimp, S. (2024). A systematic review of the evolution of food system resilience assessment. *Global Food Security, 40*, 100744. https://doi.org/10.1016/j.gfs.2024.100744

Purba, N. H., & Krishnaswamy, K. (2025). Exploring the potentials of neglected underutilized crops (NUCs): An integrative review for developing a sustainable food system model. *NPJ Science of Food, 9*(1), 199. https://doi.org/10.1038/s41538-025-00554-0

Reardon, T., & Timmer, C. P. (2007). Chapter 55 transformation of markets for agricultural output in developing countries since 1950: How has thinking changed? In R. Evenson, & P. Pingali (Eds.), *Handbook of agricultural economics* (Vol. 3, pp. 2807–2855). Elsevier. https://doi.org/10.1016/S1574-0072(06)03055-6

Sultanbawa, Y., & Sivakumar, D. (2022). Enhanced nutritional and phytochemical profiles of selected underutilized fruits, vegetables, and legumes. *Current Opinion in Food Science, 46*, 100853. https://doi.org/10.1016/j.cofs.2022.100853

Tamburini, G., Bommarco, R., Wanger, T. C., Kremen, C., van der Heijden, M. G. A., Liebman, M., & Hallin, S. (2020). Agricultural diversification promotes multiple ecosystem services without compromising yield. *Science Advances, 6*(45), eaba1715. https://doi.org/10.1126/sciadv.aba1715

Tendall, D. M., Joerin, J., Kopainsky, B., Edwards, P., Shreck, A., Le, Q. B., et al. (2015). Food system resilience: Defining the concept. *Global Food Security, 6*, 17–23. https://doi.org/10.1016/j.gfs.2015.08.001

Tuggle, C. K., Clarke, J. L., Murdoch, B. M., Lyons, E., Scott, N. M., Beneš, B., et al. (2024). Current challenges and future of agricultural genomes to phenomes in the USA. *Genome Biology, 25*(1), 8. https://doi.org/10.1186/s13059-023-03155-w

Chapter 3
Climate Change and the Water–Energy–Food (WEF) Nexus: Enhancing Resilience Through Neglected and Underutilized Crops (NUCs)

Mukhtar Ahmed, Muhammad Abid, Ahmed M. S. Kheir, Shakeel Ahmad, and P. V. Vara Prasad

Abstract The Water–Energy–Food (WEF) nexus is an important conceptualization that demonstrates the interdependency of the fundamental resources and how climate change disrupts the global and regional agri-food systems systemically. Climate change enhances water, energy, and food pressures, which cause cascading trade-offs that jeopardize food security, water availability, and energy reliability especially

M. Ahmed (✉)
Department of Agronomy, Pir Mehr Ali Shah Arid Agriculture University, Rawalpindi, Pakistan
e-mail: ahmadmukhtar@uaar.edu.pk

Swedish University of Agricultural Sciences, Umea, Sweden

Department of Biological Systems Engineering, Washington State University, Pullman, WA, USA

M. Abid
UNESCO Chair On Knowledge Systems for IWRM Director Interdisciplinary Research, Tarlai Kalan, Pakistan

Department of Mechanical Engineering, COMSATS University Islamabad, Wah Cantt, Pakistan

Department of Mechanical Engineering, Faculty of Engineering, Islamic University of Madinah, Madinah, Saudi Arabia

A. M. S. Kheir
Julius Kühn Institute (JKI)—Federal Research Centre for Cultivated Plants, Institute for Strategies and Technology Assessment, Kleinmachnow, Germany

Soils, Water and Environment Research Institute, Agricultural Research Center, Giza, Egypt

S. Ahmad
Institute of Agronomy Faculty of Agricultural Sciences and Technology, Bahauddin Zakariya University, Multan, Pakistan

P. V. V. Prasad
Department of Agronomy, Kansas State University, Manhattan, KS, USA

M. Ahmed (ed.), *Climate Change and Neglected Underutilized Crops*,
https://doi.org/10.1007/978-3-032-21005-0_3

in climatic sensitive areas like the Sub-Saharan Africa (SSA) and South Asia. Traditional monoculture-based agriculture, which is highly reliant on water, energy, and external resources, is becoming exposed to higher extremes of temperature, changes in precipitation, and extreme weather, which intensifies the shortage of resources and socio-economic imbalance. Neglected underutilized crops (NUCs), drought- and heat-tolerant, low-input, and rich in nutrients, can be used as a feasible solution to improve the resilience of the WEF nexus. Resource intensity can be decreased, food production can be stabilized during climatic stress, nutritional results can be enhanced, and local livelihood and food sovereignty can be promoted by their incorporation into diversified agri-food systems. This chapter analyzes climate change as a systemic cause of WEF vulnerabilities, interactions and trade-offs between water, energy and food systems and identifies NUCs as a strategic course of adaptation. The policy implications are based on integrated governance, investment in climate-resilient infrastructure, the encouragement of the low-resource system of cropping, and alignment with Sustainable Development Goals (SDGs) to attain sustainable, resilient, and inclusive WEF systems. Using NUCs as a component of a nexus-driven system provides an avenue to creating adaptive, nutritionally sensitive, and climate-resilient food systems that can help solve the challenges in SSA and South Asia and other vulnerable locations of the future.

Keywords Water–Energy–Food (WEF) Nexus · Climate change · Neglected and underutilized crops (NUCs) · Food security and resilience

3.1 Introduction

The WEF nexus has become a key concept to explain the complexity of the interdependence of food, water, and energy systems as a collective factor of human well-being and environmental sustainability (Javan et al., 2024). The sole management of such resources tends to lead to inefficiencies, trade-offs and increased vulnerability, especially in the current context of the increasing effects of climate change. In the light of this complexity, the Food and Agriculture Organization of the United Nations (FAO) has given the WEF nexus a global priority as a way of enhancing sustainable agriculture and securing the world food security. The nexus approach highlights the importance of concerted action and planning to deal with increasing resource demands in a climate of indecision (Bamgboye et al., 2025; Ningi & Mullen, 2025).

The WEF nexus is conceptualized as being organized around four large interfaces determining how these systems interact. The interface between water and food shows the prevailing position of agriculture that contributes almost 70% of all freshwater withdrawals globally (Ghosh et al., 2024). It is projected that climate change will create greater competition over water resources, as the regimes of precipitation as well as temperature and evapotranspiration rise, which will be experienced more in the agrarian and water-stressed areas (Ali et al., 2025). The water-for-energy interface also points out the reliance of energy systems on water to extract, mine, refine and generate electricity, both fossil and renewable energy technologies. Contrarily, the energy-for-water interface represents the energy needs of abstraction, treatment, distribution, and conveyance of water, and the energy-for-food interface represents the huge energy demands of the food system, within production and processing, as well as transportation and distribution. These interdependent routes reveal how a fluctuation in climate that occurs in one sector is capable of spreading to the whole WEF system (Ghosh et al., 2024).

The nexus of the WEF became especially relevant in the 2007–2008 food crisis where the links between the prices of food and energy and the rise in energy intensity of agricultural supply chains were strongly correlated (Ringler et al., 2016). Current overlaps on people who cannot access safe drinking water, people who have food insecurity, and people who do not have access to safe electricity only go to show the extreme interdependence of food, water, and energy systems. Predictable trends indicate that population growth, income growth and climate change will demand all three resources in significant amounts and exacerbate the current trilemma with serious threats to sustainability and equity (Sani & Scholz, 2022).

The importance of good governance and sufficient financial investment is becoming a key issue in the management of the WEF nexus because it affects the distribution of resources, the development of infrastructures, technological innovation, and adaptive capacity (Jones-Crank, 2024). Good governance systems can be used to deal with trade-offs between competing resource use, efficiency, and human rights to water and food, and financial resources offer the basis of resilient and climate-adaptive WEF systems (Cavalheiro et al., 2025). When combined with inadequate financial ability and poor governance in areas with high population growth, degradation, and scarcity of resources, vulnerabilities may be increased, and the effects of climate change on food, water, and energy security have been magnified (Chingiz et al., 2025).

In this regard, NUCs present significant possibilities in increasing resilience in the food, water, and energy nexus. Usually adjusted to marginal environments and being tolerant to climatic stresses, NUCs usually need fewer external inputs, including water, energy, and agrochemicals (Purba & Krishnaswamy, 2025). Their application in agricultural and food systems could help in diversification, less reliance on resource-demanding staple crops, and climate-adaptation and mitigation measures. Based on this, this chapter discusses the effects of climate change on food, water, and energy systems in the context of the WEF nexus, the ways the increasing demand, governance challenges, and resource interdependencies determine vulnerabilities to systems, and the opportunities of NUCs as an effective strategic avenue in enhancing resource-use efficiency, enhancing system resilience, and contributing to sustainable and climate-resilient food systems (Dhar, 2025; Ioannou & Laspidou, 2022).

Figure 3.1 shows the concept of the incorporation of the NUCs into the WFE nexus and their ability to improve its sustainability, resilience, and efficiency in resource use. It demonstrates the natural interdependencies in which water is a prerequisite of food production and energy production, the food systems produce and use energy (e.g., bioenergy) and energy is needed to extract, treat, and distribute water (Ansari et al., 2023). Governance, economic efficiency, and financial mechanisms are enabling pillars that determine the performance of the nexus through best management practices, generation of more with less resources and investment finance that enhances better environmental results. The nexus is enhanced by the inclusion of NUCs, which enhances climate resilience, and uses water and nutrients optimally, on food and energy systems that are stable even in resource limitations and climate stress (Shafiq et al., 2025). By employing strategic investments, conducive policies, and green finance, NUC based systems improve in sustainability, equity, and long-term resilience, which are the drivers of inclusive economic growth, a productive environment, and safe and secure access to water, food, and energy to all (Zahedi et al., 2024).

The Fig. 3.2 shows a WEF Nexus model that allows visualizing the interrelation of water, energy, and food systems in a more extensive socio-economic, environmental, and governance environment (Flammini et al., 2014). It brings out the role played by drivers like population growth, urbanization, shifts in diets, climate change, technological innovation, and policy environments in the demand and management of the WEF resources. At the center, a stakeholder dialogue as a form of informed decision-making, based on evidence, scenario development, and response options, will help to strike a balance between conflicting goals and interests in sectors. The framework underlines the fact that sustainable results are based on effective utilization of the resource base (land, water, energy, capital and labour). The nexus approach is especially pertinent in the situation of NUCs, since NUCs typically use less water and less energy, can thrive in marginal climatic conditions, and increase diet and resilience to climate changes. This means that nexus governance-based planning development can mitigate the strain on water and energy resources, enhance climate-resilient food systems, and promote sustainable agricultural transformation in line with nexus governance goals by integrating NUCs in WEF-based planning (Purba & Krishnaswamy, 2025).

3.2 Water–Energy–Food (WEF) Nexus: Conceptual and Quantitative Framework

The Water–Energy–Food Nexus offers a conceptual and quantitative framework that offers a solid approach to analytically examining the intricate interdependences between water utilization, energy utilization, and food production in the agricultural systems (Ali & Acquaye, 2024; Drobinski et al., 2025). The equations below are systematic in revealing resource flows and efficiencies on a variety of scales, which start with aggregate water, energy and food accounting and then pull down to crop and cropping system performance indicators. The total water consumption is broken down into irrigation, effective rainfall and processing components and

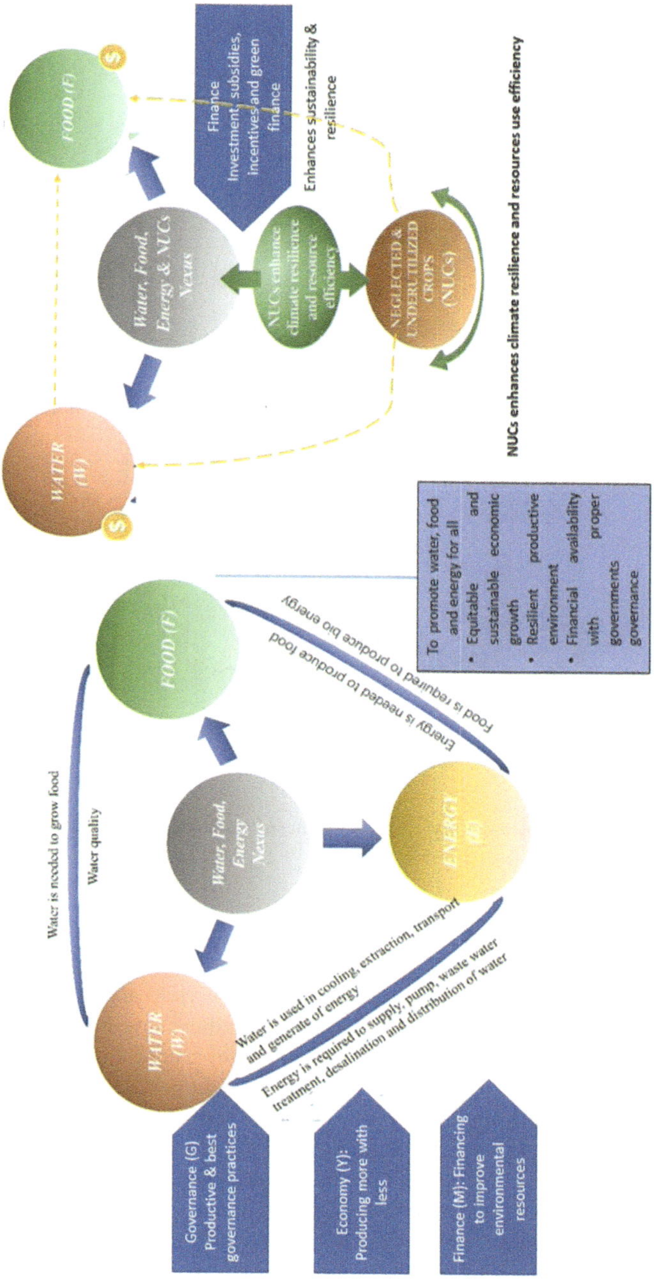

Fig. 3.1 Integrating neglected and underutilized crops (NUCs) within the water–food–energy nexus for sustainable and climate-resilient development

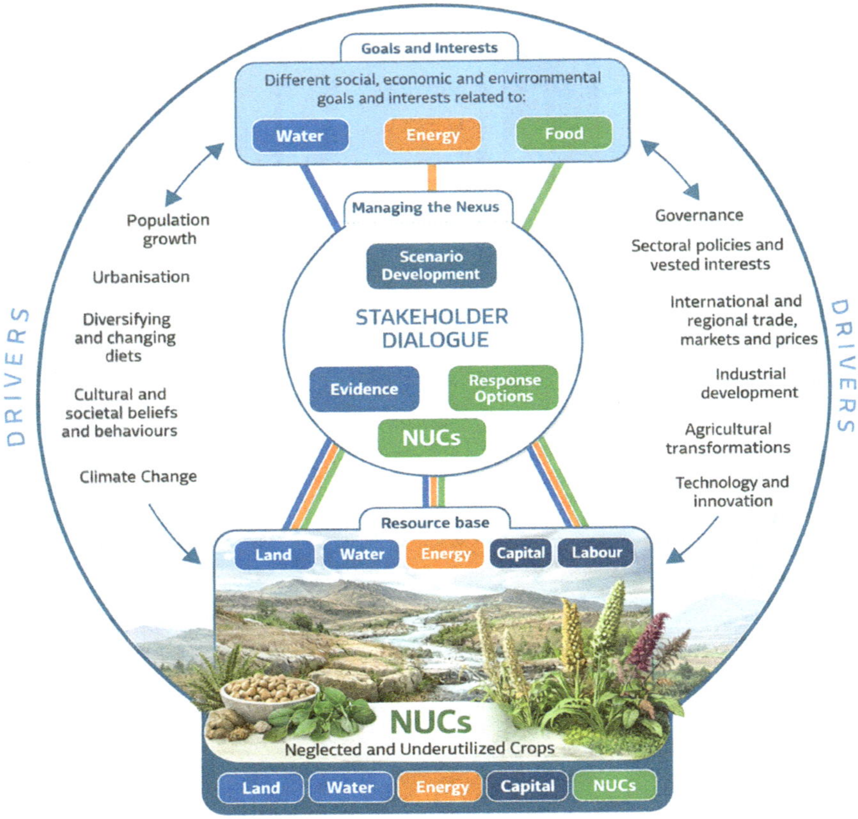

Fig. 3.2 An expanded FAO water–energy–food (WEF) nexus framework integrating neglected and underutilized crops (NUCs)

energy consumption includes irrigation pumping, fertilizers machinery and post-harvest activities, as both direct and indirect energy consumption in agriculture. To form the core output, food production is given in terms of yield and area under cultivation which connects water and energy input. WEF indices are then formulated to measure nexus efficiency and sustainability to allow comparing food production with integrated inputs of water and energy and to weight up policy or ecological priorities. At crop level, water needs through evapotranspiration and water productivity, irrigation energy needs and overall energy requirements give mechanistic understanding of resource-use efficiency whereas yield functions and water footprint metrics correlate biophysical procedures with sustainability outcomes. Lastly, crop aggregation facilitates the assessment of cropping systems and climate-smart practices and integrates carbon footprint factors to evaluate nexus performance in terms of climate change. All these equations make up a powerful analytical framework within the

WEF Nexus to determine the sustainability of agriculture, resource efficiency, and climate resilience (Albrecht et al., 2018).

3.2.1 Core Water–Energy–Food (WEF) Nexus Equations

3.2.1.1 Water Use

$$W = W_{irri} + W_{rain} + W_{process} \tag{3.1}$$

where:

W = Total water use (m^3).
W_{irri} = Irrigation water.
W_{rain} = Effective rainfall.
$W_{process}$ = Processing and post-harvest water.

3.2.1.2 Energy Use in Crop Production

$$E_{crop} = E_{pump} + E_{fert} + E_{mach} + E_{proc} \tag{3.2}$$

where

E_{crop} = Total energy use (MJ or kWh).
E_{pump} = Energy for water lifting and irrigation.
E_{fert} = Energy embodied in fertilizers.
E_{mach} = Machinery and fuel energy.
E_{proc} = Post harvest and processing energy.

3.2.1.3 Food Production

$$\mathbf{F} = \mathbf{Y} \times \mathbf{A} \tag{3.3}$$

where:

F = Total food output (kg or kcal).
Y = Crop yield (kg ha^{-1}).
A = Cultivated area (ha).

3.2.2 Integrated WEF Nexus Efficiency

3.2.2.1 Basic WEF Nexus Index

$$WEF_{index} = \frac{F}{W \times E} \text{(Represents food produced per unit of water and energy)} \tag{3.4}$$

3.2.2.2 Weighted WEF Sustainability Index

$$\mathbf{WEF_S} = \alpha\left(\frac{\mathbf{F}}{\mathbf{W}}\right) + \beta\left(\frac{\mathbf{F}}{\mathbf{E}}\right) + \gamma\left(\frac{\mathbf{E}}{\mathbf{W}}\right) \tag{3.5}$$

where: α, β, γ = Weighting coefficients reflecting policy or ecological priorities.

3.2.3 Crop-Level Water–Energy–Food (Crop-WEF) Equations

3.2.3.1 Water Component (Crop Scale)

Reference (ET_0) and Crop Evapotranspiration (ET_c)

$$\mathbf{ET_0} = \textbf{Penman Mnteith Equation}$$

$$\mathbf{ET_c} = \mathbf{K_c} \times \mathbf{ET_0} \tag{3.6}$$

where K_c = Crop coefficient (growth-stage specific).

Crop Water Requirement (CWR)

$$\mathbf{CWR} = \sum \mathbf{ET_c} \tag{3.7}$$

Crop Water Productivity (CWP)

$$\mathbf{CWP} = \frac{\mathbf{Y}}{\mathbf{CWR}}\left(\mathbf{kg\ m^{-3}}\right) \quad (3.8)$$

3.2.4 *Energy Component (Crop Scale)*

3.2.4.1 Energy for Irrigation

$$E_{irr} = \frac{\rho g H Q t}{\eta} \quad (3.9)$$

where:

P = Water density.
g = Gravity.
H = Pumping head.
Q = Discharge.
t = Time.
η = Pump efficiency.

3.2.4.2 Total Crop Energy Input

$$E_{crop} = \sum\left(E_{seed} + E_{fert} + E_{chem} + E_{mach} + E_{irr}\right) \quad (3.10)$$

3.2.4.3 Energy Productivity

$$EP = \frac{Y}{E_{crop}}\left(kg\ MJ^{-1}\right) \quad (3.11)$$

3.2.5 *Food Component (Crop Scale)*

3.2.5.1 Crop Yield

$$Y = f(W, E, S, C) \quad (3.12)$$

where:

S = Soil and nutrient status.

C = Climatic conditions.

3.2.5.2 Crop Water Footprint

$$WF_{crop} = \frac{CWU}{Y}\left(m^3 t^{-1}\right) \tag{3.13}$$

where:

WF_{crop} = Water footprint of crop production (m^3 t^{-1}).
CWU = Crop water use (m^3 ha^{-1}).
Y = Crop yield (t ha^{-1}).

3.2.6 Cropping Systems–Level WEF Nexus

3.2.6.1 Cropping System Water Use

$$W_{sys} = \sum_{i=1}^{n} W_i \tag{3.14}$$

3.2.6.2 Cropping System Energy Use

$$E_{sys} = \sum_{i=1}^{n} E_i \tag{3.15}$$

3.2.6.3 Cropping System Food Output

$$F_{sys} = \sum_{i=1}^{n} Y_i \tag{3.16}$$

3.2.6.4 Cropping System WEF Index

$$WEF_{sys} = \frac{\sum Y_i}{\sum W_i \times \sum E_i} \tag{3.17}$$

3.2.6.5 Climate-Smart Crop–WEF Nexus

$$WEF_{CSA} = \frac{Y}{W \times E \times C} \tag{3.18}$$

where: C = Carbon footprint (kg CO_2-eq).

3.3 Climate Change as a Systemic Driver of the WEF Nexus

The systems of WEF are the ones that form the well-being of the human communities, their economic growth and their sustainability. Nevertheless, the systems are becoming overburdened due to the high rate of population increase, urbanization, increased infrastructure development, and climate change. The world is estimated to be having 785 million individuals who do not have access to the basic drinking water, 2.2 billion individuals who use contaminated water, approximately 759 million people who do not have access to electricity and about 854 million individuals who are undernourished. Besides this, 2.6 billion individuals are still relying on polluting and ineffective fuels to cook. These statistics underscore the fact that WEF systems are vulnerably exposed and underscore the resultant unequal allocation of resources to regions (Bamgboye et al., 2025).

Climate change is a systemic force that transforms interaction relations in and between water, energy, and food systems and thereby fundamentally changes the equilibrium of the WEF nexus (Nunes et al., 2026). Instead of impacting each sector separately, climate change changes processes in Earth systems controlling resources availability, increases climate variability and extremes and enhances trade-offs among competing WEF demands (Torhan et al., 2022). These reciprocal effects, as pointed out in the literature on the nexus of the WEF, and synthesized by Zhang et al. (2018), undermine traditional sector management and require nexus based, nexus based responses to safeguard the sustainability of resources in the long-term.

3.3.1 Climate Change Impacts on Earth System Processes Relevant to the WEF Nexus

The processes of the Earth system, which are at the heart of the WEF nexus, include hydro-logical cycle, land-atmosphere interaction, as well as bio-geochemical flows, all of which are directly affected by climate change. Increasing temperatures enhance the rate of evapotranspiration and the moisture demand in the atmosphere and alter precipitation regimes changing run-off patterns, ground water recharge, and the availability of soil moisture. These processes redefine freshwater supply to food and energy production systems (Wang et al., 2023).

Water is the main coupling agent of the WEF nexus. Agriculture is the most important component of freshwater consumptive use in the world due to irrigation water, yet energy systems use water as a significant source of fuel, energy in thermal power plants, and as the source of electricity generation in hydropower (Siyal & Gerbens-Leenes, 2022). Changes in water availability as a result of climate trickle down the bottom of the nexus. Less water is lost during runoff, and the water that is lost during runoff has decreased groundwater recharge thereby limiting the production of crops and the availability of food. At the same time, these transformations limit hydropower production and decrease the performance of the energy systems that consume a lot of water (Zhang et al., 2017). Maize, soy, rice, and wheat yield has been decreased by about 4.1, 4.5, 1.8, and 1.8%, respectively due to climatic disturbances to water and temperature regimes on food production. The world water demand is expected to exceed 20–30% by 2050, which is partially due to the energy production, agricultural and industrial development. All these tendencies highlight the fact that climate change is a structural phenomenon that alters physical boundaries within which WEF systems carry out their functions and compels interconnected resources toward ecological and planetary extremes (Dinar et al., 2019).

3.3.2 Climate Variability, Extremes, and Long-Term Trends in the WEF Nexus

Climate change is manifested by the long-term trends as well as increasing climate variability that bring instability to the WEF systems. Drought, flooding and heat waves interfere with agricultural production, destroy water and energy infrastructure, lower the dependability of supply systems. In a nexus WEF point of view, such shocks are rarely unilateral (Zenda, 2024). Drought decreases water supply and potential hydropower generation while heat waves raises electricity demand because of cooling that leads to decrease crop productivity. The International energy agency argues that climate change would raise energy needs in the world by 25–58% by 2050 mainly because of the escalating cooling requirements. Concurrently, the predictability, upon which food, water and energy planning have been formerly based, is eroded by the long-term climate changes, including gradual warming, changed seasonality of rainfall, and more frequent extreme events (Szinai et al., 2024).

Climate risks has been increased in water-stressed and agrarian areas, especially in developing countries, using tightly coupled WEF systems. Unpredictability of the rain and escalation of the rate of evapotranspiration efficiency decrease the water-use efficacy, augment the irrigation demand, and lighten the yield of the crops. These ripple effects increase susceptibility throughout the nexus, endangering food security, energy security, and fair access to water (Shayanmehr et al., 2022).

3.3.3 *Climate Change as a Multiplier of WEF Risks and Trade-Offs*

Climate change is a highly effective risk multiplier in the nexus of WEF, which makes the existing pressures more intense and introduces new trade-offs between water, energy, and food systems. Changing climatic conditions increase water scarcity, which increases competition between water used by agricultural irrigation and water used by the energy sector, especially in areas that generate power by using a lot of water and irrigate extensive land (Mperejekumana et al., 2024).

The adaptation strategies meant to address the climate change usually introduce new pressures in the nexus. Increase in irrigation facilities to stabilise food production leads to a rise in energy requirement to power pumps and water transportation. Likewise, advanced water treatment technologies and desalination make water securer but consume a lot of energy. The nexus trade-offs might also be made worse by climate mitigation strategies; e.g. the increased use of bioenergy to minimise greenhouse gas emissions directly competes with food production on land and water, further intensifying food-energy interactions (Rosińska et al., 2024).

Examples of these interactions are those identified by Li et al. (2025) as part of the "WEF three-legged model" where the decision to focus on one area of climate pressure may put the sustainability of the others at risk. Climate change, therefore, cannot be solved by single water policies, single energy policies, and single food policies. Rather it requires integrated WEF governance frameworks which directly consider cross sectoral feedbacks, cumulative resource limitations and socioeconomic vulnerabilities (Javan et al., 2024).

In this regard, the overall resource intensity reduction strategies, including improving the efficiency of water-use, diversifying the cropping systems, planting climate-resilient and low-input crops, and energy efficiency, prove to be the key leverage points of increasing the resilience of the WEF nexus (Tian et al., 2025). In totality, changes in climate are systemic stressors that enhance interactions and trade-offs between water, energy and food systems, amplifying risks beyond single sectors and supporting the necessity to use integrated nexus-based methods of assessment and management (Table 3.1).

Table 3.1 Climate change impacts across the water–energy–food (WEF) nexus

WEF component	Climate change driver	Direct impact	Cross-sectoral nexus effect
Water	Rising temperature, altered precipitation	Reduced runoff, groundwater stress, higher evapotranspiration	Limits irrigation (food) and hydropower/cooling (energy)
Food	Heat stress, drought, floods	Yield variability, crop failure, nutritional decline	Increases water demand and energy inputs
Energy	Water scarcity, extreme events	Reduced hydropower, cooling water shortages	Competes with agriculture for freshwater
WEF nexus	Climate variability and extremes	System instability and cascading risks	Amplified trade-offs and reduced resilience

3.4 Climate Change Impacts on the Food System Within the WEF Nexus

3.4.1 Food Production, Demand, and Climate Stress

Food and fiber production continue to rely heavily on agriculture which has a dramatic impact on land, water, energy, and vegetation resources. Nevertheless, to maintain agricultural productivity, the demand on natural systems is growing due to the necessity of using land, freshwater, energy, fertilizers, and other external resources more and more. Although the world food production has increased significantly in the last century, the climate change is currently becoming a sharp limitation to the increased production, especially in large staple crops which are in the vicinity of biophysical yield cliffs (Hemathilake & Gunathilake, 2022).

Production shocks are increasing in risk in the agricultural regions due to climate variability and the growing occurrence of extreme events, including drought, floods, heatwaves, and unseasonal frosts. These stresses combine with factors in structural development of food demand such as high population growth rate, urbanization and shifts in diet towards higher proportions of animal-based and processed foodstuffs, which substantially escalate with a higher level of water and energy needed on a unit of food item produced. As a result, the world food system is having a twofold challenge and that is to satisfy the increasing demand and to work within the increasingly restrictive environmental and climatic constraints (Shah et al., 2021).

Climate change is a significant cause and effect of agriculture. The industry contributes around 21–37% of the global greenhouse gas emissions, which are mainly land-use change, enteric fermentation, fertilizer application, and energy-consuming production systems. In its turn, climate change has a detrimental effect on agricultural productivity through changes in temperature regimes, precipitation, growing seasons, as well as pests and disease interactions. Such feedback loop highlights the dire need to change food production systems to become more resilient and less emitting and ensure food security in an increasingly uncertain climatic context (Hong et al., 2025).

3.4.2 Water-for-Food and Energy-for-Food Interfaces

Agriculture is placed at the heart and the most vulnerable position in the WEF nexus. It is the main freshwater consumer worldwide making food production highly sensitive to changes in water availability, quality and seasonal distribution caused by climate change. Increased temperatures promote crop evapotranspiration and irrigation water demand and increased precipitation variability and extreme events jeopardize surface and groundwater. The problem of these dynamics is especially troublesome in the areas that already experience water variability or have a small storage volume (Messaoudi et al., 2025).

Simultaneously, the contemporary agricultural systems rely heavily on the energy sources to irrigate, produce fertilizers, proceed with mechanization, processing, cold storing of goods and transportation. Climate change reinforces this dependency by raising the irrigated acreage, and, at the same time, jeopardizing the energy infrastructure with severe weather patterns and altering hydrological conditions leading to changes in hydropower reliability. Consequently, climate change exerts pressure on water–food nexus and energy-food nexus, making the competition among sectors more intense and introducing more chances of trade-offs (Raihan et al., 2024).

The WEF nexus framework has become an important analytical approach to tackle these nexus challenges. The world population will grow, urbanize, exploit its resources, and experience climate change resulting in more than 50% growth in the worldwide demand of water, energy and food by 2050. Sectoral solutions to these pressures, besides resulting in unintended consequences, like depleting resources, environmental degradation and emission increase, have proved inefficient and unsustainable. Conversely, nexus-based approach focuses on integration, systems thinking, and coordinated management of resources and, therefore, can identify synergies and minimize trade-offs under climatic stress (Yupanqui et al., 2025).

3.4.3 Food System Vulnerability, Crop Concentration, and the Role of Innovation

The global food systems are marked by high level of crop and livestock concentration where few staple crops occupy most of the caloric consumption globally. This dependence in structure is more compliant to systemic susceptibility to climate shocks, pests, diseases and market disruption. Besides, resource-intensive livestock and crop systems further increases water and energy requirements, pushing the WEF nexus even more by climate change (Haines & Frumkin, 2021).

These weaknesses are increased in those areas where agricultural production and reliance on imports are high, including the north and island settings. As well as endangering local production due to fluctuations in temperature and precipitation, climate change exposes the supply chain to risks of global supply disruptions and price inflation of food. These dangers create a concern to diversify food systems by means of improving local food production where possible and decreasing dependence on long distance imports that have large environmental as well as economic implications (Tchonkouang et al., 2024).

Agricultural innovation comes into the picture with great importance in resolving these challenges. Innovation has not just been limited to technological development, i.e., controlled environment agriculture (CEA), climate-adaptive varieties of crops, and precision irrigation, but also institutional, organizational, and managerial developments to increase system-wide resilience. In the WEF nexus, innovation provides avenues of decoupling food production with excessive water and energy consumption, minimization of greenhouse gas emissions, and food security performance (Lindow et al., 2026). Specifically, the use of agricultural and industrial by-products is an unexploited opportunity in improving food system sustainability. Increasing the utilization of waste streams—e.g. biomass residues, organic sludges, waste heat can enhance soil health, decreasing reliance on fertilizers, decreasing emissions due to waste disposal, and bridging material loops which is in line with the principles of circular economy. Such innovations have the potential to support the goals of mitigation and adaptation of climate, as well as food security when combined using a WEF-climate nexus perspective (De Corato et al., 2024).

In general, climate change is a high-level stressor that transforms the relationships of water, energy, and food systems. The application of a WEF-Climate Change (WEF-CC) nexus approach, which is based on systems thinking, offers a holistic approach to the vulnerability of the food system and an innovative and context-specific solution to the issue. Through the transcendence of sectoral silos and integrated approaches, food systems would become stronger, use fewer resources and more able to withstand economic and environmental disruptions that are likely to occur in the climate scenarios of the future (Alhassan et al., 2026).

3.5 Climate Change Impacts on Water Systems Within the WEF Nexus

3.5.1 Water Scarcity Under Climate Change

Climate change and water have a close relation because climate change interferes with the water cycle due to unpredictable precipitations, ice melting, rising sea level, floods and drought, which exacerbate the water scarcity and water-related risks. More than 2 billion individuals do not have access to safe drinking water today, and close to 50% of the global population faces severe water shortage at some point during the year, a phenomenon that is likely to increase with the climate change and population increase (Karimi et al., 2024). Despite the Earth being largely covered by water, of 100% of the water is available as fresh water, only 0.5% of that can be utilized. The water storage of the terrestrial water (soil moisture, snow and ice) has decreased at a rate of 1 cm each year in the last 20 years big threat to water security (UN, 2023). Irrigation, hydropower, and water supply and ecosystems have been impacted by the melting glaciers, snow and permafrost. Climate change, land use and pollution are leading causes of the degradation of freshwater ecosystems and the disappearance of freshwater-dependent species (Ahmed, 2025). The water quality is also declining whereby rise in temperatures and occurrence of more frequent floods and droughts are augmenting pollution by sediments, pathogen and pesticides (Abalasei et al., 2025). Climate change has contributed to floods and droughts, with more than 70% of deaths associated with natural disasters in the last 50 years, and flood related calamities are on the rise by 134% since 2000 and drought frequency and duration have escalated by 29% (Ahmed, 2023). Farming, consuming approximately 70% of all worldwide freshwater. It is using 2000–5000 L of water a day per person of food. It is experiencing increased pressure thus posing a threat of food security (Oki & Kanae, 2006). Reducing global warming to 1.5 °C rather than 2 °C can reduce the number of people facing water shortages by one half or more. Measures like wetland protection, climate-resilient water systems, early warning systems are capable of cutting the damage by 30% with a 24-h warning. Climate-smart agriculture can greatly lower the risks and increase the water security of the world population (Hansen et al., 2025).

3.5.2 *Water-for-Energy and Energy-for-Water Interfaces*

Water and climate change have a very strong interrelation since the changes in the precipitation and temperature as well as extremes of floods and droughts directly increase the water shortage and the risks associated with water. This further enhancing the pressure on the energy systems. Due to the decreased freshwater availability and a worsening of water quality, a greater amount of energy is needed to pump, treat, desalinate, irrigate, and distribute water, which reinforces the energy-water nexus. Meanwhile, the increasing energy requirements feed back into the water systems via the water-for-energy interface because energy production, especially hydropower, thermal, nuclear production, among others, relies heavily on the availability of water. Water and energy requirements in the world will grow by 55 and 80%, respectively, with respect to 2015, which highlights the need to combine the management of these interconnected systems (Bai et al., 2026). Today approximately 8% of the total electricity production is consumed by water pumping, treatment and transport and almost 15% of the world water withdrawals is linked to energy production. Climate change also intensifies these combinations by changing water supply, causing pressure on cooling-water by power plants and energy demands in already water-strained areas, especially in Asia and Africa, where development demands are already demanding (Naderi et al., 2024).

In this respect, it is the integrated management of the water–energy and energy–water interface that plays the key role in the accomplishment of SDG 6 (clean water and sanitation) and SDG 7 (affordable and clean energy) (Ahmad et al., 2025). These issues can be demonstrated using the example of Pakistan, where close to 80% of the energy is produced by fossil fuels, but water demands are extremely high, with the per capita water demand dropping to <1000 m^3/year in 2016 compared to 5260 m^3/year in 1951, with little storage capacity of approximately 100 m^3 per capita. Approximately 3.6 billion are already dwelling in water-scarce areas worldwide, which are expected to be 5.7 billion by 2050, and safe sanitation is not distributed equally (Adnan et al., 2024). In order to deal with these pressures, various sophisticated methods of analysis, like life-cycle assessment (LCA), input–output analysis (IOA)/multi-regional input–output (MRIO), and integrated long-range energy alternatives planning (LEAP)-water evaluation and planning system (WEAP) have been more commonly employed to estimate cross-sectoral connections, emissions, and resource efficiency (Ahmad et al., 2025). The connection of water shortage, climate change, and the water–energy nexus with such integrated frameworks allows the more consistent design of policies, facilitates the implementation of renewable energy, promotes the preservation of water and energy, and minimizes the greenhouse gas emission, which, in fact, increases the sustainability of the situation regarding long-term sustainability and climate resiliency.

3.5.3 Governance, Allocation, and Climate Adaptation

The global food systems are marked by high level of crop and livestock concentration where few staple crops occupy most of the caloric consumption globally. This dependence in structure is more compliant to systemic susceptibility to climate shocks, pests, diseases and market disruption. Besides, resource-intensive livestock and crop systems further increases water and energy requirements, pushing the WEF nexus even more by climate change (Agostoni et al., 2023).

These weaknesses are increased in those areas where agricultural production and reliance on imports are high, including the north and island settings. As well as endangering local production due to fluctuations in temperature and precipitation, climate change exposes the supply chain to risks of global supply disruptions and price inflation of food. These dangers create a concern to diversify food systems by means of improving local food production where possible and decreasing dependence on long distance imports that have large environmental as well as economic implications (Tchonkouang et al., 2024).

Agricultural innovation comes into the picture with great importance in resolving these challenges. Innovation has not just been limited to technological development, i.e., controlled environment agriculture (CEA), climate-adaptive varieties of crops, and precision irrigation, but also institutional, organizational, and managerial developments to increase system-wide resilience. In the WEF nexus, innovation provides avenues of decoupling food production with excessive water and energy consumption, minimization of greenhouse gas emissions, and food security performance (Lindow et al., 2026).

Specifically, the use of agricultural and industrial by-products is an unexploited opportunity in improving food system sustainability. Increasing the utilization of waste streams—e.g. biomass residues, organic sludges, waste heat can enhance soil health, decreasing reliance on fertilizers, decreasing emissions due to waste disposal, and bridging material loops which is in line with the principles of circular economy. Such innovations have the potential to support the goals of mitigation and adaptation of climate, as well as food security when combined using a WEF-climate nexus perspective (De Corato et al., 2024).

In general, climate change is a high-level stressor that transforms the relationships of water, energy, and food systems. The application of a WEF-Climate Change (WEF-CC) nexus approach, which is based on systems thinking, offers a holistic approach to the vulnerability of the food system and an innovative and context-specific solution to the issue. Through the transcendence of sectoral silos and integrated approaches, food systems would become stronger, use fewer resources and more able to withstand economic and environmental disruptions that are likely to occur in the climate scenarios of the future (Alhassan et al., 2026).

3.6 Climate Change Impacts on Energy Systems Within the WEF Nexus

Environmental degradation and climate change threaten the world food security in a way never encountered before. The challenges that the twenty-first century has set are to fulfill the increasing demands of water, energy and food and stay within the planetary boundaries under the rapidly changing climate. The effects of global temperature increase, changed patterns of precipitations and greater occurrence of extreme situations is already being felt in food systems around the world and this poses a threat to the stability in yield and the supply chain. It has been estimated that every 1 degree centigrade rise in global temperature will cut the average yield of key staple foods by several %, leading to increased hunger and food insecurity threats, especially in poor countries and in climate-prone areas (Fanzo et al., 2021).

In that context, energy systems are an essential but insufficiently explored factor in defining the resilience of food systems. Food systems, including production, processing, storage, transport and consumption are significant sources of global greenhouse gas emissions, estimated at one-quarter to one-third, and large consumers of water and energy. These feedbacks result in food systems both as contributors to climate change and as extremely vulnerable to climatic change. To achieve food security in climate change, it is thus necessary to convert energy systems in a manner that maximizes the resilience as opposed to shifting vulnerabilities among different sectors (Tamasiga et al., 2025).

3.6.1 *Water Dependence of Energy Production*

The production of energy is also closely tied to the availability of water, and so energy systems are highly vulnerable to hydrological change caused by climate. Most of the energy options, such as hydro, thermoelectric, bio-energy, and unconventional fossil fuel, require stable water resources to extract, cool, and process. These dependencies are under threat due to climate change that has been varying in precipitation patterns, evapotranspiration, snowpack, and drought and heatwaves.

Hydropower is especially susceptible to alterations in streamflow magnitude and timing, and thermal power plants are limited in their operation through lower cooling water supply and increased temperatures of the water. The dangers become even greater when there are extreme heat events and the demand on electricity is the highest whereas the water supply is minimal. Consequently, climate change is directly causing new bottlenecks on energy security to food systems that rely on water and energy.

In the case of agriculture, these interdependences are particularly important. Water scarcity can lead to energy disruptions that may compromise the viability of irrigation, mechanization, cold storage and processing to drive up losses post-harvest, and decrease productivity. In developing nations where over 6 and more populations are left to rely on agriculture, such disturbances can easily be converted into food insecurity and livelihood loss. It is based on these dynamics that there is a need to have water constraints instilled directly into the energy planning as part of the WEF nexus (Guo & Hasani, 2024).

3.6.2 *Food–Energy Competition Under Climate Mitigation*

More trade-offs exist between food and energy systems due to climate mitigation efforts, especially those based on energy transition. Policies of bioenergy and biofuels, which are characterized as low-carbon solutions, have heightened rivalry on land, water, and food products which were originally used in food production. Develop and underdeveloped countries have to produce large volumes of edible farmland to produce fuels, compromising the availability of food and water challenges due to climate change (El-Araby, 2024).

In developing areas, especially, such tradeoffs are especially dire, as climate effects already cut yields, weakening by e.g. 6–7% per degree of warming the stapues, and food spending is a big portion of household income. Greater price competition between food and fuel also has potential to increase the prices of food, promote land-use change, and increase the rate of environmental degradation, compromising food security and environmental integrity (Ali et al., 2017).

Although there is a large body of literature on the effects of energy transition at the macroeconomic level, the immediate and secondary consequences on food security and agricultural sustainability have not been fully investigated. Nonlinear impacts of energy transitions on agricultural production, which accumulate as evidenced by the cost of inputs and land redistribution and water demand. Devoid of incorporated governance and sustainability guarantees, climate mitigation initiatives would be susceptible to maladaptation—whereby the measures implemented in order to curb the emission levels will actually compromise the water supply, food safety, or the condition of the ecosystem. To assess mitigation pathways in a comprehensive manner and react to trade-offs, a WEF-Environment (WEF-E) nexus approach is thus a necessity (Azimi et al., 2025).

3.6.3 Energy Transitions and Agricultural Implications

The important aspect of energy transitions is that they transform the agricultural systems in terms of cost of production, land-use and efficiency of value chain. Agriculture is power-intensive in all its phases, such as the cultivation and irrigation, as well as in processing, storage, and transportation, and intensive agriculture has augmented the power requirement and environmental damage. Increase in energy prices which has already been witnessed in most parts of the world have led to food price volatility and this further points to the vulnerability of food systems to the dynamics of energy markets (Tamasiga et al., 2025).

The replacement of fossil fuel with renewable energy is necessary in climate mitigation, but it poses serious effects to agricultural sustainability especially in developing nations. The inadequately handled transitions may interfere with the farming systems relying on fossil fuel, raise the costs of production, and lead to the loss of jobs in rural regions, which can affect the SDGs 1 (no poverty) and SDG 2 (zero hunger). The presented risks demonstrate that it is necessary to harmonize energy transitions and food-system resilience instead of seeking unitary solutions (Majeed et al., 2023).

One of the major ways of minimizing vulnerability in the transition is diversification of energy supply. The combination of several sources of energy—solar, wind, hydro, biomass, and new technologies—will help in making the energy more resilient, decreasing the reliance on one source, and cushion against shocks created by the climate and the market. When these systems are supported by proper policies and infrastructure, there is also a possibility of enhancing the productivity of agriculture and lessening on emissions due to decentralized and renewable energy systems (Ur Rehman et al., 2026).

Low-resource and input-efficient agricultural systems also enhance resilience in circumstances with constraints of energy and climate. Most of the NUCs are drought tolerant, use less fertilizer, and are favorable to marginal environments, which well match the ever-increasing unpredictability of water and energy. They can be integrated in diverse agricultural systems to decrease the use of fossil fuels and water-intensive inputs and increase nutritional diversity (Shafiq et al., 2025).

Other ways that can be used to decouple energy production and food crop competition include innovative production systems like cultivation of microalgae. Growth Microalgae can be cultured with no arable land (or on relatively low amounts of freshwater), can be cultured to generate both clean energy feedstocks and high-value food or feed products. These systems demonstrate the potential of integrated, nexus-based systems to create co-benefits in energy, food, water and environmental systems. The climate change can be viewed as a WEF-Environment nexus stressor that enhances the interdependences within the energy, water, food, and ecosystem nexus. To cope with these challenges, it is important to abandon sectoral silos and integrate the system-based approaches in planning that offers both synergies and trade-offs in managing the issue. It is necessary to match transitions between energy and purposes in sustainable agriculture, protect the ecosystem, and achieve food security to prevent mal-adaptation, and to prepare resilient food systems that would not let climate shocks disorient them but must carry their operations within the boundaries imposed on the planet (Olarewaju et al., 2025).

3.7 WEF Nexus Interactions and Trade-Offs Under Climate Change

Well-being, sustainable development, and economic stability of human beings depend on WEF systems. However, these resources are under increasing pressures of swift population increase, climatic changes, economic crises and inadequate investment in essential infrastructure. Global Risk Reports (2007–2019) by the World Economic Forum have always occupied the leading positions on the lists of the global risks, focusing on the importance of their strategic role in the stability of the society. By 2035, the quantity of water utilized in energy generation is expected to grow by 85% due to the 35% growth in energy demand. More intensive food and bioenergy generation will strain land, water, and energy resources. Millions of people may experience compounded WEF insecurity without effective interventions especially in those areas that are highly vulnerable to changes in seasonal rainfall and climatic vagaries (Collins, 2019).

The WEF nexus underscores the complex interconnection and interdependence between these sectors. Hence managing the resources alone can no longer work. These interdependencies are fueled by climate change, systemic impacts of which include restriction of energy production in countries with reduced water. Shortage of water complicating the production of energy services, and elevated costs of energy or bioenergy-related policies impacting food production and affordability. These interrelationships create trade-offs, synergies and possible feedbacks, necessitating governance which has to be adaptive, cross-sectoral, and context-specific (Javan et al., 2024).

3.7.1 Food–Water Interactions

Climate change, which is characterized by high temperature and changed weather patterns, has great threats to food safety and security. It causes climate variability, which may enhance food-related illnesses as well as chemical risk factors leading to pesticides, mycotoxins, and heavy metals. Severe weather conditions, drought, and flooding also pose a risk to the adequate water supply, the protection of crops and the management of cold chains, which result in the exposure, intoxications and accumulations of chemicals in humans. It is crucial to be aware of these risks and put in place mitigation and adaptation measures that can help to protect world food supply (Duchenne-Moutien & Neetoo, 2021).

The use of diverse and low water needs crops such as NUCs and water efficient technologies can help to ease the strain on water systems and improve food security. At the nexus scale, holistic planning is necessary to prevent cases of maladaptation of the short-term food security measures, leading to water sustainability or healthy ecosystems (Drobinski et al., 2025; Mondal et al., 2025).

3.7.2 Water–Energy Interactions

Production of energy is progressively becoming water-intensive, specifically hydropower, thermoelectric power plants, bioenergy, and unconventional fossil fuel. Water shortages brought about by climate may limit power production even as increasing energy usage increases the amount of water required by energy systems. On the other hand, the lack of energy, or an increase in its prices, can undermine the water supply, treatment, irrigation, and sanitation services (Guo & Hasani, 2024).

Extreme climatic events increase the possibility of cascading failures especially in areas where infrastructure is old or centralized. The key elements to minimize vulnerability and guarantee the resilience of service provision are integrated WEF governance, energy diversification, and the deployment of energy-efficient technologies that consume less water. To be able to manage such interdependencies, cross-sectoral coordination is in demand to prevent the presence of reinforcing feedback loops that contribute to systemic risk (Drobinski et al., 2025).

3.7.3 Food–Energy Interactions

The food and energy systems are interrelated as agricultural production, processing, storing and transportation are intensive energy-consuming processes. The strategies aimed at curbing climate change like bioenergy development cause land, water and crop competition, which may further aggravate food insecurity. The higher energy costs or supply shocks also add to production costs, jeopardizing affordability and access especially to vulnerable groups.

Shocks related to energy can be mitigated by diversified systems of cropping and low-input agricultural systems, such as NUCs and sustainable farming techniques. New strategies, including the use of microalgae or other non-edible biomass to generate energy, minimize competition on food crops, but contribute to the objectives of renewable energy. These strategies are a model of nexus thinking in that they bring about co-benefits in food, energy, water, and environmental systems (Said et al., 2025).

3.7.4 Governance and Data Challenges

A successful WEF nexus management is impossible without multi-level governance, collaboration of various actors, including the public, the private, and the civil-society, and strong data systems. Nevertheless, there is still a significant issue of governance because of disjointed policies, compartmentalized decision-making, and a small number of frameworks to implement nexus strategies at local, regional, and national levels. Key gaps include:

(1) Information constraints: The partial, irreconcilable, or uncoupled datasets on real-time sector and scale limitations limit the planning and monitoring of nexus tools.
(2) Measures and indicators: WEF nexus has no universal agreement on measures, unit of analysis, or comparative models, which restricts the performance evaluation and international benchmarking.
(3) Multi-actor coordination: To ensure good nexus governance, it needs to coordinate the various actors and reconciles between conflicting priorities that prevail as well as cooperative decision-making. The absence of combined policy frameworks especially when interventions are involved will only serve to strengthen trade-offs instead of achieving synergies.

All-encompassing governance approaches are expected to specify and measure interconnections, involve the participation of all stakeholders, and facilitate multi-scale planning. The creation of the globally comparable indexes and analytical platforms has the potential to provide evidence-based decisions and enhance the transfer of the nexus theory into the policy practice (Rahman & Natcher, 2025).

3.7.5 *Integrated Nexus Planning and SDG Alignment*

The climate change approach to WEF issues involves integrated, and systems-wide solutions that take into consideration physical resource interactions, socio-economic backgrounds, and new forms of vulnerability or so-called hotspots. Nexus-based planning has been shown to enhance climate resilience, equal distribution of resources and sustainable development and directly lead to SDGs, such as:

- SDG 1 (no poverty) and SDG 2 (zero hunger): controlling food and energy in the climatic areas.
- SDG 6 (clean water and sanitation): by using efficient energy and agriculture conception of water.
- SDG 7 (affordable clean energy): it achieves through the encouragement of diverse clean energy systems that are renewable.
- SDG 13 climate action (climate action): reducing maladaptive response, and building cross sectoral resilience.

An approach of WEF-Environment nexus allows identifying trade-offs, synergies, and co-benefits among these objectives so that interventions within one sector do not inadvertently make other areas more vulnerable. Nexus-based strategies offer a way to attain sustainable, equitable, and resilient WEF systems in the context of accelerating climate change by connecting the governance, data, and integrated resource management (Yupanqui et al., 2025).

3.8 The WEF Nexus and SDGs

WEF are vital resources to human livelihood and sustainable development. The increasing population, climate change, and economic strains are stretching the resources and it is estimated that the demand of energy, food, and water will grow by about 50, 35, and 40% respectively in 2030 than in 2010 (Ojeda-Matos & Jones-Crank, 2025). The disparity in the distribution of these resources in the world, in conjunction with climatic stresses underscores the necessity of integrated management. In 2015, the United Nations established 17 SDGs to determine sustainable development that is operationalized into 169 targets and 232 indicators, of which SDG 2 (zero hunger), SDG 6 (clean water and sanitation), as well as SDG 7 (affordable and clean energy), are directly linked with WEF security. Other SDGs, such as SDG 8 (decent work and economic growth), SDG 12 (responsible consumption and production), SDG 13 (climate action), SDG 14 (life below water), and SDG 15 (life on land) are also highly correlated with WEF components as it is a symbiotic phenomenon of sustainable development (Roy & Moradkhani, 2025). The WEF nexus focuses on integrated governance and management of intra- and inter-sector governance and management at a sector-scale and scale synergies, and reduction of trade-offs. Indicatively, increasing the agricultural productivity with the exclusion of water and energy limits could contribute to the issue of water stress or greenhouse gases, which are identified as harmful to SDG 6 and SDG 13 (Mutanga et al., 2024). The nexus approach is practiced in different regions that includes Mediterranean, Africa (Kenya, South Sudan, Uganda, and Ethiopia), Asia (Indonesia) and South America. In earlier work, the WEF nexus has been emphasized as a way of maximizing resource interdependencies, enhancing the efficiency of food production, sustainable agriculture, soil health, irrigation and nutrient management, as well as the integration of renewable energy (Yupanqui et al., 2025). Climate change, greenhouse gas emissions, land-use change, biodiversity, pollution, and human health associates WEF strategies with more than one SDG. Combination with the principles of the circular economy also increases resource efficiency through reuse and recycling, resolving supply and demand, and sustainable development. Water-Employment-Migration (WEM) framework provides a socio-economic perspective, including water security and employment, entrepreneurship, gender equality, and youth involvement, which are SDGs 6, 8, and 10. Although there has been progress but still difficulties in the form of gaps in governance, lack of cross-sectoral coordination, limitations on data, lack of infrastructure, and global crises like the COVID-19 pandemic prevails. WEF nexus thinking offers the avenue to speed up the SDG realization process through integrated policies, trade-offs identification and management, data-driven decision-making instruments exploration, sustainable technologies and renewable energy use, as well as stakeholder engagement at various levels. The full implementation of the WEF nexus along with circular economy principles and climate adaptation measures are the key to the sustainable management of resources, their correlative increase in resilience, and the attainment of the SDGs at the local, regional, and global scales (Strippoli et al., 2024).

3.9 Globalization, Climate Change, and the WEF Nexus

Globalization has also been spreading water, energy, and food (WEF) nexuses over long distances, establishing complex forms of interdependence due to international trade in agricultural commodities, land investments and migration patterns (Rhouma et al., 2025). Although trade may help in mitigating the effect of local climatic shocks by mitigating this through the redistribution of food and water resources between the regions of excess to those of deficit, it also creates systemic vulnerability. The shortage of food in a specific area encourages imports or acquisition of foreign lands, which, in turn, keeps the population of the area with a food deficit growing, which contributes to the further increase in dependence on non-local production (Tamasiga et al., 2025). These dynamics result in the lack of connection between consumers and the surrounding environment, which might negatively affect the sustainability of agricultural landscape management. In addition, the environmental impacts such as land-use change, nitrogen and phosphorus pollution, and freshwater depletion can be displaced to the exporting countries because of agricultural trade. The extent of reliance on the external resources may be depicted in the term of the water and ecological footprint that involves internal and external elements of resource utilization. International trade displaces approximately 24% of the land footprint in the world, and the trade between low- and high-income countries contribute about 6% to the overall trade (Javan et al., 2024). These prone areas are enhanced by climate change causing extreme weather to be more frequent and severe, skewed towards food-dependent areas, which have a disproportionate concentration in the areas insensitive to climate. Globalization of WEF systems therefore reveals inequities, like disturbances in key exportation regions may permeate in the chain of world supply, affecting food, water and energy security at the same time. The solutions to these challenges are to implement the WEF nexus, combined with the trade policies of climate resilience, sustainable resource management, and international collaboration to minimize vulnerabilities, facilitate equal access, and increase adaptability to changes in the importing and exporting areas (Behnassi et al., 2021).

3.10 Resilience in the WEF Nexus and the Role of NUCs

The chance of resilience in WEF nexus depends on diversity, redundancy, and adaptive capacity based on resources. In places like Sub-Saharan Africa (SSA) and South Asia, the current agri-food systems are very susceptible because they depend on climatic-sensitive, monocultural processes of producing few high yielding crops with low nutritional content. The weakness of these systems has become even more apparent during the COVID-19 pandemic, global crises, climate crises and an increase in malnutrition. Such agri-food systems suppress localized crops, native knowledge, and various agricultural methods, decreasing the resilience of the system and countering climate and socio-economic impacts (Kyambade et al., 2025).

NUCs or sometimes orphan crops, can play crucial role to enhance WEF nexus resilience. NUCs are generally customized to marginal and low-input settings, uses less water and energy than the traditional staples and delivers nutrient-dense sources of food. They can be integrated into agricultural systems to lower the pressure on water and energy resources and increase the climate, food security, and nutrition outcomes. NUCs help to maintain dietary diversity, enhance the quality of their diet, and strengthen social-cultural outcomes, especially in rural and marginalized communities where women smallholders are at the center of the conservation and cultivation of these types of crops (Purba & Krishnaswamy, 2025).

NUCs are involved in various food security pillars including availability, access, stability and utilization. They protect against climatic shocks naturally, contribute to local livelihoods, and enhance the food sovereignty on a community level by allowing farmers to enhance production. It has been demonstrated that communities that grow NUCs can stay food secure even under the unfavorable climatic conditions, proving that they can be a strong food source in the environment with limited water and energy.

NUCs contain high levels of proteins, vitamins, minerals, antioxidants, dietary fiber, and other bioactive compounds, which can offset increasing rates of hidden hunger, micronutrient deficiencies, and non-communicable diseases in SSA and South Asia. Introduction and reintroduction of NUCs in local diets can enhance access to a wide variety of nutritional foods, increase the quality of diets, and have a beneficial effect on health outcomes at different life stages. NUCs play a direct role in the human aspect of the WEF nexus because they are connected to nutrition security, as well as resilient agri-food systems (Purba & Krishnaswamy, 2025).

NUCs bring about food sovereignty by enabling local people to establish their own food systems. This inclusivity increases equity, promotes cultural heritage, and boosts local agri-food systems and enhances dietary diversity and successful nutrition outcomes.

The strategic use of NUCs as part of the WEF Nexus can be crucial in improving the resilience of areas that have been highly impacted by climate change like South Asia and SSA. NUCs are at the centre of the WEF Nexus because they help lessen reliance on the traditional, water and energy-intensive crops. As Fig. 3.3 shows, NUCs are a bridging system between water, energy and food systems, as their low-input and drought-tolerant traits decrease water and energy requirements, and their nutrient-dense components improve food systems and nutrition outcomes. In addition, NUCs enhance food sovereignty through community empowerment to use native knowledge, preserve native seeds, and establish their own food systems towards equitable, sustainable and context-specific agricultural development. NUCs enhance the resiliency of diversified agri-food systems, diminishing the negative effects of extreme weather, and assuring sustainable management of resources, which are their strategic value in integrated WEF Nexus planning in the SSA and South Asia. This theoretical framework drives the point that integrating the role of NUCs into the WEF Nexus is an approach that does not only guarantee short-term food and nutrition security but also complements longer-term environmental and socio-economic progress.

Figure 3.3 shows the part played by NUCs in improving resilience in the WEF nexus in SSA and South Asia when under the pressure of climate change. The diagram places NUCs in the middle, showing that they can greatly contribute to improving water security by means of drought-tolerant cultivation systems, energy security by means of bio-mass and bioenergy potential, and food and nutrition security by means of crops with a high nutrient density and local adaptation. Arrows show the relationships between the water, energy, and food system, with the importance of NUCs decreasing the pressure on resources, diversifying, and adapting to climate conditions. It is also worth mentioning that the figure highlights the general impact of climate change as stressor and how NUCs can serve as a strategic instrument to keep the systems resilient and in line with the sustainable development objectives in such climate-sensitive areas. The Fig. 3.4 shows that how the NUCs can be used as a lever point towards increasing the resilience within WEF nexus as a result of climate change pressures in Sub-Saharan Africa (SSA) and South Asia. All in the middle, NUCs can also play a part in the water, energy, food and nutrition security due to their inherent characteristics: drought-tolerant properties, low inputs, and high nutrition value. Enhanced water security is ensured through efficiency and drought resilience in water consumption and sustainable and renewable energy through low-input and

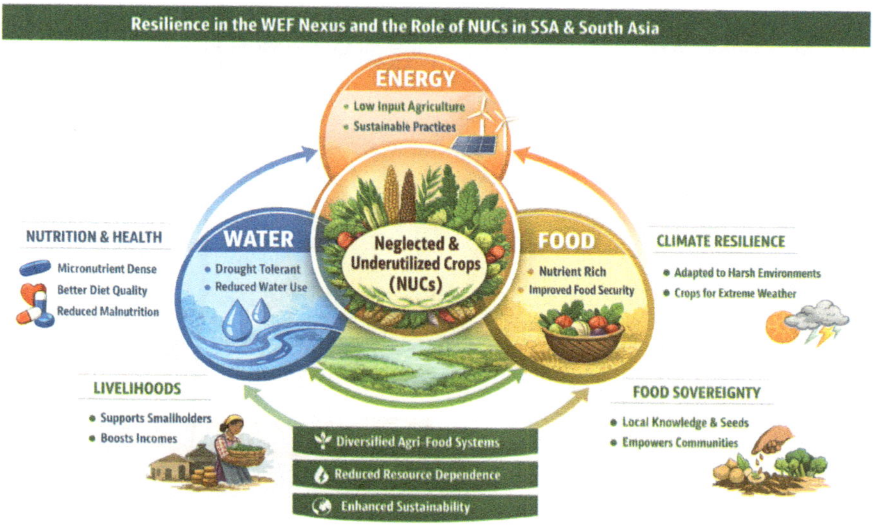

Fig. 3.3 Resilience in the WEF nexus and the role of NUCs in Sub-Saharan Africa (SSA) and South Asia

climate-smart production systems. Simultaneously, the presence of diversified diets and enhanced food availability enhances the food and nutrition security. The inter-linkages in circular ways underline the synergetic character of the WEF nexus with gains in one area strengthening the others. In general, the framework identifies adaptation, diversification, and sustainability as the main avenues by which NUCs can mitigate climate risks, enhance livelihood and develop resilient agri-food systems in climate-prone areas.

3.11 Conclusions and Policy Implications

The WEF nexus has been changing due to climate change, which is worsening resource scarcity, trade-offs, vulnerabilities, and amplified food, water, and energy systems across most food, water and energy systems. Additional extreme weather changes threatening agriculture, water, and energy security in climate-sensitive regions, like SSA. South Asia-centered harsh weather events can be viewed as critical threats to multiple facets of the economic framework of the act. The conventional mono-culture agri-food systems have increasingly become vulnerable to climatic shocks, socio-economic shocks and global crises, and there is great demand to develop resilient, integrated and adaptive management strategies.

NUCs could improve the resilience of WEF nexus as they are naturally acclimatized to marginal conditions. They use less water, energy and no agro chemicals have been required for its production. NUCs contributes to the dietary diversity and

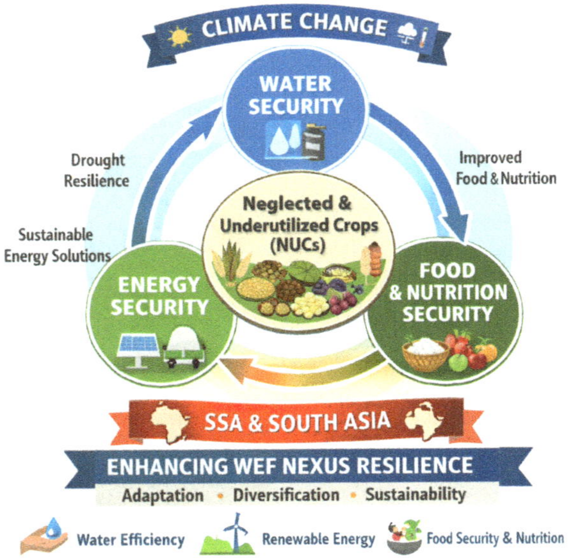

Fig. 3.4 Strengthening water–energy–food (WEF) nexus resilience through neglected and underutilized crops under climate change

enhances nutritional security. NUCs like millets, teff, sorghum, bambara groundnut, cassava, yam, and leafy vegetables have shown the ability to stabilize agri-food systems under climatic variability in addition to reducing pressures among water and energy systems, in SSA and South Asia. WEF nexus perspective emphasizes the interrelationship of water, energy, and food. It has been observed that sectoral policies are not good enough to mitigate systemic risks. Climate change is a multiplier of WEF stresses, and actions in one sector, like increasing irrigation, increasing biofuel production through energy-intensive methods, or increasing monoculture, may potentially increase another sector vulnerability. Consequently, to increase resilience, we recommends cross-sectoral governance, adaptive infrastructure investment, and low-resource climate-resilient agriculture.

Policy Implications

Promote NUC Integration in National and Regional Agri-food Policies

Governments in SSA and South Asia should focus on NUCs in agricultural planning, extension and research agendas. The promotion of the development of drought, heat-tolerant NUCs can diversify production, increase nutrition security, and lower resources intensity throughout the WEF nexus.

Strengthen WEF Nexus Governance

Cross sectoral, integrated governance structures are needed to deal with trade-offs between water, energy, and food. Where there is climate variability, the policies are supposed to allow coordination of multiple stakeholders, adaptive management, and data-informed decision-making, which will best balance the allocation of resources.

Invest in Climate-Resilient Infrastructure and Technology

The weaknesses of the WEF nexus can be minimized by investing in water-efficient irrigation, renewable energy, storage, and processing technologies. Financial assistance on the adoption of climate-resistant practices and NUC-based farming systems by smallholders is important to establishing local adaptive capacity.

Enhance Food Sovereignty and Community Participation

Policies should allow the local communities to participate actively. Women who are smallholders, should be trained so that they can take control of seeds and choice of crops.

Link NUC Promotion to Nutrition and Health Policies

Hidden hunger and micronutrient deficiencies can be tackled through the incorporation of NUCs in school feeding programs, social nutrition programs, and dietary guidelines and can result in an increased quality of diet. Agricultural policies and nutrition plans contribute to the human aspect of WEF nexus.

Foster Research, Innovation, and Market Development

The enhancement of the market viability, raising the level of farmers incomes, and the provision of climate-resilient value chains may be improved by investing in the R&D of NUC improvement, post-harvest handling, processing, and commercialization. Co-benefits can be realized through the use of innovations that can be applied in food, energy, water and environmental systems.

Integrate WEF Nexus Approaches with SDGs and Circular Economy Principles

The management of resources should be clearly associated to SDG targets. This mean that climate action, food security, clean energy and water sustainability will receive co benefits. Resource efficiency can also be maximized by incorporating the concept of the circular economy in place, which encourages reuse, recycling, and waste valorization; these concepts can facilitate long-term sustainability.

In general, the integration of NUCs, together with combined WEF nexus governance and climate-adaptive solution sets, creates an avenue towards the construction of resilient, diverse, and sustainable agri-food nexus. These strategies can be used by policymakers in SSA and South Asia to reduce risks caused by climate change, improve nutrition and food security, maximize resources utilization, and inclusive economic growth amidst the escalating global change.

References

Abalasei, M. E., Toma, D., Dorus, M., & Teodosiu, C. (2025). The impact of climate change on water quality: A critical analysis. *Water, 17*(21), 3108.

Adnan, M., Xiao, B., Bibi, S., Xiao, P., Zhao, P., & Wang, H. (2024). Addressing current climate issues in Pakistan: An opportunity for a sustainable future. *Environmental Challenges, 15*, 100887. https://doi.org/10.1016/j.envc.2024.100887

Agostoni, C., Baglioni, M., La Vecchia, A., Molari, G., & Berti, C. (2023). Interlinkages between climate change and food systems: The impact on child malnutrition—narrative review. *Nutrients, 15*(2), 416.

Ahmad, R., Liu, G., Ur Rehman, S. A., Gao, Y., Meng, F., Zhou, X., et al. (2025). Integrated systems modeling for assessing the water–energy nexus in Pakistan: Lessons learned from coupling LEAP–WEAP planning approaches. *Engineering, 50*, 72–87. https://doi.org/10.1016/j.eng.2025.02.013

Ahmed, M. (2023). The science of climate change. In R. Brinkmann (Ed.), *The Palgrave handbook of global sustainability* (pp. 195–222). Springer. https://doi.org/10.1007/978-3-031-01949-4_22

Ahmed, R. (2025). Decoding the fate of himalayan glaciers under climate change: Impacts, challenges, research gaps, and policy pathways. *Evolving Earth, 3*, 100082. https://doi.org/10.1016/j.eve.2025.100082

Albrecht, T. R., Crootof, A., & Scott, C. A. (2018). The Water–Energy–Food Nexus: A systematic review of methods for nexus assessment. *Environmental Research Letters, 13*(4), 043002. https://doi.org/10.1088/1748-9326/aaa9c6

Alhassan, A.-L., Perez, M. A., & Galagedara, L. W. (2026). Integrating climate change, food security, and innovative agriculture in Newfoundland and Labrador (NL): A Water–Energy–Food (WEF) nexus approach. *Global Transitions, 8*(1), 22–36. https://doi.org/10.1016/j.glt.2025.08.002

Ali, M. H., Biswas, P., Hassan, S. M. Q., & Islam, M. A. (2025). Climate change and its impacts on hydrological regimes over the Bengal delta. *Results in Engineering, 25*, 103861. https://doi.org/10.1016/j.rineng.2024.103861

Ali, S. M., & Acquaye, A. (2024). An examination of water–energy–food nexus: From theory to application. *Renewable and Sustainable Energy Reviews, 202*, 114669. https://doi.org/10.1016/j.rser.2024.114669

Ali, S., Liu, Y., Ishaq, M., Shah, T., Abdullah, S., Ilyas, A., & Din, I. U. (2017). Climate change and its impact on the yield of major food crops: Evidence from Pakistan. *Foods, 6*(6), 39. https://doi.org/10.3390/foods6060039

Ansari, A., Wuryandani, S., Pranesti, A., Telaumbanua, M., Ngadisih, S., Hardiansyah, M. Y., et al. (2023). Optimizing Water–Energy–Food Nexus: Achieving economic prosperity and environmental sustainability in agriculture. *Frontiers in Sustainable Food Systems, 7*, 1207197. https://doi.org/10.3389/fsufs.2023.1207197

Azimi, M. N., Rahman, M. M., & Maraseni, T. (2025). Powering progress: The interplay of energy security and institutional quality in driving economic growth. *Applied Energy, 378*, 124835. https://doi.org/10.1016/j.apenergy.2024.124835

Bai, W., Wang, J., Muttaqi, K. M., Sutanto, D., Raad, R., & Truran, A. (2026). Energy-integrated water cycles: Critical review of optimisation strategies and renewable energy synergies. *Renewable and Sustainable Energy Reviews, 226*, 116457. https://doi.org/10.1016/j.rser.2025.116457

Bamgboye, T. T., Avellán, T., Klöve, B., & Haghighi, A. T. (2025). Compounding impacts of climate change and urbanisation on water–energy–food nexus in global south countries: A systematic review. *Environmental and Sustainability Indicators, 27*, 100791. https://doi.org/10.1016/j.indic.2025.100791

Behnassi, M., Rahman, M. A., D'Silva, J., Ramachandran, G., Gupta, H., Pollmann, O., & Ramachandran, N. (2021). Enhancing resilience for food and nutrition security within a changing climate. In M. Behnassi, M. Barjees-Baig, M. El-Haiba, & M. R. Reed (Eds.), *Emerging challenges to food production and security in Asia, middle east, and Africa: Climate risks and resource scarcity* (pp. 1–42). Springer. https://doi.org/10.1007/978-3-030-72987-5_1

Cavalheiro, E. A., Oliveira, I. R., Leandro, D., & Kontz, L. B. (2025). Governance, development, and environment: Pathways to a sustainable future. *Sustainable Futures, 10*, 100813. https://doi.org/10.1016/j.sftr.2025.100813

Chingiz, D., Liu, J., & Malikov, M. (2025). The role of environmental poverty in shaping green growth in the one belt one road countries. *Sustainable Futures, 10*, 100935. https://doi.org/10.1016/j.sftr.2025.100935

Collins, A. J. R. F. (2019). *The global risks report* (14th ed., pp. 978–971). Insight Report.

De Corato, U., Viola, E., Keswani, C., & Minkina, T. (2024). Impact of the sustainable agricultural practices for governing soil health from the perspective of a rising agri-based circular bioeconomy. *Applied Soil Ecology, 194*, 105199. https://doi.org/10.1016/j.apsoil.2023.105199

Dhar, A. R. (2025). Building climate-resilient food systems through the water–energy–food–environment nexus. *Environments, 12*(5), 167.

Dinar, A., Tieu, A., & Huynh, H. (2019). Water scarcity impacts on global food production. *Global Food Security, 23*, 212–226. https://doi.org/10.1016/j.gfs.2019.07.007

Drobinski, P., Rivera-Ferre, M. G., Monem, M. A., Hatab, A. A., Behnassi, M., Chfadi, T., et al. (2025). Nexus approach to enhance water–energy–food security and ecosystems resilience under climate change in the Mediterranean. *NPJ Climate Action, 4*(1), 115. https://doi.org/10.1038/s44168-025-00308-4

Duchenne-Moutien, R. A., & Neetoo, H. (2021). Climate change and emerging food safety issues: A review. *Journal of Food Protection, 84*(11), 1884–1897. https://doi.org/10.4315/JFP-21-141

El-Araby, R. (2024). Biofuel production: Exploring renewable energy solutions for a greener future. *Biotechnology for Biofuels and Bioproducts, 17*(1), 129. https://doi.org/10.1186/s13068-024-02571-9

Fanzo, J., Bellows, A. L., Spiker, M. L., Thorne-Lyman, A. L., & Bloem, M. W. (2021). The importance of food systems and the environment for nutrition. *American Journal of Clinical Nutrition, 113*(1), 7–16. https://doi.org/10.1093/ajcn/nqaa313

Flammini, A., Puri, M., Pluschke, L., & Dubois, O. (2014). *Walking the nexus talk: Assessing the Water–Energy–Food Nexus in the context of the sustainable energy for all initiative*. FAO.

Ghosh, B., Gubareva, M., Ghosh, A., Paparas, D., & Vo, X. V. (2024). Food, energy, and water nexus: A study on interconnectedness and trade-offs. *Energy Economics, 133*, 107521. https://doi.org/10.1016/j.eneco.2024.107521

Guo, Q., & Hasani, R. (2024). Assessing the impact of water scarcity on thermoelectric and hydroelectric potential and electricity price under climate change: Implications for future energy management. *Heliyon, 10*(17), e36870. https://doi.org/10.1016/j.heliyon.2024.e36870

Haines, A., & Frumkin, H. (2021). Food systems and land use. In A. Haines & H. Frumkin (Eds.), *Planetary health: Safeguarding human health and the environment in the anthropocene* (pp. 310–359). Cambridge University Press. https://doi.org/10.1017/9781108698054.010

Hansen, J. E., Kharecha, P., Sato, M., Tselioudis, G., Kelly, J., Bauer, S. E., et al. (2025). Global warming has accelerated: Are the United Nations and the public well-informed? *Environment: Science and Policy for Sustainable Development, 67*(1), 6–44.

Hemathilake, D. M. K. S., & Gunathilake, D. M. C. C. (2022). Chapter 31—Agricultural productivity and food supply to meet increased demands. In R. Bhat (Ed.), *Future foods* (pp. 539–553). Academic Press. https://doi.org/10.1016/B978-0-323-91001-9.00016-5

Hong, C., Zhong, R., Xu, M., He, P., Mo, H., Qin, Y., et al. (2025). Interactions among food systems, climate change, and air pollution: A review. *Engineering, 44*, 215–233. https://doi.org/10.1016/j.eng.2024.12.021

Ioannou, A. E., & Laspidou, C. S. (2022). Resilience analysis framework for a water–energy–food nexus system under climate change. *Frontiers in Environmental Science, 10*, 820125. https://doi.org/10.3389/fenvs.2022.820125

Javan, K., Altaee, A., BaniHashemi, S., Darestani, M., Zhou, J., & Pignatta, G. (2024). A review of interconnected challenges in the water–energy–food nexus: Urban pollution perspective towards sustainable development. *Science of the Total Environment, 912*, 169319. https://doi.org/10.1016/j.scitotenv.2023.169319

Jones-Crank, J. L. (2024). The contribution of Water–Energy–Food Nexus governance to sustainability: A case study of Singapore. *Environmental Science and Policy, 160*, 103849. https://doi.org/10.1016/j.envsci.2024.103849

Karimi, M., Tabiee, M., Karami, S., Karimi, V., & Karamidehkordi, E. (2024). Climate change and water scarcity impacts on sustainability in semi-arid areas: Lessons from the South of Iran. *Groundwater for Sustainable Development, 24*, 101075. https://doi.org/10.1016/j.gsd.2023.101075

Kyambade, M., Mwesigwa, R., Alinda, K., Tumwine, S., & Lwanga, F. (2025). Drivers of sustainability performance in health and sanitation projects: A systematic literature review. *Social Sciences and Humanities Open, 12*, 102046. https://doi.org/10.1016/j.ssaho.2025.102046

Li, W., Ward, P. J., & van Wesenbeeck, L. (2025). A critical review of quantifying Water–Energy–Food Nexus interactions. *Renewable and Sustainable Energy Reviews, 211*, 115280. https://doi.org/10.1016/j.rser.2024.115280

Lindow, L., Campbell, C., Longwater, C., Zhang, Y., Martin-Ryals, A., & Boz, Z. (2026). Diffusion of innovation in controlled environment agriculture: A mixed-methods study of digital decision support tool adoption. *Journal of Innovation and Knowledge, 13*, 100882. https://doi.org/10.1016/j.jik.2025.100882

Majeed, Y., Khan, M. U., Waseem, M., Zahid, U., Mahmood, F., Majeed, F., et al. (2023). Renewable energy as an alternative source for energy management in agriculture. *Energy Reports, 10*, 344–359. https://doi.org/10.1016/j.egyr.2023.06.032

Messaoudi, B., Kabiri, L., Ait Lahssaine, I., El-Hafyani, M., Essafraoui, B., Kadiri, A., & Albergel, J. (2025). Assessing the impact of climate change on the Water–Energy–Food Nexus in Southeastern Morocco: A case study of the Wadi Guir watershed. *European Journal of Remote Sensing, 58*(1), 2540105. https://doi.org/10.1080/22797254.2025.2540105

Mondal, K., Chowdhury, M., Dutta, S., Satpute, A. N., Jha, A., Khose, S., et al. (2025). Synergising agricultural systems: A critical review of the interdependencies within the Water–Energy–Food Nexus for sustainable futures. *Water–energy Nexus, 8*, 167–188. https://doi.org/10.1016/j.wen.2025.04.004

Mperejekumana, P., Shen, L., Zhong, S., Muhirwa, F., Gaballah, M. S., & Nsigayehe, J. M. V. (2024). Integrating climate change adaptation into Water–Energy–Food-environment nexus for sustainable development in East African Community. *Journal of Cleaner Production, 434*, 140026. https://doi.org/10.1016/j.jclepro.2023.140026

Mutanga, S. S., Mantlana, B. K., Mudavanhu, S., Muthige, M. S., Skhosana, F. V., Lumsden, T., et al. (2024). Implementation of water energy food-health nexus in a climate constrained world: A review for South Africa. *Frontiers in Environmental Science, 12*, 1307972. https://doi.org/10.3389/fenvs.2024.1307972

Naderi, L., Karamidehkordi, E., Badsar, M., & Moghadas, M. (2024). Impact of climate change on water crisis and conflicts: Farmers' perceptions at the ZayandehRud Basin in Iran. *Journal of Hydrology: Regional Studies, 54*, 101878. https://doi.org/10.1016/j.ejrh.2024.101878

Ningi, T., & Mullen, J. D. (2025). A literature review of the food–energy–water nexus trade-offs and synergies research at the household level. *Energy Nexus, 20*, 100562. https://doi.org/10.1016/j.nexus.2025.100562

Nunes, F. C., Prasad, M. N. V., & Alves, L. D. J. (2026). Chapter 1—An interrelationship of water, energy, and food nexus to deal with climate mitigation and adaptation: An Introduction. In T. A. Kurniawan & M. N. Vara Prasad (Eds.), *Climate change impacts on the water–energy–food nexus* (pp. 1–10). Elsevier. https://doi.org/10.1016/B978-0-443-22320-4.00011-6

Ojeda-Matos, G., & Jones-Crank, J. L. (2025). The potential of the Water–Energy–Food Nexus approach in advancing the sustainable development goals: A PRISMA-based systematic review. *Environmental Science and Policy, 163*, 103961. https://doi.org/10.1016/j.envsci.2024.103961

Oki, T., & Kanae, S. (2006). Global hydrological cycles and world water resources. *Science, 313*(5790), 1068–1072.

Olarewaju, O. O., Fawole, O. A., Baiyegunhi, L. J. S., & Mabhaudhi, T. (2025). Integrating sustainable agricultural practices to enhance climate resilience and food security in Sub-Saharan Africa: A multidisciplinary perspective. *Sustainability, 17*(14), 6259.

Purba, N. H., & Krishnaswamy, K. (2025). Exploring the potentials of neglected underutilized crops (NUCs): An integrative review for developing a sustainable food system model. *NPJ Science of Food, 9*(1), 199. https://doi.org/10.1038/s41538-025-00554-0

Rahman, H. M. T., & Natcher, D. (2025). Addressing gaps in integrative water–energy–food–forest (WEFF) nexus governance. *Environmental Science and Policy, 172*, 104195. https://doi.org/10.1016/j.envsci.2025.104195

Raihan, A., Ridwan, M., & Rahman, M. S. (2024). An exploration of the latest developments, obstacles, and potential future pathways for climate-smart agriculture. *Climate Smart Agriculture, 1*(2), 100020. https://doi.org/10.1016/j.csag.2024.100020

Rhouma, A., Seitfudem, G., El-Jeitany, J., Pacetti, T., Brouwer, F., & Gil, J. M. (2025). Connecting the water footprint with the water–energy–food-ecosystems nexus concept and its added value in the Mediterranean. *Environmental and Sustainability Indicators, 26*, 100640. https://doi.org/10.1016/j.indic.2025.100640

Ringler, C., Willenbockel, D., Perez, N., Rosegrant, M., Zhu, T., & Matthews, N. (2016). Global linkages among energy, food and water: An economic assessment. *Journal of Environmental Studies and Sciences, 6*(1), 161–171. https://doi.org/10.1007/s13412-016-0386-5

Rosińska, W., Jurasz, J., Przestrzelska, K., Wartalska, K., & Kaźmierczak, B. (2024). Climate change's ripple effect on water supply systems and the water–energy nexus: A review. *Water Resources and Industry, 32*, 100266. https://doi.org/10.1016/j.wri.2024.100266

Roy, A., & Moradkhani, H. (2025). Toward an integrated sustainability assessment of water–energy–food nexus indicators. *Journal of Hydrology X, 28–29*, 100205. https://doi.org/10.1016/j.hydroa.2025.100205

Said, Z., Vigneshwaran, P., Shaik, S., Rauf, A., & Ahmad, Z. (2025). Climate and carbon policy pathways for sustainable food systems. *Environmental and Sustainability Indicators, 27*, 100730. https://doi.org/10.1016/j.indic.2025.100730

Sani, Y., & Scholz, M. (2022). Interplay of water–energy security and food consumption patterns towards achieving nutrition security in Katsina State, North-Western Nigeria. *Sustainability, 14*(8), 4478.

Shafiq, S., Zia-Ul-Haq, M., Naqvi, S. A. R., Sarfaraz, W., Ali, H., Islam, M. M., et al. (2025). Integrating neglected and underutilized crops (NUCs) in South Asian cropping systems and diets: Challenges and prospects. *Regional Sustainability, 6*(4), 100242. https://doi.org/10.1016/j.regsus.2025.100242

Shah, H., Hellegers, P., & Siderius, C. (2021). Climate risk to agriculture: A synthesis to define different types of critical moments. *Climate Risk Management, 34*, 100378. https://doi.org/10.1016/j.crm.2021.100378

Shayanmehr, S., Porhajašová, J. I., Babošová, M., Sabouhi-Sabouni, M., Mohammadi, H., Rastegari-Henneberry, S., & Shahnoushi-Foroushani, N. (2022). The impacts of climate change on water resources and crop production in an arid region. *Agriculture, 12*(7), 1056.

Siyal, A. W., & Gerbens-Leenes, P. W. (2022). The water–energy nexus in irrigated agriculture in South Asia: Critical hotspots of irrigation water use, related energy application, and greenhouse gas emissions for wheat, rice, sugarcane, and cotton in Pakistan. *Frontiers in Water, 4*, 941722. https://doi.org/10.3389/frwa.2022.941722

Strippoli, R., Gallucci, T., & Ingrao, C. (2024). Circular economy and sustainable development in the tourism sector: An overview of the truly-effective strategies and related benefits. *Heliyon, 10*(17), e36801. https://doi.org/10.1016/j.heliyon.2024.e36801

Szinai, J. K., Yates, D., Sánchez-Pérez, P. A., Staadecker, M., Kammen, D. M., Jones, A. D., & Hidalgo-Gonzalez, P. (2024). Climate change and its influence on water systems increases the cost of electricity system decarbonization. *Nature Communications, 15*(1), 10050. https://doi.org/10.1038/s41467-024-54162-9

Tamasiga, P., Dzingai, V. M., Onyeaka, H., Tchonkouang, R. D., Siyanbola, K. F., Genesis, U., & Mudimu, G. T. (2025). Energy transition effects on food security amidst climate change and progress toward sustainable development goals. *Energy and Climate Change, 6*, 100222. https://doi.org/10.1016/j.egycc.2025.100222

Tchonkouang, R. D., Onyeaka, H., & Nkoutchou, H. (2024). Assessing the vulnerability of food supply chains to climate change-induced disruptions. *Science of the Total Environment, 920*, 171047. https://doi.org/10.1016/j.scitotenv.2024.171047

Tian, G., Li, J., Wu, Z., Zhao, Q., Xia, Q., Zhang, J., & Niu, X. (2025). Unveiling the resilience of the water–energy–food nexus: A context-specific framework and evolutionary insights for the Yellow River Basin. *Agricultural Water Management, 315*, 109514. https://doi.org/10.1016/j.agwat.2025.109514

Torhan, S., Grady, C. A., Ajibade, I., Galappaththi, E. K., Hernandez, R. R., Musah-Surugu, J. I., et al. (2022). Tradeoffs and synergies across global climate change adaptations in the food–energy–water nexus. *Earth's Future, 10*(4), e2021EF002201.

UN. (2023). *Transforming our world: The 2030 agenda for sustainable development.*

Ur-Rehman, A., Sanjari, M. J., Elavarasan, R. M., & Jamal, T. (2026). Sustainability-aligned pathways for energy transition: A review of low-carbon energy network solutions. *Renewable and Sustainable Energy Reviews, 226*, 116428. https://doi.org/10.1016/j.rser.2025.116428

Wang, Q., Deng, H., & Jian, J. (2023). Hydrological processes under climate change and human activities: Status and challenges. *Water, 15*(23), 4164.

Yupanqui, C., Dias, N., Goodarzi, M. R., Sharma, S., Vagheei, H., & Mohtar, R. (2025). A review of water–energy–food nexus frameworks, models, challenges and future opportunities to create an integrated, national security-based development index. *Energy Nexus, 18*, 100409. https://doi.org/10.1016/j.nexus.2025.100409

Zahedi, R., Yousefi, H., Aslani, A., & Ahmadi, R. (2024). Water, energy, food and environment nexus (WEFEN): Sustainable transition, gaps and covering approaches. *Energy Strategy Reviews, 54*, 101496. https://doi.org/10.1016/j.esr.2024.101496

Zenda, M. (2024). A systematic literature review on the impact of climate change on the livelihoods of smallholder farmers in South Africa. *Heliyon, 10*(18), e38162. https://doi.org/10.1016/j.heliyon.2024.e38162

Zhang, C., Chen, X., Li, Y., Ding, W., & Fu, G. (2018). Water–energy–food nexus: Concepts, questions and methodologies. *Journal of Cleaner Production, 195*, 625–639. https://doi.org/10.1016/j.jclepro.2018.05.194

Zhang, Y., Huang, K., Yu, Y., & Yang, B. (2017). Mapping of water footprint research: A bibliometric analysis during 2006–2015. *Journal of Cleaner Production, 149*, 70–79. https://doi.org/10.1016/j.jclepro.2017.02.067

Chapter 4
Environmental Protection and Ecosystem Services Through Cultivation of NUCs

Mukhtar Ahmed, Muhammad Abid, Ahmed M. S. Kheir, Mustafa Kan, and P. V. Vara Prasad

Abstract Intensive agricultural practices have increased world food production. However, this resulted to the accelerated environmental degradation, loss of biodiversity and destruction of ecosystem services upon which long-term resilience of food system depends. One of the possible solutions to have sustainable agricultural productivity versus environmental protection, climate resilience, and nutrition security in future is to includes Neglected and Underutilized Crops (NUCs) in the food system. This chapter provides information about potential of NUCs in improving ecosystem services and enhancing socio-ecological resilience to climate change and resource limitation. The chapter elaborates ecosystem services theory, and measures how the NUC-based diversification facilitates the provisioning, regulation, support, and cultural services through the improvement of soil health, water

M. Ahmed (✉)
Department of Agronomy, Pir Mehr Ali Shah Arid Agriculture University, Rawalpindi, Pakistan
e-mail: ahmadmukhtar@uaar.edu.pk

Swedish University of Agricultural Sciences, Umea, Sweden

Department of Biological Systems Engineering, Washington State University, Pullman, USA

M. Abid
UNESCO Chair On Knowledge Systems for IWRM Director Interdisciplinary Research, Islamabad, Pakistan

Department of Mechanical Engineering, COMSATS University Islamabad, Islamabad, Pakistan

Department of Mechanical Engineering, Faculty of Engineering, Islamic University of Madinah, Madinah, Saudi Arabia

A. M. S. Kheir
Julius Kühn Institute (JKI)—Federal Research Centre for Cultivated Plants, Institute for Strategies and Technology Assessment, Kleinmachnow, Germany

Soils, Water and Environment Research Institute, Agricultural Research Center, Giza, Egypt

M. Kan
Department of Agricultural Economics, Agricultural Faculty, Kirsehir Ahi Evran University, Kırşehir, Türkiye

P. V. Vara Prasad
Department of Agronomy, Kansas State University, Manhattan, KS, USA

M. Ahmed (ed.), *Climate Change and Neglected Underutilized Crops*,
https://doi.org/10.1007/978-3-032-21005-0_4

regulation, carbon sequestration, biodiversity conservation, and maintenance of traditional knowledge systems. It is also the role of NUCs in assisting the agroecosystems to resist climatic variability and environmental shock and providing processes of the ecosystem that are essential to long-term productivity. The chapter also discusses about the economic incentives and policy tools like the payment of ecosystem services, market-based tools and institutional underpinnings that can be used in favor of the wider use of NUCs and an assessment of critical financial, technical and governance constraints. The United States, South Asia and Sub-Saharan Africa case studies show that the practice of NUC cultivation can contribute to improving ecosystem services and livelihood in different environments. This chapter points to the importance of the inclusion of NUCs in the sustainable agricultural systems as a pillar of climate-sensitive, resilient, and ecosystem services-driven transformation of food systems in binding the environment protection, food and nutrition security, and policy innovation.

Keywords Neglected and underutilized crops · Ecosystem services · Agrobiodiversity · Cimate resilience · Sustainable agriculture

4.1 Introduction

Modern agriculture development has led to increased food production to feed billions across the globe. However, this has damaged the environmental health. All this has resulted to the soil degradation, water pollution and loss of biodiversity. Climate change adds to these environmental strains by amplifying the occurrence of droughts, floods and other extreme weather events. Thus, threatening agricultural production and rural livelihood. As a result, the supply side of the ecosystem to provide necessary goods and services has been decreasing, and causing damage to human wellbeing and food system (IPCC, 2023; Kaur et al., 2024).

Ecosystem services (ES) are the benefits human receives from ecosystems that includes clean air/water, food and fiber, biodiversity regulation, cultural and recreational advantages (Tallis et al., 2013). ES could be categorized into four categories (i) Provisioning (food, water), (ii) Regulating (climate, flood control), (iii) Cultural (recreation, aesthetics), and (iv) Supporting (nutrient cycles, oxygen) services. These all are crucial for human well-being but due to number of factors, damage has been observed for all four categories. This highlights the need for integrated management. However, in contrast to the natural processes that take place within an ecosystem, no matter whether the human being is present or not, ecosystem services are the services that the human being values in relation to the processes that occur in the ecosystem. Healthy ecosystems are thus crucial to the survival of human beings, since they ensure the fertility of soils, control water cycles, reduce climate change, and offer many goods and services that are very vital in the livelihoods, health and development of the society (Fig. 4.1). Kim et al. (2024) classified ecosystem services

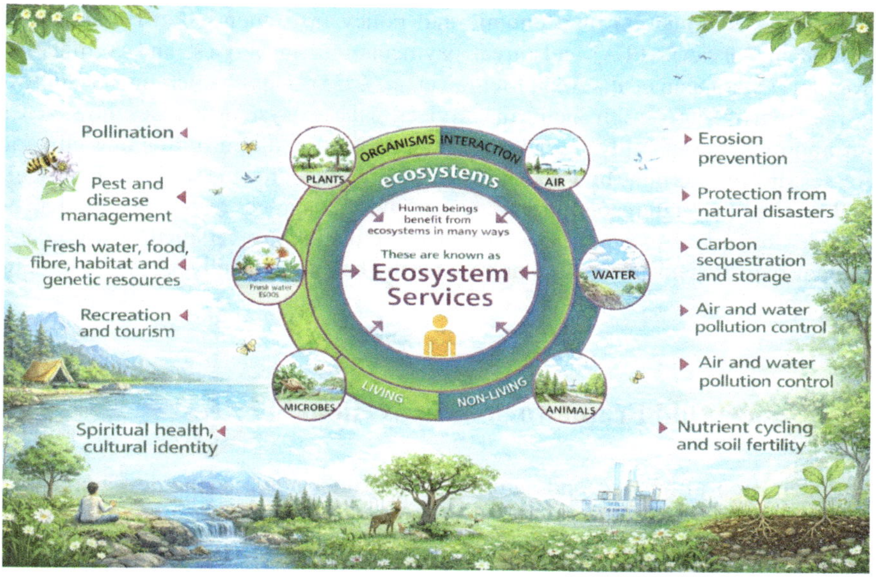

Fig. 4.1 Ecosystem services and their benefits to human well-being

into provisioning, regulating, supporting, and cultural services that can be used for the sustainable environmental management and human well-being.

Agriculture contributes to environmental change as well as to ecosystem services. Although traditional agriculture has tended to deteriorate the environment, sustainable and regenerative agriculture can create a chance to improve ecosystem services (Raza et al., 2025). Through wildlife habitat, agroforestry buffer, pollinator strip, and riparian protection areas, farmers can contribute to the biodiversity, and enhance regulating services. This could also be enhanced water infiltration, nutrient cycling, soil health, and carbon sequestration by cover cropping, no-till farming, adaptive grazing, and soil conservation (Snapp, 2017a, 2017b; Snapp & Pound, 2017). Besides this, agricultural landscapes also offer cultural ecosystem services, including the preservation of aesthetically valuable landscapes and recreational areas, as well as by providing wildlife habitats, which support tourism, education, and local traditions (Behera et al., 2026).

Integration of environmental protection and sustainable agriculture is possible by inclusion of NUCs in the system (Shafiq et al., 2025). These species are also strong against climatic stresses, low-input conditions and marginal soils thus very appropriate to sustain the ecosystem. NUCs increase the fertility of the soil by fixing nitrogen and biodiversity by encouraging pollinators. They can promote wildlife as less chemical has been required by NUCs than other staple crops. Their development is a source of a variety of ecosystem services, such as provisioning, regulating, supporting, and cultural services. Farmers will be able to break the barriers related to finances and technology and guarantee the proliferation and sustainability through the combination of economic and policy instruments like Payments for Ecosystem Services (PES) and direct payments, tax incentives, and certification programs (Marappan et al., 2025; Mehran et al., 2025). This chapter aims to address the potential of NUCs in advancing the services of the ecosystem, to assess how NUCs can improve sustainability in environmental protection and in mitigating climatic stresses, and to determine how NUCs can be incorporated into the sustainable agricultural system. Through recognition of the complex advantages of NUCs, this chapter attempts to explain its significance in ensuring resilient, diversified, and climate-resilient food systems as well as promoting the health and wellness of ecosystems and people.

4.2 Conceptual Framework of Ecosystem Services

The interaction of humans and nature has been intensified, as the development of humankind altered the ecosystems of the earth significantly (Reader et al., 2022). Ecosystem is a functional and dynamic system that consists of interacting populations of animals, microorganisms, plants and non-living factors, such as soil, water and climate. The location, soil, climate variability and water availability are some of the environmental factors that control the occurrence and operations of the ecosystems. The Convention on Biological Diversity (CBD, 1992) defines ecosystems as a complex of organisms that live in a specified place and interact with the abiotic environment. These systems are sustained by constant energy and material flows where the main sources are sunlight, nutrients, soil, and water and the main regenerative product is biomass (carbon-based life).

Living organisms that reproduce, die and grow are important to ecosystem functioning as they continuously recycle energy and matter. These cycles have been developed during millions of years due to natural disturbances (e.g., fires and floods), environmental stresses (e.g., droughts and disease), and ecological interactions (e.g., competition and predation). Nevertheless, the recent growth in the rate and severity of climatic and anthropogenic disturbances make people think about the ability of ecosystems to adapt and survive and stay robust under the pressure of global changes. Human communities have always dependent on ecosystems for physical resources like food, fuel and fibre. They have adjusted the ecosystems to increase the yield of these resources. These modifications sometimes demand the need to add external inputs such as fertilizers, pesticides and fossil fuels which may boost productivity, but can also produce negative externalities, such as nutrient runoff, water pollution and soil degradation. Urban systems also act as highly anthropogenic and highly adjusted ecosystems which rely on the constant streams of energy, water and materials and which generate processed streams of waste that needs to be ultimately absorbed or purified by natural systems. Work to increase waste reuse and recycling is indicative of working to reorganize human systems into more cyclic forms, closer to natural ecosystems (Anikwe & Ife, 2023).

Natural capital includes ecosystems, biodiversity, land, water, vegetation, and atmospheric systems that forms the ecological foundation upon which human and technical capital depends. However, the strategies of economic development have been focused on the short-term development. This neglecting ecosystem integrity leading to the extensive ecological degradation and the elimination of ecosystem resilience (Amicarelli et al., 2025). Such tendencies resulted to increase worldwide problems e.g. climatic change and loss of biodiversity which have become more and more dangerous to ecosystem stability and human health (Rac et al., 2024).

Environmental services outline the indirect and direct benefits that humans receive as a result of ecosystems. With the increased knowledge on how people rely on the natural processes at spatial and temporal levels, the necessity to identify, quantify and appreciate the services has become more evident particularly in the economic and management systems. The notion of ecosystem services denotes the benefits that are given to humans in the form of environmental assets (land, water, vegetation and atmosphere) being transformed into flows of valuable products and services like clean air, water and food (Xu & Peng, 2022). Millennium Ecosystem Assessment further conceptualized ecosystem services as the ecological properties, functions, or processes that directly or indirectly contribute to human well-being (Hori & Makino, 2018).

The most common classification of ecosystem services has been made into four categories, which are provisioning, regulating, supporting, and cultural services. Food, fibre, fuel, timber and raw materials fall under the provisioning services. Regulation of services include climatic regulation, carbon sequestration, water purification, pollination, erosion management and disease management. The ecological backbone of all the other services has been supported by supporting services of cycling nutrients, soil formation, primary production and habitat provision. Some of the non-material benefits like recreation, spiritual enrichment, aesthetic values, cultural identity and education have been considered as cultural services (Tallis et al., 2013) (Fig. 4.2).

Biodiversity is key to the maintenance of ecosystem services and can be considered as the "engine room" of ecosystem functioning. Biodiversity entails the diversity of genes, species and ecosystems including the structure of organisms into populations and community. It is the immediate source of a great deal of provisioning, e.g. food and fibre and forms the basis of controlling and sustaining services by its contribution to energy circulation, nutrient cycling and ecological interactions. Alteration in the biodiversity especially loss has a direct negative effect of limiting the ability of the ecosystems to deliver services and a negative impact on the long term ability of ecological, economic and social system to adapt to global pressures (Mace et al., 2012).

Forest ecosystems supply an essential category of ecosystem services that are known as forest ecosystem services (FES). These are the timber and non-timber forest products, carbon storage, climate moderation, hydrology control, soil security, biodiversity protection, cultural, recreational amenities. Although the role of forests in providing services has long been appreciated. The role of controlling and facilitating services as the ecosystem has been degraded due to increased climate risks,

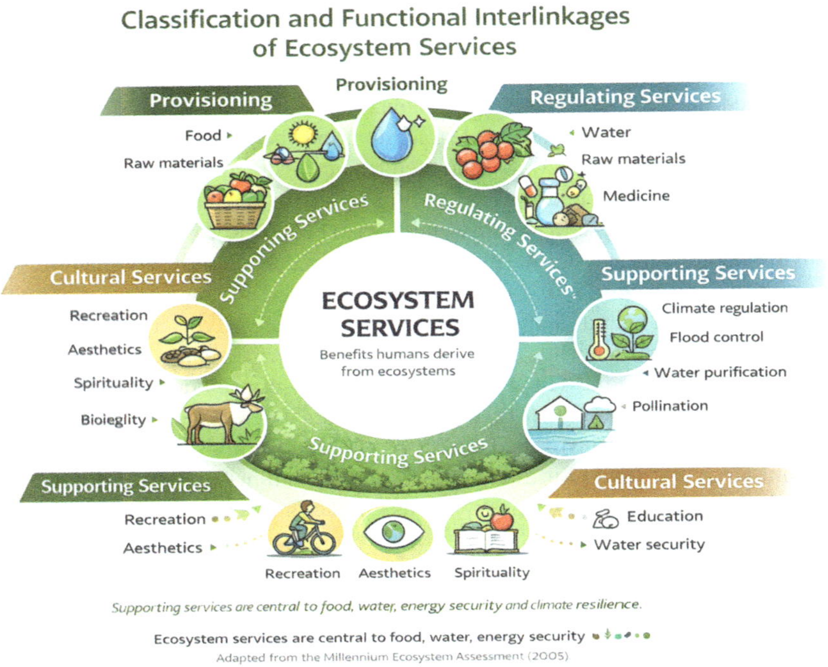

Fig. 4.2 Major categories of ecosystem services, including provisioning, regulating, supporting, and cultural services, and their key contributions to human well-being. Supporting services such as soil formation, biodiversity, and photosynthesis underpin all other ecosystem services (adapted from Millennium Ecosystem Assessment)

water shortages, and reduced biodiversity (Baral et al., 2016). There is a major distinction between ecosystem services and ecosystem processes. Photosynthesis, nitrogen fixation, nutrient cycling, soil formation and other ecosystem processes are naturally occurring processes that do not depend on human use. When these processes produce benefits that are of value to human beings, they give rise to ecosystem services. Most services do not stand out until they deteriorate or crumble, and their destruction can cause an abrupt and typically irreversible effect. Examples are fisheries collapses after long periods of stable harvests and land salinization due to changes in the hydrology cycle after the substitution of deep-rooted native vegetation with shallow-rooted crops (Danley & Widmark, 2016). Most ecosystem services have traditionally been invisible and so have not been effectively included in economic systems and often their degradation has been viewed as an externality. Under the circumstances where such externalities create great social and economic burdens like repairing rivers, climate adjustment or restoring biodiversity, the necessity to clearly identify and value ecosystem services becomes unavoidable (Owen, 2006).

Integrating ecosystem services in the decision-making process provides a well-organized system through which ecological processes, biodiversity, and human well-being can be connected. Systemically, ecosystem services hold the interface between social-ecological systems and connects human demands with nature's contributions across scales. Communication wise, the ecosystem services framework offers a user-friendly language used to communicate with the scientists, the policymakers and the stakeholders. Assessment-wise, the ecosystem services facilitate the determination of indicators and valuation methods that can be applied to incorporate ecological values in economic and policy models (Xu & Peng, 2022).

Different international projects have been launched that focus on ecosystem services as the key to sustainable development, biodiversity protection, and ecosystem stability. These includes Millennium Ecosystem Assessment (MA), Economics of Ecosystems and Biodiversity (TEEB), Common International Classification of Ecosystem Service (CICES), Convention on Biological Diversity (CBD) and the Intergovernmental Science-Policy Platform on Biodiversity and Ecosystem Services (IPBES. Payments for Ecosystem Services (PES) is another policy tool and it offers incentives to conservational and sustainable land management (Kleemann et al., 2022). Regardless of increased awareness, ecosystem services are still not well applied to policy and practice to achieve worldwide sustainability. The Ecosystem Services-based Decision-Making (ESDM) has still been confronted by the issues of uncertainty, scale mismatches, and lack of knowledge about tradeoffs among the services. An effective approach to biodiversity and ecosystem resilience and service flows into ESDM frameworks can facilitate sustainable land-use planning, forest and agricultural management (including Neglected and Underutilized Crops), and long-term human well-being (Xu & Peng, 2024).

4.3 Role of NUCs in Environmental Protection

The NUCs can contribute greatly to environmental protection because of their unique characteristics. Most NUCs are strong, have lower external needs, and are rich in nutrients hence can be used in sustainable and climate-adapted farming. Their presence enhances the health of the soil by increasing organic matter content, enhancing nutrient cycling and acting as contributors towards carbon sequestration thereby curbing greenhouse gas emissions. NUCs also promote the use of water and its conservation because lots of the species are suitable to low-rainfall or marginal soils, which decrease irrigation needs and enhance water retention in agroecosystems. Moreover, the addition of NUCs to agricultural areas increases biodiversity, which ensures the habitat of pollinators, soil life, and other wildlife, and ecological connectivity. NUCs are a multifunctional solution to agriculture enabling sustainable land management, ecosystem services, and long-term environmental sustainability through the combination of productivity and ecological benefits (Fig. 4.3) (Yesuph et al., 2025).

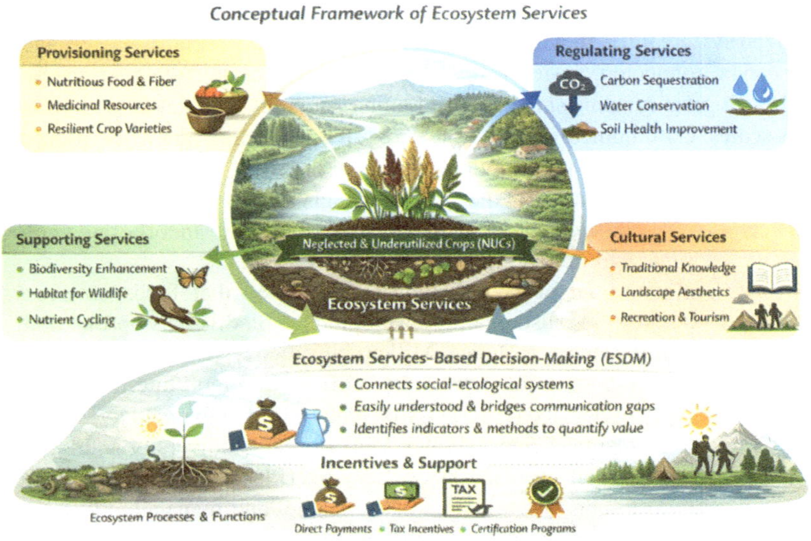

Fig. 4.3 Conceptual framework of ecosystem services and role of NUCs

4.4 Linking NUCs to Biodiversity, Ecosystem Resilience, and Ecosystem Services

NUCs occupies the strategic position at the intersection of biodiversity conservation, ecosystem resilience and ecosystem service provision. NUCs include a diverse variety of locally adapted crop species and traditional landraces which have developed in a wide variety of environmental conditions and cultural systems of management. These crops have significant part in agrobiodiversity, provision of genetic, species and functional diversity in farming landscapes. From a biodiversity perspective, NUCs enrich the species richness and genetic diversity of farming systems which would otherwise be dominated by a small range of crops grown around the world. This diversity supports the processes in the ecosystems by improving the ecological interactions like pollination, nutrient cycling, soil microbial activity and predator prey relationship. The addition of NUCs assists in ecosystem structure and not merely maximizing production, which is consistent with the fact that the provision of ecosystem services is closely tied to species composition and functional characteristics and not merely to species abundance (Shafiq et al., 2025).

NUCs are important in enhancing the resilience of the ecosystem. Most NUCs have useful characteristics including drought, heat and salinity resistance They have low nutrient demand, which allow them to reliably perform under climatic variability and environmental stress. The NUCs offer functional and response diversity in agroecosystems that can protect the food systems against yield failure, cushioning against external dependency, and serve as natural insurance against disturbances. It is specifically vital in altered ecosystems which are normally simpler in ecology and more prone to climatic stresses and system failure. Regarding ecosystem services, NUCs have a direct role in provisioning services through the supply of food, fibre, fodder, and medicinal products usually in the marginal environments where the major crops do not thrive well. They indirectly advocate the regulation and maintenance of services through strengthening of the soil structure, increasing of water infiltration, decreasing soil erosion, stabilizing nutrient cycling and promoting pollinators and useful soil organisms. Landscape-scale diversified cropping systems, which incorporate NUCs, have the potential to regulate hydrology and connect biodiversity and, therefore, maintain flows of ecosystem services beyond particular fields (Purba & Krishnaswamy, 2025).

Trade-offs between ecosystem services have also been solved by the fact that NUCs are integrated into agricultural systems. Overemphasis to maximize short-term provisioning services using a few crops may compromise supporting services like soil formation and nutrient cycling, which eventually compromises long-term productivity. Conversely, NUC-based diversification plans encourage multifunctional landscapes that can provide a wider range of the ecosystem services but still be ecologically sound. Also, NUCs facilitate social ecological resilience through the provision of local knowledge systems, cultural identity and diversification of livelihoods. Their development enhances adaptive capacity at the community level especially in areas that are susceptible to climate change, volatile market conditions and scarcity of resources. Connecting the conservation of biodiversity to food security and sustainable land management, NUCs ensure an effective avenue for ecosystem services–based decision-making (ESDM) (Power, 2010). Overall, the implementation of the NUCs into the framework of ecosystem services strengthens the core purpose of biodiversity in maintaining the normal operation of an ecosystem, resilience, and long-term human well-being. The ecosystem services delivered by NUCs are critical to sustainable agricultural practices, climate change adaptation mechanisms and conservation efforts on biodiversity. Figure 4.4 depicts the empowerment of agrobiodiversity and the ecosystem operation by the incorporation of NUCs in agricultural systems. NUCs, such as locally adapted crop species and older landraces, are beneficial in increasing genetic, species and functional diversity in agroecosystem (Amoak et al., 2025). These enriched biodiversity supports the most important ecosystem processes of nutrient cycling, soil formation, pollination and water regulation which together produce a diverse portfolio of ecosystem services. Such services are provisioning services (e.g., food, fodder and fiber), regulating services (e.g., climate control, carbon sequestration, erosion prevention and water purification), supporting services (e.g., soil fertility and habitat provision) and cultural services (e.g. traditional knowledge, cultural identity and landscape values). The continuity of the ecosystem services would lead to ecosystem and socio-ecological resilience in terms of improving adaptive capacity, stabilizing production in the face of climatic variability and livelihood security. Feedback loops indicate how resilient systems facilitates conservation, sustainable management practices and further use of NUCs, which strengthens biodiversity and service provision in the long-term. In general, the framework underlines that NUC based diversification is a systemic approach towards improvement of biodiversity, sustenance of ecosystem services, and development of resilient agricultural as well as social-ecological systems in the face of global environmental change (Mabhaudhi et al., 2022) (Fig. 4.4).

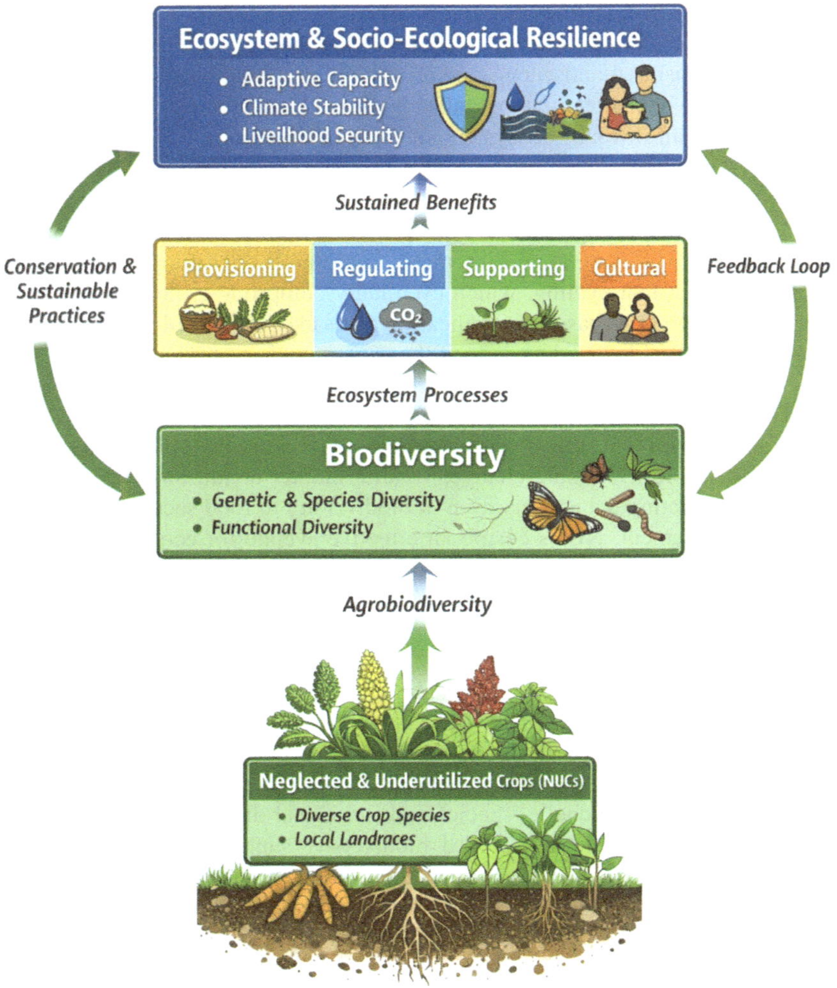

Fig. 4.4 Conceptual framework illustrating the pathway through which Neglected and Underutilized Crops (NUCs) enhance biodiversity, sustain ecosystem service flows, and build ecosystem and socio-ecological resilience through feedback mechanisms

4.5 NUCs and Resilience-The Key to Sustaining Ecosystem Services

Local species of crops and traditional landraces that add value to agrobiodiversity and functional diversity in agricultural landscapes comes under NUCs (Yesuph et al., 2025). The NUCs enrich the species and genetic diversity, therefore, enriching the ecosystems through nutrient cycling, pollination, pest management, and soil fertility. A vast array of ecosystem services (such as provisioning services (food, fodder, fibre), regulating services (climate moderation, erosion control, water purification), supporting services (soil formation, habitat provision), and cultural services (traditional knowledge, aesthetic and cultural values)) have been based on these processes. Incorporating NUCs into the agriculture system enhances diversification and counters agroecosystems to environmental pressures, climatic variability, and anthropogenic pressures. NUCs promote multifunctional and sustainable agricultural systems, stabilized production and the use of fewer external inputs, and the preservation of local biodiversity and the necessary ecological processes (Tamburini et al., 2020).

Resilience is the ability of a system to sustain structure, functions and feedbacks during disturbances both natural and caused by human beings. Biodiversity due to NUCs can absorb shock whilst maintaining vital functions. Recovery of ecosystems is determined by the overall health of the ecosystem and interaction among the species, nutrient dynamics and the abiotic factors. In the case of coral reefs, balanced ecological processes among fish populations, algae, and other organisms result in coral reefs overcoming the effects of bleaching incidences. The forces of overfishing, nutrient runoff, and climatic changes decrease resilience, and may cause degraded conditions, which are difficult or impossible to recover (Vasiliev, 2022).

Connectivity is key factor maintaining resilience and ecosystem. With the fragmentation of natural landscapes, the remnant habitats have been isolated, which decreases ecological and genetic flows. This diminishes the resilience at the landscape level. Conservation of remnant biodiversity, degraded habitat restoration, and connectivity are fundamental plans to guarantee long-term flow of ecosystem services and adaptive capacity of social ecological systems. The combination of NUCs and ecosystem resilience could be an important strategic planning of maintaining ecosystem services in the face of global environmental change. The conceptual framework of NUCs to improve biodiversity, ecosystem services and resilience have been shown in Fig. 4.5. Connectivity maintains ecological and genetic connections, which guarantee the long-term service and adaptive ability of social-ecological systems (Carof et al., 2022).

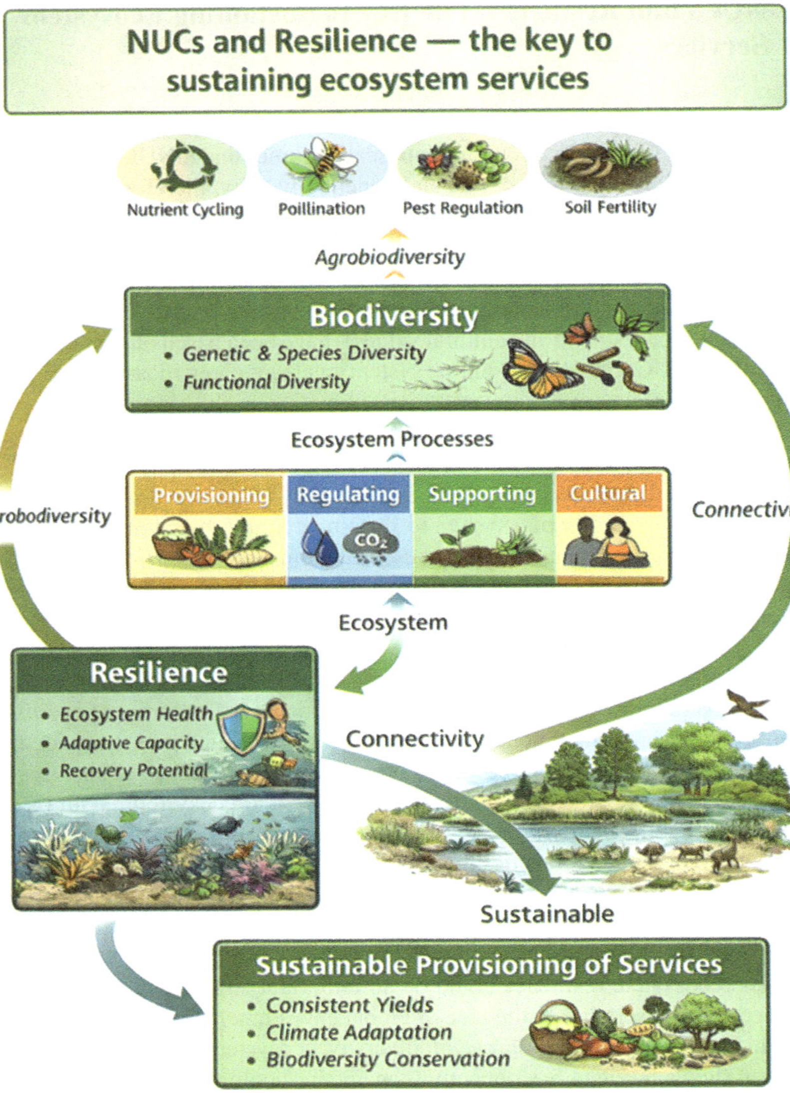

Fig. 4.5 Conceptual framework showing the role of NUCs in enhancing biodiversity, ecosystem services, and resilience for sustainable provisioning of services

4.6 Contribution of NUCs to Ecosystem Services

NUCs are important in maintaining and improving ecosystem services in both agricultural and natural landscapes. The crops do not only provide direct benefits to biodiversity but also enhance the ecological processes and offer both tangible and

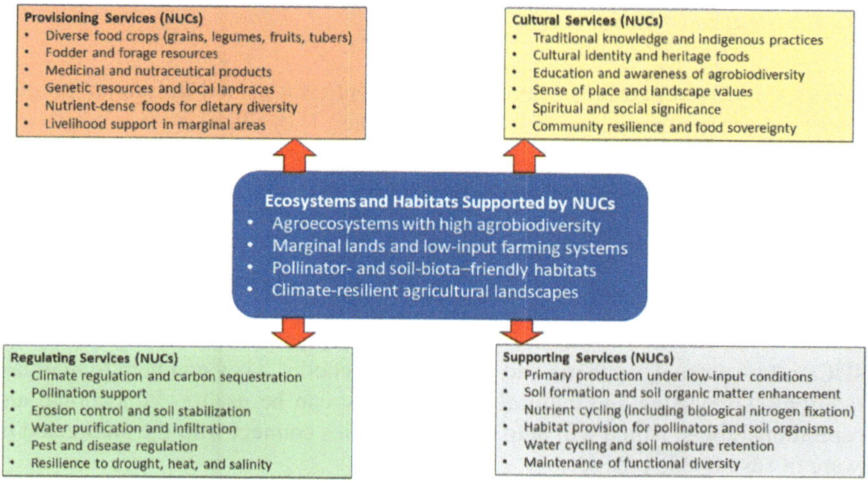

Fig. 4.6 NUCs contribution to ecosystem services

intangible benefits to the human communities. Millennium Ecosystem Assessment divides all the ecosystem services into four broad categories, which are provisioning, regulating, supporting, and cultural, and each of them can be directly traced back to the NUCs (Fig. 4.6) (Beillouin et al., 2021).

4.6.1 Provisioning Services

NUCs provide a wide range of sources of food, fiber, and pharmaceutical resources and genetic resources and vital nutrients. Their cultivation fortifies the local food security, diversification of their diet, and livelihoods in the marginal and low-input agricultural systems (Galluzzi et al., 2025).

4.6.2 Regulating Services

NCUs enhance soil fertility and water purification and erosion control by the providing vegetation cover and deep root systems. They also promote carbon sequestration, aid in climate control, aid in pollination, and natural pest and disease control thereby ensuring ecosystem stability and resilience (Kuschel et al., 2025; Nassary, 2025).

4.6.3 Supporting Services

NUCs are involved in primary production, nutrient cycling, soil formation, production of oxygen, water cycling and provide habitats. They allow the maintenance of the ecological networks and benefit other species by providing a functional diversity, which is useful in sustaining the ecological processes.

4.6.4 Cultural Services

NUCs offer spiritual, aesthetic and recreational benefits. They can help to maintain traditional knowledge and cultural identity. They can be used as educational and inspirational tools, which will help the communities connect with the nature and be aware of sustainable practices.

4.7 NUCs and Hidden Hunger

The role of NUCs in combating hidden hunger and enhancing nutrition security directly relies on the contribution of the NUCs to the ecosystem services. The condition of hidden hunger, which is the deficiency of essential micronutrients in cases where people are consuming enough calories, arises not only due to the monotony of the diet but also due to a deteriorated ecosystem that is unable to support the production of a variety of food that is rich in nutrients (Fig. 4.7) (Lowe, 2021; Weffort & Lamounier, 2024).

Green Revolution that started in the 1940s to the late 1960s was useful in increasing food production as a measure of checking hunger and food insecurity (Evenson & Gollin, 2003). It led to the production of high yielding varieties of staple crops such as wheat and rice where more irrigation, use of fertilizers as well as pesticides was applied (Fig. 4.8). These inventions increased crop production to a great extent. Wheat and rice production in the developing countries had increased more than twice. As a result, green revolution saved devastating famines, and also significantly reduced hunger in most parts of the world (Gollin et al., 2021). However, strong focus was only on the production without much consideration to the food quality and eco-friendly options (Pingali, 2012). High yield wheat and rice were found to contain less of the desirable micronutrients as compared to the traditional varieties (Fig. 4.9). This has led to loss of nutrients and has contributed to the problem of hidden hunger in which people are sufficiently fed on calories and may be lacking in the vitamins and minerals necessary to make them healthy. Green revolution led to extensive use of chemical fertilizers and pesticides as well as monoculture farming, which have brought about serious effects on the environment. Suggested practices during green revolution resulted to the increase in the crop yields but they also

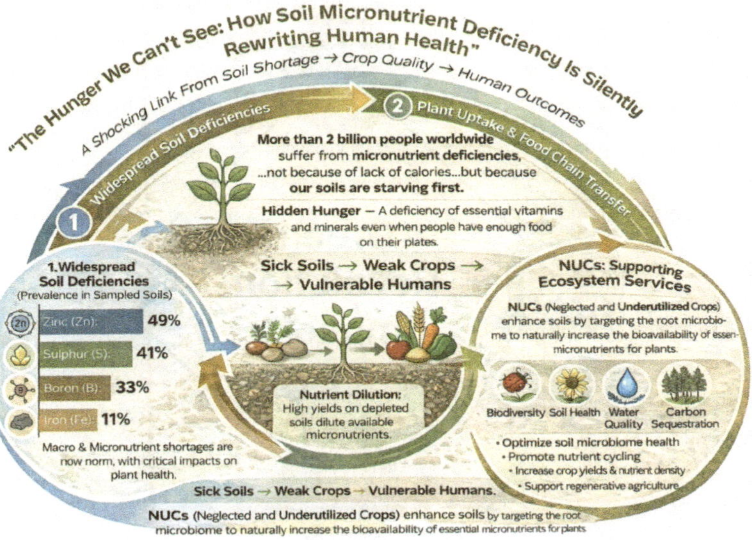

Fig. 4.7 Hidden hunger and role of NUCs

caused soil erosion, water pollution, groundwater contamination and harmful algal blooms that were hazardous to the ecosystems and human beings. The vast adoption of the monoculture suppressed the biodiversity in agriculture and exposed the crops to pests, diseases and climate stress. This resulted to more application of chemical inputs and loss of biodiversity (Fig. 4.10). At the same time, many crops e.g. NUC which were nutritionally balanced have been discarded. This led to the deterioration in the variety of crops and diets. Moreover, it has resulted in the decline of diversity that has weakened the stability of agricultural systems and contributed to the diets of few starchy staples that cause nutritionally-related health problems (Fig. 4.11) (Yilmaz & Yilmaz, 2025).

NUCs have the potential to provide the ecological and nutritional basis that is required to help fight micronutrient malnutrition in a sustainable way. The services given by NUCs to the ecosystem is central in minimizing hidden hunger by enhancing the availability of nutrient-rich foods. Many NUCs contain high amounts of essential minerals (iron, zinc, calcium), vitamins (vitamin A, folate, vitamin C), and high-quality proteins, dietary fiber, and bioactive compounds, unlike dominant staple cereals. The NUCs diversification of food systems increases the dietary diversity, enhances the nutrient adequacy, and decreases the reliance on the highly-energy-dense but nutrient-free staples. This especially applies in the marginal and low-income areas where food options are limited and this situation increases the micronutrient deficiencies in the vulnerable groups like children and women of reproductive age (Elegbeleye et al., 2025). Regulating ecosystem services also help to increase the

Fig. 4.8 Environmental impacts of the green revolution. *Source* Yilmaz and Yilmaz (2025)

Fig. 4.9 Low nutrient concentrations in a transformed agricultural landscape (HVYs: high-yielding varieties). *Source* Yilmaz and Yilmaz (2025)

nutritional value of NUCs by stabilizing food production under environmental pressure. Higher fertility of the soil, water management, natural pollination, and natural pest control assist in providing steady crop production, and food quality, even in climate variability. Climate-resistant NUCs can also be produced under drought-prone or degraded climatic conditions, thus making healthy foods available all year

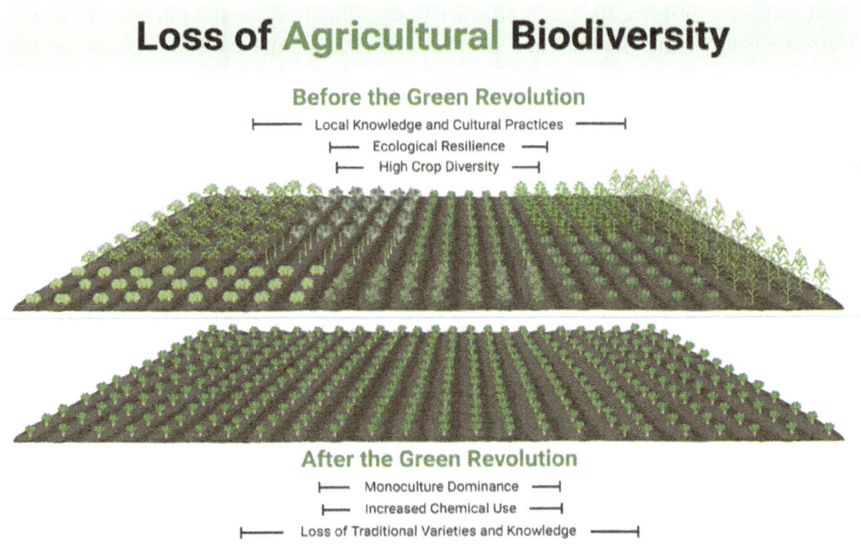

Fig. 4.10 Loss of agricultural biodiversity in the Green Revolution era. *Source* Yilmaz and Yilmaz (2025)

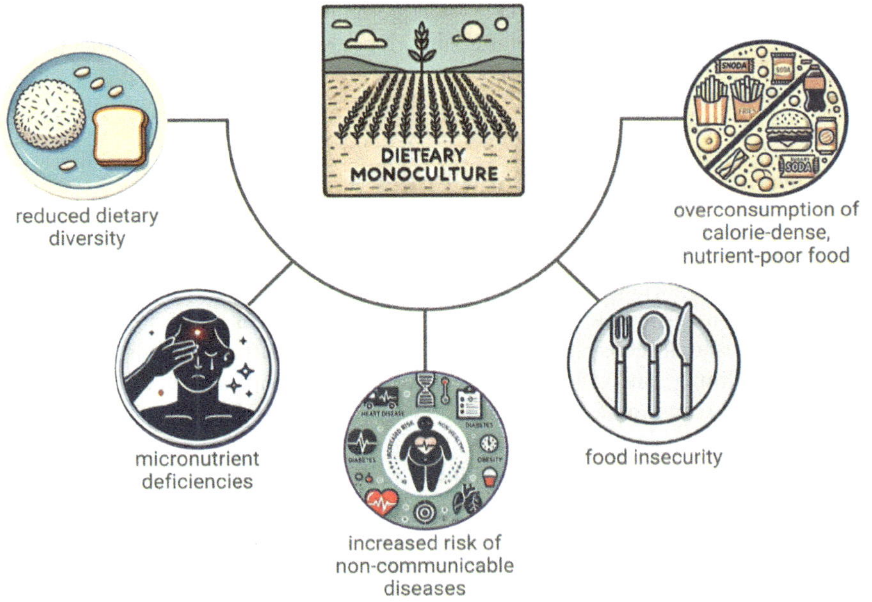

Fig. 4.11 Negative health and nutritional impacts of dietary monoculture. *Source* Yilmaz and Yilmaz (2025)

round and can minimize the susceptibility to seasonal hunger and nutrient deficiency. The reduction of the dependence on the chemical inputs by NUC-based systems also reduces the environmental degradation and, therefore, can undermine the long-term food and nutrition security. The ecological processes of producing nutrient rich crops are maintained by supporting the ecosystem services offered by NUCs like nutrient cycling, soil formation and habitat provision. Healthy soils and functional agroecosystems increase the micronutrient uptake and bioavailability in crops and ecosystem integrity which is directly related to the human nutrition outcomes. Maintaining biodiversity in these systems enhances ecosystem resilience, which would guarantee provision of nutrition-relevant ecosystem services with time. Ecosystem services related to culture are as crucial in the solution of hidden hunger. NUCs are entrenched in the traditional food systems, indigenous knowledge and local foods. The general acceptance and familiarity with a specific culture allow the frequent use of NUCs, as it allows people to better their nutrition without severe behavioral or financial barriers. The conservation of the traditional knowledge concerns cultivation, processing, and preparation of NUCs have been improved, increasing their effective use and facilitating nutrition-sensitive food systems (Purba & Krishnaswamy, 2025). In total NUCs role in fighting the hidden hunger is inherently connected with the role they can play in the ecosystem services. NUCs respond to ecological factors of food systems sustainability as well as nutritional determinants of human health through strengthening provisioning, regulating, supporting, and cultural services. Incorporation of NUCs in agricultural strategies that are built on ecosystem services hence is a holistic and resilient route of minimizing global hidden hunger and protecting the environmental health.

4.8 Economic Incentives and Policy Mechanisms

The extensive use of NUCs is usually held by economic and institutional constraints which restrict the incentives of farmers, market access, and investment opportunities rather than their agronomic or ecological capability. Whereas NUCs produce considerable services to the ecosystems like increasing soil fertility, protecting biodiversity, controlling climate, and efficiency in water use the benefits are seldom captured by market prices. Consequently, the farmers who practice NUCs are usually subjected to the opportunity costs of diversifying and not properly compensated in the extended environmental and social advantages they confer. Economic incentives and policy tools are thus necessary to internalize such positive externalities, harmonize the private decision-making with the goals of sustainability in the environment, and encourage the scaling of the NUC-based sustainable agricultural systems (El Bilali et al., 2024).

4.8.1 Payments for Ecosystem Services (PES) for Farmers Cultivating NUCs

PES Payments are one of the important policy tools which compensate farmers who proactively take care of their land in order to produce ecosystem services. PES is built upon the concept that those who benefit by the ecosystem services, i.e., governments, water authorities, carbon markets or the society in general need to pay land managers to maintain or preserve these services. PES may also give direct financial support to farmers implementing diversified cropping systems to enhance soil health, manage agrobiodiversity, trap carbon, control water movement, and attract pollinators and other beneficial organisms in the NUCs setting (Le et al., 2024).

PES mechanisms are especially compatible with NUCs as their ecological impacts are directly associated with quantifiable products, including higher soil organic carbon, decreased erosion, better soil water penetration, and better biodiversity on-farm. Indicatively, farmers growing drought resistant legumes, millets, or European landraces can be compensated in terms of nitrogen fixation, less fertilizer application, and climate resistance. PES schemes may be designed in the form of area-based payments (e.g. paid per hectare of land under NUC grow), performance-based payments based on ecosystem service indicators, or as a hybrid between practice-based and outcome-based requirements. Properly designed, PES may reduce the financial risk involved in crop diversification, promote long-term management of the natural resources, and enhance in the economic sustainability of NUC-based farm systems, especially in marginalized settings such as smallholders (Petropoulos et al., 2025).

4.8.2 Incentive-Based Programs Supporting NUC Adoption

In addition to PES, various policy tools that are based on incentives can be used to encourage the production and incorporation of NUCs into the mainstream food systems. The most utilized tools include direct payments and subsidies, which offer direct funding to the farmers in the process of switching their conventional monocropping system to a diversified system. Such payments may help to cover initial expenditure on seed purchase, transition cost of farm management or yield fluctuations in the short run which decrease barriers to adoption (Demirdogen et al., 2022).

Additionally, fiscal conditions can be used to encourage farmers and agribusinesses to invest in NUC production, processing and value addition through tax exemptions, lower taxes on land or input taxes. These incentives are especially applicable when NUCs have a role in studying the public good, specifically landscape restoration, reduction of greenhouse gases, or better watersheds (Li et al., 2024).

Market-based incentives include certification schemes and eco-labeling programs because they allow NUC producers to enter into premium markets that reward sustainability, conservation of biodiversity and nutritional quality. Organic production, climate-sensitive agriculture, biodiversity-friendly farming or traditional and indigenous foods can all be labeled, which will boost consumer awareness and lead to more purchasing, boosting farm income and strengthening environmental goals. The markets, such as carbon markets and cap-and-trade schemes, also give more prospects of monetizing the climate-controlling services of NUCs. Carpeting systems with greater sequestration of carbon in the soil or with lower carbon emissions due to lower fertilizer application and energy use can produce carbon credits, which can be traded in voluntary or regulatory markets. Incorporating NUCs into these systems can ensure the connection between local agriculture and the global climate mitigation strategy that would generate new income streams and encourage low-emission agricultural routes (Bhattacharyya et al., 2022; Harvey et al., 2014, 2025).

4.8.3 Role of Government, Private Sector, and Non-governmental Organizations

Economic incentives and policy mechanisms are only effective when these three actors - the governments, the private sector, and the non-governmental organizations (NGOs) act in coordination to promote NUCs. The governments have a key role in facilitating enabling policy environments by came up with PES schemes, establishing regulatory frameworks, offering public funding, and incorporating NUCs into national agricultural, climatic, and biodiversity plans. It is also the role of the public institutions to formulate monitoring and evaluation mechanisms, promote transparency and equity as well as make incentive programs to be aligned with the wider development agenda, such as food security, climate adaptation and livelihood in the rural areas.

The contribution made by the private sector is the integration NUCs in value chain, investing in processing and marketing infrastructure, and the certification and traceability systems. Agribusiness, retailers and financial institutions can help with accessing the markets, offer contract farming and offer innovative financing arrangements and lower the risks facing farmers adopting NUCs. Demand in sustainable and nutritious foods that are consumer driven also intensifies the involvement of the private-sector in NUC-based systems.

The space between NGOs and civil society organizations is also very crucial in terms of facilitating capacity building, farmer organization, and sharing knowledge. They tend to support participatory designs of PES, inclusive governance, and make smallholder farmers, women, and marginalized groups to enjoy the incentive programs. Another service rendered by the NGOs is the piloting of innovative incentive mechanisms, recording of the best practices, and coordination of local processes with the national and international policy processes.

Combined together, the economic incentives and policy mechanisms offer a significant framework of scaling the production of NUCs. These instruments can be used to help in aligning economic returns with environmental stewardship, develop resilient and diversified food systems, and achieve sustainable development and long-term human well-being by acknowledging and rewarding the economic services delivered by NUCs.

4.9 Barriers and Challenges

Although the environmental, nutritional, and resilience benefits of NUCs are well-documented. The extent to which they have been used is minimal as it is hindered by a variety of interrelated factors. These issues work at various levels between the farm households to markets and policy systems and tend to support each other. These limitations are critical to be overcome in order to maximize the potential of NUCs as a tool of sustainable agriculture, ecosystem service improvement, and climate-resilient food systems.

4.9.1 Financial, Technical, and Knowledge-Related Constraints

Limited financial capacity especially those farmers working in marginal environments are one of the greatest obstacles to the adoption of NUC. Cultivation of NUC is usually not subject to formal credit facilities as well as insurance programs and input subsidies, which are regularly available to major staple crops. Uncertain yields, shortage of quality seed and absence of income stability in the short-term during transition periods may increase the perceived risks to farmers. The costs of abandoning the dominant crops may be a deterrent to diversification without the financial cushions or stimulus devices.

Limitation of NUC integration is also caused by technical constraints. Numerous NUCs have been under minimal research and development, leading to a lack of agronomic practice characterisation, breeding development as well as insufficient extension services. The farmers have less access to better varieties, proper cultivation methods, techniques of handling the harvests, and processing technologies that are specific to NUCs. These technical loopholes may result in poor productivity, poor quality and increased post-harvest losses making NUC-based systems less economically attractive.

4.9.2 Market, Policy, and Land Tenure Limitations

Challenges that are associated with the markets constitute a significant bottleneck to the adoption of NUCs. NUCs value chains are usually underdeveloped, scattered or informal, and have minimal aggregation, processing, storage, and distribution infrastructure. Due to this, farmers are subjected to high transaction costs, fluctuating prices and few good markets. Lack of commercialization and scaling is further limited by the lack of standardized quality criteria and low market information systems.

These market constraints are often upheld by policy environments. Policies, subsidies and research investments in agriculture are usually biased towards only a small selection of staple crops and this has led to structural disincentives to diversification. NUCs usually do not have access to national seed systems, crop insurance, state procurement programs and food security plans. Besides this, uncertainty in regulations regarding novel foods, food standards and health claims may deter value addition and investment by the private sector towards NUC-based products.

There is also land tenure insecurity, which is especially an issue in areas where customary or informal land tenure is prevalent. Farmers who lack-title to their land could be unwilling to invest in diversified cropping system or long-term soil and biodiversity-enhancing practices related to NUCs. Credit and PES schemes may also be restricted by the insecure tenure, because in many cases, ownership of land or long-term right to use the land is a condition to become a participant of incentive-based programs.

4.9.3 Strategies to Overcome Barriers and Enable NUC Adoption

To tackle these barriers, there is need to have multi-level strategies that focus on financial, technical, institutional, and social constraints at the same time. By increasing the availability of specific financial products including PES, transition subsidies, and risk-sharing tools, and climate-related insurance, adoption risks can be mitigated. Furthermore, reward to the farmers due to the ecosystem services they have achieved by growing NUC could be good activity. Similarly, NUCs should be incorporated in the governmental investment and incentive programs.

It is also important to enhance research, extension and capacity-building activities. Participatory breeding, agronomic research, seed systems and post-harvest technologies can be invested to enhance productivity, quality and consistency of NUCs. The knowledge of science must be incorporated with traditional practices in the extension services so that the farmers are allowed to access context-specific and locally modified guidance. Knowledge among consumers can also be restored through education and awareness campaigns making them more accepting and demanding.

To facilitate commercialization and value addition, the strategies of market development are required. These are the establishment of inclusive value chains, farmer co-ops, public private alliances, certifications and market information systems. Stable demand of NUCs can be achieved by creating policies of public procurement of schools, hospitals, and social safety nets and by promoting nutrition and sustainability objectives.

These strategies are facilitated by policy and governance reforms. Long-term institutional support can be given through mainstreaming NUCs in the national policies of agriculture, nutrition, climate and biodiversity. Enhancing the rights to tenure land and the acknowledgment of customary land-use systems can also be made to promote investment in diversified and sustainable agricultural activities. Lastly participatory and inclusive governance practices involving farmers, women, indigenous, NGOs and the private sector is needed to make sure that NUC promotion leads to the achievement of equity, resilience and sustainable development.

Overall, the adoption of NUC has major challenges that limit its implementation, but they can be overcome. By applying both financial incentives, facilitating policies, market building, and systems thinking, NUCs will be able to shift their marginal crops to the central elements of resilient, ecosystem-based agricultural systems.

4.10 Case Studies and Success Stories

Evidence based on the various agroecological and socio-economic settings has shown that the NUCs cultivation when well supported by proper institutional and incentive frameworks can greatly increase the ecosystem services as well as reinforcing livelihoods and resilience of food systems. Case studies of developed and developing areas have shown how NUC-based diversification is relevant in protecting the environment, adapting to climate, and developing safely, and offer critical lessons on how to scale the diversification strategies.

4.10.1 Examples of NUC Cultivation Improving Ecosystem Services

The beneficial outcomes of NUC integration in agricultural landscapes are also detailed in many local and regional programs emphasizing the beneficial impacts on the environment. Millet, sorghum, pulse crops and local legumes have been found to enhance the soil structure, soil organic carbon and the recycling of nutrients through root systems that are deep rooted and the fixation of nitrogen in the soil by biological fixation. Through those improvements, there is a reduction in erosion, a rise in the ability of soils to withstand moisture, and an increase in the ability of soils to control and sustain ecosystem services (Al-Shammary et al., 2024; Zheng et al., 2023).

Traditional root and tubers, leafy crops, and minor cereals in diversified agricultural systems are essential in year-round ground cover. Land degradation has been reduced with increase in soil organisms and pollinators (Bhagat et al., 2024). Agro forestry systems that include under-exploited fruit trees, nuts and medicinal plants are additional effective in promoting the biodiversity through formation of multi-layered vegetation that supports wildlife, management of micro climates and stabilization of hydrology. All of this suggests that NUC cultivation can produce multifunctional landscapes that provide provisioning services as well as more extensive ecological benefits (Ndlovu et al., 2024).

4.10.2 Lessons from Payments for Ecosystem Services (PES) Programs in the USA and Other Regions

The NUCs Payments Ecosystem Services programs in the United States and other countries can offer valuable lessons on how economic incentives can be used to encourage environmentally friendly agricultural practices, including those that apply to NUCs. Conservation payment programs to cover crop, diversified rotation and habitat restoration are agri-environmental programmes that have been effective in the USA to improve the soil health, water quality and biodiversity without reducing farm incomes. Though not necessarily specifically targeting NUCs, these programmes demonstrate how the rewarding of ecosystem services has the ability to motivate farmers to shift to low-input, diversified systems that are closely related to the NUC-based approaches (Jack et al., 2008; Tyack et al., 2020; Wegner, 2016).

PES programs have proven effective in Latin America, Europe and some parts of Asia in terms of paying land managers to offer services like carbon sequestration, watershed management and biodiversity preservation. The major lessons based on these programs are that there should be proper definition of its services, proper monitoring and verification, long term funding mechanism and fair sharing of benefits. Notably, the PES schemes where local knowledge can be combined with flexible management solutions along with robust institutional underpinning are more likely to produce more sustainable environmental and social results. The lessons can be directly used to develop PES mechanisms that will encourage NUC cultivation as a way of improving ecosystem services (Berte et al., 2022; Hejnowicz et al., 2014).

4.10.3 Application in South Asia and Sub-Saharan Africa

Smallholder farming is prevalent in South Asia and Sub-Saharan Africa, and exposure to climatic events is a factor that has made NUC-based systems promising in these areas. Traditional crops, including millets, pulses, buckwheat, and native oilseeds, are being reintroduced in rainfed areas and marginal lands in South Asia to increase the fertility of the soil, lowering the water needs and diversifying the diet. Social-led programs that push these crops have enhanced the agrobiodiversity, revitalized the indigenous knowledge, and heightened the drought and extreme temperatures resistance as well as bringing about food and nutrition security (Ansari et al., 2025; Bandyopadhyay & Patnaik, 2025). Equally, native cereals, legumes and leafy vegetables are also essential in supporting livelihoods in Sub-Saharan Africa due to the fluctuating climatic conditions. NUCs accustomed to the presence of poor soils and low rainfalls have enhanced land productivity without greatly depending on external inputs hence, environmental degradation and cost of production is minimized. Integrated landscape strategies comprising NUC cultivation with soil and water management measures have complemented the ecosystem services like erosion control, water management and habitat services. In a number of instances, NUC promotions have been achieved through donor-funded and NGO-led programs as it has been able to connect NUC promotion to livelihood diversification, the empowerment of women, and the development of local markets (Olarewaju et al., 2025). The FAO has declared the 2026 as the International Year of the Woman Farmer that gives emphasis on the role of women in agri-food systems. Women is also a major participant in cultivation, conservation and value addition of NUCs which are essential in agrobiodiversity, climate resilience and nutrition security. On a global scale, women have significant proportion of the agricultural labour force, but their labour is largely informal, poorly remunerated, labour-intensive and is highly vulnerable with systemic barriers such as poor access to land, finance, technologies, extension services, education and decision making. In Pakistan, 74% of labor force in agriculture are women and girls and they contribute approximately 40 percent of household income. However, their role, protection needs, and contributions, even in the NUC-based systems, have not documented well and addressed sufficiently. According to a report issued by FAO, the federal and provincial policy changes must be made to officially acknowledge all women who are involved in agricultural and post-harvest operations as farmers. Women should be ensured with minimum wages and equal pay. Idle public land, water access and input subsidies should be given to rural women who are landless. Current FAO evaluations also indicated that 24% productivity gap have been faced by women farmers that own small land holdings. They are also having loss of 3% per day due to climate change (e.g. extreme heat) compared to men that limits the livelihood of women as well as the transformative potential of NUCs to address hidden hunger and develop resilient agri-food systems (Fig. 4.12).

The effective case studies have highlighted the need to have supportive policies, participatory government and balance between ecological objectives and livelihoods. They also point out that NUCs can only be effective when they are incorporated into larger ecosystem services and resilience frameworks as opposed to being viewed as single crops. Altogether, these case studies and success stories indicate that NUCs may be used in a transformative manner when improving ecosystem services and creating resilient food systems. The experiences of the PES-program and regional practice highlight the fact that integrating ecological understanding, economic motivation, and integrative institutions is required to replicate NUC-based solutions to a variety of socio-ecological settings (Gonza, 2025; Olawuyi & Fakhri, 2025).

Fig. 4.12 Empowering women to boost NUCs and food security

4.11 Conclusions and Recommendations

NUCs are an important but underutilized part of agricultural systems of sustainability, which can provide an unprecedented set of environmental, nutritional, and socio-economic value. In this chapter, it has been shown that NUCs increase agro-biodiversity, improve ecosystem functions, and play a significant role in providing, regulating, supporting, and cultural ecosystem services. Hence it can provide the foundation of resilient food systems. They can prosper in low-input settings and with climatic stress, which makes them especially useful under climate change, land degradation and loss of biodiversity. The use of NUCs in the mainstream agricultural systems is thus not only a diversification possibility, but a strategic requirement to protect the health of the ecosystem and the security of food in the long-term.

Policymaking has significant implications of NUCs integration regarding environmental protection and transformation of the food system. The existing policies and market organization in agriculture is mostly biased towards a short list of commodity crops at the expense of environmental implications and the stability of the ecosystem. Alternative incentives can shift towards sustainable land management by recognizing and rewarding the ecosystem services produced by the NUCs through mechanisms like Payments of Ecosystem Services, subsidies of a specific type or certification schemes and climate finance schemes. Introduction of NUCs into national agriculture, climate, biodiversity, and nutrition can be considered to achieve simultaneously the goals of environmental conservation, climate adaptation and food and nutrition security, especially in vulnerable regions. Smallholders, women, indigenous people, the private sector, and the civil society should be involved in inclusive governance structures that would promote equal distribution of benefits and implementation.

In the future, the research and development of new ideas and products should focus on bridging knowledge and innovation gaps, which limit the extrapolation of NUC-based systems. Other areas of research encompass better definition of NUCs genetic resources, participatory breeding to offer resistance to climatic changes and nutritional value, and development of context adhering agronomic and post-harvest technologies. There is also the need to put more emphasis on the quantification and valuation of the ecosystem services that NUCs offer such as their roles in carbon sequestration, ensuring soil health, preserving biodiversity and the results of nutrition. Enhancing the interdisciplinary research that incorporates ecology, agronomy, economics, and social sciences will be essential in informing the use of ecosystem services in decision-making and formulation of policies.

Finally, NUCs inclusion into sustainable farming systems has provided a potent channel on how to deal with interrelated issues of environmental degradation, climatic changes, and food insecurity. With proper coordination of ecology with economic incentives and policy alliances, NUCs can be transformed into core pillars of resilient, inclusive, and climate-smart food systems as opposed to being peripheral elements of food systems. They will need to be fully utilized through continued investment in research and policy innovation and institutional partnership to facilitate ecosystem health and human well-being.

References

Al-Shammary, A. A. G., Al-Shihmani, L. S. S., Fernández-Gálvez, J., & Caballero-Calvo, A. (2024). Optimizing sustainable agriculture: A comprehensive review of agronomic practices and their impacts on soil attributes. *Journal of Environmental Management, 364*, 121487. https://doi.org/10.1016/j.jenvman.2024.121487

Amicarelli, V., Bux, C., Del Chicca, A., Giungato, P., Tarabella, A., & Tricase, C. (2025). Toward a business-oriented concept of natural capital and its measurement in organizations. *Business Strategy and the Environment*. https://doi.org/10.1002/bse.70310

Amoak, D., Najjar, D., Nyantakyi-Frimpong, H., Amil, R. E., Maalouf, F., & Abderrazek, J. (2025). Building climate resilience in the MENA region: A feminist political ecology of agroecology and neglected and underutilized crop species. *Agriculture and Human Values, 43*(1), 4. https://doi.org/10.1007/s10460-025-10826-w

Anikwe, M. A. N., & Ife, K. (2023). The role of soil ecosystem services in the circular bioeconomy [Review]. *Frontiers in Soil Science, 3*, 2023. https://doi.org/10.3389/fsoil.2023.1209100

Ansari, M. A., Joshi, H., Mehta, P., Deo, A., Bora, K., Rao, C. A. R., et al. (2025). Climate-smart and risk-resilient adaptation strategies for potato production: a meta-analysis for South Asia. *Environmental Challenges, 20*, 101183. https://doi.org/10.1016/j.envc.2025.101183

Bandyopadhyay, U., & Patnaik, A. (2025). From green revolution to millet revolution: understanding India's transition through agri-food policies. *Food, Culture & Society*, 1–21. https://doi.org/10.1080/15528014.2025.2485505

Baral, H., Guariguata, M. R., & Keenan, R. J. (2016). A proposed framework for assessing ecosystem goods and services from planted forests. *Ecosystem Services, 22*, 260–268. https://doi.org/10.1016/j.ecoser.2016.10.002

Behera, D., Fathima, J., Saady, N. M. C., Zendehboudi, S., Albayati, T. M., Al-nayili, A., Espinoza, J. E. R. (2026). Sustainable agriculture through environmental adaptation engineering for waste management. *Green Technologies and Sustainability, 4*(1), 100242. https://doi.org/10.1016/j.grets.2025.100242

Beillouin, D., Ben-Ari, T., Malézieux, E., Seufert, V., & Makowski, D. (2021). Positive but variable effects of crop diversification on biodiversity and ecosystem services. *Global Change Biology, 27*(19), 4697–4710. https://doi.org/10.1111/gcb.15747

Berte, M. L., Mendonça, G., & Carvalho, R. (2022). Payments for environmental services integrated with the protection of biodiversity, water production and agroecology in the Paraíba Valley, in the state of São Paulo, Brazil.

Bhagat, R., Walia, S. S., Sharma, K., Singh, R., Singh, G., & Hossain, A. (2024). The integrated farming system is an environmentally friendly and cost-effective approach to the sustainability of agri-food systems in the modern era of the changing climate: A comprehensive review. *Food and Energy Security, 13*(1), Article e534.

Bhattacharyya, S. S., Ros, G. H., Furtak, K., Iqbal, H. M. N., & Parra-Saldívar, R. (2022). Soil carbon sequestration—An interplay between soil microbial community and soil organic matter dynamics. *Science of The Total Environment, 815*, 152928. https://doi.org/10.1016/j.scitotenv.2022.152928

Carof, M., Godinot, O., & Le Cadre, E. (2022). Biodiversity-based cropping systems: A long-term perspective is necessary. *Science of The Total Environment, 838*, 156022. https://doi.org/10.1016/j.scitotenv.2022.156022

CBD. (1992). *Convention on biological diversity*. http://www.cbd.int/doc/legal/cbd-en.pdf

Danley, B., & Widmark, C. (2016). Evaluating conceptual definitions of ecosystem services and their implications. *Ecological Economics, 126*, 132–138. https://doi.org/10.1016/j.ecolecon.2016.04.003

Demirdogen, A., Olhan, E., & Hasdemir, M. (2022). Heterogeneous impact of agricultural support policies: Evidence from Turkey. *Environment, Development and Sustainability, 24*(10), 12203–12225. https://doi.org/10.1007/s10668-021-01941-9

El Bilali, H., Rokka, S., Calabrese, G., Borelli, T., Grazioli, F., Tietiambou, S. R. F., et al. (2024). Conservation and promotion of neglected and underutilized crop species in West Africa: Policy and governance. *Sustainability, 16*(14), 6194. https://www.mdpi.com/2071-1050/16/14/6194

Elegbeleye, J. A., Fayemi, O. E., Agbemavor, W. S. K., Krishnamoorthy, S., Adebowale, O. J., Adeyanju, A. A., et al. (2025). Beyond calories: Addressing micronutrient deficiencies in the world's most vulnerable communities—A review. *Nutrients, 17*(24), 3960. https://www.mdpi.com/2072-6643/17/24/3960

Evenson, R. E., & Gollin, D. J. s. (2003). Assessing the impact of the Green Revolution, 1960 to 2000. *300*(5620), 758–762.

Galluzzi, G., Cruz-Garcia, G. S., Togka, K. M., Visser, B., & Mbozi, H. (2025). The role that local food plants can play in improving nutrition security and reducing seasonal scarcity in rural communities: A multi-country study. *Sustainability, 17*(10), 4683. https://www.mdpi.com/2071-1050/17/10/4683

Gollin, D., Hansen, C. W., & Wingender, A. M. J. J. o. P. E. (2021). Two blades of grass: The impact of the green revolution. *129*(8), 2344–2384.

Gonza, J. R. (2025). The role of African regional courts in shaping sustainable development: Balancing competing pillars and contributing to global jurisprudence. *Review of European, Comparative & International Environmental Law, 34*(2), 536–549. https://doi.org/10.1111/reel.70000

Harvey, C. A., Rakotobe, Z. L., Rao, N. S., Dave, R., Razafimahatratra, H., Rabarijohn, R. H., et al. (2014). Extreme vulnerability of smallholder farmers to agricultural risks and climate change in Madagascar. *Philosophical Transactions of the Royal Society B: Biological Sciences, 369*(1639), 20130089. https://doi.org/10.1098/rstb.2013.0089

Harvey, W. J., Black, N., Essaouabi, S., Petrokofsky, L., Rangan, V., Bird, M. S., et al. (2025). The impact of sustainability certification schemes and labels on greenhouse gas emissions: A systematic evidence map. *17*(2), 792. https://www.mdpi.com/2071-1050/17/2/792

Hejnowicz, A. P., Raffaelli, D. G., Rudd, M. A., & White, P. C. L. (2014). Evaluating the outcomes of payments for ecosystem services programmes using a capital asset framework. *Ecosystem Services, 9*, 83–97. https://doi.org/10.1016/j.ecoser.2014.05.001

Hori, J., & Makino, M. (2018). The structure of human well-being related to ecosystem services in coastal areas: A comparison among the six North Pacific countries. *Marine Policy, 95*, 221–226. https://doi.org/10.1016/j.marpol.2018.02.023

IPCC. (2023). Summary for Policymakers. In: *Climate Change 2023: Synthesis Report. A Report of the Intergovernmental Panel on Climate Change. Contribution of Working Groups I, II and III to the Sixth Assessment Report of the Intergovernmental Panel on Climate Change* [Core Writing Team, H. Lee and J. Romero (eds.)].

Jack, B. K., Kousky, C., & Sims, K. R. E. (2008). Designing payments for ecosystem services: Lessons from previous experience with incentive-based mechanisms. *Proceedings of the National Academy of Sciences of the United States of America, 105*(28), 9465–9470. https://doi.org/10.1073/pnas.0705503104

Kaur, S., Bedi, M., Singh, S., Kour, N., Bhatti, S. S., Bhatia, A., et al. (2024). Chapter Eight—Monoculture of crops: A challenge in attaining food security. In A. Sharma, M. Kumar & P. Sharma (Eds.), *Advances in Food Security and Sustainability* (Vol. 9, pp. 197–213). Elsevier. https://doi.org/10.1016/bs.af2s.2024.07.008

Kim, H., Lazurko, A., Linney, G., Maskell, L., Díaz-General, E., Březovská, R. J. et al. (2024). Understanding the role of biodiversity in the climate, food, water, energy, transport and health nexus in Europe. *Science of The Total Environment, 925*, 171692. https://doi.org/10.1016/j.scitotenv.2024.171692

Kleemann, J., Koo, H., Hensen, I., Mendieta-Leiva, G., Kahnt, B., Kurze, C., et al. (2022). Priorities of action and research for the protection of biodiversity and ecosystem services in continental Ecuador. *Biological Conservation, 265*, 109404. https://doi.org/10.1016/j.biocon.2021.109404

Kuschel, E., Obriejetan, M., Kuzmanić, T., Mikoš, M., Seifert, L., Conevski, S., et al. (2025). A systematic framework for assessing the temporally variable protective capacity of nature-based solutions against natural hazards. *Infrastructures, 10*(12), 318. https://www.mdpi.com/2412-3811/10/12/318

Le, T.-A. T., Vodden, K., Wu, J., Bullock, R., & Sabau, G. (2024). Payments for ecosystem services programs: A global review of contributions towards sustainability. *Heliyon, 10*(1), e22361. https://doi.org/10.1016/j.heliyon.2023.e22361

Li, T., Yue, X.-G., Qin, M., & Norena-Chavez, D. (2024). Towards Paris Climate Agreement goals: The essential role of green finance and green technology. *Energy Economics, 129*, 107273. https://doi.org/10.1016/j.eneco.2023.107273

Lowe, N. M. (2021). The global challenge of hidden hunger: Perspectives from the field. *The Proceedings of the Nutrition Society, 80*(3), 283–289. https://doi.org/10.1017/s0029665121000902

Mabhaudhi, T., Hlahla, S., Chimonyo, V. G. P., Henriksson, R., Chibarabada, T. P., Murugani, V. G., et al. (2022). Diversity and diversification: Ecosystem services derived from underutilized crops and their co-benefits for sustainable agricultural landscapes and resilient food systems in Africa [Systematic Review]. *Frontiers in Agronomy, 4*, 2022. https://doi.org/10.3389/fagro.2022.859223

Mace, G. M., Norris, K., & Fitter, A. H. (2012). Biodiversity and ecosystem services: a multilayered relationship. *Trends in Ecology & Evolution, 27*(1), 19–26. https://doi.org/10.1016/j.tree.2011.08.006

Marappan, K., Sadasivam, S., Natarajan, N., Arumugam, V. A., Lakshmaiah, K., Thangaraj, M., et al. (2025). Underutilized fruit crops as a sustainable approach to enhancing nutritional security and promoting economic growth [Review]. *Frontiers in Sustainable Food Systems, 9*, 2025. https://doi.org/10.3389/fsufs.2025.1618112

Mehran, M., Ikram, M., Ghoneim, A. M., Ghafar, S., Ashraf, M., Rehman, H. u., et al. (2025). Biodiversity Within Agroecosystems and Its Benefits to Soil. In M. Shaaban (Ed.), *Soils and Sustainable Agriculture: Interplay of Soil, Plant, Water and Environmental Systems for Sustainable Agriculture* (pp. 233–266). Springer Nature Switzerland. https://doi.org/10.1007/978-3-031-91114-9_10

Nassary, E. K. (2025). Exploring the role of nature-based solutions and emerging technologies in advancing circular and sustainable agriculture: An opinionated review for environmental resilience. *Cleaner and Circular Bioeconomy, 10*, 100142. https://doi.org/10.1016/j.clcb.2025.100142

Ndlovu, M., Scheelbeek, P., Ngidi, M., & Mabhaudhi, T. (2024). Underutilized crops for diverse, resilient and healthy agri-food systems: a systematic review of sub-Saharan Africa [Systematic Review]. *Frontiers in Sustainable Food Systems, 8, 2024*. https://doi.org/10.3389/fsufs.2024.1498402

Olarewaju, O. O., Fawole, O. A., Baiyegunhi, L. J. S., & Mabhaudhi, T. (2025). Integrating sustainable agricultural practices to enhance climate resilience and food security in Sub-Saharan Africa: A multidisciplinary perspective. *Sustainability, 17*(14), 6259. https://www.mdpi.com/2071-1050/17/14/6259

Olawuyi, D. S., & Fakhri, R. (2025). Case studies. In D. S. Olawuyi & R. Fakhri (Eds.), *Biodiversity and Nature Conservation Law and Policy in the Middle East and North Africa Region* (pp. 87–294). Cambridge University Press. https://doi.org/10.1017/9781009519663.007

Owen, A. D. (2006). Renewable energy: Externality costs as market barriers. *Energy Policy, 34*(5), 632–642. https://doi.org/10.1016/j.enpol.2005.11.017

Petropoulos, T., Benos, L., Busato, P., Kyriakarakos, G., Kateris, D., Aidonis, D., & Bochtis, D. (2025). Soil organic carbon assessment for carbon farming: A review. *Agriculture 15*(5), 567. https://www.mdpi.com/2077-0472/15/5/567

Pingali, P. L. J. P. o. t. n. a. o. s. (2012). Green revolution: impacts, limits, and the path ahead. *109*(31), 12302–12308.

Power, A. G. (2010). Ecosystem services and agriculture: Tradeoffs and synergies. *Philosophical Transactions of the Royal Society b: Biological Sciences, 365*(1554), 2959–2971. https://doi.org/10.1098/rstb.2010.0143

Purba, N. H., & Krishnaswamy, K. (2025). Exploring the potentials of neglected underutilized crops (NUCs): An integrative review for developing a sustainable food system model. *NPJ Science of Food, 9*(1), 199. https://doi.org/10.1038/s41538-025-00554-0

Rac, I., Erjavec, K., & Erjavec, E. (2024). Agriculture and environment: Friends or foes? Conceptualising agri-environmental discourses under the European Union's Common Agricultural Policy. *Agriculture and Human Values, 41*(1), 147–166. https://doi.org/10.1007/s10460-023-10474-y

Raza, M. Y., Ahmad, M., Qadir, G., Ikram, S., Ahmad, S., & Malik, A. (2025). Regenerative agriculture. In M. Ahmed (Ed.), *Climate Resilient and Sustainable Agriculture: Volume 1: Adaptation and Mitigation Strategies* (pp. 391–405). Springer Nature Switzerland. https://doi.org/10.1007/978-3-032-00190-0_15

Reader, M. O., Eppinga, M. B., de Boer, H. J., Damm, A., Petchey, O. L., & Santos, M. J. (2022). The relationship between ecosystem services and human modification displays decoupling across global delta systems. *Communications Earth & Environment, 3*(1), 102. https://doi.org/10.1038/s43247-022-00431-8

Shafiq, S., Zia Ul Haq, M., Raza Naqvi, S. A., Sarfaraz, W., Ali, H., Islam, M. M., et al. (2025). Integrating neglected and underutilized crops (NUCs) in South Asian cropping systems and diets: Challenges and prospects. *Regional Sustainability, 6*(4), 100242. https://doi.org/10.1016/j.regsus.2025.100242

Snapp, S. (2017a). Chapter 2-Agroecology: Principles and practice. In S. Snapp & B. Pound (Eds.), *Agricultural Systems (Second Edition)* (pp. 33–72). Academic Press. https://doi.org/10.1016/B978-0-12-802070-8.00002-5

Snapp, S. (2017b). Chapter 5-Designing for the long-term: Sustainable Agriculture. In S. Snapp & B. Pound (Eds.), *Agricultural Systems (Second Edition)* (pp. 123–167). Academic Press. https://doi.org/10.1016/B978-0-12-802070-8.00005-0

Snapp, S., & Pound, B. (2017). Chapter 4-Farming systems for sustainable intensification. In S. Snapp & B. Pound (Eds.), *Agricultural Systems (Second Edition)* (pp. 93–122). Academic Press. https://doi.org/10.1016/B978-0-12-802070-8.00004-9

Tallis, H., Guerry, A., & Daily, G. C. (2013). Ecosystem services. In S. A. Levin (Ed.), *Encyclopedia of Biodiversity (Second Edition)* (pp. 96–104). Academic Press. https://doi.org/10.1016/B978-0-12-384719-5.00284-7

Tamburini, G., Bommarco, R., Wanger, T. C., Kremen, C., van der Heijden, M. G. A., Liebman, M., & Hallin, S. (2020). Agricultural diversification promotes multiple ecosystem services without compromising yield. *6*(45), eaba1715. https://doi.org/10.1126/sciadv.aba1715

Tyack, N., Dempewolf, H., & Khoury, C. K. (2020). The potential of payment for ecosystem services for crop wild relative conservation. *Plants, 9*(10), 1305. https://www.mdpi.com/2223-7747/9/10/1305

Vasiliev, D. (2022). The role of biodiversity in ecosystem resilience. *IOP Conference Series: Earth and Environmental Science, 1072*(1), Article 012012. https://doi.org/10.1088/1755-1315/1072/1/012012

Weffort, V. R. S., & Lamounier, J. A. (2024). Hidden hunger—A narrative review. *Jornal de Pediatria, 100*, S10-S17. https://doi.org/10.1016/j.jped.2023.08.009

Wegner, G. I. (2016). Payments for ecosystem services (PES): A flexible, participatory, and integrated approach for improved conservation and equity outcomes. *Environment, Development and Sustainability, 18*(3), 617–644. https://doi.org/10.1007/s10668-015-9673-7

Xu, Z., & Peng, J. (2022). Ecosystem services-based decision-making: A bridge from science to practice. *Environmental Science & Policy, 135*, 6–15. https://doi.org/10.1016/j.envsci.2022.04.010

Xu, Z., & Peng, J. (2024). Recognizing ecosystem service's contribution to SDGs: Ecological foundation of sustainable development. *Geography and Sustainability, 5*(4), 511–525. https://doi.org/10.1016/j.geosus.2024.05.001

Yesuph, D. S., Dalka, D. D., & Tadesse Baye, M. (2025). Underutilized and neglected crop species and their role in enhancing household food security amid climate change, Wolaita Zone, Ethiopia. *Heliyon, 11*(3), e42345. https://doi.org/10.1016/j.heliyon.2025.e42345

Yilmaz, H., & Yilmaz, A. (2025). Hidden hunger in the age of abundance: The nutritional pitfalls of modern staple crops. *Food Science & Nutrition, 13*(2), Article e4610. https://doi.org/10.1002/fsn3.4610

Zheng, F., Liu, X., Ding, W., Song, X., Li, S., & Wu, X. (2023). Positive effects of crop rotation on soil aggregation and associated organic carbon are mainly controlled by climate and initial soil carbon content: A meta-analysis. *Agriculture, Ecosystems & Environment, 355*, 108600. https://doi.org/10.1016/j.agee.2023.108600

Part II
Cereal and Pseudocereal Crops: Climate Adaptation and Resource Nexus

Chapter 5
Quinoa (*Chenopodium Quinoa*): Climate Resilience, Nutritional Benefits and Water-Energy Efficiency

Umar Farooq, Ahmed M. S. Kheir, Amjad Malik, and Mukhtar Ahmed

Abstract Quinoa (*Chenopodium quinoa* Willd.) has been recognized as climate-resistant and nutritionally better neglected and underused crop that can survive under climate change, land degradation and water scarcity. This chapter brings recent scientific findings that suggest that quinoa has a role to play in diversifying agricultural systems, innovating climate-resilient agriculture, and enhancing nutritional security, especially in marginal and resource-limiting contexts. A multidimensional strategy was applied by synthesizing peer-reviewed studies, FAOSTAT data, crop modeling studies, genomic and physiological studies, and empirical case studies of Andes, South Asia, and Mediterranean-type areas. The findings indicate that quinoa has a high level of genetic diversity and adaptation behavior and is resistant to drought, salty environments, heat, and frost, and can yield between 0.5 and more than 5.0 t ha^{-1} under good conditions and remain stable during moderate abiotic conditions. Experimental observations in non-Andean countries, such as Pakistan, have shown that it is able to yield 1.2–2.4 t ha^{-1}. Crop growth models have shown that it responds to 40–50% evapotranspiration cuts with little yield loss, that's why it is superior to other crops in terms of water-use efficiency. Quinoa is nutritionally a high-quality protein source with a lysine protein, essential amino acids, minerals, vitamins, and bioactive compounds. It is naturally gluten-free, which makes it a valuable food crop in the fight against protein-energy malnutrition and micronutrient deficiencies. Regardless of these benefits, the adoption rates are limited by a lack of awareness among farmers, poor seed systems, bitter saponin concentration, poor quality value-chain development and poor policy support. The chapter

U. Farooq (✉) · M. Ahmed
Department of Agronomy, Pir Mehr Ali Shah Arid Agriculture University, Rawalpindi, Pakistan
e-mail: umerfaroooq42@gmail.com

A. M. S. Kheir
Julius Kühn Institute (JKI)—Federal Research Centre for Cultivated Plants, Institute for Strategies and Technology Assessment, Kleinmachnow, Germany

Soils, Water and Environment Research Institute, Agricultural Research Center, Giza, Egypt

A. Malik
Kenanow Bachelor of Education Program Faculty of Education, University College of the North (UCN), Thompson, MB, Canada

M. Ahmed (ed.), *Climate Change and Neglected Underutilized Crops*,
https://doi.org/10.1007/978-3-032-21005-0_5

recommends integrated solutions through low-saponin and stress-tolerant varieties breeding, extension and seed delivery system fortification, investment in processing and market infrastructures and national policy entrenchment of quinoa on agricultural, nutrition and climate-resilience, to achieve its potential of being a sustainable crop in future food systems.

Keywords Quinoa · Climate resilience · Nutritional security · Marginal lands · Sustainable agriculture

5.1 Introduction

Quinoa (*Chenopodium quinoa* Willd.) is a vivid example of a neglected and underutilized species (NUS) that has a real potential to enhance food and nutrition security in the world. After remaining localized to the Andean highlands, the outstanding nutritional properties of quinoa, an almost complete plant protein with a favorable balance of essential amino acids (especially lysine) in it, and large amounts of minerals and bioactive compounds, have prompted increasing scientific and consumer attention. Recent reviews point to the potential of quinoa as a gluten-free and nutrient-rich food that can be used to supplement the healthier diet and fill the micronutrient deficits in the risk groups (Angeli et al., 2020a).

In addition to nutrition, quinoa is valued due to its wide range of genetic diversity and plasticity. The collections of germplasm of quinoa currently include tens of thousands of accessions (more than 16,000 accessions in repositories around the world), the majority of which are found in the Andes—a resource which supports the effective adaptation of the crop to highly diverse climates and soils. This genetic diversity has been a key element towards making quinoa experimentally grow in over a hundred countries (Ma et al., 2025).

The agronomic adaptability of quinoa is especially applicable in the face of the current dangers to the global food system, which are the accelerating climate change, the increase of soil salinity and desertification of most areas, and the chronic malnutrition. Published in the past 2 years, field and multi-environment research records the behavior of quinoa under drought, salinity, marginal-soil conditions, and provide inconsistent but encouraging yields beyond the Andean zone of activity—demonstrating the suitability of the crop as a marginal-land climate-resilient candidate crop. All these characteristics render quinoa strategic NUS of climate-smart agricultural portfolios, particularly in areas where traditional cereals are not doing so well (Cui et al., 2024). Nevertheless, globalization of quinoa has also had both valuable lessons and threats at a rapid pace. There is a warning that commercialization without proper governance and appropriate value-chain design may negatively affect the smallholder producer and local biodiversity. Policies that safeguard the rights of farmers, on-farm diversity, and market development could help to have sustainable and equitable goals (Hirich et al., 2021). Earlier work have shown great potential of quinoa due to its high nutritional benefits, genetic potential, and adaptable to environmental changes (Dehghanian et al., 2024). In this chapter we will discuss quinoa potential as successful future crop.

5.2 Botanical Description and Taxonomy

5.2.1 Taxonomic Classification (Family: Amaranthaceae)

Quinoa (*Chenopodium quinoa Willd.*) belongs to the family Amaranthaceae, to the subfamily Chenopodioideae. It is a herbaceous, annual, dicotyledonous plant, popularly known as a pseudo-cereal because of its application as a cereal, although it does not belong to the real family, grain (Poaceae) (Angeli et al., 2020b). Studies conducted at the chromosome level shows that quinoa is allotetraploid ($2n = 4x = 36$), which means that it occurs as a result of the hybridization of two different diploid ancestors, and that is why it is genetically plastic and adaptable (Angeli et al., 2020a).

5.2.2 Origin and Domestication History (Andean Region)

The centre of origin of quinoa is the Andean highlands of South America specifically around the Lake Titicaca area of Bolivia and Peru today. The archaeological data indicate that the quinoa cultivation started approximately 7000 years ago, which is why it became one of the historical Andean domesticated crops. The highland indigenous people have preserved different landraces by herding and herding, and seed-exchange methods. This protracted history of human selection has maintained an array of germplasm that is also well adapted to very diverse microclimates (Jaikishun et al., 2019).

With the conquest by the Spaniards, the use of quinoa in colonial set ups fell down, although it stayed in local Andean farms, a factor that has contributed to the preservation of the genetic diversity of quinoa and its cultural importance (Bazile et al., 2016).

5.2.3 Morphological Characteristics

Quinoa exhibits a great level of morphological diversity, which is reflected in the characteristics of plant height, branching, inflorescence structure, seed color and phenology:

Leaves and plant habit: The plant habit and leaves are usually the alternate ones, and the surface of the lobed leaves is frequently powdery (mealy).

Inflorescence: Panicles are very different. Amaranth form panicles (long, open) occur in some of the accessions and glomerulate panicles (compact groups) occur in others.

Agronomic characteristics: Quantitative characteristics vary among genotypes, i.e., flowering, biomass, panicle length, 1000-weight of seed and seed yield are broadly varied.

Seed metabolic composition: A high-throughput analysis of 471 quinoa accessions showed that there was significant variable metabolite difference in the composition of the seed metabolites, such as essential amino acids, fatty acids, and saponins (Tabatabaei et al., 2022).

Seed colour: The seed coat colours (white, yellow, and red, black, purple) are broadly varied thus indicating genetic variation and adaptation.

5.2.4 Genetic Diversity and Ecotypes

Genetic diversity of quinoa is one of the most pronounced features and it is organized in terms of various ecotypes adapted to various agro-ecological zones:

Classification on ecotype: Quinoa is generally categorized into five major ecotypes by the type of their natural niches:

- Highland (Altiplano)
- Inter-Andean valleys
- Salt flats (salares)
- Yungas (tropical/subtropical)
- Coastal (lowland)

Genetic structure: Recent genome-wide association studies (GWAS) provide evidence of major groups of genetic groupings of quinoa accessions, such as: lowland Chilean, Peruvian highland and Bolivian highland lines.

Landrace diversity: SSR-marker evaluation of 268 landrace accessions collected in 2 time periods (1980s vs. 2014–2015) in the Ecuadorian Andes revealed 124 alleles, high anticipated heterozygosity, and a total of more than 900 distinct genotypes.

Wild relatives/breeding potential: The useful properties (e.g. heat tolerance) of wild tetraploid relatives such as *Chenopodium berlandieri* and *C. hircinum* have been tested. As a case study, *C. berlandieri* remained more viable to heat stresses, and was therefore proposed as having the potential to become an introgression into cultivated quinoa.

Table 5.1 Comparison of phenological scales of quinoa (*Chenopodium quinoa*) (Sosa et al., 2017)

Principal stage	BBCH code range	Description
0—Germination	00–09	Dry seed, water uptake, radicle emergence, cotyledon opening
1—Leaf development	10–19	Cotyledons expanded, first true leaves appear and develop
2—Branching	20–29	Formation and elongation of lateral branches
5—Inflorescence emergence	50–59	Flower buds visible, inflorescence becomes clearly distinguishable
6—Flowering	60–69	First flowers open, full flowering, end of an thesis
7—Fruit/seed development	70–79	Seed formation, early grain filling
8—Ripening	80–89	Grain filling complete, seed color changes, physiological maturity
9—Senescence	90	Final drying of plant and readiness for harvest

5.2.5 Phenological Stages

Quinoa has a normal phenological cycle, which includes germination, seedling development, vegetative development, panicle development, flowering, filled grains and physiological maturity (Table 5.1). Phenological growth stages of quinoa (*Chenopodium quinoa*) according to the BBCH scale have been shown in Figs. 5.1 and 5.2. Genotype, temperature and photoperiod are influential factors in determining the length of each stage. The modern researches show that photoperiod sensitivity is one of the major limitations when introducing Andean varieties to subtropical or tropical conditions, which tends to influence flowering time and stability of yields (Angeli et al., 2020b).

5.3 Global Status of Quinoa

5.3.1 Historical Neglect and Limited Cultivation

Quinoa was left out over centuries and is only grown by indigenous Andean people who have preserved it but given less attention to improve its genetic diversity. The reason behind its marginalization was the domination of European cereals that were brought in during the colonial expansion and the status of quinoa as a low-status food, which limited its spread outside of conventional systems of farming. It was not until the recent decades that scientific interest began with studies illustrating quinoa as a better nutritional composition with the capacity to endure extreme climatic

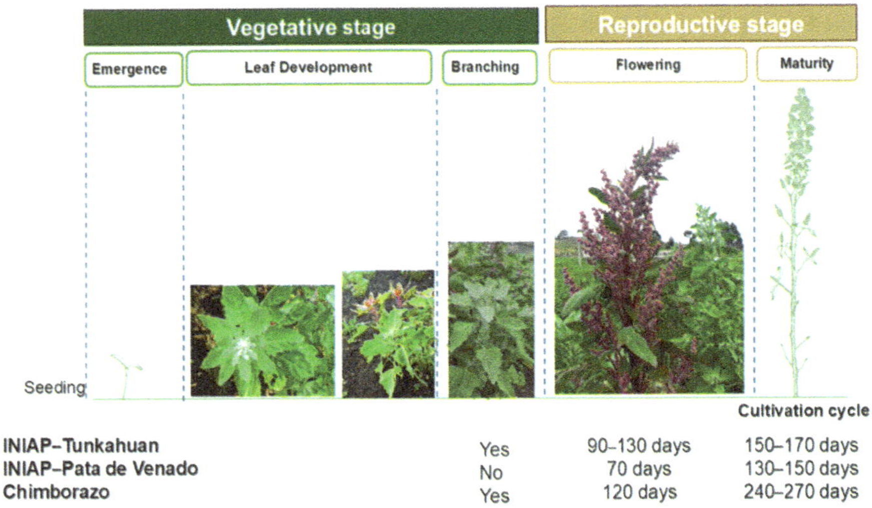

Fig. 5.1 Phenology growth of quinoa plant (Sosa et al., 2017)

conditions, making it a potential emerging crop of relevance in food and nutrition security (Cui et al., 2024).

5.3.2 *Global Distribution*

Quinoa cultivation has experienced rapid globalization in the last twenty years (Figs. 5.3 and 5.4). Its flexibility to various areas of agro-ecological locations such as salt-affected, drought-prone, and high-altitude areas has seen the crop planted or experimented in over a hundred countries (Bazile, 2023). Even though the production of quinoa is growing in Asia, Africa, the Mediterranean region, and Europe, the Andes, especially Peru and Bolivia, still hold the bulk of the world production, which has the centuries of experience of farmers and a rich variety of adapted landraces (Alandia et al., 2020). The new production areas are now assessing genotypes based on heat tolerance, photoperiod sensitivity, and biomass production, as the contemporary studies show that genotype-environment interaction is strong (Mabrouka et al., 2023).

5.3.3 *International Year of Quinoa*

Quinoa grain could have significant impact on anti-hunger fight as reported in 2013, the International Year of Quinoa (IYQ). In this year, a significant progress was made

Fig. 5.2 Phenological growth stages of quinoa (*Chenopodium quinoa*) and inflorescence emergence (50), full flowering (60a) and thick grain at the main inflorescence level (85a) according to the BBCH scale (Sosa et al., 2017)

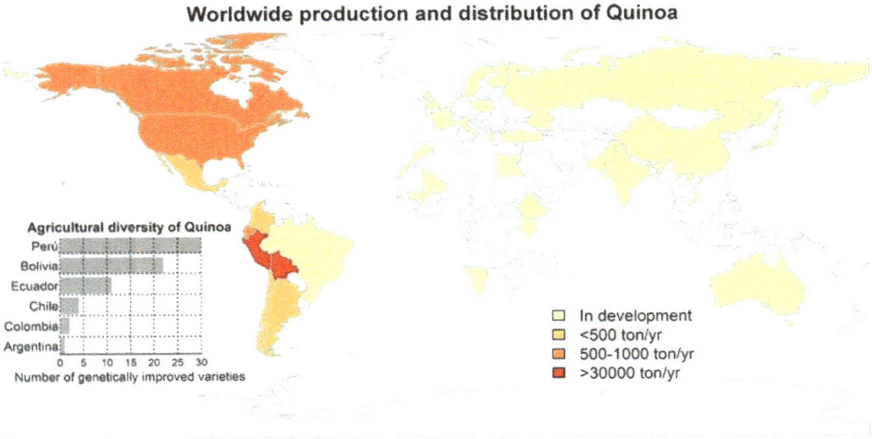

Fig. 5.3 Global distribution of quinoa production. Bolivia and Peru are the main producers, followed by Ecuador, USA, and Canada. Chile, Argentina, Colombia, and Mexico produce mostly for local consumption. The other countries are currently developing quinoa projects (field trials). Bar plot on the left shows productive varieties developed by genetic improvement in countries within the original distributional range (FAO, 2024)

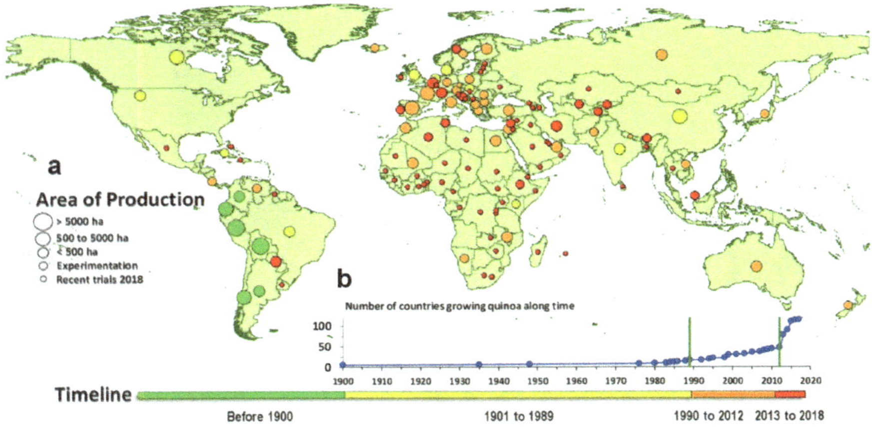

Fig. 5.4 Time series global distribution of quinoa (Alandia et al., 2020)

to aware world about potential of the crop. The program has highlighted the possible benefits of quinoa in sustainable food production, nutritional value improvement and the resilience to climate. This resulted to the organization of international research trials, germplasm exchange schemes, and policy discussions. The IYQ also spurred the motivation of investment in value chain development and enhanced the international discourse of quinoa as a climate-sensitive crop that would be very important future food systems.

5.4 Agronomic Characteristics

5.4.1 Climate Requirements

Quinoa has become quite popular in terms of its extensive climatic adaptability to grow in the lowlands along the coast to the high-altitude Andesian areas. The optimum temperature is 15–20 °C. However, some ecotypes can survive much lower or higher temperatures. In semi-arid to temperate climates, the crop thrives well with low humidity leading to low disease occurrence. Recent work have elaborated the ability of quinoa to survive in areas of climate variability and show that it grows well and races well within changes in temperature regime, and non-uniform rainfall distribution (Akram et al., 2024).

5.4.2 Soil Adaptability

5.4.2.1 Salinity

Among the unique agronomic characteristics of quinoa, its extreme level of salinity tolerance is very important. It is able to survive in soils with salinity levels equivalent to the seawater, primarily because of physiological processes like osmotic adaptation, sodium binding and efficient ion compartmentalization (Tabatabaei et al., 2022). Studies in saline habitats indicate that some genotypes are able to retain chlorophyll levels, stomata activity and biomass under high conductivity levels of electricity, and quinoa is a model crop in saline agriculture (Fig. 5.5) (Sun et al., 2025).

5.4.2.2 Poor Soils and Marginal Lands

In addition to tolerance to salinity, quinoa can grow in low quality soils such as sandy, rocky or soil lacking organic matter. It has a deep and fibrous root system that increases the foraging of nutrients and morphological plasticity that enables adaptation to different soil textures. Recent articles focus on the potential of quinoa in the marginal areas where the conventional cereals do not work, and it is important in enhancing food production in the resource-poor agricultural systems (Bazile, 2023).

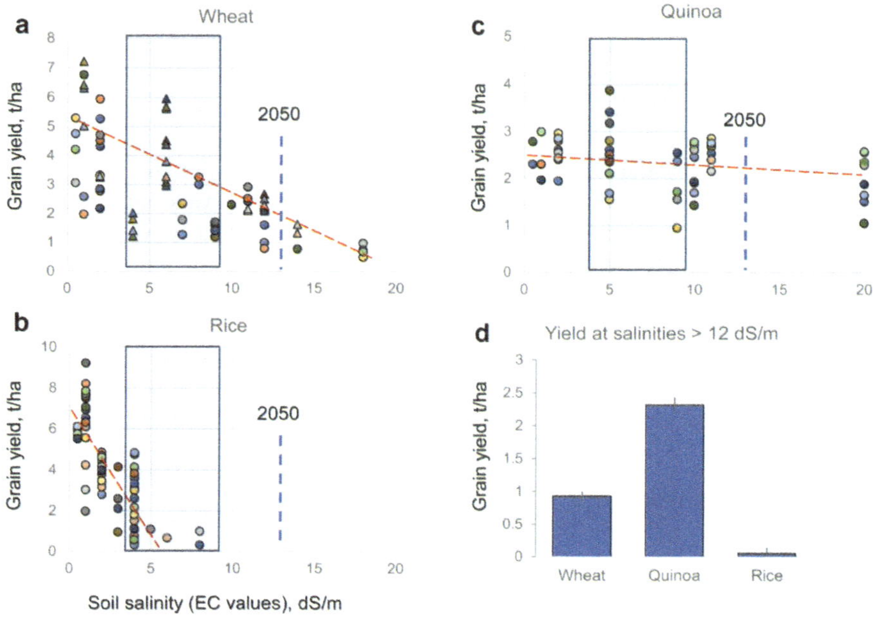

Fig. 5.5 Effect of soil salinity on grain yield in wheat (**a**), rice (**b**) and quinoa (**c**) species. Data is collated based on 11 papers describing performance of 38 genotypes for wheat; 9 papers/31 genotypes for rice; and 12 papers/35 genotypes for quinoa (Sun et al., 2025). Each symbol represents a specific genotype. In **a** circles represent bread wheat (*Triticum aestivum*) while triangles define durum (*Triticum turgidum* ssp. *durum*) wheat genotypes. The slopes of the red dotted lines illustrate differences in sensitivity of species to increasing soil salinity. Shaded boxes represent the current range of salinities experience by crops in major grain producing countries experiencing salinity issues (China, India, USA, Mexico, Pakistan, Australia). The bold vertical dotted line shows predicted values of soil salinity in 2050 derived from trends in changes in EC values in irrigation water (Liu et al., 2020). **d** grain yield of three species grown at salinities predicted for 2050

5.4.3 *Tolerance to Abiotic Stresses*

5.4.3.1 Drought

Several physiological processes such as stomatal control, osmoprotectant storage and quick phenological adaptation make quinoa highly drought resistant. Comparative tests conducted in irrigated, rainfed and very severe drought condition reveals that quinoa remains relatively stable in terms of yield by adjusting the level of leaf area, root depth and water-use efficiency. It has been observed that there exist considerable genotype-specific variations, a fact that supports the need to use drought-adapted lines in arid areas (Grenfell-Shaw & Tester, 2021).

5.4.3.2 Frost

Quinoa is also resistant to short-time frost especially during the early stages of vegetation, which has led to its popularity in high-altitude Andean habitats. Ecotypes have a great variety of frost tolerance and can be correlated with the activity of antioxidant enzymes, stability of the membrane, and the accumulation of osmolytes produced by cold. Cold climate Winter-sown quinoa has demonstrated a high germination and establishment rate in cold climates but reproductive stages are more sensitive (Jacobsen et al., 2007).

5.4.3.3 Heat

Heat tolerability is also gaining a lot of concern as quinoa finds its way to low latitude and dry areas. Increasing temperatures influence flowering, pollen viability and grain filling, although there are genotypes that show a high degree of resistance to heat. Hot-arid trials revealed the ability to maintain a steady level of yield by relying on such characteristics as early flowering, well-controlled transpiration, and increased activity of heat-shock proteins (Ain et al., 2023) (Fig. 5.6).

Quinoa can tolerate a wide range of temperatures (from − 8 to 35 °C) and relative humidity conditions (from 40 to 88%), depending on genotype characteristics and phenological stage. Despite the adaptation of quinoa outside the Andes, a high temperature during flowering and seed set can significantly reduce the yield and is one of the major barriers to the global expansion of quinoa. For example, studies in Italy, Morocco, Germany, Portugal, India, Egypt, Mauritania, and the United States have reported that high temperatures reduce the quinoa seed yield. Most research has focused on understanding the effect of temperature on quinoa seed germination. However, few studies have focused on understanding the physiological changes in quinoa under high temperature during the phenological stages (Alandia et al., 2020; Sun et al., 2025).

5.4.4 Potential Yield Under Various Environment

Quinoa has diverse genotypes and yield potentials as the yield of quinoa in different environments lies between 0.5 to over 5 t/ha under optimal conditions. The components of yield like panicle size, seed density, and weight of grains are sensitive to environmental stress that is dynamic. Research carried out in the period of 2022–2024 indicates that even during the moderate drought or salinity, quinoa yields are rather stable, although lines adapted to non-Andean areas perform well (Golabi et al., 2024). This makes quinoa one of the key crops in food production that is resistant to climate change.

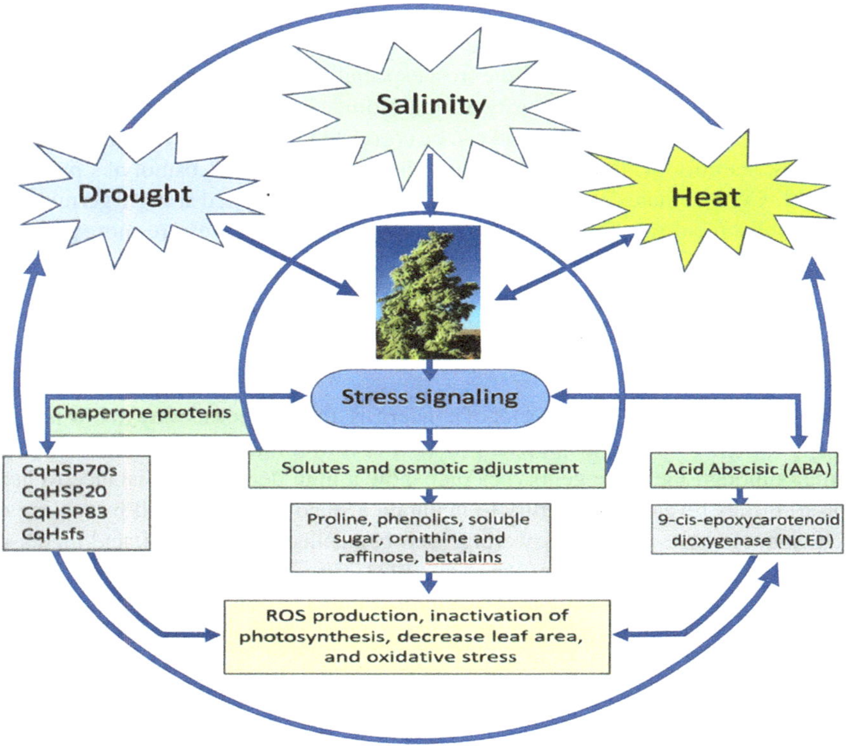

Fig. 5.6 Primary physiological responses to drought, heat, and salinity on quinoa. The expression of NCED genes is upregulated to drought and salinity and chaperone proteins are upregulated to drought and heat stress (Hinojosa et al., 2018)

5.5 Nutritional Profile and Health Benefits

Grains play an important role in human diet by meeting approximately half of an individual's need for energy and protein intake. Wheat, corn, rice, barley, oat, rye, and sorghum are the most crucial foods in the world in human and animal diets. A comparison of the nutritional values of these grains in relation to quinoa is given in Table 5.2. Quinoa's superiority over other grains results from its richer protein, lipid, and ash content (Navruz-Varli & Sanlier, 2016).

Table 5.2 Comparison of the nutritional values of grains and quinoa (edible 100 g) (Navruz-Varli & Sanlier, 2016)

Composition	Quinoa	Rice	Barley	Wheat	Corn	Rye	Sorghum
Lipid (g)	6.07	0.55	1.3	2.47	4.74	1.63	3.46
Protein (g)	14.12	6.81	9.91	13.68	9.42	10.34	10.62
Ash (g)	2.7	0.19	0.62	1.13	0.67	0.98	0.84
Fiber (g)	7.0	2.8	15.6	10.7	7.3	15.1	6.7
Carbohydrate (g)	64.16	81.68	77.72	71.13	74.26	75.86	72.09
Energy (kcal)	368	370	352	339	365	338	329

5.5.1 Macronutrient Composition

5.5.1.1 Protein

Quinoa seeds contain a high protein content (12–22%) according to the genotype and the culture. The proteins are predominantly albumins and globulins, which are highly digestible. Very low levels of prolamins that are less digestible are present. This top-quality protein is necessary in the development of the muscle as well as the restoration of tissue. The content of the protein also depends on the soil fertility and climate to a certain extent (Xi et al., 2024).

5.5.1.2 Carbohydrates

Quinoa is rich in carbohydrates (50–64% of the dry weight), which is the primary constituent of the grain. Most of these are starch which supplies slow release energy. Quinoa also has traces of maltose, D-xylose, and glucose. It has a low GI, which qualifies it to be a diabetic diet. Carbohydrate component contributes to energy, as well as fiber consumption (Ogungbenle, 2003; Zhang & Li, 2024).

5.5.1.3 Fats

Quinoa has 4–9% of fat, which consists predominantly of unsaturated fatty acids (linoleic and a-linolenic acids). These fatty acids are vital to cardiovascular health and brain functioning. Quinoa also has a minor portion of saturated fat that has minimal health hazards. Lipid fraction also adds to energy density and improves the bioavailability of fat-soluble vitamins (Guo et al., 2025).

5.5.2 *High-Quality Protein and Essential Amino Acids (Lysine-Rich)*

Essential amino acids rich in lysine and other essential amino acids are present in quinoa. Quinoa has high proteins and low amounts of carbohydrates. Essential amino acids rich in lysine and other essential amino acids are dominant in quinoa. Quinoa is the only grain that has all the nine essential amino acids. It contains lysine especially in high amounts and this is usually lacking in other vegetarian proteins. This renders quinoa a great addition to the vegetarian or vegan food. The biological value of quinoa protein is high to support the proper growth, immune system and the production of enzymes. Its amino acids constitution is comparable to milk and egg (Guo et al., 2025).

5.5.3 *Micronutrients (Iron, Magnesium, Zinc, Vitamins)*

Quinoa contains the necessary minerals, which include iron, magnesium, zinc, and phosphorus. Iron is useful in the production of hemoglobin and prevents anemia. Magnesium facilitates nerve and muscle functioning. Zinc is very important in immunity and growth. Vitamin E is also found in quinoa, as well as traces of carotenoids, which are antioxidants. Combination of these minerals and vitamins fights micronutrient deficiencies (Guo et al., 2025).

5.5.4 *Bioactive Compounds (Polyphenols, Flavonoids)*

Quinoa contains a large number of bioactive compounds, such as polyphenols, flavonoids, saponins, and peptides. They are antioxidant and anti-inflammatory, which helps to protect the cells against oxidative stress. Research indicates that quinoa extracts have the ability to counteract free radicals and improve the performance of antioxidant enzymes in the body. The frequent intake can be used to decrease the risks of long-term illnesses such as diabetes, heart issues, and disorders caused by inflammation (Guo et al., 2025).

5.5.5 *Gluten-Free Properties*

Quinoa is also gluten-free by nature, which is why the representatives of celiac disease or those can consume it intolerant to gluten. Its gluten is low and thus it does not has any reactions. Quinoa that is gluten free may be used in breads, pasta and cereals. The property can also be incorporated in health foods and functional diets all over the world (Bravi et al., 2024).

5.5.6 *Role in Combating Malnutrition*

Quinoa is a nutritious crop that is capable of resolving both protein-energy malnutrition and micronutrient deficiencies. It contains protein that is of high quality and enhances protein consumption in vegetarian diets. Iron and zinc are some of the minerals that prevent anemia and aid in immunity. The use of bioactive compounds and antioxidants will improve the health of the body. This renders quinoa of great importance in areas with low nutritional diversity. Its production and use can have a profound effect on food and nutrition security in the world (Angeli et al., 2020a). Overall benefits of quinoa crop have been shown in Fig. 5.7.

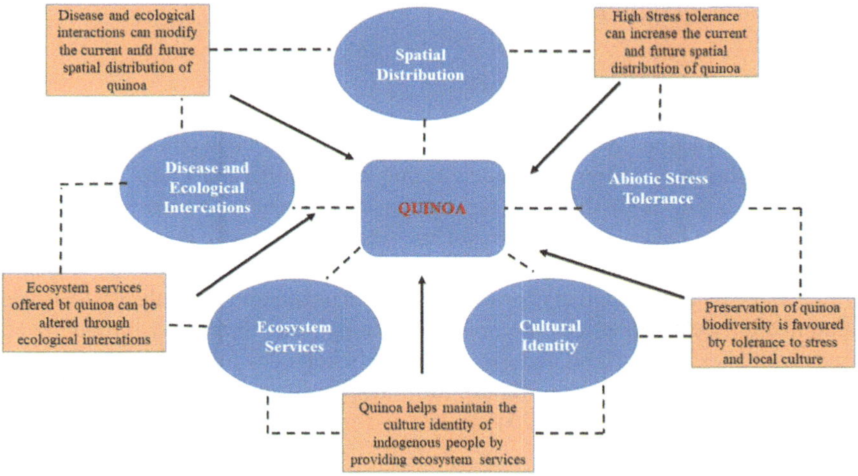

Fig. 5.7 Conceptual framework illustrating the interactions among spatial distribution, stress tolerance, ecosystem services, and cultural identity of quinoa

5.6 Traditional and Modern Uses

5.6.1 Traditional Andean Uses

Quinoa has been a part of the Andean food culture and it has been eaten thousands of years as a staple dish alongside other Andean product. It has been prepared in traditional families as whole grain in soups, porridges and stews including mazamorra, atamalada and in festive and ritual dishes that have cultural values. The Bolivian and Peruvian communities have preserved a variety of landraces with each having been selected on the basis of taste, cooking characteristics and local adaptation. The methods of traditional processing, including the washing, rubbing and winnowing were created to eliminate saponins of the seed coat, to make the preparation of quinoa safe and effective (Angeli et al., 2020a).

5.6.2 Food Products

5.6.2.1 Flour and Bakery Products

The use of quinoa flour in the food system is becoming popular in modern food system because of the protein level, balanced amino acid profile and gluten-free quality. Milling yields fine flour that may be combined with wheat or incorporated on its own in recipes of bakeries. Research indicates that quinoa flour enhances pastries and bread by increasing their nutritional value, in terms of protein, minerals and essential amino acids. Its physiological characteristics such as the ability to absorb water and gelatinization behavior affect the process of dough handlings and provide baked goods with a less firm crumb and a more intense smell. The increasing popularity of quinoa-based products is due to both nutritional sensitivity and gluten-free bakery products demand (Sun et al., 2025).

5.6.2.2 Beverages

Quinoa is also called upon to make both traditional and modern drinks. Quinoa has been used as a base ingredient of fermented beverages in Andean regions, and is regularly used on community gatherings and seasonal festivals. However, quinoa has gained interest all over the world, with it now being processed into plant-based milk substitutes with a mild taste and high nutritional content. They are lactose-free and gluten-free drinks that can be enriched with vitamins or probiotics. Digestibility and bioactive properties of quinoa beverages increase with fermentation processes, which make them functional-foods products (Sun et al., 2025).

5.6.3 Snacks

The growth of quinoa into ready-to-eat snack products sees on the increase of the popularity of quinoa in international health-food markets. Snack bars, breakfast cereals and energy bites use popped and puffed quinoa, which gives the product crisp texture and a high nutritional level. The extrusion technology also allows the creation of baked or shaped snacks that are capable of preserving the integrity of proteins and minerals and provide attractive sensory characteristics to the snacks. Quinoa snacks are clean-label high-nutrient products substituting traditional cereal-based products because of the large number of product innovations in the near past (Muñoz-Pabon et al., 2022).

5.6.4 Industrial Uses

5.6.4.1 Cosmetics

Quinoa has become significant in the field of cosmetics because it is composed of a high level of proteins, essential fatty acids and saponins. The quinoa protein that is extracted improves the ability to retain moisture and assists healing of the skin, whereas the oil provides the benefits of emolliency and antioxidants. Natural surfactants in seed coats, saponins, find their uses in cleansing and exfoliating products, and meet the interest of consumers in botanical and chemical-free formulation. They are mildly foaming and cleaning and have prompted cosmetic manufacturers to consider quinoa-based components in creams, lotions, conditioners, and so on (Xi et al., 2024).

5.6.4.2 Nutraceuticals

Quinoa ingredients are also being included in the nutraceutical industry due to their rich polyphenols, flavonoids, peptides and mineral contents. Grain and bran extracts are created into antioxidant supplements and functional capsules that aim at enhancing metabolism, immune enhancement, and reduction of levels of oxidative-stress. Quinoa protein isolates and bioactive peptides are incorporated in special dietary items among the athletes, children, and malnourished population. Mineral and phytochemical-enriched bran fractions have become functional ingredients to improve the quality of the food (Xi et al., 2024).

5.6.5 *Use as Forage and Animal Feed*

In addition to the use of quinoa as human food and industrial, quinoa can also be used as a forage crop in livestock systems. Young leaves and shoots are rich in protein, calcium and iron; thus they are good in feeding ruminants during the vegetative growth periods. By-products of grains like those that bran and husk are also added to the poultry and livestock feed after washing or heat treatment to eliminate saponins. Studies have shown that quinoa forage has the potential to enhance feed ratio and weight gain of animals but when there is excess saponin levels then quinoa should not be used without proper management. Quinoa is a versatile alternative feed source since it can be grown in arid and marginal locations due to its ability to tolerate poor soils and low-input systems (Sun et al., 2025).

5.7 Barriers to Large-Scale Adoption

5.7.1 *Limited Awareness Among Farmers*

The lack of awareness among the farmers is still one of the main barriers to further implementation of quinoa in most parts of the world beyond the Andes. Farmers in some countries consider quinoa a new crop and do not know the basics of its cultivation, water, and possibility to sell in the market which is a setback in the introduction of quinoa. Research indicates that despite its ability to thrive in marginal conditions, a large proportion of the growers do not have the hands-on information on which cultivar to plant, when and how to plant plants, and how to manage plants after harvesting. Consequently, even in the areas where the climatic conditions are good adoption is still slow (Sun et al., 2025).

5.7.2 *Insufficient Extension Services and Market Linkages*

The poor linkages between farmers, markets and institutional buyers and weak extension systems also hamper the growth of quinoa cultivation. In most developing areas, the extension agents have minimal training in new crops and quinoa is not always a part of any demonstration trials or farm field schools. This is a disconnection between scientific improvements and practice at the field level. The low market connections also demoralize farmers since not all of them have secure buyers, price assurance, and processing facilities. In the absence of effectively structured supply chains, farmers might be unwilling to invest in an unknown returning new crop (Sun et al., 2025).

5.7.3 Problems with Availability of Seed

The availability of the seed is still a big issue especially with the lack of certified, locally adapted, and high yielding varieties. Most nations continue to rely on seed imported which might not fit regional photoperiod, temperature or even soil conditions. There is also the limitation of available access by weak seed multiplication systems and the absence of registered quinoa-driven seed companies. Studies point out that the quality production of seed in the area is important since the interaction of the genotype x environment does have significant effects on crop performance. Large-scale cultivation will not be an easy task without viable chain of seed supply (Sun et al., 2025).

5.7.4 Bitter Saponins and Processing Requirements

Most varieties contain high levels of saponins leading to extra burdens to farmers and processors. Saponins have to be washed away or abraded or removed by special machines—all of which add labor and expense to the harvest. Traditional washing is not possible in the water-scarce areas, and mechanical de-saponification is frequently not there. Research highlights that incomplete removal of saponin bitterness is among the primary factors that make consumer acceptance in new markets reduce. It is therefore necessary to develop low-saponins varieties and low processing technologies (Sun et al., 2025).

5.7.5 Lack of Policy Support

In most areas, quinoa has not been incorporated in agricultural policies fully. The institutions have not done much to support quinoa as most national development programs, subsidies, insurance plans and credit facilities are centered on major cereals. Even though FAO has recognized the world significance of quinoa, policy actions have not been undertaken by many countries (Ahmadzai, 2020).

5.7.6 Limited Value Chain Development

Quinoa value chains are not properly developed in the countries that are not located in South America. The existence of poor processing industries, certification systems, branding programs, and transport infrastructure results to high post-harvest losses, poor product diversification and fluctuating market prices. Smallholders are restricted by the fact that the milling, de-saponification, and packaging have not been invested in. An effective value chain has to have aligned gains in production, processing, distribution and marketing (Hirich et al., 2021).

5.8 Opportunities for Scaling Up

5.8.1 Quinoa in Sustainable Farming Systems

Quinoa is also sustainable and low-input agriculture crop because of its low fertilizer and irrigation requirements and environmental stress resistance. It increases the resilience of cropping by introducing a species with a stress-tolerant crop to the farming systems (Sun et al., 2025).

5.8.2 Role in Diversification and Crop Rotation

Quinoa favors the practice of diversification in cereal-based rotations, through the reduction of the pressure of pests and diseases, and also through the enhancement of the soil structure in terms of its deep roots. Rotations that are made of quinoa enhance productivity and stability in semi-arid areas (Bazile, 2023).

5.8.3 Potential for Organic Cultivation

Quinoa is also natural and very resilient to organic farming systems in which there is restricted use of synthetic inputs. The high prices of organic quinoa at the international markets make farmers consider going organic (Bazile, 2023).

5.8.4 Value Addition and Processing Potential

Quinoa has many opportunities in value added products—in flours and baked goods, nutraceuticals and puffed grains. Another enhancement of taste by processing is the elimination of saponins, which increase consumer acceptability (Sun et al., 2025).

5.8.5 Demand from Health-Food Markets

The nutritional benefits of quinoa such as complete amino acid profile, gluten-free, and low glycemic index have seen its continued demand increase globally. This provides good export opportunities (Cui et al., 2024).

5.9 Breeding and Improvement Efforts

5.9.1 Genetic Variability and Breeding Goals

The large genetic diversity of quinoa makes it possible to select adaptation, tolerance to stress and grain quality. Stable yield, lodging resistance and uniform seed size are some of the major breeding priorities (Sun et al., 2025).

5.9.2 Development of Low-Saponin Varieties

Saponin-reduced varieties have been breeding programs, which lower the processing demand and enhances user acceptability (Bazile, 2023).

5.9.3 Stress-Tolerant Lines

Continuous breeding is on varieties that are resistant to drought, salinity and frost. These screening in ecotypes reveal appropriate lines to be used in Asia and Africa (Sun et al., 2025).

5.9.4 Molecular Breeding, Genomics, and Modern Tools

Genomic technologies, such as marker-assisted breeding, SNP-based breeding, and CRISPR are rapidly increasing the rate of stress tolerance and nutrition quality improvement, but this is still lagging behind major cereals (Sun et al., 2025).

5.9.5 Challenges in Germplasm Access

Breeding is hampered by the fact that the access to quinoa germplasm is still limited by the international system plus the absence of well-developed gene banks in the new areas of growth (Bazile, 2023).

5.10 Policy and Socioeconomic Perspectives

5.10.1 Global Initiatives by FAO and CGIAR

FAO and CGIAR have worked on quinoa research by conducting trials around the globe, technical assistance, and awareness like the International Year of Quinoa 2013 yet many countries have not developed robust follow-up policies (Bazile, 2023).

5.10.2 Market Development Challenges

New producers are hampered by the existence of strong demand, weak processing infrastructure, and lack of quality assurance and competition with traditional exporters (Cui et al., 2024).

5.10.3 Socioeconomic Benefits for Smallholder Farmers

Quinoa will be able to offer income diversification, low cost of inputs as well as increased profitability in marginal environment to enhance livelihood of smallholders (Sun et al., 2025).

5.10.4 Gender and Community-Based Production Models

The production of quinoa in Andes has empowered the economic status of the women. Equivalent community frameworks can be utilized to promote gender equity in developing manufacturing areas (Sun et al., 2025).

5.11 Crop Growth Modelling for Quinoa

AquaCrop, DSSAT, and APSIM models are used to determine the stress performance of quinoa. Indicatively, it was simulated that quinoa could withstand 40–50% of reduced ETc with minimal yield loss (Kheir et al., 2021). Recent efforts found a Crop Water Stress Index (CWSI) value of 0.44 to keep the yields at approximately 2.5 t/ha and save 29% of water (Mendoza-Márquez et al., 2025).

5.12 Case Studies

5.12.1 Case Study 1 Andean Countries (Peru, Bolivia)

In 2023, a project conducted in the high Andes of Peru and Bolivia examined the sustainability and competitiveness of smallholder quinoa production through the study "Performance Evaluation of Quinoa Producer Associations in Andean Highlands." The research focused on quinoa producer associations in Apurímac, where farmers cultivate organic varieties such as Blanca de Junín, Choclito, Pasancalla, Negra, and Roja. While associations reported adequate financial and non-financial performance, their competitiveness—particularly in terms of organizational capacity, resources, and strategic alliances—remains limited, especially in rural districts. The findings highlight challenges such as weak institutional support, limited training, and poor documentation, indicating low sustainability of quinoa producer associations in the rural Andean region (Villegas-Casaverde et al., 2025). The structured questionnaires were used to survey 23 quinoa farmer associations, interviews and the participatory workshops, and the data obtained were analyzed using ANOVA, Pearson correlation and PCA. The research established the relationship generated between 17 and 40.8 tonnes of organic quinoa over a year basing on land access, inputs, and market resources availability. Despite the satisfactory levels of production, general competitiveness was rated as moderately inadequate and the challenges of urban based associations were more as a result of low levels of technical training, poor institutional support as well as by the level of government participation. The researchers have made the conclusion that the enhancement of collective marketing structures, enhancement of extension services, and safeguarding the local landrace diversity are essential measures that need to be taken to ensure the long-term sustainability and empowerment of the Andean quinoa producers (Villegas-Casaverde et al., 2025).

5.12.2 Case Study 2 South America

In 2024, scientists performed a field test called Adaptation Potential of Quinoa Accessions under Arid Mediterranean-Like Conditions in the Peruvian coastal desert, which is a setting similar to the temperate zones in Europe. In order to evaluate the morphological performance of twenty-five quinoa accessions, disease performance (AUDPC), saponin contents, and grain yields, replicated field trials were used. The findings demonstrated that there was significant genetic variation with grain yields of 0.35 to 8.80 t/ha, and saponin content of 0.04 to 0.21%, showing great possibilities of finding sweet and high yielding genotypes. The research also found that after selecting several accessions with a high yield and low saponin content, they could be used as the rightful candidates in quinoa growth and effective adaptation in the European agro-climatic regions. The next step in the development of genetically engineered genetics should be to have the actual testing in the European fields and the breeding programs that should be collaborative to speed up the enhancement of the genotypes to be used in temperate cultivation (Alania-Choque et al., 2024).

5.12.3 Case Study 3 South Asia

To determine the flexibility of quinoa to the semi-arid climate of Punjab, field experiment called Assessment of Phenological, Physiological, and Nutritional Traits in Elite Quinoa Genotypes Under Local Agronomic Conditions was conducted. The findings showed that there was great genotypic variation in the tested materials. Q4, Q11, Q15 and Q126 genotypes had the best biomass and were highly efficient about their canopy, implying that they would be suitable in high yield production due to water limitation. Nutritional profiling showed that the highest phenolic compounds and anthocyanins were accumulated in the genotypes Q9 and Q52 that led to better antioxidant value and enhanced nutritional quality. Root analysis further revealed that Q15, Q24 and Q82 had a wide and profound root system that guarantees higher water and nutrient absorption and resistance to semi-arid ecological stress. Overall, this research has validated that the well-identified quinoa genotypes have high potential to be successful in South Asian agro ecologies, in which the crop would be diversified and will enhance food and nutritional security of Pakistan. The next step in the research should involve multilocation testing of the most successful genotypes and demonstrations at the farmer-level to increase the production of quinoa on a commercial scale (Nazeer & Akram, 2025).

5.12.4 *Case Study 4 Quinoa—A Model Crop for Understanding Salt-Tolerance Mechanisms in Halophytes*

Variation in salinity tolerance of four lowland genotypes of quinoa (*Chenopodium quinoa* Willd.) was assessed by Ruiz-Carrasco et al. (2011). Results showed that quinoa have the ability to endure severe salinity (400–750 mM NaCl) and maintain high leaf K^+ concentration (400–650 mM) to maintain a high K+/Na+ ratio. Molecular multiplication showed that there was high induction of SOS1 and NHX1 antiporters, which allowed Na^+ to be deposited in vacuoles. The existence of epidermal bladder cells was pointed out as one of the significant structures to store and protect the leaf tissues with respect to ion. It is interesting to note that there are no changes or even improvements in seed protein and minerals quality during the salt stress. The authors recommended a broader field-testing of salt-tolerant genotypes beyond the Andes and greater incorporation of these systems into breeding programs.

5.12.5 *Case Study 5: Modelling Quinoa Growth Under Saline and Water-Limiting Conditions*

Estrella Delgado et al. (2025) used SWAP-WOFOST model to assess the potential of quinoa to irrigation salinity and drought in Laayoune, Morocco and applied the field data to calibrate the model. The authors simulated the soil salinity, evapotranspiration, biomass and yield by combining a salt uptake functionality with the Maas-Hoffman salinity stress functionality. The model was correct in projecting salinity dynamics and trends in physiological responses, including a decrease of salinity in soil through root salt uptake of about 40% during crop growth, a real reduction of transpiration of 25–45% in high salinity and biomass losses of more than 30% in the case of 20 dS m^{-1}. The findings revealed that quinoa is able to endure active salt addition, but they also revealed that the salt concentration of soil also continues to increase despite the presence of irrigation. The SWAP-WOFOST model has shown that major stress interactions were obtained accurately and therefore it should be applied in the salinity-sensitive agriculture.

5.13 Conclusion

Quinoa (*Chenopodium quinoa* Willd.) represents a promising neglected and underutilized species with significant potential to contribute to global food security, nutritional adequacy, and climate resilience. Its special flexibility to varied agro ecological requirements, such as marginal lands, salty soils and harsh climates, makes it a climate smart crop that can fit in sustainable agricultural systems. The nutritional composition of the species, such as quality protein rich in lysine and amino acids, vitamins, minerals, and bioactive compounds, highlight the importance of the use of the species in treating malnutrition and health-conscious diets. The local and global food systems of quinoa can be considered through traditional Andean farming methods, as well as, modern food processing and industrial usage in production. Though it holds potential, lack of awareness among the farmers, issues with seed supply, bitter saponins, lack of policy support and value chain underdevelopment are some of the barriers whereby large-scale adoption is no longer achievable. However, breeding systems, which concentrate on low- saponin and stress-tolerant is a solution to boost productivity and adaptability, as well as, the current genetic technology. In the Andes, South Asia and emerging African and European areas, case studies have indicated that quinoa can be used to reach high yields that can sustain smallholder livelihoods when its management and technical support is done appropriately. The adoption and market development of global programs has also been precipitated by the efforts of FAO, CGIAR and national programs. In the future, to achieve greater quinoa production, I believe that it will be necessary to implement coordinated strategies to achieve this goal through research, extension services, gender-based community frameworks, and value-chain development. When these strategies are properly applied, quinoa will be an exemplary crop that will also solve climate resilience, food security and sustainable nutrition, making a significant contribution to the realization of several Sustainable Development Goals such as Zero Hunger, Climate Action and Good Health and Well-being.

References

Ahmadzai, H. (2020). Trends in Quinoa adoption in marginal areas, economic viability, and policy outlook. *Journal of Agribusiness and Rural Development*, *57*, 235–247. https://doi.org/10.17306/J.JARD.2020.01351

Ain, Q. T., Siddique, K., Bawazeer, S., Ali, I., Mazhar, M., Rasool, R., Mubeen, B., Ullah, F., Unar, A., & Jafar, T. H. (2023). Adaptive mechanisms in quinoa for coping in stressful environments: An update. *PeerJ*, *11*, e14832. https://doi.org/10.7717/peerj.14832. PMID: 36883058; PMCID: PMC9985901.

Akram, M. Z., Libutti, A., & Rivelli, A. R. (2024). Drought stress in quinoa: effects, responsive mechanisms, and management through biochar amended soil: A review. *Agriculture*, *14*(8), 1418. https://www.mdpi.com/2077-0472/14/8/1418

Alandia, G., Rodriguez, J. P., Jacobsen, S. E., Bazile, D., & Condori, B. (2020). Global expansion of quinoa and challenges for the Andean region. *Global Food Security*, *26*, 100429. https://doi.org/10.1016/j.gfs.2020.100429

Alania-Choque, J., Vásquez-Espinoza, L. G., Anculle-Arenas, A., Bustamente-Muñoz, J. L., Jellen, E. N., Gutiérrez-Rosales, R. O., & Mayta-Anco, M. E. (2024). Characterization and agronomic evaluation of 25 accessions of chenopodium quinoa in the Peruvian coastal desert. *Agronomy*, *14*(9), 1908. https://www.mdpi.com/2073-4395/14/9/1908

Angeli, V., Miguel Silva, P., Crispim Massuela, D., Khan, M. W., Hamar, A., Khajehei, F., …Piatti, C. (2020a). Quinoa (*Chenopodium quinoa* Willd.): An overview of the potentials of the "golden grain" and socio-economic and environmental aspects of its cultivation and marketization. *Foods (Basel, Switzerland)*,*9*(2), 216. https://www.mdpi.com/2304-8158/9/2/216

Angeli, V., Miguel Silva, P., Crispim Massuela, D., Khan, M. W., Hamar, A., Khajehei, F., Graeff-Hönninger, S., Piatti, C. (2020b). Quinoa (*Chenopodium quinoa* Willd.): An overview of the potentials of the golden grain and socio-economic and environmental aspects of its cultivation and marketization. *Foods, 9*(2), 216. https://doi.org/10.3390/foods9020216. PMID: 32092899; PMCID: PMC7074363.

Bazile, D. (2023). Global trends in the worldwide expansion of quinoa cultivation. *Biology and Life Sciences Forum, 25*(1), 13. https://www.mdpi.com/2673-9976/25/1/13

Bazile, D., Jacobsen, S. E., & Verniau, A. (2016). The global expansion of quinoa: Trends and limits. *Frontiers in Plant Science, 7*, 622. https://doi.org/10.3389/fpls.2016.00622

Bravi, E., Sileoni, V., & Marconi, O. (2024). Quinoa (*Chenopodium Quinoa* Willd.) as functional ingredient for the formulation of gluten-free shortbreads. *Foods (Basel, Switzerland), 13*(3), 377. https://www.mdpi.com/2304-8158/13/3/377

Cui, H., Yao, Q., Xing, B., Zhou, B., Shah, S. S., & Qin, P. (2024). The performance of agronomic and quality traits of quinoa under different altitudes in northwest of China. *14*(6), 1194. https://www.mdpi.com/2073-4395/14/6/1194

Dehghanian, Z., Ahmadabadi, M., Asgari Lajayer, B., Gougerdchi, V., Hamedpour-Darabi, M., Bagheri, N., Sharma, R., Vetukuri, R. R., Astatkie, T., & Dell, B. (2024). Quinoa: A promising crop for resolving the bottleneck of cultivation in soils affected by multiple environmental abiotic stresses. *Plants (Basel), 13*(15), 2117. https://doi.org/10.3390/plants13152117. PMID: 39124236; PMCID: PMC11313704.

Estrella Delgado, D. C., De Swaef, T., Vanderborght, J., Laloy, E., Cnops, G., De Boever, M., …Garré, S. (2025). Modeling quinoa growth under saline and water-limiting conditions using SWAP-WOFOST. *Agricultural Water Management, 309*, 109356. https://doi.org/10.1016/j.agwat.2025.109356

FAO. (2024). FAOSTAT database, Food and Agriculture Organization of the United Nations. Available at http://faostat.fao.org/ (Accessed February 15, 2024).

Golabi, M., Lak, S., Gilani, A., Fazel, M. A., & Egdernezhad, A. (2024). Effect of sowing date and planting method on yield and yield components of three varieties of quinoa (Chenopodium quinoa Willd). *Italian Journal of Agronomy, 19*(2), 100013. https://doi.org/10.4081/ija.2024.2212

Grenfell-Shaw, L., & Tester, M. (2021). Abiotic stress tolerance in Quinoa. In S. M. Schmöckel (Ed.), *The Quinoa Genome* (pp. 139–167). Springer International Publishing. https://doi.org/10.1007/978-3-030-65237-1_9

Guo, Z., Deng, X., Ping, C., Li, X., Li, D., Wu, X., …Kong, R. (2025). Quinoa: Nutritional and phytochemical value, beneficial effects, and future applications. *Applied Food Research,5*(1), 100766. https://doi.org/10.1016/j.afres.2025.100766

Hinojosa, L., González, J. A., Barrios-Masias, F. H., Fuentes, F., & Murphy, K. M. (2018). Quinoa abiotic stress responses: A review. *Plants,7*(4), 106. https://www.mdpi.com/2223-7747/7/4/106

Hirich, A., Rafik, S., Rahmani, M., Fetouab, A., Azaykou, F., Filali, K.,…Choukr-Allah, R. (2021). Development of Quinoa value chain to improve food and nutritional security in rural communities in Rehamna, Morocco: Lessons learned and perspectives. *Plants, 10*(2), 301. https://www.mdpi.com/2223-7747/10/2/301

Jacobsen, S. E., Monteros, C., Corcuera, L. J., Bravo, L. A., Christiansen, J. L., & Mujica, A. (2007). Frost resistance mechanisms in quinoa (Chenopodium quinoa Willd.). *European Journal of Agronomy, 26*(4), 471–475. https://doi.org/10.1016/j.eja.2007.01.006

Jaikishun, S., Li, W., Yang, Z., & Song, S. (2019). Quinoa: In perspective of global challenges. *Agronomy,9*(4), 176. https://www.mdpi.com/2073-4395/9/4/176

Kheir, A. M. S., Alkharabsheh, H. M., Seleiman, M. F., Al-Saif, A. M., Ammar, K. A., Attia, A., …Schillaci, C. (2021). Calibration and validation of AQUACROP and APSIM models to optimize wheat yield and water saving in arid regions. *Land, 10*(12), 1375. https://www.mdpi.com/2073-445X/10/12/1375

Liu, M., Pan, T., Allakhverdiev, S. I., Yu, M., & Shabala, S. J. T. i. P. S. (2020). Crop halophytism: An environmentally sustainable solution for global food security. *25*(7), 630–634.

Ma, J., Du, J., Zhang, W., Zhao, H., Ma, Y., Hong, H., Huang, W., & Li, J. (2025). Analysis of flavor metabolism characteristics of quinoa (*Chenopodium quinoa* Willd.) sprout plant beverage by GC-IMS combined with intelligent sensory technology. *Food Research International, 221*, 117373. https://doi.org/10.1016/j.foodres.2025.117373

Mabrouka, O., Smail, M., Mohammed Tahar, H., & Chenchouni, H. (2023). Yield, growth development and grain characteristics of seven Quinoa (Chenopodium quinoa Willd.) genotypes grown in open-field production systems under hot-arid climatic conditions. *Scientific Reports, 13*. https://doi.org/10.1038/s41598-023-29039-4

Mendoza-Márquez, B., Chuchon-Remon, R., Ramos-Fernández, L., Falconí-Palomino, J., Garcia-Lopez, Y., & Gómez-Pando, L. (2025). Water stress thresholds and irrigation management via infrared thermography and aquacrop model in quinoa cultivation in arid zones. *Agricultural Research*. https://doi.org/10.1007/s40003-025-00869-0

Muñoz-Pabon, K. S., Roa-Acosta, D. F., Hoyos-Concha, J. L., Bravo-Gómez, J. E., & Ortiz-Gómez, V. (2022). Quinoa snack production at an industrial level: effect of extrusion and baking on digestibility, bioactive, rheological, and physical properties. *Foods (Basel, Switzerland), 11*(21), 3383. https://www.mdpi.com/2304-8158/11/21/3383

Navruz-Varli, S., & Sanlier, N. (2016). Nutritional and health benefits of quinoa (Chenopodium quinoa Willd.). *Journal of Cereal Science, 69*, 371–376. https://doi.org/10.1016/j.jcs.2016.05.004

Nazeer, S., & Akram, M. Z. (2025). Elucidating genotypic variation in quinoa via multidimensional agronomic, physiological, and biochemical assessments. *Plants (Basel), 14*(15). https://doi.org/10.3390/plants14152332

Ogungbenle, H. N. (2003). Nutritional evaluation and functional properties of quinoa (Chenopodium quinoa) flour. *International Journal of Food Sciences and Nutrition, 54*(2), 153–158. https://doi.org/10.1080/0963748031000084106

Ruiz-Carrasco, K., Antognoni, F., Coulibaly, A. K., Lizardi, S., Covarrubias, A., Martínez, E. A.,…Zurita-Silva, A. (2011). Variation in salinity tolerance of four lowland genotypes of quinoa (Chenopodium quinoa Willd.) as assessed by growth, physiological traits, and sodium transporter gene expression. *Plant Physiology and Biochemistry, 49*(11), 1333–1341. https://doi.org/10.1016/j.plaphy.2011.08.005

Sosa, V., Brito-Salentim, V. H., Fuentes, F., & Steinfort, U. (2017). Phenological growth stages of quinoa (Chenopodium quinoa Willd.) based on the BBCH scale. *Annals of Applied Biology, 170*, 8.

Sun, H., Khan, W. u. D., Tanveer, M., Ijaz, U., Lu, Z., & Shabala, S. (2025). Can quinoa (Chenopodium quinoa) replace traditional cereals under current climate scenarios? [Review]. *Frontiers in Plant Science, 16*, 2025. https://doi.org/10.3389/fpls.2025.1636565

Tabatabaei, I., Alseekh, S., Shahid, M., Leniak, E., Wagner, M., Mahmoudi, H., Thushar, S., Fernie, A. R., Murphy, K. M., Schmöckel, S. M., Tester, M., Mueller-Roeber, B., Skirycz, A., & Balazadeh, S. (2022). The diversity of quinoa morphological traits and seed metabolic composition. *Scientific Data, 9*(1), 323. https://doi.org/10.1038/s41597-022-01399-y

Villegas-Casaverde, M., Prado-Canchari, A., Choque-Quispe, K., Aguirre-Landa, J. P., Agreda-Cerna, H. W., Reynoso-Canicani, D. D., Mescco Cáceres, E., Coral Surco, R. G., Valverde Mamani, G., Tapia Tadeo, F., & Loayza-Céspedes, J. C. (2025). High andean association producers of organic quinoa: A sustainability study based on competitiveness and performance. *Sustainability, 17*(9), 3929. https://www.mdpi.com/2071-1050/17/9/3929

Xi, X., Fan, G., Xue, H., Peng, S., Huang, W., & Zhan, J. (2024). Harnessing the potential of quinoa: nutritional profiling, bioactive components, and implications for health promotion. *Antioxidants (Basel), 13*(7). https://doi.org/10.3390/antiox13070829

Zhang, H., & Li, R. (2024). Effects of quinoa on cardiovascular disease and diabetes: A review. *Frontiers in Nutrition, 11*, 1470834. https://doi.org/10.3389/fnut.2024.1470834

Chapter 6
Pendant Amaranth (*Amaranthus Caudatus*): Sustainable Production and Role in Soil Health

Laiba Noor, Umar Farooq, Mukhtar Ahmed, and Shakeel Ahmad

Abstract *Amaranthus Caudatus* is an underutilized, climate-resilient pseudo-cereal crop that contains excellent nutritional, physiological, and biological characteristics for sustainable agriculture and soil management. It contains good proteins (14–17%), high lysine levels, and major nutrients such as minerals, hence a good food source to individuals with protein-energy malnutrition. The crop is a good to be planted under water stress conditions, high temperature, and low nutrient content in the soil due to its high C4 photosynthetic and robust antioxidant system. This makes the crop to show high-water use efficiency of up to 35%. The plant has a deep taproot system, which is approximately up to 1.5 m long, with lateral roots. This enhances the nutrient and water uptake as well as physical properties of soil by enhancing aggregation. The crop also possesses high diversity of microbes due to rhizodeposition, and therefore supports large quantities of the Plant growth-promoting rhizobacteria (PGPR) and Arbuscular mycorrhizal fungi (AMF). However, crop productivity can be made stable by 20 to 40% by the implementation of sustainable methods such as organic farming, integrated nutrient management, conservation agriculture and intercrops, which will reduce overreliance on agrochemicals. Potential grain crop productivity of 1.5–2 t ha^{-1} under low input agriculture conditions has been estimated by the use of crop modeling software such as HYDRUS-1D, DSSAT, and APSIM. Crop phenology, productivity, and resource use efficiency was simulated under various agroclimatic conditions (altitudes ranging between 0 to 3000 m). *Amaranthus caudatus* could enhance the livelihood of small farmers, guarantees the sovereignty of their seeds, and augments the security of people to consume it.

Keywords *Amaranthus caudatus* · Sustainable agriculture · Soil health · Climate resilience · Nutritional security

L. Noor (✉) · U. Farooq · M. Ahmed
Department of Agronomy, Pir Mehr Ali Shah Arid Agriculture University, Rawalpindi, Pakistan
e-mail: laibanoorkhalid@gmail.com

S. Ahmad
Institute of Agronomy Faculty of Agricultural Sciences and Technology, Bahauddin Zakariya University, , Multan, Pakistan

M. Ahmed (ed.), *Climate Change and Neglected Underutilized Crops*,
https://doi.org/10.1007/978-3-032-21005-0_6

6.1 Introduction

Amaranthus caudatus (also called Pendant Amaranth) is receiving increasing attention because of its promising potential in terms of facilitating sustainable agriculture and improving soil health. It is potential future smart crop. It has been traditionally grown throughout the Andean region. It has excellent nutritional value, remarkable physiological resistance, and the ability to thrive in low-input agricultural systems (Rastogi & Shukla, 2013). *Amaranthus caudatus* could help to promote crop diversification, ecological intensification, and regenerative soil management practices. *Amaranthus caudatus* is good option for sustainable and resilient agricultural systems (Chivenge et al., 2022).

Amaranthus caudatus is considered one of the three greatest grain amaranth species that are grown in the world. During the past few years, *A. caudatus* has attracted the interest of scientific and agricultural circles due to its high nutritional value and a good amino acid composition with a high lysine level and a high concentration of minerals and antioxidants (Singh & Punia, 2020). It has potential to withstand high degree of drought resistance and ability to survive changes in climatic conditions. It has been observed that it has a good capacity to absorb nutrients and thrives well in low-fertility soils (Rastogi & Shukla, 2013). It is also gaining importance due to its role in food and nutritional security, particularly in the areas that have few resources or low income (Stallknecht & Schulz-Schaeffer, 1993). This crop was first domesticated in the Andean region, but now has been grown under diverse agroecological conditions (Mukuwapasi et al., 2024). The crop enjoys an outstanding nutritional composition, especially in its high-quality protein content and lysine-rich amino acid composition. All these properties render *Amaranthus caudatus* an effective dietary supplement in places where malnutrition (chronic) and protein-energy malnutrition have been widespread (Nazeer et al., 2022). As the world populations are growing, and climate issues are becoming more challenging, the Food and Agriculture Organization (FAO) has suggested that resilient crops that have not been fully utilized should be diversified to ensure food systems. *Amaranthus caudatus* is in a good situation to satisfy this requirement (Stallknecht & Schulz-Schaeffer, 1993).

The crop is highly adaptable to poor soil and requires low levels of agricultural inputs. It has stable yields in drought and temperature stress and is therefore suitable in climate-vulnerable areas such as sub-Saharan Africa and South Asia (Mng'omba, 2025). *Amaranthus caudatus* has become popular in health food markets in Europe, North America, and Asia due to increased consumer demand for gluten-free and antioxidant-rich foods. This pattern has created an additional economic potential for smallholder farmers and facilitated the diversification of the source of income (Mng'omba, 2025). *Amaranthus caudatus* grows quickly and has a strong root system, which has positive effects on the soil structure, its porosity, and aggregation, increasing water penetration and helping control the erosion process in weak agroecosystems. Release of organic acids and exudates in amaranth roots helps to support the proliferation of beneficial microbial associations, including PGPR and AMF (Mlakar et al., 2010). These exchanges increase nutrient availability and uptake efficiency and promote ecological intensification by decreasing the synthesis and agrochemical requirements, and still keeping the production levels high. Crops adapted to intercropping and rotation systems increase the functional biodiversity, natural pest control, and nutrient recycling, and work to improve the agroecosystem in response to climate variability (Mukuwapasi et al., 2024).These characteristics render *A. caudatus* an essential crop to enhance the soil health and ecological-based intensive crop systems through natural processes, as opposed to chemical-based input systems (Pauline Chivenge et al., 2015). This chapter focuses on *Amaranthus caudatus* potential as a future agroecological smart crop to be grown under various agroecological systems.

6.2 Botanical, Ecological, and Agronomic Characteristics of *Amaranthus Caudatus*

6.2.1 Taxonomy and Botanical Classification

The *Amaranthus caudatus* is an annual flowering plant that belongs to the Amaranthaceae family. It is also known by a variety of common names, such as pendant amaranth, tassel flower, quilete, and khaddar (Jimoh et al., 2019a, 2019b) (Table 6.1). Taxonomic position of *A. caudatus* includes Kingdom Plantae; Order Caryophyllales; Family Amaranthaceae; Genus Amaranthus; Species Amaranthus caudatus (Jimoh et al., 2019a, 2019b). There are 70–75 species of the genus Amaranthus worldwide (Iamonico, 2023). Historically, the genus was classified as subgenera on the basis of such characteristics as monoecy or dioecy. As an example, an earlier system of classification named two subgenera (one of monoecious and the other of dioecious plants), although more recent systems (such as Mosyakin and Robertson's) have given a system of three subgenera (Jimoh et al., 2019a, 2019b). Nevertheless, genus Amaranthus taxonomy remains a thorn in the flesh primarily due to the sheer number of distinct morphological varieties, the allocation of traits to species, as well as hybridization and the resulting increase in morphological variants or phenotypic plasticity (Iamonico, 2023). Particularly, in the case of *A. caudatus*, the recent shift in naming of plants has demonstrated that, although some varieties or subspecies (e.g., *A. caudatus* var. albiflorus, var. atropurpureus, sub) are not considered different taxa but are instead referred to as being identical to *A. caudatus* (Iamonico, 2023).

Also, the same reexamination indicated that a few taxa that were formerly treated as part of *A. caudatus* (such as mantegazzianus, gibbosus) should be recognized as distinct species, and that one of them, formerly regarded as *A. caudatus* var. gibbosus, was renamed Amaranthus baileyanus (Iamonico, 2023), as shown in Table 6.2. The strict sense of *A. caudatus* L. is therefore a widely accepted and valid species; however, its infraspecific taxonomy is undergoing revision. With this in mind, its taxonomic status can be said to be simplified as what has always been the case with synonyms and delimiting the boundaries of a species, we need only sum up by saying

Table 6.1 Taxonomic classification of *Amaranthus caudatus* by (Linnaeus, 1753)

Rank	Classification
Kingdom	Plantae
Clade	Tracheophytes
Clade	Angiosperms
Clade	Eudicots
Order	Caryophyllales
Family	Amaranthaceae
Genus	Amaranthus
Species	*Amaranthus caudatus* L.

Table 6.2 Accepted synonyms and nomenclatural status by Iamonico (2023)

Synonym/former name	Current status
A. caudatus var. albiflorus	Synonym of *A. caudatus*
A. caudatus var. atropurpureus	Synonym of *A. caudatus*
A. caudatus var. gibbosus	Elevated to Amaranthus baileyanus
A. caudatus subsp. saueri	Synonym of *A. caudatus*
Amaranthus mantegazzianus	Distinct species
Amaranthus edulis	Sometimes distinct in molecular studies

that the genus Amaranthus is a genus whose taxonomic status, despite its simplicity, is in fact, some complexities as many have been and continue to be in history and today, because it is highly diversified and morphologically variable (Iamonico, 2023).

6.2.2 Morphology and Growth Habit

Amaranthus caudatus is an upright C4 herbaceous perennial plant with considerable variability in shape and size across the several types of the plant and locations. The plants usually grow to a height of 0.5 to 2.5 m, and the stem is strong and can branch slightly or extensively depending on the nature of the genotype and the spacing between the plants. The leaves are simple and come out alternately. Forms of the leaves are ovate-rhomboid-ovate; the lamina is largely hairless, having smooth margins. The variety of colors is truly impressive, as it is green and dark purple, because the plant has different quantities of pigments, as shown in the Fig. 6.1.

This species is most distinguished by the presence of huge flowering plants with huge, soft-tipped terminal inflorescences, which may be as long as 0.9–1.5 m (Stetter et al., 2016). As illustrated in Fig. 6.2 the inflorescences consist of thick bundles of small, unisexual flowers. The fruit is an utricle with a single seed (semi- Hard, grain

Fig. 6.1 Amaranth variation in inflorescence colors (Mukuwapasi et al., 2024)

types) and a darker seed (vegetable or ornamental type), that is 1–1.5 mm across, round-shaped, and covered with a usually whitish coating.

A. caudatus have deep and robust taproot system with laterals branches that proves quite useful in the extraction of water and nutrients under stress conditions. Good root system is one of the primary factors of its quality to adapt under marginal agroecosystems (Brenner et al., 2000). Characteristics of *A. caudatus*, such as the high biomass production, large flowers, variation of colors, and long root system,

Fig. 6.2 Flower morphology: The inflorescence of *A. caudatus* consists of flower clusters in which several female flowers surround a male flower in the center (Stetter et al., 2016)

render it an attractive plant to be considered as a pseudo-cereal, leafy vegetable, and multipurpose crop in various environmental conditions.

6.2.3 Physiological Attributes Contributing to Stress Resilience

Amaranthus caudatus shows a range of physiological traits that allow the plant to survive in the most hostile environments, such as drought, heat, poor soil, and change of temperature (Brenner et al., 2000). It is a C_4 photosynthetic species, which can capture maximum light and sustain high temperature. It has high CO_2 fixation feature with high rate of assimilation, and reduced rate of photorespiration. These features are highly important for good water-use efficiency and biomass production under unfavorable environments (Sage & Zhu, 2011). More mesophyll conductance and enhanced control of internal CO_2 concentration also enable the C_4 strategy to allow *A. caudatus* normal photosynthesis even under a limited supply of water (Anuradha et al., 2023).

The plant species mechanically accumulate cytoprotective substances, including proline, glycine betaine, and soluble sugars, which maintain the cells inflated. These proteins aids membranes to survive under drought or salinity stress. Plant also have antioxidative stress enzymes include: SOD (Superoxide dismutase), CAT (Catalase), and POD (Peroxidases). The enzymes fight against photosynthesis system damage by eliminating the reactive oxygen species that have been generated due to the stress (Rastogi & Shukla, 2013). These antioxidant reactions are closely correlated with the capacity of the plants to maintain chlorophyll active state and high chlorophyll fluorescence efficiency (Fv/Fm) under the conditions of abiotic stress.

Furthermore, *A. caudatus* has an effective nutrient absorption from low-fertility and degraded soils, which is facilitated by its deep root penetration and capability to regulate nutrient transporters in a nutrient-limited environment (Brenner et al., 2000). In addition, its adaptive stomatal conductance to ensure transpiration under lower water, without compromising adequate CO_2 uptake makes it drought tolerance (Anuradha et al., 2023). It has been realized as well that high protein content of the seed (14–17%) alongside the high levels of lysine and other essential amino acids may be linked to the effective nitrogen assimilation and, thus, the performance of plants in low-input agriculture.

6.2.4 Adaptation to Marginal Environments

Amaranthus caudatus is quite good to grow under harsh environments and in such farming systems where low inputs have been used. Water scarcity, soil fertility, and external inputs are the elements that restrain the productivity of the usual plants (Stallknecht & Schulz-Schaeffer, 1993). Its roots, which are deep and much branched, aid in the plant obtaining not only water but also nutrients very effectively, both from the layers of the soil that are deeper, and enable the plant to grow steadily even under poor water and nutrients stress (Brenner et al., 2000). High water-use-efficiency of this plant has been supported by the C_4 pathway of the photosynthesis of the plant. Simultaneously, carbon assimilation under higher temperatures enables it to continue producing biomass and yield under rainfed semi-arid conditions. These characteristics, together with low fertilizer responsiveness and compatibility with the biological-based nutrient cycling, make *A. caudatus* particularly suitable to resource limited farming systems (Mlakar et al., 2010).

Concerning the environmental conditions, *A. caudatus* has a wide range of adaptation in different agroecological conditions. It is a summer crop, and its optimal growth temperature is between 20 and 30 °C, although it can also endure short, hot, and damp periods (Mlakar et al., 2010). It is a highly adaptive species and can be cultivated in just about any type of soil, such as sandy and loam soils, as long as where there is effective drainage, and the pH level is below 6 to 7. Its average rainfall demand, as well as flexibility to respond to day length, enable it to be planted and cultivated under variable climatic conditions (Stallknecht & Schulz-Schaeffer, 1993). All these characteristics lead to the realization that *A. caudatus* can be a tough crop that may be cultivated sustainably and help in improving soil health despite the shift in climatic conditions (Stallknecht & Schulz-Schaeffer, 1993).

6.2.5 Phenology of Amaranthus Caudatus

Amaranthus caudatus passes through distinct stages of the developmental process, starting with germination and eventually maturing physiologically (Fig. 6.3). These stages greatly depend on the environment and the genetic structure of the plant (Brenner et al., 2000). The *A. caudatus* seeds grow under the most favorable climatic conditions with adequate water and temperatures between 20 and 30 °C, and can sprout very fast, within 7 to 10 days of germination. Nevertheless, it may change depending on the soil properties and the depth of the planting of the seeds (Travlos et al., 2020). The phenology of *A. caudatus* shows that an early germination was followed by a long vegetative period during which the leaves and stems of the plants developed at a rapid rate. Typically, this vegetative stage lasts longer in *A. caudatus* than in some other species of amaranth, and this aids the plant in accumulating a high biomass and also in accessing resources (Jimoh et al., 2019a, 2019b).

A. caudatus typically flowers 68–79 days after sowing. This timing may be highly influenced by various characteristics, i.e., the quantity of fertilizer, photoperiod, and temperature, which were observed in controlled experiments when early and late flowering were also determined under various conditions (Martínez-Núñez et al., 2019). *A. caudatus* shows its characteristic drooping inflorescences progressively, and transition between flowering and maturity may extend the period of planting the crop, particularly where the conditions are not so favorable (Anuradha et al., 2023), as illustrated in Fig. 6.3. This protracted developmental period shows that overall

Fig. 6.3 Phenological growth stages of Amaranthus caudatus under controlled/pot conditions, showing key growth stages (Martínez-Núñez et al., 2019)

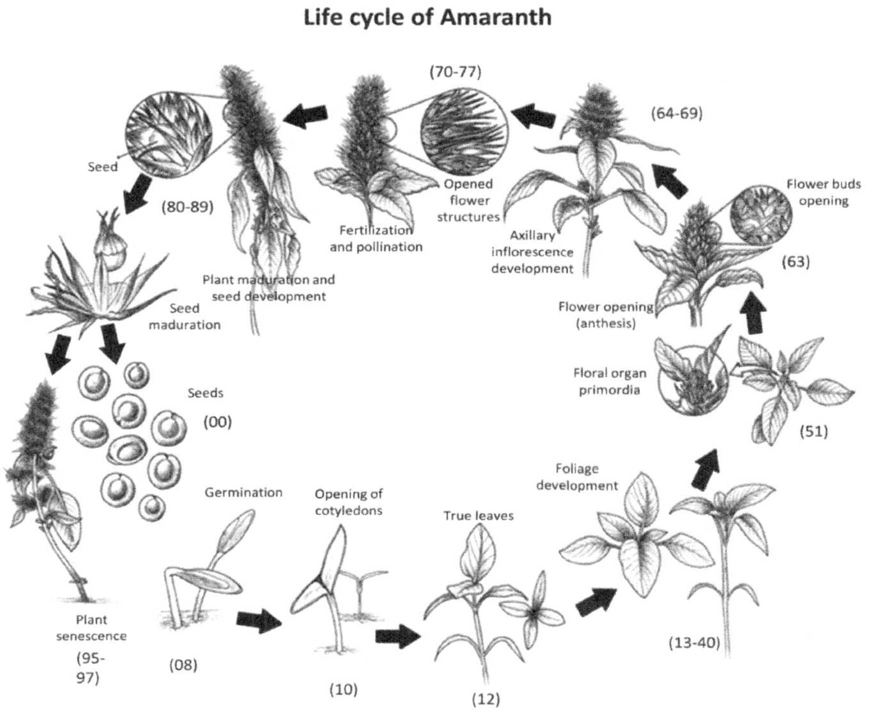

Fig. 6.4 Life cycle of amaranth under greenhouse conditions, showing key growth stages from germination to seed maturation and senescence (BBCH codes)(Martínez-Núñez et al., 2019)

time to maturity is between 110 to more than 150 days according to the BBCH scale, as illustrated in Fig. 6.4. Climatic conditions and management demonstrates the capacity of *A. caudatus* to become accustomed to a broad environment and systems of cultivation (Travlos et al., 2020).

6.2.6 *Production Zones and Agroecological Distribution of Pendant Amaranth*

Amaranthus Caudatus is a plant that has been observed growing in various agroecological areas across the globe. It is capable of adjusting to various climates and soils and this is really amazing. Its native place is the high Andean mountains of South America, comprising of Peru, Bolivia, Ecuador, and Argentina, and the valleys surrounding the countries as well (Falasca et al., 2014). Andean region has been cultivating this plant as a source of food. Agroclimatic zoning studies in Argentina revealed that *A. caudatus* can be grown in humid, subhumid and semiarid countries, but is particularly productive in semiarid and subhumid environment where its low water demands and ability to tolerate hydric stress give it an advantage (Mukuwapasi et al., 2024). In addition to South America, *A. caudatus* has been cultivated in other locations under different degree. It can be used as leafy vegetable, a minor grain crop, or a health-food crop. It has also been reported in Africa, where both grain and leaf morphotype have been used, but despite lack of formal production statistics, it has been reported that *A. caudatus* is well adapted and can be grown under rainfed, low-input systems and particularly in regions where regular cereals have not be growing because of drought or inappropriate soil. Asia, is favorable for its adaptation under both the mixed cropping and subsistence agriculture systems (Mukuwapasi et al., 2024). *A. caudatus* is also highly ecologically flexible as it can grow in both tropical low land areas and also in higher altitudes (e.g. up to approximately 3000 m above sea level) in Andes indicating a broad spectrum of agroecological adaptability. *A. caudatus* is a climate-resilient crop with promise to be used in sustainable and marginal farming environments due to its drought tolerance, phenotypic plasticity, and low input needs in all these regions (Falasca et al., 2014).

6.3 *Amaranthus Caudatus* Production Systems

6.3.1 Preparation of Sustainable Land and Soil Conservation

One of the key pillars of an environment-sensitive production system of Amaranthus caudatus is its growth and survival in low-input and climatically vulnerable regions (Pretty et al., 2018). It has reduced tillage requirements and conservation agriculture could be good as this crop requires low inputs (Fig. 6.5).

Conservation tillage systems are best to promote soil moisture retention. It can provide aggregate stability and prevents nutrient losses as compared to conventional plough-based systems (Chivenge et al., 2015). *A. caudatus* has a shallow root position and active initial development, that's why plant may grow well in areas with low tillage. Therefore, plant does not conflict with the conservation agriculture models, which have been established to ensure the long-term sustainability of the soil (Pretty et al., 2018).

6.3.2 Soil Fertility and Nutrient Management

The use of sustainable soil fertility management is the key to achieving the maximum productivity of the *A. caudatus* and the maintenance of soil health and ecosystem functioning (Yadav & Yadav, 2024). Organic materials like farm yard manure, compost, green manure and vermicompost enhances the physical properties of soil, cation exchange capacity and nutrient availability besides promoting microbial

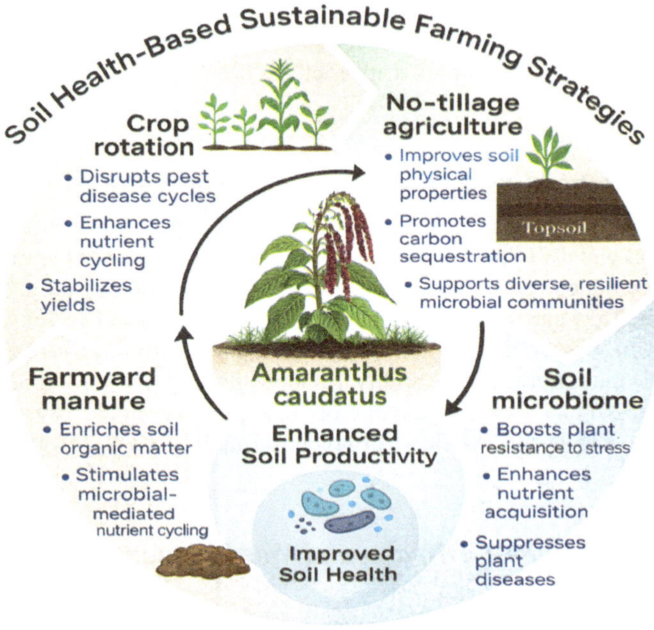

Fig. 6.5 Overview of soil health-based sustainable farming strategies (Liang et al., 2025)

activity (Vessey, 2003). Various investigations have indicated that organic amendments are highly beneficial in increasing the carbon and enzyme functions of soil microbial biomass and nutrient cycling relative to inorganic fertilization alone. Added to this are biofertilizers, which are bacteria able to fix nitrogen, microorganisms capable of solubilizing phosphate, and microorganisms capable of mobilizing potassium. This all enhances nutrient-use efficiency and diminish dependence on chemical fertilizers (Vessey, 2003). Combined nutrient application, including organic, biological, and judicious inorganic applications have been reported to be able to maintain crop production and enhance the biological quality of the soil (Yadav & Yadav, 2024).

6.3.3 Water-Smart and Climate-Smart Agronomic Practices

A. caudatus has C4 photosynthetic pathway that makes it to have high water and radiation use efficiency (Sage & Zhu, 2011). However, optimized irrigation timing during critical vegetative phases like flowering and grain filling, along with the mulching and soil moisture preservation, could results in a high level of water productivity and yield. Furthermore, practices like sowing dates, stress-tolerant genotypes, and diversification of the cropping systems could be adopted to grow *A. caudatus* successfully under extreme weather conditions (Lipper et al., 2014).

6.3.4 Diversified Cropping Systems and Rotations

Crop rotations and diversified systems of crop production increase the sustainability of amaranth production systems through better nutrient cycles, lower occurrence of pests and diseases, and stable yields (Smith et al., 2017). The rotation of legumes or cereals with *A. caudatus* helps to enhance the levels of nitrogen in the soil, the activity of the microflora, and the ability to suppress weeds and soil-borne pathogens (Sage & Zhu, 2011). Diversification of crops also promotes agroecosystem stability and resilience, and sustenance of small-holder farming systems (Vessey, 2003).

6.3.5 Integrated Pest, Disease, and Weed Management

Integrated pest, disease, and weed management method should be used to ensure the stability of the yield (Sage & Zhu, 2011). *A. caudatus* is comparatively resistant to most of the pests and diseases, preventative cultural methods. However, methods like crop rotation, clean seed, timely planting, and post-plant sanitation, are useful in reducing the populations of pests and disease (Pimentel & Pimentel, 2007). The use of mechanical weeding, mulching, and increased crop competitiveness heighten reduction in weed pressure without extensive use of herbicides. Limited, specific chemical treatments as part of an integrated pest management program, combined with biological control, reduces the risk to the environment and health.

6.3.6 Yield Performance and Resource-Use Efficiency

Amaranthus caudatus has a positive yield behavior and resource-use efficiency in different agroecological conditions (Yadav & Yadav, 2024). The reported grain yields are moderate to high with low-input conditions, which point to the efficient use of water and nutrients (Rastogi & Shukla, 2013). Amaranth can also be utilized as climate-smart and resource-efficient crop since sustainable production systems can enhance the stability of yields and lessen reliance on outside inputs (Pretty et al., 2018).

6.4 Rhizosphere Interactions and Soil Biological Processes

6.4.1 Root System Architecture and Rhizodeposition

The rhizosphere, which is known as the zone of soil that is directly affected by the roots of the plants, is at the core of the process of controlling the biological processes and dynamics of nutrients in the soil. *Amaranthus caudatus* has a well-established root system that can cover a wide area in soil exploration, which increases the uptake of nutrients and water in low-input and stressful environments (Rastogi & Shukla, 2013; Vadez et al., 2015). Root exudation or rhizodeposition refers to the release of organic molecules, including sugars, amino acids, organic acids, and secondary metabolites, which promote microbial growth and/or change microbial community structure. These rhizodeposits serve as energy sources to the soil microorganisms, and they enhance nutrient mobilization processes, thus enhancing soil fertility and the availability of nutrients to the plants (Vessey, 2003).

6.4.2 Associations with Plant Growth-Promoting Rhizobacteria and Arbuscular Mycorrhizal Fungi

The growth and development of plants are positively correlated with plant PGPR and AMF which greatly improve the functional performance of the amaranth rhizosphere. Amaranth has also been reported to have Pseudomonas, Rhizobium, Streptomyces, and Bacillus as PGRPs in its rhizospheres, which play a role in the fixation of nitrogen, solubilization of phosphates, the production of phytohormones, and the inhibition of pathogen growth (Vessey, 2003; Zaib et al., 2023). AMF become symbionts of the plant roots and their extraradical hyphae increase the effective root surface area, therefore, improving phosphorus absorption, water uptake, and tolerance to drought and nutrient stress (Zaib et al., 2023).

6.4.3 Effects on Soil Enzyme Activities, Microbial Biomass, and Diversity

The processes that take place in the rhizosphere have a drastic impact on soil enzyme activities, microbial biomass, and microbial diversity, which are the main indicators of biological health of the soil. Dehydrogenase, phosphatase, urease, and b-glucosidase are enzymes in the soil that are crucial in the carbon, nitrogen, and phosphorus cycles (Geisseler & Scow, 2014). Among the soil-related changes, organic amendments and biologically active rhizospheres (related to amaranth) can increase the enzyme activity and the carbon content of the biomass in the microbes as compared to the conventional soils (Mäder et al., 2002). This is due to the importance of an increased microbial diversity in increasing functional redundancy and carrying capacity of soil ecosystems, providing them with the ability to efficiently recycle nutrients and maintain themselves against environmental stress (Kumar et al., 2023).

6.5 Contributions to Soil Health and Ecosystem Services

6.5.1 Soil Organic Matter Dynamics and Carbon Sequestration

The soil organic matter (SOM) is a key factor that determines the soil fertility, structural stability and functioning of the ecosystem. Sustainable cropping systems with *Amaranthus caudatus* will lead to SOM accumulation as a result of root biomass, rhizodeposition, and crop residue inputs (Lal, 2015; Yadav & Yadav, 2024). Carbon inputs of root origins are especially significant to stabilize carbon over a long period, as they are known to increase microbial processing and stabilize soil organic carbon pools (Trivedi et al., 2018).Enhanced SOM enhances nutrient retention, water-binding ability, and microbial activity, among other things, and enhances agroecosystems water retention and soil resilience to natural calamities.

6.5.2 Soil Structure, Aggregation, and Erosion Control

Biologically active cropping systems are necessary to be able to provide ecosystem services of improved soil structure and aggregate stability. Through its root development and relationship with soil microorganisms, especially arbuscular mycorrhizal fungi, *Amaranthus caudatus* helps to stabilize soil structure and aggregation (Soler et al., 2011).Microbial exudates and fungal hyphae serve as adhesive agents to hold aggregates of soil together, which makes them less vulnerable to erosion and crusting. The greater aggregation will enhance pore connectivity, infiltration and aeration, which are essential for crop production and soil conservation in the long term.

6.5.3 Nutrient Cycling and Fertility Enhancement

Sustainable agroecosystems rely heavily on biological processes to mediate nutrient cycling processes that include soil microorganisms and plant-soil interactions. Systems with amaranth encourage active cycling of nutrients by providing inputs of organic matter and activating the microbial activity that triggers the processes of nutrient mineralization and immobilization (Soler et al., 2011). In symbiotic relationships with PGPR and AMF, nutrient acquisition efficiency (especially in terms of nitrogen and phosphorus) improves, and the losses of nutrients are minimized by leaching and volatility (Begum et al., 2019; Vessey, 2003). These biological processes enhance the fertility of soils with minimum reliance on external inputs, which would enhance the environmentally sustainable management of nutrient management.

6.5.4 Biological Indicators of Soil Health and System Resilience

Microbial biomass, enzyme activities, functional diversity and communities of microorganisms are examples of biological indicators that are increasingly being utilized as a measure of soil health and ecosystem resilience. Organic and integrated management systems of Amaranth have been linked to increased biomass carbon of microbes, enzyme activity and therefore, microbial diversity relative to conventional management systems (Mäder et al., 2002). The benefit of high biological activity is that soil systems can withstand climatic and management-induced stresses and remain productive and perform ecosystem services because it enhances functional redundancy and buffering capacity of soil systems (Tilman et al., 2011; Pretty et al., 2018).

6.6 Socio-Economic Relevance

6.6.1 Importance for Smallholder and Resource-Poor Farming Systems

Amaranthus caudatus is specifically useful when it comes to smallholder and resource-poor farmers because it is tolerant to marginal soils, a low-input food crop, and drought and heat stress (Rastogi & Shukla, 2013; Yadav & Yadav, 2024). The fact that it yields reasonably when conditions are not optimal renders it a dependable crop towards improving food security and decreasing production hazard in low resource settings.

6.6.2 Economic Viability Under Low-Input Sustainable Production

The economical examination of amaranth production in low-input and organic management systems has shown that the crop is economically viable since it does not need to heavily rely on synthetic fertilizers and pesticides, which are expensive (Yadav & Yadav, 2024). It has been demonstrated that integrated nutrient management and sustainable soil management systems yield higher yield stability and better resource-use efficiency, which generates competitive yields compared to staple cereal-based systems of the smallholder. The use of the crop as a leafy vegetable and as a grain also adds value in terms of the economy and the market potential (Rastogi & Shukla, 2013).

6.6.3 Contributions to Rural Livelihoods, Food Security, and Seed Sovereignty

One direct contribution made by amaranth is because of grain and leafy vegetable cultivation, which help in care, food, and income of the rural populations (Yadav & Yadav, 2024). It contains many essential amino acids, minerals, and vitamins that enhance the dietary diversity and food security among the vulnerable groups (Rastogi & Shukla, 2013). The benefit of genetic variation in local varieties and the most enhanced types is also the fact that farmers are able to save and share seeds, which facilitates their seed sovereignty and less reliance on the commercial seed markets and suppliers.

6.7 Selected Case Studies Highlighting Sustainable Cultivation Outcomes

6.7.1 Case Study 1: Value Chain Potential of Grain Amaranth for Food and Nutrition Security in Africa

Possible applications of underutilized grain amaranth (*Amaranthus spp.*) in the value chain with a focus on food and nutrition security, especially in Africa have been elaborated earlier (Aderibigbe et al., 2022). Since staple crops do not meet the daily macro and micro nutrient requirements, especially in the vulnerable population of women and under-fives. Hence grain amaranth has high potential of providing high protein, high-calorie, and essential micronutrient materials, However, its potential was not being realized due to poor seed systems and an underdeveloped value-chain. Policy tool was given in scaling up production, processing, and marketing of the grain amaranth as a strategic plan to fight food security. *Amaranthus caudatus* sustainable production methods and soil health-oriented management can help to enhance the productivity and the value-chain integration.

6.7.2 Case Study 2: Grain Amaranth as a Climate-Resilient and Nutrient-Rich Underutilized Crop

Grain amaranth (*Amaranthus* spp.) potential for nutritional security and climate resilience. Have been elaborated by (Murthy & Kumar, 2017) The authors given the historical significance of grain amaranth and its distribution over the whole world, with an emphasis on the fact that this food is gaining popularity due to its high nutritional value. It has been claimed that amaranth grain is composed of higher protein content (14–16%), which has better protein quality in terms of lysine level when compared to major cereals, including rice, wheat, and maize. The findings also indicated that amaranth is a good source of essential minerals, b-carotene, unsaturated oils, and squalene, which makes it useful in food and industrial applications. Being a C4 crop, the grain amaranth exhibits a high level of biomass and plasticity in climatic conditions influenced by climatic changes.

6.7.3 Case Study 3: Functional, Medicinal, and Sustainable Utilization Potential of Amaranth

Amaranth (*Amaranthus* spp.) has emerged as a multifunctional crop with significant potential for improving human nutrition, health, and agricultural sustainability, as highlighted in the comprehensive review by (Terzieva et al., 2026). Work demonstrated that amaranth is a source of high-quality proteins, essential amino acids, dietary fiber, vitamins, and minerals, which makes it an affordable food item in efforts to curb malnutrition and enhance food security. It contains bioactive compounds that have antioxidant, anti-inflammatory and hypocholesterolemic effects. It also contains flavonoids, phenolic acids and peptides that indicates that amaranth can be used to prevent non-communicable diseases and healthy dietary practices. Even though some anti-nutritional compounds such as saponins, phytates, and oxalates are present, they can be alleviated by opting to use proper methods of processing foods. Besides nutrition- and functionality-related advantages, Amaranthus species are highly adaptable to various climatic environments thus good under stress conditions (Terzieva et al., 2026). Previous work highlighted multifunctionality of amaranth as a health-promising, nutrient-rich, and environmentally friendly crop aimed at fostering the establishment of a climate-resilient farming and diversification of any agriculture system (Terzieva et al., 2026) (Fig. 6.6).

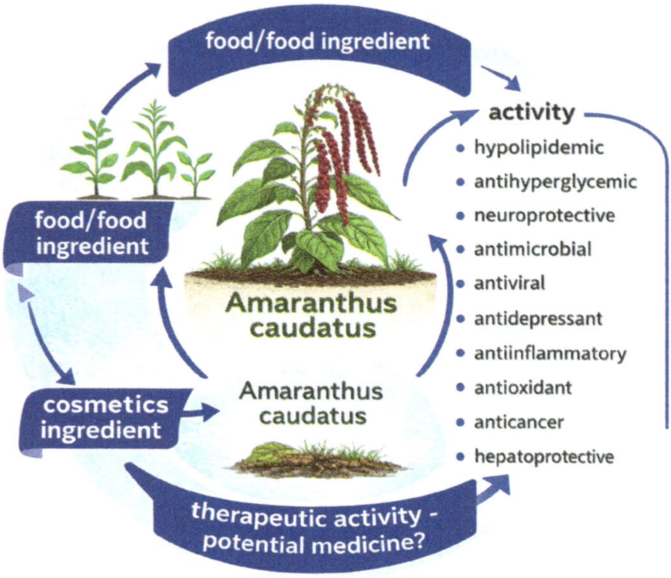

Fig. 6.6 Multiple benefits of amaranth (Terzieva et al., 2026)

6.8 Emerging Tools and Future Perspectives

6.8.1 Applications of Crop Modelling in Sustainable Amaranth Production

Crop modelling is an effective method to predict growth, development, and yield of crops that are not utilized across the globe e.g. amaranth (*Amaranthus spp.*). Through crop models, crop could be subjected to different environmental and management factors (Karimov et al., 2025). Models of phenology and biomass accumulation, including CROPGRO, LINTUL, and HYDRUS 1D, can be used to predict the effects of various agronomic practices, including planting density, the timing of irrigation, and regimes of fertilization without necessarily using time-consuming field experiments (Gautam & Khedkar, 2024). Moreover, crop models are used in planning sustainable production, assist in the combination of soil health management with productivity optimization, and predict crop performance during climate change (Karimov et al., 2025). Modelling frameworks would be especially useful in under-utilized crops such as amaranth, that typically have limited empirical data. This could give a scientific foundation of productivity, resilience, and ecological sustainability.

6.8.2 Crop Modelling and Irrigation Efficiency in Amaranth

Agroecosystem services are a major challenge to climate change, which limits agricultural activity as well as leading to scarcity of water in arid zones such as those in Uzbekistan (Table 6.3). The paper by (Karimov et al., 2025) is an evaluation of the performance of full (FDI) and a deficit (DDI) drip irrigation regime on amaranth cultivation in Tashkent based on a HYDRUS-1D simulation model. The model was validated using field experiments in two growing seasons alongside measuring soil moisture, analysis of root zone, and crop performance. The findings showed that although FDI generated better yields (2004 kg/ha), DDI yielded better in terms of water productivity (an increase of 56.5%) because of lower unproductive evaporations and deeper root penetrations, which is an indication of amaranth adaptation to water stress (Karimov et al., 2025). Strong correlations between simulated and observed soil water content ($R^2 = 0.862$ for FDI, $R^2 = 0.936$ for DDI) confirmed the model's reliability. The findings demonstrate that by incorporating accurate irrigation technologies with the HYDRUS-1D crop modelling it is possible to maximize water consumption and run sustainable amaranth production in arid areas under the pressure of climate extremes (Karimov et al., 2025).

Table 6.3 Water use of the Amaranth crop under full and deficit irrigation (Karimov et al., 2025)

Parameters	Index	Unit	Full drip irrigation (FDI)	Deficit drip irrigation (DDI)
Potential transpiration	Tp	cm^3/yr	41.58	25.18
Evaporation	E	cm^3/yr	4.23	2.63
Crop evapotranspiration potential	ETp	cm^3/yr	45.81	27.81
Precipitation	P	cm^3/yr	12.35	12.35
Irrigation, net	I	cm^3/yr	18.10	9.05
Transpiration	T	cm^3/yr	24.68	12.17
Actual crop evapotranspiration	Eta	cm^3/yr	28.91	14.802
Yield maximum	Ym	kg/ha	3000	3000
Yield actual	Ya	kg/ha	2004	1604
Water productivity (Ya/Eta)	WP	kg/m^3	0.69	0.108

6.8.3 Breeding and Rhizosphere-Focused Innovations for Soil Health

Breeding strategies targeting *A. caudatus* in the future, are being directed towards the enhancement of stress tolerance, nutrient-use efficiency, and positive rhizosphere interactions (Rastogi & Shukla, 2013). Deep and fibrous root systems, improved exudation phenotypes and symbiotic responsiveness to carry out the nutrient acquisition and the soil microbial activity could be the traits required in the targeted plant (Begum et al., 2019).The incorporation of rhizosphere-oriented characteristics into breeding pipelines are also agronomic and ecological in nature and it is in line with the objectives of sustainable and regenerative agriculture (Tedersoo, 2017).

6.8.4 Integration into Regenerative and Climate-Smart Agriculture

Planting amaranth in regenerative and climate-smart food systems has assisted in enhancing the health of the soil, biodiversity, and ecosystem benefits. Organic amendments, crop diversification, and conservation agriculture are all examples of sustainable management practices that promote carbon sequestration in the soil, greenhouse gas emissions reduction, and the resilience of the system to climate extremes (Lal, 2015). Amaranth grows well under marginal soil and extremes climatic conditions such as droughts, thus it is highly suitable crop to scale regenerative functionality within the parameters of smallholder and low-input integration systems (Yadav & Yadav, 2024).

6.8.5 Policy, Extension, and Scaling-Up Strategies

The ability of succession in fostering sustainable amaranth production in the regions and country-wise depends on the existence of effective policy frameworks and extension services. Climate-smart and regenerative practices can be adopted with the help of policies that conserve seeds, promote low-input sustainable agriculture, market access, and training of farmers (Pretty et al., 2018). Smallholder farmers can learn how to apply scientific knowledge into practice to improve productivity, profitability, and environmental sustainability by using demonstration trials, capacity building, and participatory approaches.

6.9 Conclusion

Amaranthus caudatus is a resistant, nutritionally valuable, and environmentally friendly crop, and it has great potential for low-input agriculture. It has high resource-use efficiency and potential to improve soil health. The rhizosphere interactions with the beneficial microbes also enhance nutrient uptake, soil enzyme activity, microbial diversity, and general soil ecosystem processes. Amaranth provides benefits to the ecosystem by boosting soil organic matter, improves carbon capture, improves soil structure, and aids in nutrient cycling, that all make agroecosystem resilient. In addition to the ecological advantages, it sustains the livelihoods of the smallholders, enhances food security, and offers economic feasibility on low-input management. The contemporary technologies, such as crop modeling, remote sensing, and breeding with a rhizosphere in mind provide the opportunities to maximize the use of resources, improve the sustainable production, and increase the scale of the regenerative practices. Altogether, the introduction of *A. caudatus* into agricultural systems is a complex solution to enhance productivity, soil health, and socio-economic performance as well as promote the development of climate-smart and regenerative agriculture. The further activities should be concentrated on the adoption scaling, rhizosphere efficiency, and technological methods of management to achieve the full potential of the crop in the food system in the form of sustainable food.

References

Aderibigbe, O., Ezekiel, O., Owolade, S., Korese, J., Sturm, B., & Hensel, O. (2022). Exploring the potentials of underutilized grain amaranth (Amaranthus spp.) along the value chain for food and nutrition security: A review. *Critical Reviews in Food Science and Nutrition, 62*(3), 656–669.

Anuradha, Kumari, M., Zinta, G., Chauhan, R., Kumar, A., Singh, S., & Singh, S. (2023). Genetic resources and breeding approaches for improvement of amaranth (Amaranthus spp.) and quinoa (Chenopodium quinoa). *Frontiers in Nutrition, 10*, 1129723.

Begum, N., Qin, C., Ahanger, M. A., Raza, S., Khan, M. I., Ashraf, M.,... Zhang, L. (2019). Role of arbuscular mycorrhizal fungi in plant growth regulation: Implications in abiotic stress tolerance. *Frontiers in Plant Science, 10*, 1068.

Brenner, D., Baltensperger, D., Kulakow, P., Lehmann, J., Myers, R., Slabbert, M., & Sleugh, B. (2000). Genetic resources and breeding of Amaranthus.

Chivenge, P., Mabhaudhi, T., Modi, A. T., & Mafongoya, P. (2015). The potential role of neglected and underutilised crop species as future crops under water scarce conditions in Sub-Saharan Africa. *International Journal of Environmental Research and Public Health, 12*(6), 5685–5711.

Chivenge, P., Zingore, S., Ezui, K., Njoroge, S., Bunquin, M., Dobermann, A., & Saito, K. (2022). Progress in research on site-specific nutrient management for smallholder farmers in sub-Saharan Africa. *Field Crops Research, 281*, Article 108503.

Falasca, S. L., Pitta Alvarez, S. I., & Bernabé, M. A. (2014). Agroclimatic Zoning for Amaranth Amaranthus CaudatusUsing a Geographical Information System in Argentina.

Gautam, K., & Khedkar, R. (2024). Amaranthus: Biology, functional potential and sustainable utilization: Pseudo Millets. In *Millets: The Multi-Cereal Paradigm for Food Sustainability* (pp. 177–231). Springer.

Geisseler, D., & Scow, K. M. (2014). Long-term effects of mineral fertilizers on soil microorganisms–A review. Soil Biology and Biochemistry, 75, 54–63

Iamonico, D. (2023). Nomenclature survey of the genus Amaranthus (Amaranthaceae): 12 questions about Amaranthus caudatus. *Plants, 12*(7), 1566.

Jimoh, M., Afolayan, A., & Lewu, F. (2019). Therapeutic uses of Amaranthus caudatus L.

Jimoh, M. O., Afolayan, A. J., & Lewu, F. B. (2019b). Germination response of Amaranthus caudatus L. to soil types and environmental conditions. *Thaiszia J. Bot, 29*, 85–100.

Karimov, A., Khalikov, B., Isaev, S., Sindarov, O., Khayitov, K., Avliyakulov, M.,... Muhammadov, Y. I. (2025). Efficiency of drip irrigation in amaranth production using the HYDRUS-1D model. *Frontiers in Sustainable Food Systems, 9*, 1612679.

Kumar, U., Raj, S., Sreenikethanam, A., Maddheshiya, R., Kumari, S., Han, S., & Gahlot, D. K. (2023). Multi-omics approaches in plant–microbe interactions hold enormous promise for sustainable agriculture. *Agronomy, 13*(7), 1804.

Lal, R. (2015). Restoring soil quality to mitigate soil degradation. *Sustainability, 7*(5), 5875–5895.

Liang, X., Yu, S., Ju, Y., Wang, Y., & Yin, D. (2025). Integrated management practices foster soil health, productivity, and agroecosystem resilience. *Agronomy, 15*(8), 1816.

Linnaeus, C. (1753). Species plantarum, vol. 2, (pp. 561–1200). *Stockholm, Sweden, Laurentius Salvius, Holmiae.*

Martínez-Núñez, M., Ruiz-Rivas, M., Vera-Hernández, P., Bernal-Muñoz, R., Luna-Suárez, S., & Rosas-Cárdenas, F. (2019). The phenological growth stages of different amaranth species grown in restricted spaces based in BBCH code. *South African Journal of Botany, 124*, 436–443.

Mäder, P., Fliessbach, A., Dubois, D., Gunst, L., Fried, P., & Niggli, U. (2002). Soil fertility and biodiversity in organic farming. *Science, 296*(5573), 1694–1697.

Mlakar, S. G., Turinek, M., Jakop, M., Bavec, M., & Bavec, F. (2010). Grain amaranth as an alternative and perspective crop in temperate climate. *Journal for Geography, 5*(1), 135–145.

Mng'omba, S. (2025). Grain amaranth, a potential and resilient food crop amenable to processing for diverse food and other products in Southern Africa. *Frontiers in Sustainable Food Systems, 9*, 1656596.

Mukuwapasi, B., Mavengahama, S., & Gerrano, A. S. (2024). Grain amaranth: A versatile untapped climate-smart crop for enhancing food and nutritional security. *Discover Agriculture, 2*(1), 44.

Muratova, A., Gorelova, S., Golubev, S., Kamaldinova, D., & Gins, M. (2023). Rhizosphere microbiomes of Amaranthus spp. grown in soils with anthropogenic polyelemental anomalies. *Agronomy, 13*(3), 759.

Murthy, N., & Kumar, J. A. (2017). Grain Amaranth (Amaranthus sp.)—An underutilized crop species for nutritional security and climate resilience. *Mesore Journal of Agricultural Science, 51*, 12–20.

Nazeer, S., Zubair Akram, M., & Ali, M. (2022). Amaranth as nutrition-rich and climatic resilient crop: A review. *Agriculturae Conspectus Scientificus, 87*(4), 281–293.

Pimentel, D., & Pimentel, M. H. (2007). Food, energy, and society: CRC press.

Pretty, J., Benton, T. G., Bharucha, Z. P., Dicks, L. V., Flora, C. B., Godfray, H. C. J.,... Morris, C. (2018). Global assessment of agricultural system redesign for sustainable intensification. *Nature Sustainability, 1*(8), 441–446.

Rastogi, A., & Shukla, S. (2013). Amaranth: A new millennium crop of nutraceutical values. *Critical Reviews in Food Science and Nutrition, 53*(2), 109–125.

Sage, R. F., & Zhu, X.-G. (2011). Exploiting the engine of C4 photosynthesis. *Journal of Experimental Botany, 62*(9), 2989–3000.

Singh, A., & Punia, D. (2020). Characterization and nutritive values of amaranth seeds. *Current Journal of Applied Science and Technology, 39*(3), 27–33.

Soler, C. T., Bado, V., Traore, K., Bostick, W. M., Jones, J., & Hoogenboom, G. (2011). Soil organic carbon dynamics and crop yield for different crop rotations in a degraded ferruginous tropical soil in a semi-arid region: A simulation approach. *The Journal of Agricultural Science, 149*(5), 579–593.

Stallknecht, G. F., & Schulz-Schaeffer, J. (1993). *Amaranth rediscovered* (pp. 211–218). New crops. Wiley.

Stetter, M. G., Zeitler, L., Steinhaus, A., Kroener, K., Biljecki, M., & Schmid, K. J. (2016). Crossing methods and cultivation conditions for rapid production of segregating populations in three grain amaranth species. *Frontiers in Plant Science, 7*, 816.

Tedersoo, L. (2017). Biogeography of Mycorrhizal Symbiosis.

Terzieva, S., Baycheva, S., Tzanova, M., Ivanova, T., Dimitrova, D., & Grozeva, N. H. (2026). Multifunctional edible amaranths: A review of nutritional benefits, anti-nutritional factors, and potential in sustainable food systems. *Foods, 15*(1), 130.

Travlos, I., Gazoulis, I., Kanatas, P., Tsekoura, A., Zannopoulos, S., & Papastylianou, P. (2020). Key factors affecting weed seeds' germination, weed emergence, and their possible role for the efficacy of false seedbed technique as weed management practice. *Frontiers in Agronomy, 2*, 1.

Tilman, D., Balzer, C., Hill, J., & Befort, B. L. (2011). Global food demand and the sustainable intensification of agriculture. *Proceedings of the national academy of sciences, 108*(50), 20260–20264.

Trivedi, P., Singh, B. P., & Singh, B. K. (2018). Soil carbon: introduction, importance, status, threat, and mitigation. In *Soil carbon storage* (pp. 1–28): Elsevier.

Vadez, V., Kholová, J., Hummel, G., Zhokhavets, U., Gupta, S., & Hash, C. T. (2015). LeasyScan: A novel concept combining 3D imaging and lysimetry for high-throughput phenotyping of traits controlling plant water budget. *Journal of Experimental Botany, 66*(18), 5581–5593.

Vessey, J. K. (2003). Plant growth promoting rhizobacteria as biofertilizers. *Plant and Soil, 255*(2), 571–586.

Yadav, A., & Yadav, K. (2024). From humble beginnings to nutritional powerhouse: the rise of amaranth as a climate-resilient superfood. *Tropical Plants, 3*(1).

Zaib, S., Zubair, A., Abbas, S., Hussain, J., Ahmad, I., & Shakeel, S. N. (2023). Plant Growth-Promoting Rhizobacteria (PGPR) reduce adverse effects of salinity and drought stresses by regulating nutritional profile of barley. *Applied and Environmental Soil Science, 2023*(1), 7261784.

Chapter 7
Canihua (*Chenopodium pallidicaule*): Adaptation to Marginal Environments and Food Security

Umar Farooq, Zameer Abbas, and Aashir Sameen

Abstract Canihua (*Chenopodium pallidicaule*) is a new climate-resilient pseudocereal with a strong reputation due to its high nutritional values and the capacity to grow under harsh conditions. The development of interest on sustainable and nutritious crops around the world has led to the need to know the agronomic behaviour of the crop, its adaptation capacity and the role it plays in food and nutrition security. This chapter will provide an overall review of the ecological, physiological, nutritional and yield performance of Canihua in different climatic settings, and its growing contribution in food and industrial purposes. Based on case-study evidence, canihua (*Chenopodium pallidicaule*) shows strong agronomic performance, resilience, and food-use potential. A field experiment in the Peruvian Altiplano recorded significant varietal differences in grain yield, with Pitojiura (976.1 kg ha^{-1}), K'ello (913.7 kg ha^{-1}), Chilliwa Rosada (886.8 kg ha^{-1}), and Puca (869.8 kg ha^{-1}) outperforming other varieties, while Cupi produced the lowest yield. Germplasm evaluations indicated that most accessions-maintained yield despite the presence of Eurysacca quinoae larvae and Peronospora disease, demonstrating inherent pest and disease tolerance. Food-based studies further confirmed the nutritional and functional value of canihua, including the development of elder-friendly mousse with antioxidant and anti-diabetic properties and fermented gluten-free bread with improved amino acid composition, mineral bioavailability, and antioxidant capacity. These findings emphasize the role of high-yielding varieties, tolerant germplasm, and value-added processing in strengthening canihua utilization.

Keywords Canihua · Food security · Marginal environments · Nutritional value · Climate resilience

---

U. Farooq (✉) · Z. Abbas · A. Sameen
Department of Agronomy, Pir Mehr Ali Shah Arid Agriculture University, Rawalpindi, Pakistan

M. Ahmed (ed.), *Climate Change and Neglected Underutilized Crops*,
https://doi.org/10.1007/978-3-032-21005-0_7

7.1 Introduction

Canihua (*Chenopodium pallidicaule*), also called kaniwa is a premodern Andean pseudo cereal, which was domesticated in the highlands of Peru and Bolivia. It is of the same botanical origin as quinoa and is still hardly used globally. It has been cultivated for a long time by indigenous populations residing above 3800 m where the tough environmental conditions do not permit the survival of the traditional food crops. Such a long history of domestication has made canihua a self-sustaining species that can survive in environments where most crops do not survive (Mangelson et al., 2019).

The growing interest in canihua in scientific and commercial circles in recent years is related to the growing demand to develop climate-resistant food systems. The species exhibits high levels of tolerance to abiotic stresses including frost, drought, salinity and soils with low amounts of nutrients. It also thrives in brief growth periods and degraded landscapes, which makes it a key crop to marginalized environments with climatic uncertainty, desertification or unsustainable cereal-based monocultures. Canihua is nutritionally defined as a high protein food with an ideal amino acid content, especially in lysine. Minerals, dietary fiber, antioxidants and bioactive compounds linked with better metabolic and cardiovascular health are also present in high levels in the grain. The natural lack of bitter saponins in the pericarp is also a significant benefit in comparison to quinoa, as it allows pericarp to be processed without having to undergo extensive post-harvest washing or even mechanical treatment. The attribute decreases the economic limitations faced by the farmers and makes consumers accept it more (Mavroeidis et al., 2022).

In addition to its physiological and nutritional benefits, canihua has great cultural value amongst the Andean rural societies. It has served in the past to provide food and livelihood security in areas where food production is limited by environmental misfortune. By increasing its growth, it would facilitate economic diversification and empower the smallholders, especially women, who usually take central roles in the process of production, processing, and marketing of traditional crops (Gomez Cahuata et al., 2022).

On the international scale, canihua can be compared with the modern trends of sustainable development. It can help make the diets healthier, the low-input farming systems more resilient, and the pressure on the increasingly susceptible ecosystems lower. Nevertheless, although it has its benefits, cultivation is still constrained by knowledge gaps, absence of better varieties, poor value chains and low research investment. The developments of modern breeding, remote sensing, and enhanced market connections are still in their early developments stages in comparison with older crops (Bruno, 2023).

The identification of canihua as an effective and viable resource entails the use of multidisciplinary strategy, which combines agronomy, nutrition, socioeconomics and technological innovations. The chapter will discuss its biological characteristics, environmental flexibility, nutritional and functional characteristics, market potential and food security. It also indicates the prospects of modelling and remote sensing that can speed up the spread of canihua into different parts of the world that are characterized by various agricultural operation (Bruno, 2023).

7.2 Botanical and Agronomic Characteristics

7.2.1 Taxonomy and Genetic Background

Canihua (*Chenopodium pallidicaule*) which is a member of the Amaranthaceae family is closely related to quinoa (*Chenopodium quinoa*) in the genus Chenopodium. It has been thought to be one of the oldest domesticated pseudo cereals of the Andean plateau. The crop possesses a high level of genetic diversity because of adaptation to strict mountainous conditions over a long period. The traditional systems of farming have preserved most of the local landraces with their desirable agronomic and nutritional characteristics (Fig. 7.1). Its genetic resources are still useful breeding programs that would improve stress tolerance, quality of grain and output in low input farming systems. Canihua has a smaller genome and lower metabolic rates of saponins than quinoa, which have led to its improved palatability and processing benefits (Mavroeidis et al., 2022).

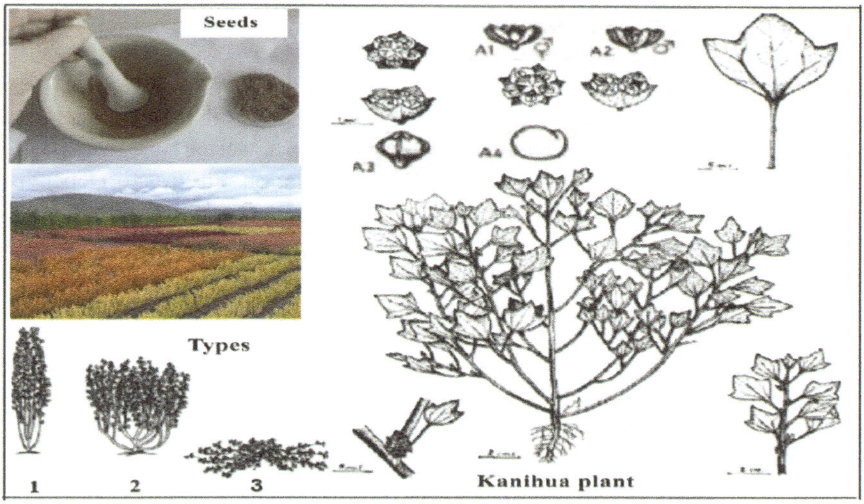

Fig. 7.1 Canihua plant (*Chenopodium pallidicaule Aellen*). Seeds (Photo: Gladys Moscoso-Mujica and Angel Mujica). Kañihua plant; A1 hermaphrodite flower; A2 flower masculine; A3 fruit; A4 seed. Types (plant growth habit): 1 Sayhua; 2 Lasta; 3 Pampa Lasta (Moscoso-Mujica et al., 2024)

7.2.2 Morphology

Canihua has a short stature, with most of the plants being shorter than quinoa. The plant has a height that is usually between 20 and 60 cm thus being more resistant to lodging in a high-altitude windy environment. The stem is straight and can be branched about genotype and nutrient content. Leaves are small, alternate and differ in shape between lanceolate and a little triangular with a green to reddish pigmentation determined by the amount of anthocyanin.

The infloescences are generated on the terminal or axial panicles, which are usually thick and tight. The flowering is apetalous (lacks a petal) and the reproductive system is wind pollinated (Fig. 7.2). The size of seeds is significantly smaller in comparison with quinoa, and the diameter of the seed coat may range between under 1 mm. and deep red or black color, depending on flavonoid and phenolic pigment content in the seed. The grains are smooth, do not contain a bitter pericarp and do not need to be subjected to the removal of saponin, which makes the process of harvesting, and post processing easy (Rodriguez et al., 2020).

7.2.3 Growth Habit and Photoperiod Sensitivity

Canihua has a versatile growth pattern that can be utilized in short growing periods and elevations. It is very freezing tolerant both at the vegetative and reproductive stages, and can survive down to 10 °C. This strength enables it to be planted earlier compared to majority of other crops. The species is moderately sensitive to photoperiod with most of the landraces being adapted to day lengths that are characteristic of highlands in the equators. Genotypes greatly differ in flowering time, which helps them adapt to the ecological environment at varying elevation and climate. Depending on the traditional production conditions, the crop has a life cycle of 120 to 150 days but better lines take shorter time to mature. It can yield very well under low irrigation and low and low inputs of nutrients, which makes it adaptable to resource-poor agricultural systems as well as stress-prone areas. (Mangelson et al., 2019).

Canihua is a nutritional food that makes it a functional food. The seeds also have quality proteins that have balanced essential amino acids and especially lysine, which is usually reduced in cereal grains. Lipid fraction contains the healthy unsaturated fatty acids with the carbohydrates being mainly made up of low digestible starches, which also make the glycemic response lower. Canihua contains a natural zero level of bitter saponins as opposed to quinoa where it is a major post-harvest demerit. Other significant concentrations of minerals in the grain include iron, calcium, magnesium and zinc, which is particularly useful to populations who have micronutrient deficiencies (Table 7.1). Bioactive compounds such as flavonoids, betalains, phenolic acids, and dietary fiber that adds to its health-promoting benefits improve its antioxidant activity. There is a positive relationship between dark seed coat pigments

Fig. 7.2 Morphological growth habit variation in semi-domesticated and wild cañahua plants, **a** Last'a (semiprostrate growth plant with several branches), **b** Saiwa (erect growth type with fewer branches and upright), and **c** Pampa Last'a or Illamanku (plant with many branches and pros trate shape), **d** close up of little flowers with stamens in the small panicle, **e** fruits with peri carp, **f** achene or seed with colored perisperm, and **g** diversity in colors showing two differentiated plant morphotypes (Rodriguez et al., 2023)

and the higher polyphenol level content and provides an additional nutritional and technological significance (Gomez Cahuata et al., 2022).

Table 7.1 Proximal chemical composition of Cañihua (g/100gdry weight) (Gomez Cahuata et al., 2022)

Variety	Protein	Fat	Carbohydrates	Crude fibre	Ashes
NR	15.3	7.8	66.4	7.0	3.5
Cupi	14.4	5.7	63.6	11.2	5.0
Ramis	14.9	7.0	65.7	8.2	4.3
NR	15.4	6.2	73.1	6.1	2.6
Ilpa INIA 406	18.8	7.2	64.6	7.1	4.1
NR	14.4	7.9	67.2	–	3.5

7.3 Phenology of Canihua (Chenopodium pallidicaule)

Canihua has a phenological pattern of high adaptation to the agro climatic conditions of the Andeans highlands, which are harsh. The crop is relatively short seasonal, and it is well adapted to short-season farming and areas where it is likely to be affected by early frost. It has the normal growth stages of germination, vegetative growth, flowering, grain filling, physiological maturity, but the length of each developmental period is different in various areas with variation in altitude, temperature and soil fertility (Fig. 7.3). The seedling development is usually fast, once sown and this facilitates good establishment even when there is low moisture content (Quevedo-Olaya et al., 2025).

There is a significant phenological plasticity of the crop especially under temperature of elevation. In the high altitude, the vegetative and reproduction period is prolonged due to the cooler temperatures, at lower altitudes where the temperatures are warmer; the crop is quicker to reach maturity. The conditions of the habitat, which typically result in optimal development, are cold and semi-arid conditions with controlled photo thermal cues that control the growth progression. The capacity

Fig. 7.3 Phenology of Canihua (*Chenopodium pallidicaule*) during growing (Rodriguez et al., 2023). **a** Germination, **b** two leaves, **c** branching, **d** branches with inflorescences, **e** small flowers distributed in small panicles, **f** seed shattering, **g** grains (achenes), **h** plant in physiological maturity, **i** manual harvest, **j** drying of plants for manual threshing

of canihua to survive the lowest temperatures at the developmental phases, such as initial growth and grain filling, allows cultivating the product with high productivity even on mountainous land with low quality (Scherberger et al., 1975).

Even under environmental pressure like drought, exposure to wind and low soil fertility, flowering and seed formation are reliable. The flowers are also normally inconspicuous and self-pollinating, which guarantees consistent reproductive success during times of pollinator limitation. Nevertheless, the probability of partial cross-pollination can arise in good conditions, which gives a possibility of genetic variation and local adaptation. The flowering to grain filling is effective, which helps to ensure high seed set and quality even in the extreme climate changes (Patiño Fernández & Cruz Mamani, 2025). Together with the phenological traits of canihua, such as brief crop duration, frost-resistance, stress-induced reproductive stability, and adaptability to changes in altitude, the phenological traits of canihua can be classified as one of the most climate-resilient pseudo cereals, which are grown in high-altitude environments (Bruno, 2023).

7.4 Adaptation to Marginal Environments

7.4.1 Performance Against the Biotic Stresses

Canihua is highly resistant to biotic stress and its grain productivity does not fail when subjected to significant pests and diseases. Recent analyses of Eurysacca quinoae larval infestations and Peronospora species infections were not frequently damaging, and majority of accessions were still able to produce normally in field conditions (Fig. 7.4). This tolerance is facilitated by morphological and physiological characteristics, which include decreased foliar area, compact growth form, and irregular architecture of the plant, which assist in limiting the effects of insects and pathogens on the essential structures, including the panicle. Overall, existing information suggests that canahua has intrinsic biotic-stress resilience in a variety of accessions and can therefore continue to be productive despite the presence of pests and diseases (Rodriguez et al., 2023).

7.4.2 Performance in Drought, Frost, Salinity, and Poor Soils

Canihua also shows extraordinary ability to withstand severe abiotic conditions enabling it to be produced on a long-term basis in conditions that are not favorable to the majority of cereal crops. It does not survive drought because of its ability to control stomata, deep rooting and high water-use efficiency. It is also highly frost resistant, and can be kept in frosty environments without experiencing any serious yield loss in its early growth or reproductive life cycle. Also, canihua is highly tolerant

Fig. 7.4 Pest and diseases reported in the field, **a** and **b** moth larvae, "Kcona-Kcona" (Eurisacca spp.) attacking in the infrutescences, **c** noctuid moth larvae, "Ticona" (Helicoverpa spp.), **d**, **e**, and **f** leaves infected with downy mildew (Pheronospora spp.) (Rodriguez et al., 2023)

of soil salinity and alkalinity by selective uptake of ions and osmotic adaptation. It develops in soils that are deficient in nutrients, those with low organic content and low fertility, and is an invaluable crop to marginalized agricultural systems (Rodriguez et al., 2023).

7.4.3 Physiological Tolerance Mechanisms

Physiological characteristics of stress tolerance have been shown to be the buildup of protective secondary metabolites, antioxidative defenses and preservation of membrane stability during the occurrence of cold and drought. Canihua balances the osmotic pressure and aids in avoiding cellular injuries in a salty and dehydrating environment. It has an enhanced root-to-shoot communication, efficient nutrient uptake and adaptive morphology that helps it to survive in degraded soils and changeable climatic situations (Limachi et al., 2023).

7.4.4 *Soil-Climate Suitability Mapping and Agro Ecological Zones*

Agro ecological evaluations show that canihua is adapted naturally to the high altitude, cold, and semi-arid plateaus, especially the Andean highlands that are above 3800 meters. Nevertheless, according to environmental suitability models, the area where it is grown may expand to other climatic conditions with more land degradation. Solar-climatic mapping suggests that it can expand to areas with a risk of frost, intrusion of saltwater, and lack of water, and grow into new areas of agricultural diversification in areas where the technique has traditionally been practiced (Merida-Lopez et al., 2024).

7.4.5 *Comparative Resilience vs. Quinoa and Cereal Crops*

Canihua is usually more resistant to stress, particularly in cold and dry climates as compared to quinoa, wheat, barley, and maize. It also has a stable reproduction success even when temperatures and moisture are extreme unlike most cereal crops, which lose a lot of productivity. This comparative advantage makes canihua effective as one of the stable elements of climate-resilient agricultural systems (Bonifacio-Callisaya & Bonifacio-Flores, 2022) (Fig. 7.5).

7.4.6 *Potential for Expansion into South Asia, MENA Region, and Sub-Saharan Africa*

New climate issues such as water shortage, salinization of the soil, and temperature changes have rendered canihua an excellent candidate to be implemented in South Asia, the Middle East and North Africa (MENA) and Sub-Saharan Africa. The areas are especially susceptible to food insecurity because of the erosion of the environment and the lack of variety of crops. The low input needs, stress adaptability and nutritional value of Canihua make it very effective in resilience building of the low-input smallholder systems (Fig. 7.6).

7.5 Nutritional and Functional Properties

One of the grains with the highest content of nutrients in the Andes has been known as canihua (*Chenopodium pallidicaule*). Its special nutritional characteristics and functional food nature make it very appropriate to the traditional and the contemporary health-conscious markets. The grain has outstanding protein quality, great mineral

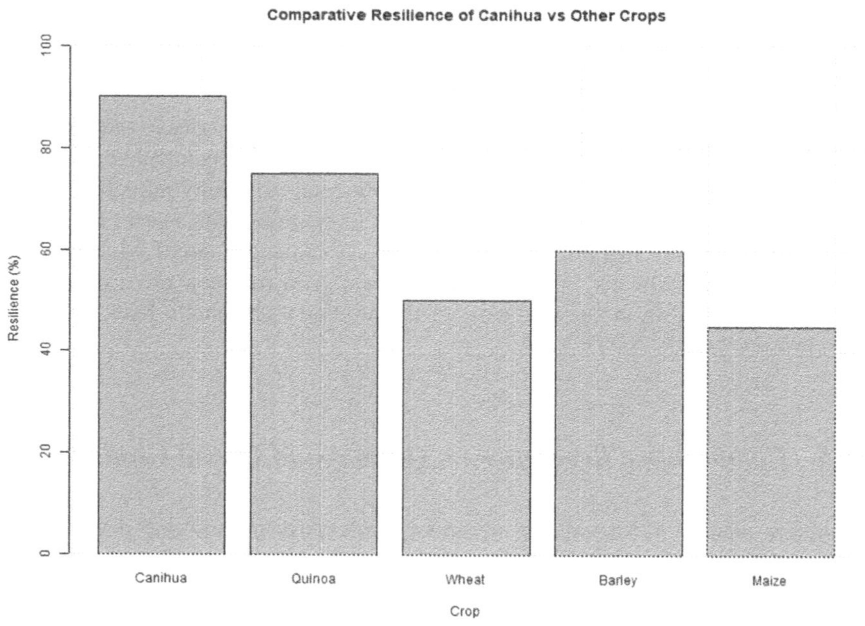

Fig. 7.5 Resilience of Canihua versus Quinoa and Cereal crops (Bruno, 2023)

density, bioactive substances, and significant technological properties that underpin a variety of food processing processes (Mérida-López et al., 2023).

7.5.1 Protein Quality and Amino Acid Composition

Canihua is having a good balance of essential amino acids with special attention to lysine that is normally lacking in cereal grains like wheat, rice and maize. It possesses a high protein digestibility and the amino acid content concurs with the recommended human dietary needs. Based on this better nutrition balance, canihua can be used as a good source of protein among vulnerable groups, such as children and people with a low dietary diversity (Peñarrieta et al., 2008) (Fig. 7.7).

7.5.2 Dietary Fiber and Mineral-Rich Composition

The dietary fiber contained in the grain also helps in enhancing the overall gastrointestinal health and delayed glucose release, which are effective in metabolism maintenance. Canihua is greatly appreciated due to high levels of minerals such as iron,

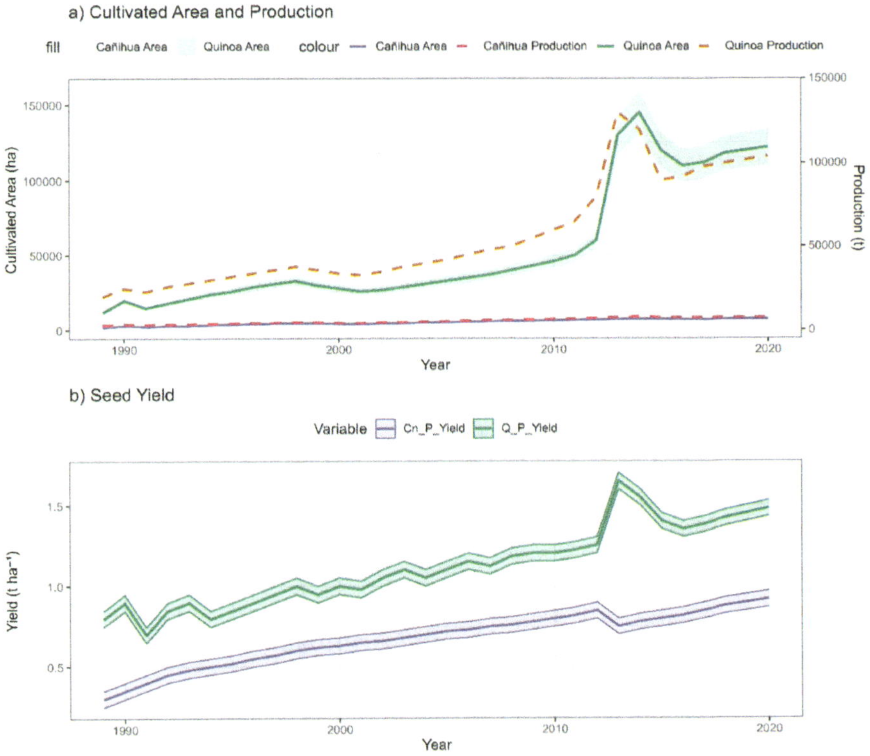

Fig. 7.6 Cultivated area, production (**a**) and yield (**b**) of cañahua in Peru and Bolivia (Rodriguez et al., 2023)

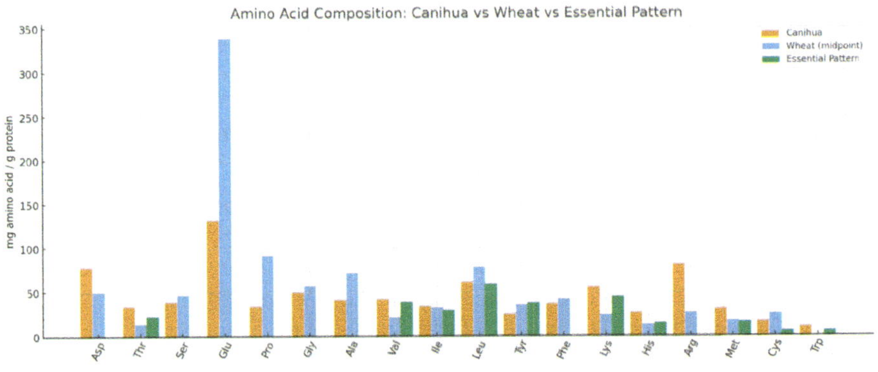

Fig. 7.7 Amino acid composition of cañihua and wheat grains (mg amino acid/g protein) (Gomez Cahuata et al., 2022)

Table 7.2 Nutrition facts for Cañihua seeds (Healwith food) (Mérida-López et al., 2023)

Macronutrients	Vitamins	Minerals
Protein: 15.4 g (31%)	Thiamin (B1): 0.7 mg (48%)	Calcium: 134 mg (13%)
Crude fat:6.2 g	Riboflavin (B2): 0.4 mg (25%)	Iron: 15 mg (86%)
Carbohydrates: 61 g (20%)	Niacin (B3): 1.4 mg (7%)	Phosphorus: 424 mg (42%)
Crude fiber: 7.7 g (31%)		

calcium, magnesium and zinc that are concentrated there. These are nutrients that help in maintaining bone, providing oxygen, and the immune system and make canihua a functional choice in areas where micronutrient deficiencies are common (Limachi et al., 2023). Since kaniwa is a unexplored food, there is little research examining the nutritional value of the food. This means that the nutrition facts of kaniwa can only be found on some nutrients. Table 7.2 gives the data regarding the approximate nutrient composition of kaniwa not only in absolute (in grams or milligrams) but also in Percent Daily Values (in brackets). One hundred grams of kaniwa seeds (3.5 oz) determine the values (Mérida-López et al., 2023).

7.5.3 Polyphenols, Flavonoids, and Antioxidant Potential

Canihua is also rich in amounts of polyphenolic compounds and flavonoids with much of it being found in the seed coat, which give it a high antioxidant capacity. These biomolecules are significant in the mitigation of oxidative stress and preventing disorders associated with inflammation. It also has added natural pigments, including dark anthocyanin-rich pericarps in certain cultivars that increases its functional value of health promotional foods (Serena-Romero et al., 2023).

7.5.4 Low Saponin Content and Sensory Acceptability

In contrast to quinoa, canihua contains very low levels of saponin naturally, which means that it does not require laborious washing or rough post-harvest treatment of canihua. This attribute enhances customer acceptance since it has a more neutral and sweet taste. Low processing needs also make it more economically viable to smallholder communities and less use of water in Andean arid zones (Huamán Tapara et al., 2024).

7.5.5 Functional Properties for Food Applications

Canihua has desirable techno-functional characteristics such as good water absorption, emulsifying ability, soldify, as well as foaming. Its properties are suitable to be used in baked products, infant foods, gluten-free formulations, fermented drinks and nutraceutical foods. The product has a fine grain size that allows it to be milled into flour easily, which contributes to a wide range of product development opportunities in both the traditional and industrial food industries (Gomez Cahuata et al., 2022).

7.6 Food Application and Technological Potential

As a dietary component, Canihua has long been an important food crop in the Andean highlands, which has provided dietary diversity, as well as a source of nutrition security. Its healthy nutrition, low processing needs, and functionality make it versatile about the conventional food systems, as well as the contemporary industrial use. Due to the increasing popularity of climate-resilient and health-promoting grains, canihua has promising innovations to be implemented in the food industry (Quispe et al., 2024).

7.6.1 Traditional Andean Food Products

Canihua is also made in different culturally meaningful ways in its indigenous territories. The grain is roasted and milled into a very nutritious flour called kaniwaco used to prepare porridges, drinks, and high-energy foods to children and workers. Popped canihua is a snack that is cooked like puffed cereals and it is the snack commonly eaten on festivals and at meals. Canihua is a dietary protein and micronutrient-enriched fermented drink and soup in rural areas (Luque-Vilca et al., 2024).

7.6.2 Modern Processing and Product Diversification

Cannihua is gaining commercial appeal, as it is gluten-free and of very high nutritional value. Fine milling is done to give flour, which is used to make bakery products, extruded snacks, breakfast cereals, infant food formulations and ready-to-cook mixes. Its amino acid composition may be considered as the benefit of the grain going into protein fortification in special diets like vegan, high-protein, and food that can be consumed by diabetics. In addition, bioactive compounds enable canihua to be applied in nutraceuticals, functional drinks, and health-focused supplements to aid in antioxidant and metabolic wellbeing (Zegarra et al., 2019).

7.6.3 Technologies for Improving Digestibility and Sensory Quality

A number of processing technologies are being put into use to make canihua more palatable and digestible. Controlled roasting enhances flavor and anti-nutritional factors besides maintaining nutrient density. To enhance the digestibility of proteins, bioavailability of minerals, and texture to enhance widespread consumer acceptance, soaking, germination, and extrusion are becoming more frequent. The inherent low saponin levels reduce bitterness and do away with harsh washing, which means that canihua is more appropriate in areas with little water resources (Kim & Iida, 2023).

7.6.4 Shelf-Life and Processing Quality Enhancements

Canihua has good oxidative stability during storage since it contains a high percentage of polyunsaturated fatty acids and natural antioxidants. Nevertheless, moisture and environmental pollution defense is crucial in preserving food grains. The packaging, dehulling and heat-stabilizing procedures are supporting the enhancement of shelf life and minimization of post-harvest losses. The basic requirements will be quality control systems and standardized processing specifications, which will enable the scaling of canihua into competitive global markets (Mankotia et al., 2025) (Fig. 7.8).

7.7 Contributions to Food Security and Sustainable Diets

7.7.1 Role in Nutrition Security for Low-Income Communities

Canihua is a fortified pseudocereal that has potential to significantly increase the quality of the diet of low-income, rural communities. Its seeds are a source of good quality protein (usually 15–19%) with a balanced essential amino acid composition particularly lysine balancing cereal-based diets, which lack some essential amino acids. Canihua can be used as a resource-poor household to enhance the biological value of daily protein consumption and promote growth and recovery after undernutrition (Moscoso-Mujica et al., 2024).

Fig. 7.8 Traditional uses of Canihua (*Chenopodium pallidicaule*) (Rodriguez et al., 2023). **a** Local seed grinding with a grit stone, **b** flour from toasted grain, "pito" in Bolivia and "cañihuacu" in Peru, **c** "llip't'a," complement product for chewing coca leaves, **d** "K'ispiño" traditional dumplings of quinoa and cañihuaco in plastic bags, **e** bread with cañihuaco, **f** cake, **g** muesli, **h** muffing, and **i** pancake

7.7.2 Suitability for Food Fortification Programs

Canihua flour possesses functional properties such as good water-binding, emulsification and gelling capacity and thus it can be used as a vehicle to fortify and formulate complementary foods. It has a relatively low saponin content in most cultivars, which lowers processing burdens and enhances palatability, which facilitates easier incorporation of institutional feeding and mass fortification programs. Further processing techniques like germination, fermentation, and extrusion, increase nutrient bioavailability and minimize anti nutritional effects, making it more useful in child-feeding, maternal nutrition, and products that are elderly-friendly products (Peñarrieta et al., 2008).

7.7.3 Addressing Malnutrition and Micronutrient Deficiencies

Canihua is a source of reasonable amounts of bioavailable iron, zinc, calcium and magnesium, which are often lacking in the populations that are at risk. It is possible to add canihua to regular diets to decrease incidences of anemia and other micronutrient-related illnesses without having to depend on imported supplements. Moreover, polyphenols and antioxidant compounds present in it add quality to the nutritional diet and can have ancillary health effects that are applicable to both undernutrition and the double burden of malnutrition (Mérida-López et al., 2023).

7.7.4 Climate-Resilient Diversification in Smallholder Systems

In addition to the direct nutrition benefits, canihua elevates food security through the agro ecological resilience. The crop also endures drought, frost, high radiation and bad soils, which restrict most of the staples and hence provides a steady production solution to the smallholder farmers in the marginal environments. The required low input levels enable resource-poor farmers to adopt it, whereas its integration with other crops enhances agrobiodiversity and it minimizes the economic risk of complete crop failure during a climatic stress. Canihua, as a locally produced and low-cost source of nutrients, contributes to the food supply in households as well as income diversification (Kim & Iida, 2023).

7.8 Environmental Sustainability and Ecosystem Benefits

7.8.1 Low Water and Fertilizer Input Requirements

Canihua (Chenopodium pallidicaule) is very acclimatized to marginal conditions and does not need a lot of external inputs to be cultivated successfully. It has a natural ability to survive drought and low fertility soils which means that it can grow with minimal irrigation and minimum use of fertilizers. This feature renders canihua especially appropriate to resource-limited farmers and water-scarce areas so that learners can rely less on expensive irrigation systems and synthetic manure. Canihua helps to make agricultural activities more sustainable, as it needs fewer inputs, and reduces environmental strain metrics linked to intensive crop production (Moscoso-Mujica et al., 2024).

7.8.2 Biodiversity Conservation and Agro Ecology Benefits

Canihua is also a conventional Andean pseudocereal that is rather significant in terms of agrobiodiversity conservation. It is grown alongside other native crops and this makes the agroecosystems resilient as well as diverse. Genetic diversity can be beneficial in the resistance to pests, diseases and climatic variations that contribute to the long-term sustainability of the ecological system (Rosales-Espinoza et al., 2022). The positive aspects of canihua to mixed cropping systems include that it maintains cultural heritage, promotes agro ecological systems, and the local varieties of crop varieties, which is needed in sustainable food systems.

7.8.3 Soil Health and Carbon Sequestration Potential

Canihua has a positive effect on the health of soil, as it has low external nutrient requirements, reduces disturbances caused by soil, and provides organic matter. It has better root system that increases soil structure, water infiltration, and erosion, especially on the degraded highland fields. Vicia sativa cover crops and organic fertilizer (guano) are all crop management strategies that can enhance soil fertility and yield in C. pallidicaule. Such practices can also contribute to an increase in the carbon sequestration of soil, and canihua can be seen as a good option in climate-smart agriculture.

7.8.4 Reduced Ecological Footprint Compared to Conventional Cereals

Compared to major cereals like wheat or maize, canihua has a lower ecological footprint. Its tolerance to abiotic stresses reduces the need for chemical fertilizers, pesticides, and irrigation, thereby lowering energy inputs and greenhouse gas emissions. Because it can grow on marginal soils unsuitable for conventional crops, canihua helps avoid land-use competition and minimizes environmental degradation.

7.9 Socio-economic Perspectives

7.9.1 Market Potential and Value Chain Development

Canihua has a potential of a strong niche-market since it has the characteristics of functional food, it is gluten-free, and has high nutrient content. Its products are marketed in local consumption and international specialty market. They are in the process of developing value chains, small-scale processors are manufacturing flour, snacks, and fortified products.

7.9.2 Role in Rural Livelihood Improvement

The production of canihua gives subsistence and income to the smallholder farmers, especially in the low-fertile or high-altitude areas. With the creation of a durable and enhancing crop, households will have surplus to sell to strengthen food security and economic support.

7.9.3 Women Participation and Community-Based Production

Women tend to dominate canihua production, processing and marketing particularly in cooperative or community-based context. These models promote gender equity, traditional knowledge and resilience in the household (Rosales-Espinoza et al., 2022).

7.9.4 Barriers: Seed Supply, Awareness, Policy Support, Commercialization Gaps

Even though it has its positive side, the use of canihua is hampered by inaccessibility to quality seed, lack of awareness about the health benefits, low policy, and ineffective market infrastructure.

7.9.5 Case Studies from Andean Production Systems

Canihua is sensitive to rising temperature and humidity, with studies in Peru indicating sensitivity of canihua to climatic factors, which needs climate-adaptive management. ISSR markers of genetic variation help in conservation and breeding plans of Puno-region accessions (Rosales-Espinoza et al., 2022). In Bolivia and Ecuador, the income of the household has risen and the food products rich in nutrients are more accessible due to cooperative-based production.

7.10 Modelling and Remote Sensing for Assessment and Scaling

7.10.1 Crop Growth Modelling for Yield Prediction and Climate Adaptation

Crop growth models are becoming more important in the prediction of crop yield such as canihua particularly in conditions of climate variability. These models recreate growth in various agro ecological conditions by integrating physiological information, phenology, and environmental condition and aid in estimating yield response to environmental stresses e.g. drought or frost. They are also useful in identifying the best genotypes and planting date of the smallholder systems, hence making adaptive agronomic decisions.

7.10.2 UAVs and Satellite Remote Sensing for Stress Monitoring

With help of remote sensing using satellite and UAV systems, there can be continuous monitoring of the health of crops and thus the detection of stressors early (such as water deficit or nutrient imbalance). These systems have the ability to observe the canopy growth, biomass gain, and phenological phases and therefore allow the implementation of specific, precise, and timely interventions and less input.

7.10.3 Geographic Suitability Mapping and Digital Agronomy

Machine learning combined with GIS-based suitability modelling can provide highly effective predictive mapping of lands that can be utilized to grow canihua. Suitability models can be used to outline areas where canihua can possibly grow by combining soil, climate and topographic data with environmental variables derived by satellites.

7.10.4 *Use of AI in Expansion Planning for Marginal Regions*

Machine learning and, in particular, artificial intelligence (AI) are currently applied to uncover multi-dimensional, large datasets (climate, soil, crop performance) to forecast where crops can be expanded. According to recent reviews, yield prediction and land suitability are tied to the use of machine learning and remote sensing with great possibilities of scaling neglected crops.

7.11 Case Studies

7.11.1 *Case Study 1: Yield Trial in the Andes, Peru*

A field experiment was done on ten canihua (*Chenopodium pallidicaule*) varieties that had been grown at the Agrarian Experimental Station ILLPA in Puno by Huaman Tapara, Flores Huarco, Valverde Rodriguez, and Jara Claudio. The research was carried out between October 2018 and April 2019 and consisted of a randomized complete block study having four blocks and fifty experimental units with an altitude of 3750 m above the sea level. The varieties that were tested were Pitojiura, K'ello, Puca, Chilliwa, and Cupi among others. Their results revealed a high difference in the yield of grains: Pitojiura variety performed the highest yield of 976.1 kg/ha, then K'ello (913.7 kg/ha), Chilliwa Rosada (886.8 kg/ha) and Puca (869.8 kg/ha); Cupi recorded the lowest yield of 477.4 kg/ha, which was about 50% lower than that of Pitojiura variety. The intermediate at the year of production was toncco q'ello at about 684.2 kg/ha. These variations underscore the need of varietal selection to help in maximizing yield in high altitude marginal environments. The authors suggest that increased emphasis should be placed on local breeding initiatives using high yielding landraces such as Pitojiura and K'ello in enhancing food production in the Andean smallholders.

7.11.2 *Case Study 2. Pest and Disease Resilience in the Peruvian Altiplano*

A study by Betsabe Leon Ttacca, Bravo Portocarrero, Llanos Nina, Leon Tacca and Medina Espinoza tested variations of key insect pests and disease severity by 27 accessions of a germplasm bank (Camacani Experimental Center, Universidad Nacional del Altiplano, and Puno) and three commercial varieties in the Peruvian Andean Altiplano. Three development stages of the crop were followed through a season of a crop to determine the levels of pests, especially the Eurysacca quinoae larvae, and the disease, including the Peronospora (mildew). They discovered that rather than the economic damage threshold (6 larvae/plant) being met, the larvae in the accession (up to 5.5 larvae/plant in accession 03 21 26, 4.5 in variety Rami's) were below the level at which economic damage would occur. Peronospora was the worst disease; some Accessions (03-21-267, 03-21-218, 03-21-124, 03-21-26, 03-21-64) had high disease area under the disease progress curve but others (03 21 03, 03 21 24) were not affected. This inherent resistance/tolerance in many of the accessions and varieties was observed even though the overall grain yield did not reduce significantly despite the presence of pests and pathogens. The authors imply that these tolerant lines may be prioritized to be bred and integrated pest-management practices designed with respect to the resistant cultivars may be used to reinforce canihua production within the Altiplano.

7.11.3 Case Study 3: Elder-Friendly Canihua Mousse in Japan

Researchers Dah Sol Kim and Fumiko Iida have published a study in Tokyo, Japan (Kaniwa (*Chenopodium pallidicaule*) its Nutritional Composition and its Applicability as an Elder-Friendly Food with Gelling Agents). They also examined the nutritional character of canihua and experimented in creating a mousse that would be appropriate to the elderly population according to the Korean Industrial Standard (KS) of food hardness. They obtained regression equations with the help of various gelling agents (guar gum, locust bean gum, xanthan gum) in order to acquire three hardness levels of tongue, gum, and teeth chewing. The findings indicated that canihua mousse containing a concentration of less than 1.97% guar gum (or corresponding proportions of other gums) was capable of fitting into the tongue-chewable texture; greater concentrations allowed it to be used in gum or teeth chewing. The mousse made of canihua was nutritionally favorable (good in unsaturated fat), had antioxidant activity, and was anti-diabetic. The paper offers a promising avenue of implementing canihua in elderly nutritional programs, particularly in the aging societies. Future studies are required in consumer acceptance, and scale-up of product development in old age people in different cultural environments.

7.11.4 Case Study 4: Gluten-Free Fermented Canihua Bread in Peru

In Lima, Peru, Genny Isabel Luna Mercado developed and optimized a gluten-free bread made of fermented dough with chia seed as canihua (Chenopodium pallidicaule). The experiment compared the techno-functional, thermo-mechanical and physicochemical characteristics of two types of canihua (Cupi and Illpa) and potato starch. Optimization was done by using 7% chia replacement, as well as sourdough starter (*Lactobacillus plantarum*), which gave the bread a specific volume of 2.96 cc/g. The product shown had better amino acid composition, mineral bio accessibility, and antioxidant properties than unfermented canihua. The authors recommend that such findings can be used to scale up bakery production in community or commercial environments. Subsequent efforts ought to be directed towards the validation of rheological stability and shelf life under actual commercial environments, as well as how consumers to be taken on a routine basis in both local and international markets can accept the products.

7.12 Correlation Heatmap of Canihua

The Canihua (*Chenopodium pallidicaule*) correlation heatmap contains a visual representation of the relationship between its important nutritional, functional, and agronomic traits. There were also strong positive associations between protein content, essential amino acids and dietary fiber indicating the complementary relationship of these characteristics in improving the nutritional value of the grain. Protein and fiber content were also significantly positively correlated with various minerals (iron, calcium and magnesium) indicating the potential of Canihua as a nutrient-dense functional food. On the other hand, the negative or weaker correlations were found among fat content and certain carbohydrate and fiber parameters, which shows that there was a differentiation between fat and other macronutrients in the seed. By and large, the heatmap provides a clear and concise picture of the interaction between different traits to enable focused breeding, food fortification and value-added product development. The visualization can also be an effective method enabling the researchers and policymakers to find out the optimal trait combinations that will yield the most nutritional advantages and functional characteristics of Canihua (Fig. 7.9).

7.13 Conclusion

Canihua (Chenopodium pallidicaule), a neglected Andean pseudo cereal, and offers a good chance to enhance food security, sustainable diets and a resolute agricultural system in marginal environments. Its outstanding nutritional value in terms of high quality protein, essential amino acids, dietary fiber and bioavailable minerals make it a strategic crop to counter malnutrition and micronutrient deficiencies, especially in the low-income and resource-starved populations. The practical attributes of the grain, which together with contemporary food processing methods have introduced the grain into a multiplicity of food items, such as gluten-free bakery, aging-friendly formulations, and fortified foods, make it more palatable to the various population groups.

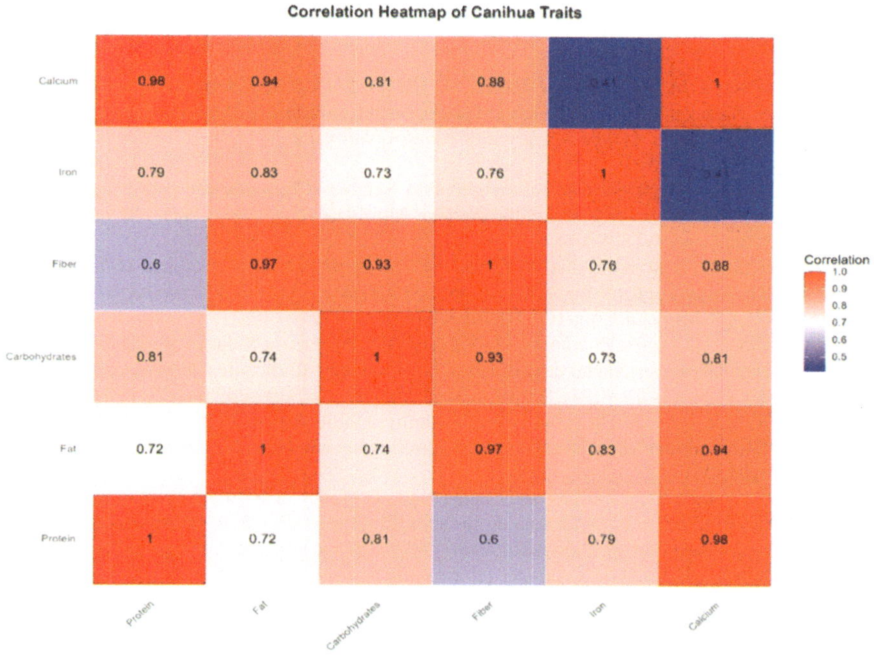

Fig. 7.9 Correlation heatmap of Canihua traits

The organism has a natural ability to withstand abiotic stresses (drought, frost, salinity, low-fertility soils, etc.) which justify its use in the development of high-altitude and semi-arid areas with harsh climates. Smallholder agroecosystem integration leads to agro ecological diversification, soil health and biodiversity, as well as reducing ecological footprint compared to conventional cereals. Socio-economic evaluation shows that cultivation of canihua sustains rural livelihoods, as women significantly contribute to the production system of the community and there are new market opportunities because of value chain development. The opportunities to apply sitespecific adaptation, maximization of yield, and expansion into non-traditional areas are further increased by the developments in the area of digital agronomy, remote sensing, and crop modelling with the support of AI-based planning. Strategic In order to maximize the potential of canihua, strategic there should be investment in breeding programs, policy support, better seed systems and knowledge dissemination. All of these properties make canihua a climate-resilient, nutritious, and environmentally sustainable crop and have enormous potential in terms of both food security and sustainable development around the world.

References

Bonifacio-Callisaya, M., & Bonifacio-Flores, A. (2022). Selección y evaluación agronómica en morfotipos Saiwa y Last 'a de qañawa (Chenopodium pallidicaule Aellen). *Revista de Investigación e Innovación Agropecuaria y de Recursos Naturales, 9*(3), 16–27. https://doi.org/10.53287/nuuh5152mg75i

Bruno, M. C. (2023). Archaeobotanical insights into Kañawa (Chenopodium pallidicaule Aellen) domestication: A rustic seed crop of the Andean Altiplano. *Agronomy, 13*(8), 2085. https://doi.org/10.3390/agronomy13082085

Gomez Cahuata, J. F., Rosas-Quina, Y. E., & Pachari Vera, E. (2022). Cañihua (*Chenopodium pallidicaule* Aellen) a promising superfood in food industry: A review. *Nutrition & Food Science, 52*(6), 917–928. https://doi.org/10.1108/NFS-09-2021-0277

Huamán Tapara, J. C., Flores Huarco, N. H., Valverde Rodríguez, A., & Jara Claudio, F. R. (2024). Rendimiento de diez variedades de Cañihua (Chenopodium pallidicaule Allen) en Chumbivilcas, Cusco-Perú. *Revista Alfa, 8*(24), 930–938. https://doi.org/10.33996/revistaalfa.v8i24.314

Kim, D.-S., & Iida, F. (2023). Kaniwa (Chenopodium pallidicaule)'s nutritional composition and its applicability as an elder-friendly food with gelling agents. *Gels, 9*(1), 61. https://doi.org/10.3390/gels9010061

León Ttacca, B., Bravo Portocarrero, R., Llanos Nina, E., León Tacca, J., & Medina Espinoza, J. (2020). Variabilidad en la incidencia de insectos plaga y severidad de enfermedades en accesiones de cañihua (*Chenopodium pallidicaule*) del Altiplano peruano. Universidad Nacional del Altiplano.

Limachi, J., Huanca, S., Castañeta, G., Miranda-Flores, D., Palma, V., Yañiquez, J., Tejeda, L., Mollinedo, P., Dimov, K., Popova, T., & Peñarrieta, M. (2023). Antioxidant potential and fatty acid profile of different canihua (Chenopodium pallidicaule) cultivars, raised in Bolivian Altiplano. *Food Science and Applied Biotechnology, 6*(2), 383. https://doi.org/10.30721/fsab2023.v6.i2.284

Luna Mercado, G. I. (2020). *Optimización de pan sin gluten a base de cañihua (Chenopodium pallidicaule) fermentada con masa madre y chía*. Master's thesis, Universidad Nacional Agraria La Molina.

Luque-Vilca, O. M., Paredes-Erquinigo, J. Y., Quille-Quille, L., Choque-Rivera, T. J., Cabel-Moscoso, D. J., Rivera-Ashqui, T. A., & Silva-Paz, R. J. (2024). Utilization of sustainable ingredients (Cañihua flour, whey, and potato starch) in gluten-free cookie development: Analysis of technological and sensorial attributes. *Food, 13*(10), 1491. https://doi.org/10.3390/foods13101491

Mangelson, H., Jarvis, D. E., Mollinedo, P., Rollano-Penaloza, O. M., Palma-Encinas, V. D., Gomez-Pando, L. R., Jellen, E. N., & Maughan, P. J. (2019). The genome of *Chenopodium pallidicaule*: An emerging Andean super grain. *Applications in Plant Sciences, 7*(11), e11300. https://doi.org/10.1002/aps3.11300

Mankotia, R., Singh, H., Varsha, N., Ghritlahare, M. K., Krishna, M. S., Meena, R., Murmu, D. K., & Mani, A. (2025). Nutritional, functional, and environmental benefits of pseudocereals—A review. *International Journal of Research in Agronomy, 8*(4S), 111–121. https://doi.org/10.33545/2618060X.2025.v8.i4Sb.2785

Mavroeidis, A., Roussis, I., & Kakabouki, I. (2022). The role of alternative crops in an upcoming global food crisis: A concise review. *Food, 11*(22), 3584. https://doi.org/10.3390/foods11223584

Mérida-López, J., Pérez, S. J., Morales, R., Purhagen, J., Bergenståhl, B., & Rojas, C. C. (2023). Comparison of the chemical composition of six Canihua (Chenopodium pallidicaule) cultivars associated with growth habits and after Dehulling. *Food, 12*(8), 1734. https://doi.org/10.3390/foods12081734

Mérida-López, J., Rojas, C. C., Bergenståhl, B., & Purhagen, J. (2024). Functional properties of starch cultivars of two Andean grains grown in Bolivia: Amaranth (Amaranthus caudatus)

and canihua (Chenopodium pallidicaule). *Heliyon, 10*(15), e35140. https://doi.org/10.1016/j.heliyon.2024.e35140

Moscoso-Mujica, G., Mujica, Á., Chura, E., Begazo, N., Jayo-Silva, K., & Oliva, M. (2024). Kañihua (Chenopodium pallidicaule Aellen), an ancestral Inca seed and optimal functional food and nutraceutical for the industry: Review. *Heliyon, 10*(14), e34589. https://doi.org/10.1016/j.heliyon.2024.e34589

Patiño Fernández, M. A., & Cruz Mamani, J. N. (2025). Sistematización de Investigaciones sobre la Cañahua en la Facultad de Agronomía de la Universidad Mayor de San Andrés. *Sapiens International Multidisciplinary Journal, 2*(1), 51–65. https://doi.org/10.71068/vp5vm630

Peñarrieta, J. M., Alvarado, J. A., Åkesson, B., & Bergenståhl, B. (2008). Total antioxidant capacity and content of flavonoids and other phenolic compounds in canihua (*Chenopodium pallidicaule*): An Andean pseudocereal. *Molecular Nutrition & Food Research, 52*(6), 708–717. https://doi.org/10.1002/mnfr.200700189

Quevedo-Olaya, J. L., Schmiele, M., & Correa, M. J. (2025). Potential of Andean grains as substitutes for animal proteins in vegetarian and vegan diets: A nutritional and functional analysis. *Food, 14*(17), 2987. https://doi.org/10.3390/foods14172987

Quispe, A. J., Moreno, M. C., Leon, A. M., Bouchon, P., & Medina, W. T. (2024). Design of cañihua-rice: Development and characterization of an analogue of rice by warm-extrusion of cañihua (Chenopodium pallidicaule Aellen) and rice (Oryza sativa) flours. *Food and Humanity, 2*, 100193. https://doi.org/10.1016/j.foohum.2023.12.001

Rodriguez, J. P., Bonifacio, A., Gómez-Pando, L. R., Mujica, A., & Sørensen, M. (2023). Cañahua (Chenopodium pallidicaule Aellen). In *Neglected and underutilized crops* (pp. 45–93). Elsevier. https://doi.org/10.1016/B978-0-323-90537-4.00011-9

Rodriguez, J. P., Jacobsen, S.-E., Andreasen, C., & Sørensen, M. (2020). Cañahua (Chenopodium pallidicaule): A promising new crop for arid areas. In A. Hirich, R. Choukr-Allah, & R. Ragab (Eds.), *Emerging research in alternative crops* (Vol. 58, pp. 221–243). Springer International Publishing. https://doi.org/10.1007/978-3-319-90472-6_9

Rosales-Espinoza, E., Jimenez-Dávalos, J., Sota-Cano, A. F., & Cáceres-Vizarreta, A. I. (2022). Caracterización molecular de la cañihua (chenopodium Pallidicaule aellen). *Ciencia Latina Revista Científica Multidisciplinar, 6*(6), 9149–9165. https://doi.org/10.37811/cl_rcm.v6i6.4063

Scherberger, R. R., Kaess, H., & Brückner, S. (1975). Studies on the action of an anticholinergic agent in combination with a tranquilizer on gastric juice secretion in man. *Arzneimittel-Forschung, 25*(9), 1460–1463.

Serena-Romero, G., Ignot-Gutiérrez, A., Conde-Rivas, O., Lima-Silva, M. Y., Martínez, A. J., Guajardo-Flores, D., & Cruz-Huerta, E. (2023). Impact of in vitro digestion on the digestibility, amino acid release, and antioxidant activity of Amaranth (Amaranthus cruentus L.) and Cañihua (Chenopodium pallidicaule Aellen) proteins in Caco-2 and HepG2 cells. *Antioxidants, 12*(12), 2075. https://doi.org/10.3390/antiox12122075

Zegarra, S., Muñoz, A. M., & Ramos-Escudero, F. (2019). Elaboración de un pan libre de gluten a base de harina de cañihua (Chenopodium pallidicaule Aellen) y evaluación de la aceptabilidad sensorial. *Revista Chilena de Nutrición, 46*(5), 561–570. https://doi.org/10.4067/S0717-75182019000500561

Chapter 8
Fiindi (*Digitaria exilis*): Drought Tolerance and Water Use Efficiency

Muhammad Asif, Umar Farooq, Wajahat Hussain, Ahmed M. S. Kheir, Mukhtar Ahmed, and Shakeel Ahmad

Abstract Fiindi (*Digitaria exilis*), an under-explored African cereal, has outstanding endurance to adverse conditions as it has a short growth period, extensive rooting system, and has high water-use-efficiency. It has the potential to grow grain on marginal soils where large grain crops tend to fail and this makes it a potential candidate of climate-smart agriculture. This chapter is justified by the fact that Fiindi has a potential to enhance food security in drought-prone areas with the growing climate variability. The goal is to combine botanical, physiological, biochemical, and molecular information about the drought adaptation of Fiindi, with the help of superior modelling, remote sensing, and breeding techniques that can justify its enhancement and broad usage. The methods are multidisciplinary in this chapter as they apply physiological methods, crop simulation modelling, remote sensing-based stress detection and participatory breeding. Case studies indicate that Fiindi was measurably performing: in Mali, drought-tolerant landrace JI1 maintained 78% relative water content and had 0.92 t ha^{-1} under extreme stress; in Burkina Faso, NDVI-based modelling forecasts yields with less than 20% error; and in Nigeria, participatory breeding had established 18–25% increased water-use efficiency and high-preference genotypes. On the whole, Fiindi is a crop with a strong set of adaptative characteristics, which makes it a strategic member of the dryland agro-ecosystem. The emphasis of future work should be put on the growth of genomic resources and digital agriculture to boost breeding and consider the use of water-efficient production methods.

M. Asif (✉) · U. Farooq · W. Hussain · M. Ahmed
Department of Agronomy, Pir Mehr Ali Shah Arid Agriculture University, Rawalpindi, Pakistan

A. M. S. Kheir
Julius Kühn Institute (JKI)—Federal Research Centre for Cultivated Plants, Institute for Strategies and Technology Assessment, Kleinmachnow, Germany

Soils, Water and Environment Research Institute, Agricultural Research Center, Giza, Egypt

S. Ahmad
Institute of Agronomy Faculty of Agricultural Sciences and Technology, Bahauddin Zakariya University, Multan, Pakistan

M. Ahmed (ed.), *Climate Change and Neglected Underutilized Crops*,
https://doi.org/10.1007/978-3-032-21005-0_8

Keywords *Digitaria exilis* · Water-use efficiency · Water drought tolerance · Remote sensing · Modelling

8.1 Introduction

Digitaria exilis (Fiindi or white fonio) is an ancient African cereal crop that has been praised due to its outstanding ability to adapt to harsh climates, poor soils, and that the grain of the crop grows quickly. It has been regarded as one of the oldest domesticated cereals in West Africa and it continues to be the focus of the traditions, food systems and livelihoods of the rural communities in the region (Dansi et al., 2012).

Fiindi has attracted interest as a species of strategic implementation in enhancing climate-resilient agriculture since it is a resilient C_4 crop with special morphological and physiological detail in drought-prone areas. Drought, heat stress, soil degradation, and water scarcity have continued to affect the productivity of agriculture globally endangering food security of low-income and arid countries. According to the climate models, the frequency and intensity of droughts will rise in most regions of sub-Saharan Africa, which will increase production risks of key cereals, including maize, rice, and wheat (FAO, 2023).

In this regard, non-exploited crops possessing an intrinsic resistance are becoming part and parcel of diversified, climate-adaptive agriculture systems. Crops not fully utilized like Fiindi have profound benefits since they are environmentally resilient as well as being nutritionally and culturally beneficial. Smallholder farmers are vulnerable to climate variability which means that these crops are instrumental where the better or high input cereals perform poorly. Fiindi is not only abundantly spread in West Africa, specifically in Guinea, Mali, Burkina Faso, Senegal, and Nigeria, but also grown in low-fertility and sandy soils with very little agronomic requirements (Cruz et al., 2016a). Although Guinea is the leading in the production within the region but it offers more because of widespread cultivation but not excellent yielding output. Some of the nations that produce smaller areas such as Mali and Burkina Faso show the same or even greater yield showing yield gaps existing all over the state. These gaps indicate that there can be substantial productivity increase by means of better varieties and management of soil fertility and use of agronomic practices that are climate adaptable. Reducing the yield gap would potentially raise substantially the quantity of fonio produced without expanding the extent of the agricultural area, which would be part of sustainable intensification and regional food security (Fig. 8.1). The plant has great agronomic importance because it matures early, can be cultivated on poor soils and provides a nutritional diversity into the diet due to its high mineral and amino acid content. Fiindi is a very good crop to use in dry agriculture because of its deep root system, fast phenological growth and physiological characteristics that allow it to survive severe water shortage. The fact that it has a short crop cycle means that farmers can plant it at short rainy seasons, eliminating the exposure to terminal drought. Furthermore, Fiindi has high-performance in terms of water consumption, so it can be used in areas with a fluctuating rainfall and low moisture levels in the soil (Mabhaudhi et al., 2019). These qualities make it a major climate resistant crop that can withstand environmental extremes. The main elements of the adaptation strategy of Fiindi are drought tolerance and water-use efficiency (WUE). The mechanisms that enable the crop to continue with its metabolic activity during water stress include stomatal regulation, osmotic adjustment, antioxidant activity, leaf rolling, and root proliferation. The knowledge of such adaptations is essential in the enhancement of crops and in the establishment of agronomic strategies that would yield maximum production under the struggle of limited water supplies.

This chapter aims at delivering a synthesis of adaptive characteristics of Fiindi in details with its possible contribution to climate-smart agriculture. The chapter includes botanical and agronomic traits, phenology, drought-tolerance, WUE strategies, modelling, remote sensing applications, and breeding methods that can be used to achieve more resilience. Through this combination of the existing scientific understanding, the chapter illuminates how Fiindi can play an important role in sustainable dryland agriculture and the food security of the world.

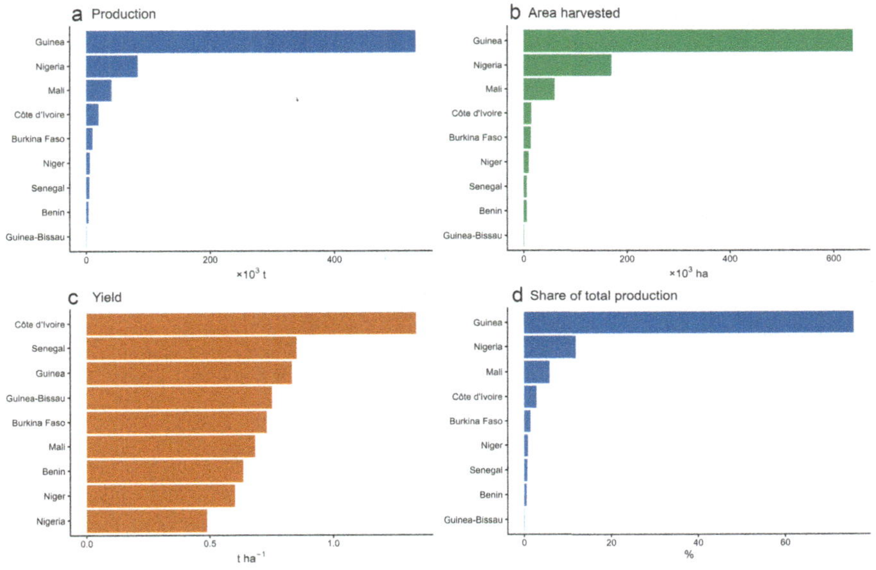

Fig. 8.1 Production (**a**), harvested area (**b**), yield (**c**) and contribution to total production (**d**) of fonio (*Digitaria exilis* and *D. iburua*) across maor West African producing countries (FAO, 2024)

8.2 Botanical and Agronomic Overview of *Digitaria exilis*

8.2.1 Taxonomic Position and Morphology

Digitaria exilis belongs to the family Poaceae and to the subfamily Pinacodeae, the tribe Paniceae and is taxonomically related to the drought-tolerant C_4 grasses (Abrouk et al., 2020). It is an annual cereal with slender culms, thin linear leaf, and digitate inflo World trade bonded with small grains of about 1–1.5 mm in diameter (Table 8.1). It has an C_4 photosynthetic pathway which helps to enhance its photosynthetic efficiency in high temperatures and in low water conditions. The grain size and fine husk indicates its being under-domesticated crop, and characteristics such as lodging and grain shattering remain in most of the landraces (Mabhaudhi et al., 2018).

8.2.2 Ecological Distribution and Environmental Adaptability

Digitaria exilis is a native plant species of West Africa used in the cultivated land of Guinea, Mali, Burkina Faso, Senegal and Nigeria (Animasaun et al., 2023) (Fig. 8.2). The species can survive in semi-arid and sub-humid climate with annual rainfall of 300–1200 mm and temperatures between 25–35 °C. This flexibility to degraded and low-fertility soils sandy, gravelly, and lateritic soils provides it with an edge over other major cereals, such as maize, sorghum and rice (Schillinger, 2011). Such ecological adaptability renders Fiindi to be applicable in drought-prone areas with the agricultural productivity being constrained by the lack of water and soil erosion (Mabhaudhi et al., 2018).

8.2.3 Growth Requirements and Soil Preferences

Fiindi has been found to have low agronomic input requirements. It thrives on low organic matter and low nutrient marginal soils and does not need or need minimal fertilizer to go through the growth cycle (Mier & van den Hurk, 2014). The majority of varieties are able to grow within a period of 60–90 days, which allows the crop to avoid droughts in late seasons. It has a short growth cycle and requires low moisture level; hence farmers can produce Fiindi in places where rainfall patterns are short and unreliable. Though it is tolerant of the poor soils, its performance is reduced in the heavy clay soils where the drainage is limited (Martin et al., 2016).

Table 8.1 Morphological comparison between two major cultivated species of fonio (*Digitaria* species)

Organs	Traits	*Digitaria exilis*	*Digitaria iburua*
Roots	Type	Fasciculated (hundreds of fine roots)	Fasciculated
	Dimension	Diameter (1 mm), length ≤ 228 cm	–
Stubble	Size	30–80 cm	45 cm–1.4 m
	Aspect	Fine, hollow, erected	Glabrous, upright
Tillers		2–8 tillers with 5–9 nodes	4–5 nodes
Leaves	Width	0.3–0.9 cm	1 cm
	Length	5–15 cm	30 cm
	Pubescence	Glabrous	Glabrous
Straw	Aspect	Erected/simple	Erected
	Height	About 45 cm	>50 cm
Racemes	Number	2–5 fingers	4–10 sous-doigts
	Aspect	Sessile or not	–
	Length	5–12 cm/3–15 cm	12–13 cm
Spikelets	Length	1.5–2 mm	2 mm
	Pubescence	Glabrous	Courts poilsoupresque glabrous
Pedicels	Number	Grouped by 2–4	Often 5
	Aspect	Very frail, glabrous	Glabrous, sometimes hairy
	Length	0.5–1.8 mm	2.5 mm
Caryopsis	Base	Narrow	Pointed
	Shape	Ovoid to ovo-ellipsoid	Ellipsoid
	Dimension	Diameter 0.5–1 mm, length 0.75–2 mm	Length 1.5–1.75 mm
	Color	White to yellow, pink, purple wine-red	Blackish

Source Adoukonou-Sagbadja (2010)

8.3 Phenology, Distribution, and Production Zones

8.3.1 Phenology

The life cycle of fiindi is short and normally takes a period of between 60 and 90 days depending on the variety and environmental factors. Germination takes place fast when the soil is moist normally in 3–7 days following sowing (Table 8.2). Early tillering occurs, generating more than one shoot per plant, the effect of which increases the canopy formation and the acquisition of resources (Bucens & Pain, 2000). Flowering is very plastic and is usually induced by photoperiod and water

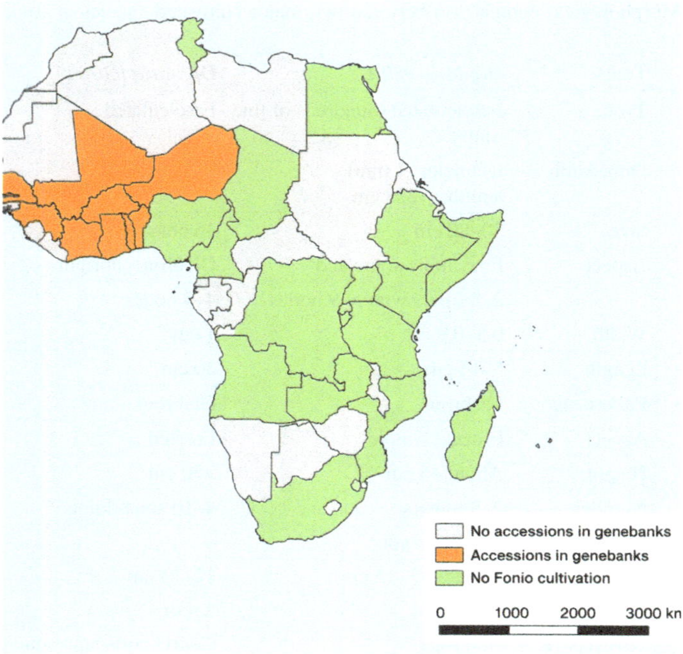

Fig. 8.2 Spatial distribution of White and Black Fonio (*Digitaria exilis* and *Digitaria iburua*) cultivation in West Africa. *Source* Blench RM (2012)

supply, so that early flowering landraces can avoid terminal drought. Grain-filling phases are not long and usually take 20–25 days and are responsive to water shortage where stress at this stage influences the weight and quality of the grains (Fig. 8.3).

Table 8.2 Phenological stages of *Digitaria exilis*

Stage	Approx. timing	Key features
Germination and emergence	0–5 days	Rapid sprouting, emergence
Seedling stage	5–10 days	Leaf formation, root establishment
Tillering	10–25 days	Multiple shoots develop
Stem elongation	25–35 days	Height increases, transition to reproduction
Booting	35–40 days	Panicle enclosed in flag leaf sheath
Heading	40–45 days	Panicle emerges
Flowering	45–50 days	Pollination and fertilization
Grain filling	50–60 days	Grain weight increases
Maturity	60–70 days	Grains dry, ready for harvest

Source Fonio (Digitaria Exilis) in West Africa (2015)

Fig. 8.3 **a** Typical *Digitaria* sp. plant; **b** a filed of *D. exilis*; **c** mature *D. iburua* (black fonio) sp.; **d** spikelet arrangement of *D. iburua*; **e** threshed seeds of *D. exilis* (white fonio); and **f** threshed seeds of *D. iburua*. *Source* Farooq and Siddique (2023)

The quick phenology of the crop permits it to be used in semi-arid and sub-humid areas with unpredictable rainfall, which means that the crop can be used to produce the products despite unpredictable dryland conditions (Mabhaudhi et al., 2018).

8.3.2 Geographical Distribution

Fiindi is indigenous to West Africa (Fig. 8.4) and it has been grown over centuries in Guinea, Mali, Burkina Faso, Senegal, Nigeria, and sections of Cote d Ivoire (Dansi et al., 2010). It grows in agroecological areas with low to moderate rainfall (300–1200 mm/year) and good drainage soils characterized by sandy and loamy soils. Fiindi has been adapted especially to semi-arid savannas, low fertility soils and degraded land that is not conducive to the major cereals. It has an ecological plasticity with a range of growth of between 200 and 1500 above sea level, which is both a local adaptation and genetic variation of landraces (Abrouk et al., 2020).

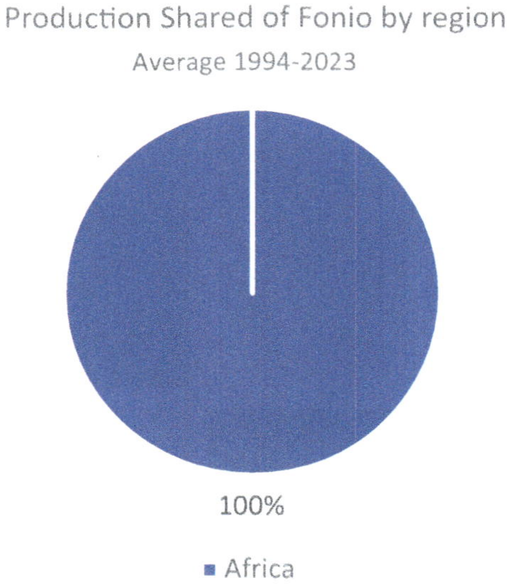

Fig. 8.4 Production share of Fonio by region from 1994–2023. *Source* FAO (December 10, 2024)

8.3.3 Agronomic Importance

8.3.3.1 Nutritional Quality

Fiindi grains are rich in nutrients and contain important amino acids, such as methionine and cysteine (Table 8.3), which are lacking in the major cereals (Temegne et al., 2016). The grain is free of gluten, well digested, and contains a lot of dietary fiber and micronutrients, including iron, zinc, and magnesium, which makes it a good candidate to counteract the sign of micronutrient deficiencies (Wang et al., 2021).

Table 8.3 Nutrient composition of Digitaria exilis (per 100 g dry grain)

Nutrient/Component	Amount (per 100 g)
Energy (kcal)	350–380
Carbohydrates	76.9–78.1 g
Protein	7.1–11 g
Fat (total lipids)	1.7–4.2 g
Dietary fiber	2.2–8.0 g
Sugar	0.6–0.9 g

Source FAO/INFOODS (2023)

8.3.3.2 Cultural and Economic Value

Fiindi has a cultural value throughout the West African region, where it is consumed in cultural meals, rituals, and tribal events (Karl et al., n.d.). The crop is also economical as it gives income to small holder farmers, most of whom are still manual processors causing it to be labor intensive. Having been a hungry-season crop is especially critical, because of its early maturity, which allows filling food gaps (Cruz et al., 2016b).

8.4 Physiological Mechanisms of Drought Tolerance

Digitaria exilis has drought tolerance that is a synthesis of physiological, biochemical, and structural mechanisms that enable the plant to maintain its water balance, photosynthetic and transpiration regulation in limited moisture (Fig. 8.5). All these processes also make it resilient to semi-arid and sub-humid West Africa regions. These adaptive processes are described in the following subsections.

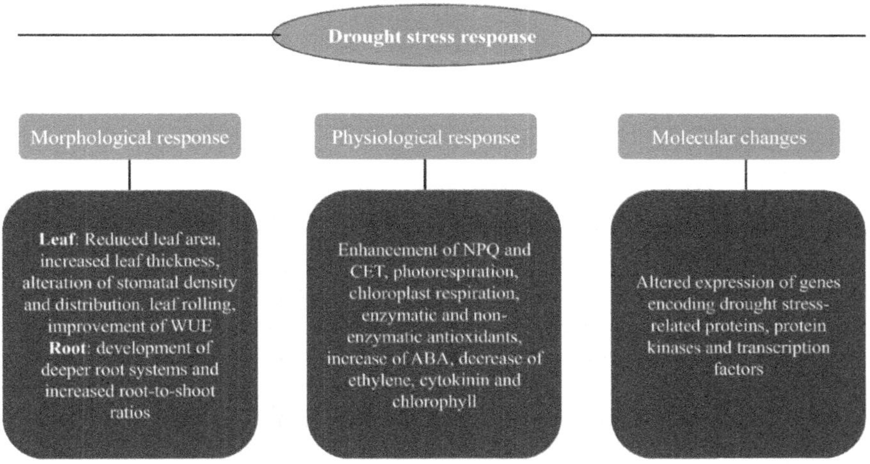

Fig. 8.5 Drought stress responses in major C_4 crops, with an emphasis on photosynthesis. Plants respond to drought at various levels to protect photosynthesis and other aspects of plants. NPQ, non-photochemical quenching. CET, cyclic electron transfer. WUE, water use efficiency

8.4.1 Water Relations and Cellular Homeostasis of Fonio (Digitaria exilis)

8.4.1.1 Water Potential

When plants are exposed to drought, the water potential of leaves and roots are reduced, and it has a direct effect on the maintenance of turgor and metabolism. Digitaria exilis reverses this decrease by raising the concentration of cell sap so as to sustain an adequate turgor pressure to facilitate enzyme activity and cell growth (Hsiao & Xu, 2000). A decrease in water potential of leaves also leads to premature morphological reactions in Fiindi, including leaf rolling and a decrease in leaf expansion that aids in lowering water loss.

8.4.1.2 Relative Water Content (RWC)

RWC is a strong measure of the hydration of plants. The use of drought resistant cereal species has allowed them to have a high RWC during water famine due to better osmotic adjustment and efficient absorption of roots. The same trends have been reported in the African millets, whereby the preservation of RWC positively affects stability and recovery of the membrane following the rehydration process. The capacity to maintain RWC even in the presence of extremely high moisture stress is one of the main properties which help Fiindi to become drought resilient (Zhao et al., 2006).

8.4.1.3 Osmoregulation Strategies

The gain of compatible solutes like proline, soluble sugars and other low-molecular-weight metabolites makes osmotic adjustment in Fiindi. These compounds stabilize the proteins, preserve the turgor of the cells, and eliminate the damage of membranes caused by dehydration. Osmotic adjustment is important in helping the C_4 grass to continue its photosynthetic activity in drought conditions, keeping the chloroplasts hydrated and stomatal functioning (Farooq et al., 2009). Similar osmolyte-mediated processes are probably used by Fiindi which have been seen in some of the drought-tolerant cereals.

8.4.2 *Photosynthetic Responses Under Drought of Fonio (*Digitaria exilis*)*

8.4.2.1 Stomatal vs Non-stomatal Limitations

Drought inhibits photosynthesis by allowing stomatal and non-stomata limitations. The non-stomatal causes of reduced carbon assimilation include reduced chlorophyll content and electron transport inhibition, and the process of stomatal closure also lowers the amount of CO_2 diffusion. The C_4 grasses generally have a better stomatal control because of the CO_2-concentrating mechanism in the bundle sheath cells so partial photosynthetic activity may continue even at low stomatal conductance (Flexas et al., 2004). This natural drought-buffering effect is enjoyed by the C_4 species Fiindi.

8.4.2.2 Chlorophyll Degradation

The loss of chlorophyll is a typical response to drought, which is evidence of disrupted pigment stability and augmented photooxidative strain. The drought stability of chlorophyll is needed to sustain biomass build up and yield formation. There is a high relationship between the resistance to chlorophyll degradation and improved membrane stability and lesser lipid peroxidation in drought adapted cereals (Ng et al., 2010). The greenness of Fiindi throughout the late season moisture deficit indicates comparable mechanisms of protection.

8.4.2.3 Energy Dissipation Mechanisms (NPQ, Xanthophyll Cycle)

Drought causes the excitation pressure of the photosystem II (PSII) to heighten and leads to the amplification of the requirement of excess energy dissipation. Photoinhibition is counteracted by non-photochemical quenching (NPQ) and xanthophyll cycle which transform excess radiations into heat. These photoprotective mechanisms are usually induced in C_4 grasses undergoing drought (Demmig-Adams & Adams, 2006). To ensure PSII integrity during periods of limited soil moisture, Fiindi probably makes use of an increased NPQ activity.

8.4.3 *Water Uptake Dynamics*

8.4.3.1 Role of the Root System

The major element of drought avoidance is root architecture. The fibrous root structure of Fiindi also penetrates soil layers fast and pushes the early acquisition of soil moisture to the maximum and allows root penetration even in unpredictable rain. Expansive root systems are beneficial in maintaining transpiration and enhancing water uptake by drought-adapted cereal under stress induced by moisture deficit (Lynch, 2013).

8.4.3.2 Deep Rooting vs Lateral Spread

There is variability in the root depth and lateral spread in the different landraces of Fiindi. Deep-rooting phenotypes enable access to deeper water pools whereas lateral roots increase the utilization of surface moisture after the intermittent rainfall. Drought-tolerant grasses studies have shown that this form of architectural plasticity correlates with a high-yield under terminal drought (Uga et al., 2013).

8.4.3.3 Root Hydraulic Conductance

The one that defines the ability of the plant to move water between root and shoots is hydraulic conductance. When drought occurs, most cereal plants increase the activity of aquaporins to enhance the movement of water across root membranes. Enhanced root hydraulic conductivity has been closely attributed to improved drought performance of millet and sorghum (Heil & Land, 2014). The fact that Fiindi survived in extreme water scarcity implies the use of similar physiological mechanisms.

8.4.4 Cuticle Conductance and Transpiration Regulation

8.4.4.1 Epicuticular Wax Deposition

Epicuticular wax is a very important barrier that restrains non-stomatal water loss. Drought triggers the production of wax leading to increased layers of cuticle which curbs cuticular transpiration. The development of better wax deposition is also one of the characteristics of drought-tolerant cereals like pearl millet and sorghum. Similar increases in the wax load would be expected to help Fiindi survive long dry intervals (Dumarey et al., 2009).

8.4.4.2 Reduced Stomatal Density

Grasses adapted to drought often have a lower stomatal density, which means that they lose less water through transpiration water loss but maintain the necessary gas exchange. Less stomatal density leads to the enhancement of water use efficiency (WUE) in the case of low moisture conditions. Even though there are no particular studies on Fiindi stomatal density, its drought-avoidance behavior indicates similar structural adaptations (Liu et al., 2007).

8.5 Biochemical and Molecular Responses to Drought of Fonio (*Digitaria exilis*)

Drought stress provokes a multifaceted system of biochemical and molecular reactions in Digitaria exilis, which helps the plant to endure water shortage and avoid metabolic homeostasis (Fig. 8.6). Among these responses, there are the regulation of reactive oxygen species (ROS), the activation of enzymatic and non-enzymatic antioxidant systems, the accumulation of compatible solutes, membrane stabilization, and transcriptional reprogramming.

8.5.1 Reactive Oxygen Species (ROS) Generation under Drought

Reactive oxygen species (ROS), such as hydrogen peroxide (H_2O_2), superoxide radicals (O^{2-}), and hydroxyl radicals (OH^{-1}) are usually produced in excess in chloroplasts, mitochondria and peroxisomes due to water shortage (Gill & Tuteja, 2010). Unregulated ROS may cause the destruction of proteins, lipids, and nucleic acids, which disrupt cell processes. Controlled production of ROS is a shared signal that is used to determine stress-responsive responses in drought-adapted cereals that mediate the acclimation to low water supplies (Mittler et al., 2004). The data provided by Fiindi (e.g., JI1 accession) indicates that there was moderate accumulation of ROS during stress which could be interpreted as the effective redox systems in the tolerant genotypes.

8.5.2 Antioxidant Defense Systems

Plants use both enzymatic and non-enzymatic antioxidants to counter the effects of ROS to avoid oxidative stress.

8.5.2.1 Enzymatic Antioxidants

Superoxide dismutase (SOD), catalase (CAT), ascorbate peroxidase (APX), and glutathione reductase (GR) are also important enzymes that collectively work to scavenge the ROS and restore the cellular antioxidants. These enzymes have been

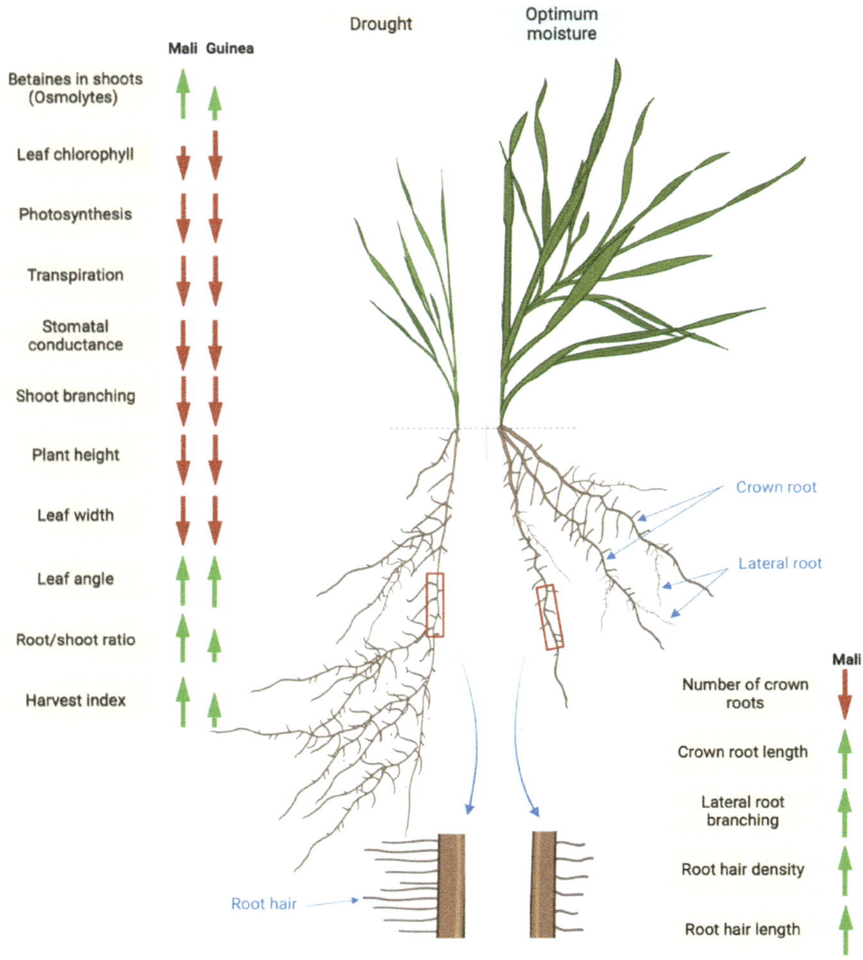

Fig. 8.6 Summary of biochemical, physiological and morphological acclimation responses to drought by two accessions of White fonio. *Source* Pudasaini et al. (2025)

reported to increase their activities in drought-tolerant cereals, and this is in line with elevated membrane stability and photosynthetic activity (Apel & Hirt, 2004). Field and controlled studies in Fiindi also reveal improved SOD and CAT processes in JI1 and comparable accessions, which explain their adaptation to extended water droughts.

8.5.2.2 Non-enzymatic Antioxidants

These non-enzymatic molecules are flavonoids, carotenoids, phenolics, and anthocyanins, which help in detoxification of ROS and photoprotection. These compounds have been detected to be high in various orphan plants during drought stress (Järvisalo & Saris, 2014). These metabolites, in addition, have synergistic interactions with the enzymatic system to increase the total antioxidant activity and prevent chloroplast membranes against peroxidative damage.

8.5.3 Compatible Solutes and Osmo Protectants

In compatibility to water deficit, compatible solutes (e.g. proline, glycine betaine, soluble sugars, etc.) are deposited to aid in osmotic adjustment, protein stabilization and membrane protection (Ashraf & Foolad, 2007). Proline is especially important in drought adaptive accessions of Fiindi, which helps them to withstand turgor and stabilize enzymes. The soluble sugars are also considered Osmo protectants and as the carbon stores that help in post-rehydration recovery.

8.5.4 Lipid Peroxidation and Membrane Stability

Drought stress has a critical effect on the membrane integrity. A major indicator of oxidative damage is lipid peroxidation which is usually determined by the content of malondialdehyde (MDA). Low MDA and decreased electrolyte leakage in drought-tolerant cereals are linked to an increase in the water retention capacity and membrane stability (Scherberger et al., 2020). Accessions of Fiindi with the larger capacity to antioxidant show reduced accumulation of MDA, which means that the cellular membranes are well-guarded in case of long-term drought.

8.5.5 Gene Expression and Transcriptional Regulation

The drought-responsive genes are central in the coordination of physiological and biochemical adoptions. DREB, NAC, bZIP, and WRKY families of transcription factors control downstream stress-responsive genes that are engaged in osmolyte production, antioxidant protection and water movement (Shapira & Nadler, 1975). The recent findings in orphan cereals indicate that the epigenetic changes, such as the DNA methylation and histone acetylation play a role in stress memory and drought tolerance (Chinnusamy & Zhu, 2009). These molecular perspectives can be used in Genomics-assisted breeding in Fiindi to select genotypes that have a better antioxidant system, osmotic adjustment, and stress-responsive gene expression.

8.6 Morphological and Anatomical Adaptations of Fonio (*Digitaria exilis*)

The morphological and anatomical characteristics affecting drought tolerance in *Digitaria exilis* are highly effective as the morphological characteristics of the plant make it maximize the water intake, reduce the loss of water and the accomplishment of its life cycle in inadequate moisture conditions. The combination of adaptation in roots, leaves, and reproductive organs increase the ability to survive and be productive in the dry land.

8.6.1 Root Architecture Traits

Water and nutrient uptake are mostly done through a root system, whose structure is directly related to adaptation to drought. Digitaria exilis usually forms a root system that is dense and fibrous and can easily probe soil moisture (Uga et al., 2013). Deep-rooting phenotypes utilize water found in the subsoil layers, whereas lateral roots are used to utilize the transient surface water. Differential root depth, density, and branching structure, either amongst land rises or to local rainfall regimes, is evidenced by variation among landraces, including DI, JI1, and JI2 (Matusik & Gibson, 2013). Deeper and more prolific root systems continue to have higher relative water content and better stomatal conductance in times of drought leading to better growth and yield stability.

8.6.2 Leaf Anatomical Responses

Leaf anatomy is important in controlling the loss of water and the photosynthetic efficiency of leaves under conditions of water scarcity. Cells become specialized as bulliform cells, which are large epidermal cells on the adaxial surface, which makes the plants roll and fold their leaf in times of water deficit, and this way, they reduce the area in direct sunlight and transpiration (Fahlgren et al., 2015). Secondly, the Fiindi landraces can survive drought conditions due to its drought tolerant, which is the thick cuticle layers, which restrict cuticular transpiration and increase water retention (Dumarey et al., 2009). Intercellular airstairs in mesophyll tissue additionally adds to the efficiency of diffusion of CO_2 and reduces water loss. All of these leaf anatomical characteristics aid in the plants adaptation to withstand drought photosynthesis.

8.7 Water Use Efficiency (WUE) of Fonio (*Digitaria exilis*)

One of the characteristics used to define the performance of plants in the conditions of water scarcity is water use efficiency (WUE). It is characterized as the ratio of biomass or carbon accrued per unit of water lost by the transpiration process and is an integration of the responsiveness of physiological, morphological and anatomical characteristics (Condon, 2004). The knowledge of WUE in *Digitaria exilis* helps understand how it adapts to arid and marginal soils and helps design the strategies of crop improvement.

8.7.1 Concept and Physiological Basis of WUE

The ratio of carbon assimilation to stomatal conductance is intrinsic WUE which is important in sustaining photosynthesis during droughts and reduced water wastage (Niesel & Gallinat, 2006). C_4 types like Fiindi have a CO_2-concentrating apparatus which promotes increment of carbon at low stomatal conductance, thereby inherently having high inner WUE than C_3 cereals. Whole-plant WUE is a ratio of the sum of the biomass formed divided by water used during the growth time which is calculated as the products of root uptake efficiency, canopy architecture, and leaf-level gas exchange (Blum, 2009). The intrinsic and whole-plant WUE have a role in the survival and productivity of plants in water-limited environments in Fiindi.

8.7.2 Stomatal Conductance and Photosynthetic Capacity

One of the major factors of WUE is the stomatal behavior. Fiindi controls stomata to respond to both soil moisture and vapor pressure deficit modulating the CO_2 uptake against the transpiration water loss. Drought induced, reduced stomatal conductance is substituted with the increased capacity of photosynthesis in the C_4 metabolism which enables carbon fixation to proceed efficiently (Beck et al., 1975). Diversity between accessions in density and responsiveness of stomata also contribute to WUE, as accessions with deep root systems are able to keep stomata open more during temporary drought.

8.7.3 Carbon Isotope Discrimination as a Proxy for WUE

Carbon isotope discrimination (D13C) is a combined measure of long-term WUE as it is a combination of intercellular and atmospheric concentration of CO_2 over the period of growth. There is a positive correlation between low D13C and high intrinsic WUE in drought tolerant cereals (Vatner & McRitchie, 2008).

8.7.4 Influence of Root Traits on WUE

Root system architecture has a strong influence on the WUE in that it dictates the degree of exploitation of soil water. Deep and proliferative roots improve access to the subsoil moisture, which promotes the continued transpiration and carbon assimilation in dry seasons (Matusik & Gibson, 2013). In Fiindi, the accessions like JI1 and JI2 are deeply rooted with a high degree of lateral proliferation of roots which are associated with increased WUE. Water transport between soil and shoot is further controlled by root hydraulic conductivity, which relies on the activity of aquaporins, thereby enhancing optimal water use in that of the stress conditions (Kudoyarova et al., 2014).

8.8 Modelling Approaches for Understanding Drought Response and WUE

8.8.1 Role of Crop Simulation Models

Crop simulation models have a role to play since they can give an insight on the relationship between spectral vegetation and spectral albedo to the chemical composition of the vegetation. The applicability of the use of crop simulation models is common to quantify the growth of plants, water consumption, and yields to various environmental and management conditions, which include Agricultural Production Systems sIMulator (APSIM), Decision Support System for Agrotechnology Transfer (DSSAT), and Aqua Crop (Holzworth et al., 2014).APSIM (Agricultural Production Systems simulator) can be used to give a detailed modelling of soil-water interactions, crop growth, and biophysical interactions which is very helpful in analyzing the impacts of drought. DSSAT combines genetic coefficients with environmental factors to determine the crop performance; meanwhile, Aqua Crop is a more specific model of water-limited environment, with water productivity in the spotlight (Kersten, 2017). The models have effectively been implemented in the orphan crops like millets and sorghum, in which empirical data is constrained, by fitting landrace-specific parameters (Steduto et al., 2009).

8.8.2 Modelling Fiindi Drought Resilience

Evapotranspiration models and soil-plant-atmosphere continuum (SPAC) models can be used to simulate the drought resilience of Fiindi. Evapotranspiration considers crop water requirements that are calculated using data on meteorology and crop-specific coefficients and thus water deficits are estimated as well as irrigation time (Nemecek et al., 2011). The SPAC models combine the soil moisture dynamics, root water uptake, and plant hydraulic conductance to predict both the canopy and whole-plant levels of plant responses to stress. The models can be used to assess how characteristics of root architecture, stomatal conductance, and WUE influence growth and yield in case of varying drought conditions (Heil & Land, 2014).

8.8.3 Predictive Modelling for Yield under Climate Variability

Climate variability brings some uncertainty in the rainfall and temperature which affect production of crops. Predictive models can also be used to incorporate drought indices like Standardized Precipitation Index (SPI) and Crop water stress index (CWSI) to simulate the impact of water shortages on the yield of Fiindi (Vicente-Serrano et al., 2010). Simulation-generated crop growth trajectories can also be used to tell the key growth phases that are susceptible to water stress, thus providing a management intervention. These forecasting abilities are especially helpful to the smallholder farmers who have Fiindi planted on rainfed systems.

8.9 Remote Sensing and Geospatial Technologies for Drought Monitoring

The application of remote sensing and geospatial technologies has transformed the drought assessment and crop monitoring which provides rapid and massive subdivision of water stress and crops performance. The tools are important in offering vital information in both the study and the management of under-utilized crops like Digitaria exilis, especially where there is a scarcity of field-based measurements.

8.9.1 Use of Remote Sensing in Drought Assessment

The use of satellite imagery to measure vegetation indices such as the Normalized Difference Vegetation Index (NDVI) and the Enhanced Vegetation Index (EVI) are common techniques in determining the health of plants under drought environments and the development of their canopy. They are indices which measure chlorophyll composition and photosynthetic activity, which are indirect measures of water stress (Tucker et al., 2005). The data of land surface temperature (LST) measured by thermal devices allows one to determine the dynamics of heat stress and evapotranspiration in crops and the Vegetation Health Index (VHI) combines both NDVI and LST data to create real-time measurements of vegetation state in extensive areas (Kogan, 2001).

8.9.2 Mapping Fiindi Fields in West Africa

Spatial visualization of the Fiindi cultivation is important to inform about the drought vulnerability in the regions and development of interventions. Satellite data, e.g. Landsat, Sentinel-2, MODIS, etc., in high resolution enables the ability to discriminate fonio fields on the basis of canopy structure, phenology, and spectral characteristics (Tucker et al., 2005). Geospatial mapping is used to provide the baseline data to model water availability and predict the possible yield to be used in different climatic conditions.

8.9.3 Monitoring Crop Stress and Predicting Yield

With remote sensing, constant observation of crop stress can be made through satellite-based proximal sensing of soil moisture (e.g., SMAP, SMOS) or UAV proximal sensor steps. The UAV platforms enable high-resolution phenotyping, such as canopy temperature, leaf area index, and NDVI, which allows determining the preciseness of the performance of Fiindi in water-restricted conditions (Atzberger, 2013). Soil moisture data can estimate drought intensity and timing, and the UAV platforms are capable of doing this. These methods favor the yield prediction models and early-stage detection of regions that need agronomic interventions.

8.10 Future Prospects and Challenges

8.10.1 Climate Change and Fiindi's Importance

Hypothetical warming and rainfall fluctuation are likely to increase the drought stress in West African agroecosystems (IPCC, 2022). The changing conditions under which fiindi is inherently drought tolerant and has a short life cycle make fiindi especially suitable to climate-smart agriculture. It has a fast phenology, which enables it to avoid terminal droughts sooner and low input needs, which can be grown in marginal soils that major cereals cannot produce on (Ke et al., 2019). Increasing its cultivation would make the smallholder farming systems more resilient to climate variability.

8.10.2 Need for Genetic Improvement

Regardless of its adaptive qualities, Fiindi has a number of domestication constraints, such as small grains, grain shattering, and irregular grain maturity (Abrouk et al., 2020). To resolve these characteristics by traditional and molecular breeding is necessary to enhance productive and stable conditions during drought conditions. The genetic enhancement on WUE, root structure, and physiological stress responses will give it a competitive edge over the large cereals (Saroj et al., 2015).

8.10.3 Integration of Digital Agriculture

Digital agriculture devices, such as high-throughput phenotyping, the use of Unmanned Aircraft Vehicles, and remote sensing, can provide the opportunity to measure growth, stress, and make management decisions in Fiindi (Araus & Cairns, 2014). Interaction with crop models, including APSIM and Aqua Crop, makes genotype x environment x management interactions stimulable, which is useful in both the targeted breeding of specific genotypes that perform well and in the agronomic precision of the practices (Van Ittersum et al., 2003).

8.10.4 Policy and Commercialization Opportunities

The development of policies to facilitate seed systems, market access, and value chain development is important in scaling the Fiindi cultivation. Economic incentives of farmers can be promoted through investments in mechanization in terms of sowing and dehiscing, and the promotion of nutrient-rich fonio products (Karl et al., n.d.). Partnerships between governments and businesses can speed up the process of commercialization but provide fair access to smallholder farmers.

8.11 Case Studies

Case Study 1: Variation in Drought Tolerance of Southern Mali Fiindi Landraces

An experiment in Southern Mali compared drought-tolerance skills of identified Digitaria exilis landraces in smallholder rainfed systems (Dansi et al., 2012). It aimed at comparing physiological and agronomic characteristics of five divergent accessions in a season with uneven rainfall and enormous evaporative stress. The parameters of measurements were relative water content, stomatal conductance, root length density and grain yield in mild and severe drought. The JI1 genotype was the most successful genotype in terms of performance, as it retained the highest relative water content (78%) and deep-rooted ability (up to 65 cm) and could then access the subsoil moisture during periods of drought growth. It was also the best in terms of grain yield (0.92 t ha^{-1}). Architecture of the root systems was a key contributing factor where drought-tolerant accessions exhibited better lateral spread and increased root bio-mass and led to better efficiency in water uptake. The research proposed the significance of tapping natural genetic diversity in Fiindi to breed drought resistant varieties. It suggested giving high priority to landraces which have high root proliferation, early flowering and high water-use efficiency to be incorporated in regional crop-improvement programs to improve food security in the drylands.

Case Study 2: Remote Sensing and Modelling to predict the yield of fiindi in Burkina Faso

A study in Northern Burkina Faso examined the application of satellite remote sensing and Aqua Crop modelling in the evaluation of Fiindi growth, water stress and yield variability over small-scale fields (Tucker et al., 2005). The research involved the use of NDVI, land surface temperature (LST) and indices of soil moisture using Sentinel-2 and MODIS images. The measurements that were done under the field consisted of plant height, canopy cover, phenology, and final grain yield, which served to calibrate and validate the model of Aqua Crop. The findings revealed that the NDVI values at the booting phase were highly associated with the final yield of the grain (R2 = 0.82) which implies the consistency and reliability of vegetation indices in the prediction of early yields. LST identified both heat and drought stress two weeks before it was visible, and satellite-acquired soil moisture imagery was valid in identifying areas of extreme water shortage. When comparing the remote sensing with the Aqua Crop simulations, the prediction was of yield within the range of ± 10% accuracy among various positions of the landscape. The research has concluded that remote sensing should be used hand in hand with crop modelling to improve the monitoring of drought and also to make better decisions on Fiindi production. It suggested the development of site-specific advisory around the sowing of dates, application of fertilizers, and moisture conservation measures in order to enhance productivity in the semi-arid areas.

Case Study 3: Participatory breeding of high WUE Fiindi in Northern Nigeria

Northern Nigeria had a participatory breeding program to determine Fiindi accessions that have better water-use efficiency (WUE) and low-input farming adaptation (Ceccarelli & Grando, 2007). It entailed the use of groups of farmers in five villages testing 12 genetically diverse Fiindi lines under rainfed and moisture-stressful environments. The parameters that were evaluated were key intrinsic WUE, canopy temperature depression, days to flowering, grain yield and traits that were desired by farmers like grain appearance and processing quality. Two accessions, FI-07 and JI2 had an extraordinarily high performance in terms of WUES with a 18–25% higher WUES than local checks and lower temperature in their canopies in the peak drought periods. Their early maturation also allowed them to beat terminal drought and their farmers also liked their quality of grain, dehiscing ease and good taste. Trials of adoption the next season demonstrated that over 60 percent of households that tried it found the lines to be better than the traditional ones. The paper showed that a mixture of scientific selection and farmer expertise hastens the process of creating and adopting climate resilient Fiindi varieties. It suggested to scale up participatory breeding initiatives in West Africa as well as incorporate farmer-selected genotypes that were drought-tolerant into national seed systems.

8.12 Conclusion

Digitaria exilis (Fiindi) is a superb entity with a combination of morphological, physiological, biochemical, and molecular features that allow it to survive and be productive under water-limited and marginal environments. It has an effective system of root up taking water through the deep and proliferative root systems, and has anatomical features of the leaves such as bulliform cells and thick cuticles that minimize the loss of moisture through the leaf and retain photosynthetic activity during drought conditions. The adaptations in reproductive processes, early flowering and rapid increase in the growth of the species enable it to avoid terminal drought so that seed set and yield remain consistent in the event of unachievable rainfall. On the physiological level, Fiindi is very efficient in water use, as it combines stomatal control, osmotic adjustment and efficient control of transpiration. Its resilience is additionally promoted by biochemical and molecular processes, such as antioxidant defense systems and stress-responsive expression of genes to prevent oxidative damage of the cell structures and metabolic homeostasis. The combination of these multi-layered adaptations in combination with the C_4 photosynthetic pathway allows Fiindi to be sustained in growth and productivity during adverse environmental circumstances. Fiindi is not only a potentially powerful animal due to its physiological resilience. It is a valuable crop to the smallholder farmers and the food-insecure areas due to its nutritional value, cultural value, and the ability to grow on marginal soils. Modern breeding strategies, genomics, high-throughput phenotyping, crop modelling, and remote sensing can be used to improve its productivity, improve its water utilization, and speed up the production of drought-tolerant varieties. Moreover, adoption, commercialization, and higher social-economic impact can be enhanced through participatory methods, value-chain enhancements. To conclude, Fiindi is a prototype climate-smart agriculture orphaned drug. It has a rare combination of adaptation characteristics to make it a strong crop in helping to achieve food and nutrition security in an increasingly unpredictable climatic environment. The strategic incorporation of science, agronomic management and policy will play a crucial role in providing the key to unlocking its full potential and making Fiindi a crop that can be used as a staple crop in water-limited areas and not the traditional and under-utilized millet.

References

Abrouk, M., Ahmed, H. I., Cubry, P., Šimoníková, D., Cauet, S., Pailles, Y., Bettgenhaeuser, J., Gapa, L., Scarcelli, N., Couderc, M., Zekraoui, L., Kathiresan, N., Čížková, J., Hřibová, E., Doležel, J., Arribat, S., Bergès, H., Wieringa, J. J., Gueye, M., et al. (2020). Fonio millet genome unlocks African orphan crop diversity for agriculture in a changing climate. *Nature Communications, 11*(1), 4488. https://doi.org/10.1038/s41467-020-18329-4

Adoukonou-Sagbadja, H. (2010). *Genetic Characterization of Traditional Fonio Millets (Digitaria exilis, Digitaria iburua Stapf) Landraces from West Africa: Implications for Conservation and Breeding* (PhD thesis, Justus-Liebig University Giessen).

Animasaun, D. A., Adedibu, P. A., Olawepo, G. K., & Oyedeji, S. (2023). Fonio millets. In *Neglected and underutilized crops* (pp. 201–219). Elsevier. https://doi.org/10.1016/B978-0-323-90537-4.00028-4

Apel, K., & Hirt, H. (2004). REACTIVE OXYGEN SPECIES: Metabolism, oxidative stress, and signal transduction. *Annual Review of Plant Biology, 55*(1), 373–399. https://doi.org/10.1146/annurev.arplant.55.031903.141701

Araus, J. L., & Cairns, J. E. (2014). Field high-throughput phenotyping: The new crop breeding frontier. *Trends in Plant Science, 19*(1), 52–61. https://doi.org/10.1016/j.tplants.2013.09.008

Ashraf, M., & Foolad, M. R. (2007). Roles of glycine betaine and proline in improving plant abiotic stress resistance. *Environmental and Experimental Botany, 59*(2), 206–216. https://doi.org/10.1016/j.envexpbot.2005.12.006

Atzberger, C. (2013). Advances in remote sensing of agriculture: Context description, existing operational monitoring systems and major information needs. *Remote Sensing, 5*(2), 949–981. https://doi.org/10.3390/rs5020949

Beck, M. L., Freihaut, B., Henry, R., Pierce, S., & Bayer, W. L. (1975). A serum haemagglutinating property dependent upon polycarboxyl groups. *British Journal of Haematology, 29*(1), 149–156. https://doi.org/10.1111/j.1365-2141.1975.tb01808.x

Blench RM (2012) Vernacular names for African millets and other minor cereals and their significance for agricultural history. *Arabidopsis halleri. Archaeol Anthropol Sci.*https://doi.org/10.1007/s12520-012-0104-5

Blum, A. (2009). Effective use of water (EUW) and not water-use efficiency (WUE) is the target of crop yield improvement under drought stress. *Field Crops Research, 112*(2–3), 119–123. https://doi.org/10.1016/j.fcr.2009.03.009

Bucens, D., & Pain, M. C. (2000). Influence of hematocrit, blood gas tensions, and pH on pressure-flow relations in the isolated canine lung. *Circulation Research, 37*(5), 588–596. https://doi.org/10.1161/01.res.37.5.588

Ceccarelli, S., & Grando, S. (2007). Decentralized-participatory plant breeding: An example of demand driven research. *Euphytica, 155*(3), 349–360. https://doi.org/10.1007/s10681-006-9336-8

Chinnusamy, V., & Zhu, J.-K. (2009). Epigenetic regulation of stress responses in plants. *Current Opinion in Plant Biology, 12*(2), 133–139. https://doi.org/10.1016/j.pbi.2008.12.006

Condon, A. G. (2004). Breeding for high water-use efficiency. *Journal of Experimental Botany, 55*(407), 2447–2460. https://doi.org/10.1093/jxb/erh277

Cruz, C. D., Peterson, G. L., Bockus, W. W., Kankanala, P., Dubcovsky, J., Jordan, K. W., Akhunov, E., Chumley, F., Baldelomar, F. D., & Valent, B. (2016a). The 2NS translocation from *Aegilops ventricosa* confers resistance to the *Triticum* Pathotype of *Magnaporthe oryzae. Crop Science, 56*(3), 990–1000. https://doi.org/10.2135/cropsci2015.07.0410

Cruz, J.-F., Béavogui, F., Dramé, D., & Diallo, T. A. (2016b). *Fonio, an African cereal.* Cirad [UMR Qualisud] IRAG, Institut de recherche agronomique de Guinée.

Dansi, A., Adoukonou-Sagbadja, H., & Vodouhè, R. (2010). Diversity, conservation and related wild species of Fonio millet (Digitaria spp.) in the northwest of Benin. *Genetic Resources and Crop Evolution, 57*(6), 827–839. https://doi.org/10.1007/s10722-009-9522-3

Dansi, A., Vodouhè, R., Azokpota, P., Yedomonhan, H., Assogba, P., Adjatin, A., Loko, Y. L., Dossou-Aminon, I., & Akpagana, K. (2012). Diversity of the neglected and underutilized crop species of importance in Benin. *The Scientific World Journal, 2012*, 1–19. https://doi.org/10.1100/2012/932947

Demmig-Adams, B., & Adams, W. W. (2006). Photoprotection in an ecological context: The remarkable complexity of thermal energy dissipation. *New Phytologist, 172*(1), 11–21. https://doi.org/10.1111/j.1469-8137.2006.01835.x

Dumarey, C., Sket, D., Joseph, D., & Boquet, P. (2009). Basic phospholipase of Naja nigricollis venom. *Comptes Rendus Hebdomadaires Des Seances De l'Academie Des Sciences. Serie D: Sciences Naturelles, 280*(13), 1633–1635.

Fahlgren, N., Gehan, M. A., & Baxter, I. (2015). Lights, camera, action: High-throughput plant phenotyping is ready for a close-up. *Current Opinion in Plant Biology, 24*, 93–99. https://doi.org/10.1016/j.pbi.2015.02.006

FAO. (2023). *Revealing the true cost of food to transform agrifood systems*. Food and Agriculture Organization of the United Nations.

FAO. (2024). FAOSTAT database, Food and Agriculture Organization of the United Nations. Available at http://faostat.fao.orgaccessed 15 December 2025)

Farooq, M., & Siddique, K. H. M. (Eds.). (2023). *Neglected and underutilized crops: Future smart food*. Academic Press, an imprint of Elsevier.

Farooq, M., Wahid, A., Kobayashi, N., Fujita, D., & Basra, S. M. A. (2009). Plant drought stress: Effects, mechanisms and management. *Agronomy for Sustainable Development, 29*(1), 185–212. https://doi.org/10.1051/agro:2008021

Flexas, J., Bota, J., Loreto, F., Cornic, G., & Sharkey, T. D. (2004). Diffusive and metabolic limitations to photosynthesis under drought and salinity in C_3 plants. *Plant Biology, 6*(3), 269–279. https://doi.org/10.1055/s-2004-820867

Gill, S. S., & Tuteja, N. (2010). Reactive oxygen species and antioxidant machinery in abiotic stress tolerance in crop plants. *Plant Physiology and Biochemistry, 48*(12), 909–930. https://doi.org/10.1016/j.plaphy.2010.08.016

Heil, M., & Land, W. G. (2014). Danger signals—"damaged-self recognition across the tree of life". *Frontiers in Plant Science, 5*, 578. https://doi.org/10.3389/fpls.2014.00578

Holzworth, D. P., Huth, N. I., deVoil, P. G., Zurcher, E. J., Herrmann, N. I., McLean, G., Chenu, K., Van Oosterom, E. J., Snow, V., Murphy, C., Moore, A. D., Brown, H., Whish, J. P. M., Verrall, S., Fainges, J., Bell, L. W., Peake, A. S., Poulton, P. L., Hochman, Z., et al. (2014). APSIM—Evolution towards a new generation of agricultural systems simulation. *Environmental Modelling & Software, 62*, 327–350. https://doi.org/10.1016/j.envsoft.2014.07.009

Hsiao, T. C., & Xu, L. (2000). Sensitivity of growth of roots versus leaves to water stress: Biophysical analysis and relation to water transport. *Journal of Experimental Botany, 51*(350), 1595–1616. https://doi.org/10.1093/jexbot/51.350.1595

IPCC. (2022). Climate Change 2022: Impacts, Adaptation, and Vulnerability. In: Contribution of Working Group II to the Sixth Assessment Report of the Intergovernmental Panel on Climate Change [H-O. Pörtner, D.C. Roberts, M. Tignor, E.S. Poloczanska, K. Mintenbeck, A. Alegría, M. Craig, S. Langsdorf, S. Löschke, V. Möller, A. Okem, B. Rama (eds.)]. *Cambridge University Press. Cambridge University Press, Cambridge, UK and New York, NY, USA, 3056*. https://doi.org/10.1017/9781009325844

Järvisalo, J., & Saris, N. E. (2014). Action of propranolol on mitochondrial functions—Effects on energized ion fluxes in the presence of valinomycin. *Biochemical Pharmacology, 24*(18), 1701–1705. https://doi.org/10.1016/0006-2952(75)90009-x

Karl, K.A., MacCarthy, D., Chimwaza, G., Fredenberg, E., Guarin, J., Méndez Leal, E., Kozlowski, N., Narh, S., Sheikh, H., Valdivia, R., Wesley, G., Yang, M., Van Deynze, A., & van Zonneveld, M. (n.d.). *Opportunity crop profiles for the vision for adapted crops and soils (VACS) in Africa*. https://doi.org/10.7916/7MSA-YY32.

Ke, Y., Li, C., McKenzie, A. M., & Liu, P. (2019). Risk transmission between Chinese and U.S. agricultural commodity futures markets—A CoVaR approach. *Sustainability, 11*(1), 239. https://doi.org/10.3390/su11010239

Kersten, E. (2017). 6th international symposium on maritime medicine 1974. *Bulletin of the Institute of Maritime and Tropical Medicine in Gdynia, 26*(2), 131–132.

Kogan, F. N. (2001). Operational space technology for global vegetation assessment. *Bulletin of the American Meteorological Society, 82*(9), 1949–1964. https://doi.org/10.1175/1520-0477(2001)082<1949:OSTFGV>2.3.CO;2

Koréissi, Y. (2015). *Fonio (Digitaria exilis) in West Africa: Towards improving nutrient quality*. Wageningen University.

Kudoyarova, G. R., Korobova, A. V., Akhiyarova, G. R., Arkhipova, T. N., Zaytsev, D. Y., Prinsen, E., Egutkin, N. L., Medvedev, S. S., & Veselov, S. Y. (2014). Accumulation of cytokinins in roots and their export to the shoots of durum wheat plants treated with the protonophore carbonyl cyanide m-chlorophenylhydrazone (CCCP). *Journal of Experimental Botany, 65*(9), 2287–2294. https://doi.org/10.1093/jxb/eru113

Liu, H., Li, G., Chang, H., Sun, D., Zhou, R., & Li, B. (2007). Calmodulin-binding protein phosphatase PP7 is involved in thermotolerance in *Arabidopsis*. *Plant, Cell & Environment, 30*(2), 156–164. https://doi.org/10.1111/j.1365-3040.2006.01613.x

Lynch, J. P. (2013). Steep, cheap and deep: An ideotype to optimize water and N acquisition by maize root systems. *Annals of Botany, 112*(2), 347–357. https://doi.org/10.1093/aob/mcs293

Mabhaudhi, T., Chibarabada, T. P., Chimonyo, V. G. P., Murugani, V. G., Pereira, L. M., Sobratee, N., Govender, L., Slotow, R., & Modi, A. T. (2018). Mainstreaming underutilized indigenous and traditional crops into food systems: A South African perspective. *Sustainability, 11*(1), 172. https://doi.org/10.3390/su11010172

Mabhaudhi, T., Chimonyo, V. G. P., Hlahla, S., Massawe, F., Mayes, S., Nhamo, L., & Modi, A. T. (2019). Prospects of orphan crops in climate change. *Planta, 250*(3), 695–708. https://doi.org/10.1007/s00425-019-03129-y

Martin, J. K., Luthra, M. G., Wells, M. A., Watts, R. P., & Hanahan, D. J. (2016). Phospholipase A2 as a probe of phospholipid distribution in erythrocyte membranes. Factors influencing the apparent specificity of the reaction. *Biochemistry, 14*(25), 5400–5408. https://doi.org/10.1021/bi00696a003

Matusik, E., & Gibson, T. P. (2013). Fluorometric assay for N-acetylprocainamide. *Clinical Chemistry, 21*(13), 1899–1902.

Mier, P. D., & van den Hurk, J. J. (2014). Lysosomal hydrolases of the epidermis. I. Glycosidases. *The British Journal of Dermatology, 93*(1), 1–10. https://doi.org/10.1111/j.1365-2133.1975.tb06468.x

Mittler, R., Vanderauwera, S., Gollery, M., & Van Breusegem, F. (2004). Reactive oxygen gene network of plants. *Trends in Plant Science, 9*(10), 490–498. https://doi.org/10.1016/j.tplants.2004.08.009

Nemecek, T., Huguenin-Elie, O., Dubois, D., Gaillard, G., Schaller, B., & Chervet, A. (2011). Life cycle assessment of Swiss farming systems: II. Extensive and intensive production. *Agricultural Systems, 104*(3), 233–245. https://doi.org/10.1016/j.agsy.2010.07.007

Ng, W. G., Donnell, G. N., Koch, R., & Bergren, W. R. (2010). Urinary alpha-L-fucosidase. *Birth Defects Original Article Series, 11*(6), 335–339.

Niesel, A., & Gallinat, A. (2006). Neue technische Verfahren in der operativen Hysteroskopie—Wird die hysteroskopische Myomektomie einfacher? *Geburtshilfe und Frauenheilkunde, 66*(10), 976–982. https://doi.org/10.1055/s-2006-924493

Pudasaini, R., Amyotte, J., Brettingham, D. J. L., & Raizada, M. N. (2025). Physiological, metabolomic, morphological and root system architecture acclimation responses to drought in the African orphan millet white fonio (Digitaria exilis). *Plant and Soil, 512*(1–2), 1529–1558. https://doi.org/10.1007/s11104-024-07159-8

Saroj, S. K., Singh, M. N., Vishwakarma, M. K., Singh, T., & Mishra, V. K. (2015). Identification of stable restorers and genetics of fertility restoration in late-maturing pigeonpea [C *ajanus cajan* (L.) M illspaugh]. *Plant Breeding, 134*(6), 696–702. https://doi.org/10.1111/pbr.12309

Scherberger, R. R., Kaess, H., & Brückner, S. (2020). Studies on the action of an anticholinergic agent in combination with a tranquilizer on gastric juice secretion in man. *Arzneimittel-Forschung, 25*(9), 1460–1463.

Schillinger, W. F. (2011). Practical lessons for successful long-term cropping systems experiments. *Renewable Agriculture and Food Systems, 26*(1), 1–3. https://doi.org/10.1017/S1742170510000359

Shapira, E., & Nadler, H. L. (1975). Problems associated with the determination of arylsulfatase A and B using nitrocatechol sulfate substrate. *Clinica Chimica Acta; International Journal of Clinical Chemistry, 65*(1), 1–4. https://doi.org/10.1016/0009-8981(75)90327-7

Steduto, P., Hsiao, T. C., Raes, D., & Fereres, E. (2009). AquaCrop—The FAO crop model to simulate yield response to water: I. Concepts and underlying principles. *Agronomy Journal, 101*(3), 426–437. https://doi.org/10.2134/agronj2008.0139s

Temegne, N. C., Mouafor, B. I., & Ngome, A. F. (2016). Agro-morphological characterization of cassava (Manihot esculenta Crantz) collected in the humid forest and Guinea Savannah agro-ecological zones of Cameroon. *Greener Journal of Agricultural Sciences, 6*(7), 209–225. https://doi.org/10.15580/GJAS.2016.8.062716107

Tucker, C. J., Pinzon, J. E., Brown, M. E., Slayback, D. A., Pak, E. W., Mahoney, R., Vermote, E. F., & El Saleous, N. (2005). An extended AVHRR 8-km NDVI dataset compatible with MODIS and SPOT vegetation NDVI data. *International Journal of Remote Sensing, 26*(20), 4485–4498. https://doi.org/10.1080/01431160500168686

Uga, Y., Sugimoto, K., Ogawa, S., Rane, J., Ishitani, M., Hara, N., Kitomi, Y., Inukai, Y., Ono, K., Kanno, N., Inoue, H., Takehisa, H., Motoyama, R., Nagamura, Y., Wu, J., Matsumoto, T., Takai, T., Okuno, K., & Yano, M. (2013). Control of root system architecture by DEEPER ROOTING 1 increases rice yield under drought conditions. *Nature Genetics, 45*(9), 1097–1102. https://doi.org/10.1038/ng.2725

Van Ittersum, M. K., Leffelaar, P. A., Van Keulen, H., Kropff, M. J., Bastiaans, L., & Goudriaan, J. (2003). On approaches and applications of the Wageningen crop models. *European Journal of Agronomy, 18*(3–4), 201–234. https://doi.org/10.1016/S1161-0301(02)00106-5

Vatner, S. F., & McRitchie, R. J. (2008). Interaction of the chemoreflex and the pulmonary inflation reflex in the regulation of coronary circulation in conscious dogs. *Circulation Research, 37*(5), 664–673. https://doi.org/10.1161/01.res.37.5.664

Vicente-Serrano, S. M., Beguería, S., & López-Moreno, J. I. (2010). A multiscalar drought index sensitive to global warming: The standardized precipitation evapotranspiration index. *Journal of Climate, 23*(7), 1696–1718. https://doi.org/10.1175/2009JCLI2909.1

Wang, X., Chen, S., Ma, X., Yssel, A. E. J., Chaluvadi, S. R., Johnson, M. S., Gangashetty, P., Hamidou, F., Sanogo, M. D., Zwaenepoel, A., Wallace, J., De Peer, Y., Bennetzen, J. L., & Van Deynze, A. (2021). Genome sequence and genetic diversity analysis of an under-domesticated orphan crop, white fonio (*Digitaria exilis*). *GigaScience, 10*(3), giab013. https://doi.org/10.1093/gigascience/giab013

Zhao, F. J., Jiang, R. F., Dunham, S. J., & McGrath, S. P. (2006). Cadmium uptake, translocation and tolerance in the hyperaccumulator *Arabidopsis halleri*. *New Phytologist, 172*(4), 646–654. https://doi.org/10.1111/j.1469-8137.2006.01867.x

Chapter 9
Indian Barnyard Millet (*Echinochloa frumentacea*): Climate-Smart Cultivation and Nutrient Cycling

Zohaa Fatima, Umar Farooq, and Adnan Arshad

Abstract Indian barnyard millet (*Echinochloa frumentacea*) is a nutritious, climate-resilient and short-duration cereal crop with significant nutrient cycling potential. Its high tolerance to abiotic stresses and its capacity to adapt to marginal and rainfed agroecosystems makes it a good choice for climate smart cultivation. The main objectives of this chapter are to evaluate Indian barnyard millet's significance and potential in sustainable cultivation as well as nutrient recycling. For this reason, a multi-dimensional analytical approach was used, which combined peer-reviewed literature with the most recent case studies. These studies include use of remote sensing, climate change modeling, multiple site genotype analysis, nutrients use efficiency and yield prediction. The studies suggest that C_4 physiology of barnyard millet provides it with outstanding water-use efficiency and fast growth cycle that makes its yield stable under adverse environmental conditions. Studies reveal that the average barnyard dry fodder yield under Indian conditions is about 5 tons/ha. It has also been determined that barnyard millet generates 2–3 times more biomass than traditional cereals with the same amount of water. Indian barnyard millet can be transformed to a non-marginal primary crop by using the traditional, agronomic and biotechnological methods because of its sustainable agriculture use, climate resilience and nutritional security.

Keywords Barnyard millet · Nutrient recycling · Climate resilience · Sustainable agriculture · Minor millets

---

Z. Fatima (✉) · U. Farooq
Department of Agronomy, Pir Mehr Ali Shah Arid Agriculture University, Rawalpindi, Pakistan

A. Arshad
State-Key Laboratory Herbage Improvement and Grassland Agroecosystem, Key Laboratory of Grassland Livestock Industry Innovation, Ministry of Agriculture and Rural Affairs, Engineering Research Center of Grassland Industry, Ministry of Education, College of Pastoral Agriculture Science and Technology, Lanzhou University, Lanzhou, China

Department of Climate-smart Agriculture, National Cotton R&D Center, Xinjiang Academy of Agricultural Sciences, Ürümqi, China

M. Ahmed (ed.), *Climate Change and Neglected Underutilized Crops*,
https://doi.org/10.1007/978-3-032-21005-0_9

9.1 Introduction

Indian barnyard millet (*Echinochloa frumentacea*) is a climate change resilient and highly nutrient crop. It can survive marginal, low input and stress conditions because of its short growth cycle that is typically 60–90 days and low input needs. The optimal temperature for barnyard millet is 25–30 °C and it can also tolerate up to 38 °C (Renganathan et al., 2020).

Barnyard millet is mainly cultivated in India, but other parts of Asia also grow it in rainfed and marginal soils. However, it remains underutilized in most agricultural systems throughout the world despite its adaptation. The policy movements such as United Nations (UN) declaring the year "2023" as the "International Year of Millets", are helping to draw such grains back to the agendas of the world (Muskan et al., 2025).

Barnyard millet is important because of its resilience and adaptability for adverse agro-ecological areas. For having nutrient rich grain, it can fulfil dietary needs and can also be used by glucose intolerant consumers. As it is a C_4 crop, barnyard millet has high water use efficiency (WUE) and can reduce its transpiration rate in water stressed conditions. Its adaptation to erratic rainfall, low-input farming and tolerance to marginal soils makes it suitable for climate smart agriculture (Ishaq et al., 2025). Research has shown that it can withstand soil electrical conductivity of between 4 and 6 dS/m (Maithani et al., 2022).

The chapter attempts to provide an overview of the potential of barnyard millet under climate-smart agriculture cultivation. It also covers botanical traits, agronomic performance, nutritional value, nutrient cycling and tolerant behaviour to the different stresses as well as the modeling and remote sensing for its assessment. The chapter integrates the existing empirical evidence with the progressive strategies to underline the importance of Indian barnyard millet's contribution to sustainable food systems and climate resilience.

9.2 Botanical Description

9.2.1 Taxonomy and Morphology

The Barnyard millet (*Echinochloa frumentacea*) is from Poaceae family. A minor millet crop is underutilized. *Echinochloa* genus contains several species that are utilized as millets but in India, *E. frumentacea* is the most common specie that is cultivated.

Barnyard millet is an annual and climate resistant crop that has shallow and fibrous root system. In favorable conditions, it can grow to 60–120 cm in length with green stems and leaves. It has flat leaves with smooth or a little hairy surface (Fig. 9.1). The morphology normally involves terminal panicle inflorescence with extremely diversification between accessions: cylindrical, pyramidal, globose or elliptic. The

Fig. 9.1 Morphology of Indian barnyard millet (Singh et al., 2025). **a** (Inflorescence compactness): A1 compact; A2 intermediate; A3 open. **b** (Inflorescence colour): B1 green; B2 light purple; B3 dark purple. **c** (Shape of lower raceme): C1 straight; C2 curved; C3 slender. **d** (grain colour): D1 straw white; D2 light grey; D3 grey. **e** (Branching of lower raceme): E1 absent; E2 present. **f** (Pigmentation of plant): F1 A absent; F1 B present. **g** (Leaf color), pale green, light green and dark green. **h** (Inflorescence shape): H1 cylinder; H2 globose; H3 pyramidical

compactness of inflorescence is also variable (compact, intermediate, loose) and so is the inflorescence colour (green, light purple, dark purple). Its grain is of a white or yellow colour. Spikelet has two florets: the lower one is usually sterile (lemma with palea), and the upper is bisexual having a shiny lemma and partially covered palea (Renganathan et al., 2020).

9.2.2 *Phenology*

Phenology is important to improve cultivation under diverse climatic conditions. Indian barnyard millet finishes growing much earlier than the other major cereals and usually in 60–90 days under optimal conditions (Table 9.1). The phenological stages

Table 9.1 Phenological stages of barnyard millet (Echinochloa frumentacea) (Kuraloviya et al., 2022)

BBCH code	Stage description
00–05	Germination
07–09	Emergence
10	1st leaf unfolded
11–19	2–9 leaves unfolded
20–29	Tillering
30–39	Culm elongation
41–49	Booting
50–59	Panicle emergence/heading
60–69	Flowering/anthesis
70–79	Milk stage
80–89	Dough stage
90–99	Ripening/harvest maturity

include flowering timings, panicle emergence, grain filling etc. (Fig. 9.2). Which can also differ in relative to morphology, genotype, management and environmental factors such as temperature, photoperiod and water availability (Fig. 9.3).

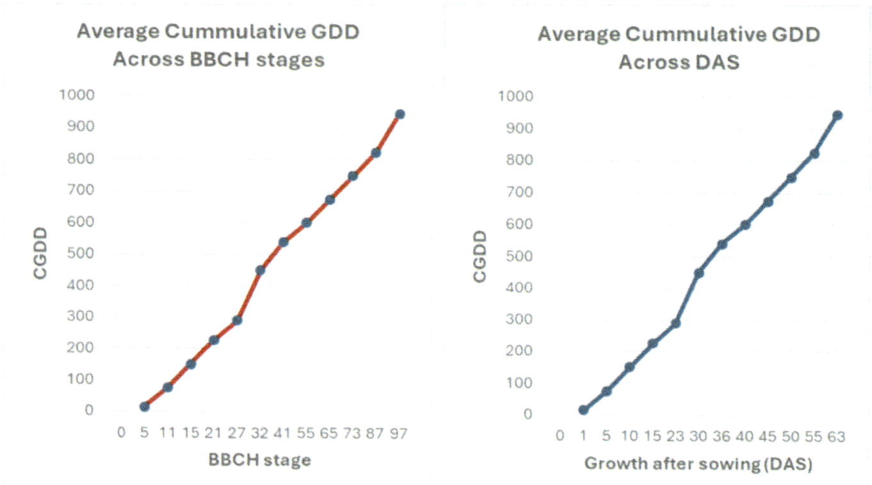

Fig. 9.2 The average CGDD across DAS necessary for Indian barnyard millet to reach the same BBCH stage Singh et al. (2025) and Kuraloviya et al. (2022). Both graphs in this figure show a positive relationship of average cumulative growing degree days (CGDD) with DAS and BBCH. As the average CGDD increases with the increase of DAS and BBCH scale, this shifts phenological stage of Barnyard millet from vegetative stage to reproductive stage

Fig. 9.3 Phenological stages with BBCH codes for Indian barnyard millet (Renganathan et al., 2020)

9.2.3 Domestication History and Global Distribution

Domestication of barnyard millet includes selection from wild relatives in marginal or stressed conditions with traits like short growth cycle, drought resistance and poor soils adaptability. Some old studies suggest that it has been cultivated since thousands of years ago. In one source, it was also said that archaeological evidence of its cultivation dates to 5000 B.C in India and 3000 B.C. in China. But these very old dates are not accepted universally and are not confirmed.

Barnyard millet is mainly cultivated in marginal soils, hilly areas or rainfed zones of South and East Asia, specifically in India, China, Japan, and Korea. It is also cultivated in some parts of Africa, tropical and subtropical regions. India is the largest producer of barnyard millet. As in a recent review, it is cultivated in India over 0.146 million hectares. The production was about 0.147 million tonnes and recent three-year average productivity was 1034 kg/ha (Renganathan et al., 2020).

9.3 Genetic Resources, Diversity and Genomic Advances

9.3.1 Germplasm Collections and Molecular Diversity

The collections of barnyard millet germplasm are massive throughout the world, with total of 8600 accessions globally and 3700 and 3200 in Japan and India, respectively. Cultivation on hilly, marginal and tribal lands in India contributed to this diversity by preserving many landraces and local varieties. The genetic reservoir can be useful in breeding stress-resistant, high-yielding or nutritionally better varieties (Mishra et al., 2022).

A genetic variability study demonstrates a large amount of variation in genotypes for traits contributing to morphology and yield (Table 9.2). This study was on 99 genotypes of barnyard millet under sodic soil conditions. The results showed high values of phenotypic coefficient of variation (PCV), genotypic coefficient of variation (GCV) and heritability of plant height, ear width, lower raceme length, flag-leaf sizes and grain yield per plant (Dhanalakshmi et al., 2019).

Such stress-induced molecular and phenotypic diversity is promising in the selection and breeding of tolerant genotypes adapted to marginal soils.

9.3.2 Genomics, Transcriptomics and Breeding Tools

Genomic research involves sequencing, mapping and gene discovery of genes that encode valuable agronomic characteristics like drought tolerance, early maturity, high grain micronutrient composition and yield stability. Entire genome sequencing of barnyard millet has made it possible to identify single nucleotide polymorphisms (SNPs), simple series repeats (SSRs) and quantitative trait locus (QTL) of stress tolerance and nutrition value (Fig. 9.4). Marker-assisted selection can be used for early maturity and drought tolerance while loci identification will be for zinc and iron accumulation in grains (Mukherjee et al., 2025).

Stress responsive genes in barnyard millet under conditions of drought, salinity, and high-temperature conditions have been revealed using RNA sequencing. The

Table 9.2 Gene banks conserving germplasm of barnyard millet worldwide (Renganathan et al., 2020)

Country	Number of accessions	Organization	Location
India	1888	National Bureau of Plant Genetic Resources	New Delhi
India	985	University of Agricultural Sciences	Bangalore
India	749	International Crop Research Institute for the Semi-Arid Tropics	Patancheru
India	300	Vivekananda Parvatiya Krishi Anusandhan Sansthan	Almora
India	1561	Indian Institute of Millets Research	Hyderabad
Japan	3671	National Institute of Agrobiological Sciences	Kannondai
United States	232	USDA Agricultural Research Service	Washington
United States	306	National Centre for Genetic Resources Conservation	Collins
United States	304	North Central Regional Plant Introduction Station	Ames
China	717	Institute of Crop Science, Chinese Academy of Agricultural Sciences	Beijing
Kenya	208	National Gene Bank of Kenya, Crop Plant Genetic Resources Centre	Muguga
Ethiopia	92	International Livestock Research Institute	Addis Ababa
Australia	66	Australian Plant Genetic Resource Information Service	Queensland
Pakistan	50	Plant Genetic Resources Program	Islamabad
Norway	44	Svalbard Gene Bank	Spitsbergen
United Kingdom	44	Millennium Seed Bank Project, Seed Conservation Department and Royal Botanic Gardens	London
Germany	36	Leibniz Institute of Plant Genetics and Crop Plant Research	Gatersleben

analysis of the differential gene expression enables breeders to select on desired genes to be used in enhancing resistance to abiotic stresses and grain quality. The Transcriptome information can also be used to perceive the pathways of nutrient uptake, efficiency of C_4 photosynthesis, and plant development to offer targets of functional genomics and genome-editing (Mukherjee et al., 2025).

Genomic and transcriptomic data are being combined with advanced breeding tools to introduce a better variety (e.g., Marker-Assisted Selection (MAS), Genomic Selection (GS), CRISPR/Cas9 Genome Editing, and Speed Breeding & Multi-Environment Trials) (Mukherjee et al., 2025).

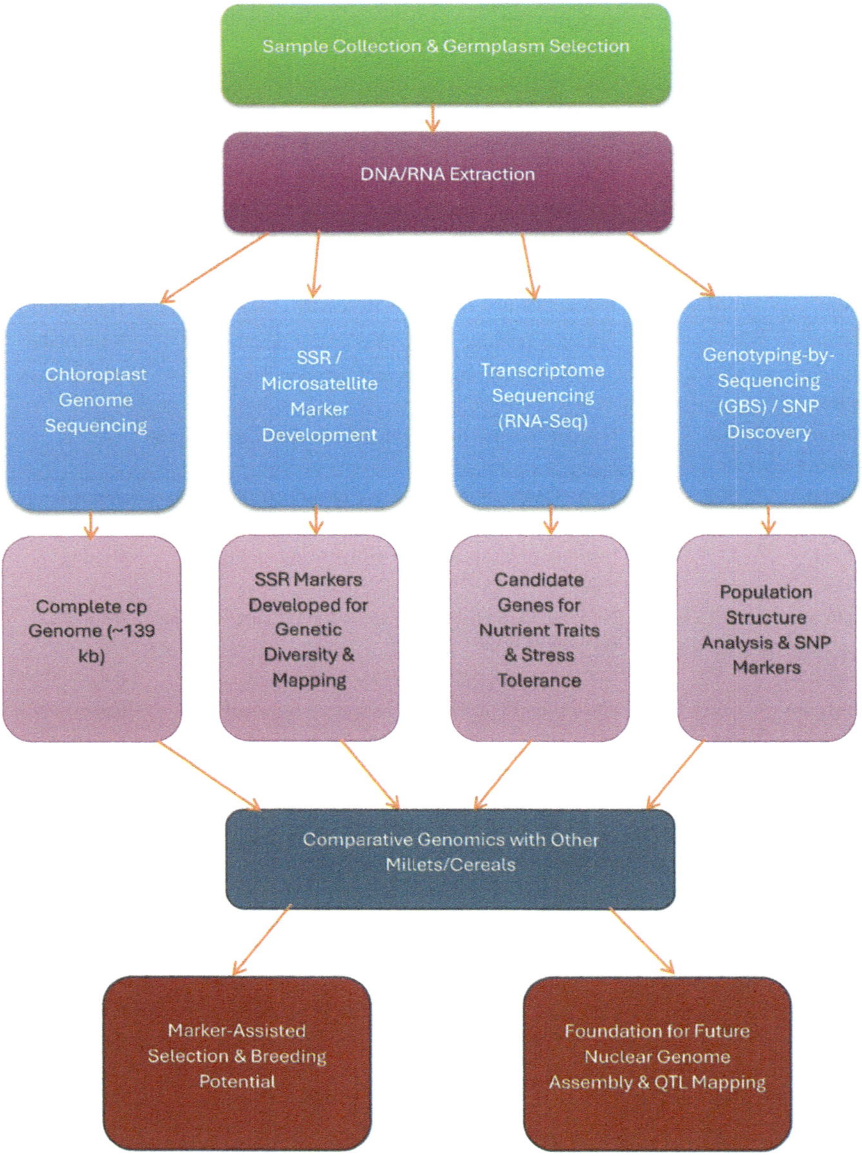

Fig. 9.4 Genomic-mapping workflow for barnyard millet (Renganathan et al., 2020)

9.3.3 Heatmap of Traits Correlation

The heatmap of correlation referred to as Fig. 9.5 gives a clear visual representation of the relationships between genetic, physiological, growth and yield traits of Indian barnyard millet crop. These traits are organized into functional categories such as genetic variability indices, stress-related traits, growth parameters and yield components. This heatmap presents both positive and negative correlations and interactions among traits. Plant height and grain yield exhibit negative relationships whereas the growth cycle, dry fodder yield and yield related characteristics have positive relationships. Physiological characteristics such as optimum temperature and salinity tolerance have different correlations with productivity traits. Genetic parameters like PCV (Phenotypic Coefficient of Variation) GCV (Genotypic Coefficient of Variance) and heritability are moderately correlated with various traits that give varying degrees of genetic control. The lower-triangle, colour-gradient design enables clear visualization and summary of complex interrelationships, which helps in trait selection and crop improvement strategies.

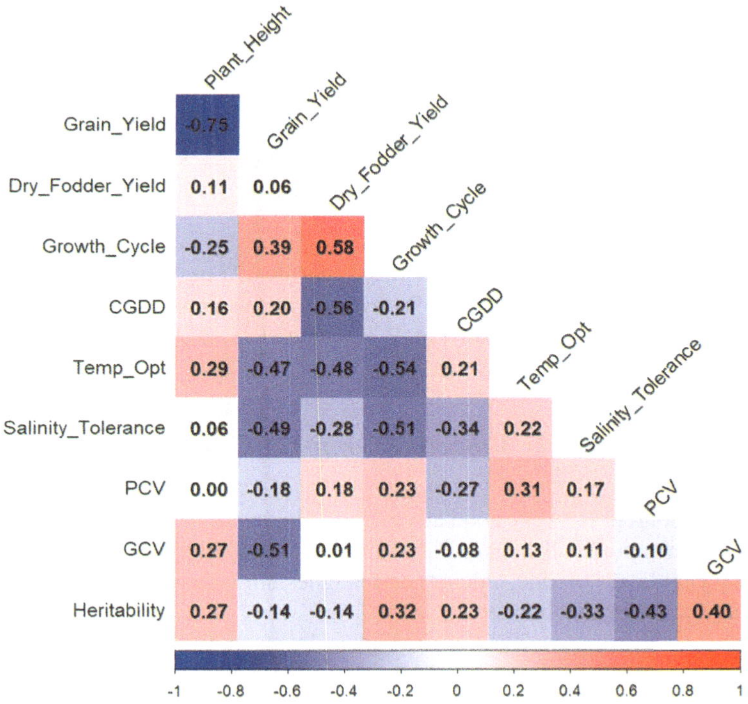

Fig. 9.5 Heatmap of trait correlations in Indian barnyard millet performance

Table 9.3 Macronutrient composition (per 100 g dry weight) of Indian barnyard millet (Bhatt et al., 2023)

Macronutrient component	Per 100 g dry weight
Energy	~398 kcal
Protein	~10.5 g
Fat	~3.6–3.8 g
Carbohydrate/starch	51.5–62 g
Dietary fiber	12.6 g (soluble ~4.2 g, insoluble ~8.4 g)

9.4 Nutritional, Nutraceutical and Functional Properties

9.4.1 Macronutrients Composition

Indian barnyard millet grain is nutritionally better than the most cereals and millets. Moderate to high slowly digestible carbohydrates are good for diabetic person as it maintains blood glucose stability. Moderate protein in it maintains muscles and repairs tissues and dietary fibre is good for bowl movement and digestion (Table 9.3). This balances energy and nutrient source that can be used in the everyday diet, especially in resource poor environments (Fig. 9.6) (Bhatt et al., 2023).

9.4.2 Essential Amino Acids

In some studies Lysine and Methionine was found 78–122 mg and 87–123 mg respectively per 100 g dry weight in Indian barnyard millet as appreciable amount, which supports growth and tissue repair (Fig. 9.7) (Bhatt et al., 2023).

9.4.3 Micronutrients, Bioactive Compounds and Antioxidants

The mineral content is a very good advantage of barnyard millet. The range of grain iron has been reported between 15.6–18.6 mg/100 g (Table 9.4). The grain also has significant levels of zinc, calcium and magnesium with other vital minerals and micronutrients (Renganathan et al., 2020).

Besides, the barnyard millet has the highest levels of bioactive compounds, such as polyphenols, flavonoids, antioxidants, and carotenoids when compared to other millets. In recent studies, the presence of total phenolic and carotenoid content in barnyard millet was reported to be two times more than in certain related small millets (Renganathan et al., 2020).

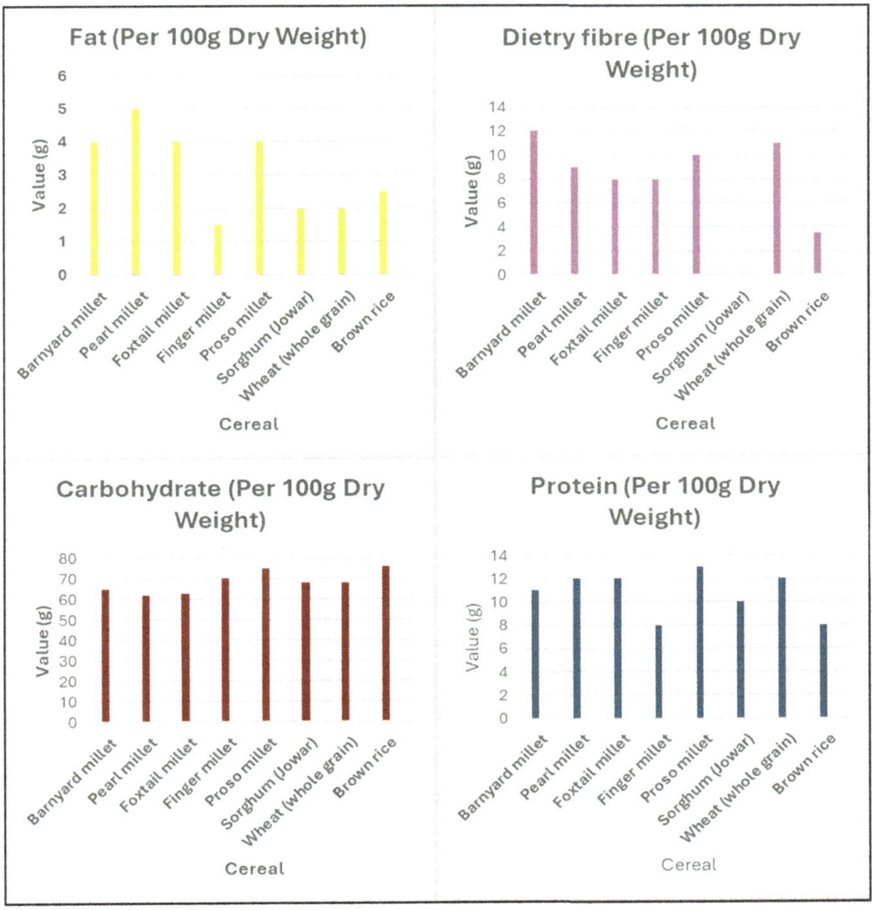

Fig. 9.6 Macronutrient (fats, dietary fibre, carbohydrates and protein) composition of Indian barnyard millet versus other cereals (Bhatt et al., 2023)

9.4.4 Processing, Value Addition and Fodder Application

Indian barnyard millet is nutritionally and functionally suitable to value addition. Its flour can be used for different products such as porridge, flat bread, bakery products and snacks fitting all modern diets and health sensitive consumers. An example is the study on pizza-base made with barnyard millet flour and with a satisfactory level of nutritional and functional quality, providing a gluten-free and low-Glucose Index (GI) alternative to wheat-based items.

The potential of value-addition is high, considering the increasing consumer demands of healthy, non-gluten and nutrient-rich foods especially in the urban or

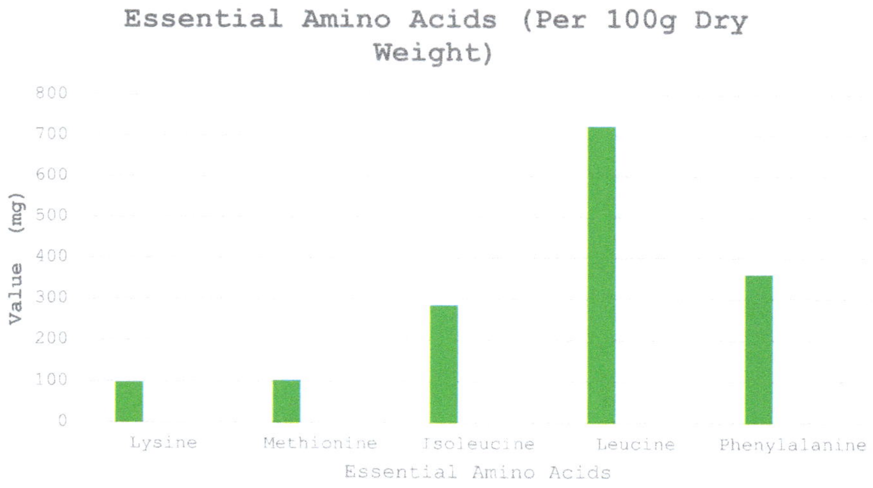

Fig. 9.7 Essential amino acids mg per 100 g of Indian barnyard millet (Bhatt et al., 2023)

Table 9.4 Micronutrients, bioactive compounds and antioxidants (per 100 g dry weight) of Indian barnyard millet (Bhatt et al., 2023)

Minerals	Per 100 g dry weight	Functional role
Iron (Fe)	15–19.5 mg	Haemoglobin synthesis, oxygen transport and metabolism
Phosphorus (P)	280–340 mg	Bone and teeth formation, energy metabolism (ATP), nucleic acids, cell membrane integrity
Calcium (Ca)	17–27 mg	Bone and teeth mineralization, muscle contraction, nerve signalling and blood clotting
Zinc (Zn)	4–5 mg	Immune function, wound healing, DNA synthesis, growth and reproductive health
Magnesium (Mg)	120 mg	Supports neuromuscular function and bone health
Bioactive compounds and antioxidants	*Per 100 g dry weight*	*Functional role*
Total phenolic content	20.3–27.8 mg GAE/g	Anti-inflammatory, anti-cancer
Flavonoids	80–120 mg	Anti-inflammatory, cardioprotective
Tannins	3.25–3.96 mg/g	Antioxidant, antimicrobial
Carotenoids	Trace	Improves eye health

health-conscious market. It offers a pathway of promotion of millet diets among non-traditional communities linking the safety of food with the market need.

It can also be used as quality fodder because of its nutritional value and great volume of fodder production. Studies reveal that the average barnyard dry fodder yield under Indian conditions is about 5 tons/ha (Sood et al., 2015).

9.5 Climate-Smart Cultivation of Barnyard Millet

As Indian barnyard millet is resilient towards climate changes, it is planted in marginal soils, intermittent rainfall areas and low-input farming systems. It is also drought resistant and usually cultivated in rainfed or low fertility conditions. It uses less amount of water and agrochemicals as compared to the major cereals. Due to its short growth cycle (typically 75–80 days) and C_4 photosynthesis, barnyard millet is water-efficient and can turn sunlight in to biomass under heat and drought stress. Because of these characteristics, it is a good fit to climate-smart agriculture, particularly in water-constrained areas or areas with uncertain rainfall (Nagaraja et al., 2023).

9.5.1 Heat and Drought Tolerance

Barnyard millet is a C_4 plant and is very efficient in photosynthesis in the conditions of high temperatures. Research has revealed that it can retain reproductive success at flowering stage even when the day temperatures are greater than 35 °C. But for cereals e.g. rice and wheat, they suffer spikelet sterility under the same conditions. Protective mechanisms such as efficient transpiration, increased cooling of the leaves and heat-shock proteins accumulation contribute to heat tolerance (Mukherjee et al., 2025).

One of the major characteristics of the barnyard millet is drought resistance which allows it to grow in rainfed and moisture-restricted conditions (Renganathan et al., 2020). It has a shallow and fibrous root system that quickly draws in any moisture available in the soil. The crop also ends its life cycle in 60–90 days without falling to terminal drought (Sood et al., 2015). Stomatal regulation, osmotic adjustment and maintenance of relative water content (RWC) are physiological adaptations which enable filling of grain at water-deficit conditions (Jayakodi et al., 2019).

9.5.2 Cultivation in Marginal, Dry Land, Saline and Erratic Rainfall Areas

Indian barnyard millet can grow in nutrient-deficient and low-fertility soils such as degraded uplands and red laterite soils (Hasan et al., 2025). Its root system is fibrous, thereby increasing nutrient uptake rate, whereas crop residues incorporated into the soil increases soil organic carbon (SOC) and structure with time. Because of this, barnyard millet can be used in sustainable crop rotations and intercropping in marginal areas (Maithani et al., 2022).

Barnyard millet is not a halophyte, but it has moderate salinity tolerance. Research has shown that it can withstand soil electrical conductivity of between 4 and 6 dS/m. It can germinate and develop seedling in saline soils which is better than rice and wheat under same soils. It can survive a moderate level of salt stress because of the osmotic adjustment and selective infiltration of ions (Maithani et al., 2022).

Indian barnyard millet is the best for irregular rainfall areas where rain is unevenly distributed because of its brief germination period and the ability to sown at any time (Renganathan et al., 2020). Depending on the arrival of monsoons, the crop can be sown in various times, such that flowering and grain fill take place in enough moisture. It has rapid early growth and this lessens its competition with weeds to form a good crop canopy despite uneven rainfall (Sukanya et al., 2022).

9.5.3 Water-Use Efficiency (WUE)

Indian barnyard millet is highly efficient in water-use (WUE), which is why it is a good crop in semi-arid and rainfed environments (Renganathan et al., 2020). WUE can be defined as the ratio of biomass or grain yield per unit of transpired water. Water use efficiency in this crop is high because it has an efficient stomatal regulation mechanism that reduces water losses when there is a shortage of soil moisture (Sood et al., 2015). The early growth leads to rapid canopy cover thus minimizing soil evaporations. This crop can absorb shallow soil moisture effectively though the fibrous root system. It has been determined that barnyard millet generates 2–3 times more biomass on the same amount of water than traditional cereals such as rice. This productivity helps in ensuring that the yields are constant even in the water-threatened regions (Jayakodi et al., 2019).

9.5.4 Short Duration and C_4 Physiology

Indian barnyard millet is a short-period cereal and is normally sown in 60–90 days depending on the genotype and the environment (Renganathan et al., 2020). The benefits of short duration are numerous as they include escape of terminal drought as the life cycle is over before severe water stress. It can be planted between the staple crops (rice and wheat) as a catch crop or relay crop (Hasan et al., 2025).

Indian barnyard millet, being a C_4 plant, is highly beneficial in terms of physiological benefits when the conditions of high light, temperature and low water are present. Its C_4 pathway has high photosynthetic efficiency which reduces photorespiration and allows the fixation of carbon even in high temperatures. Because of C_4 physiology, photosynthesis uses less Rubisco and thus it needs less nitrogen per unit of biomass. This crop also has higher drought resistance and low stomatal conductance which causes minimal water loss at same photosynthetic rate (Maithani et al., 2022).

9.5.5 Role in Climate Mitigation and Soil Carbon Sequestration Potential

Indian barnyard millet is a low input and drought tolerant crop which can also survive in rainfed areas. Hence, the crop does not emit a lot of greenhouse gases (GHG) in comparison to high input cereal which requires intensive irrigation and fertilization. Recent studies on millets (including barnyard millet) point out that they are beneficial both in terms of water- and carbon-footprint as compared to most of the major cereals and rabi crops (Ishaq et al., 2025).

Moreover, it can grow in marginal soils with minimal external intervention. This may reduce pressure on the arable soils, preserve soil health and achieve more sustainable land use. Barnyard millet can be included in the climate mitigation and adaptation strategies using correct agronomic and cropping systems (i.e. rotations, intercropping, organic management).

9.5.6 Yield Potential

Indian barnyard millet has high potential yield under different agro-climatic conditions due to its adaptability to drought tolerance, heat stress and marginal soils as well as irregular rainfall (Figs. 9.8 and 9.9). There is an important relation between genotype, environment and management practices with yield production to evaluate varieties in different stress conditions (Renganathan et al., 2020).

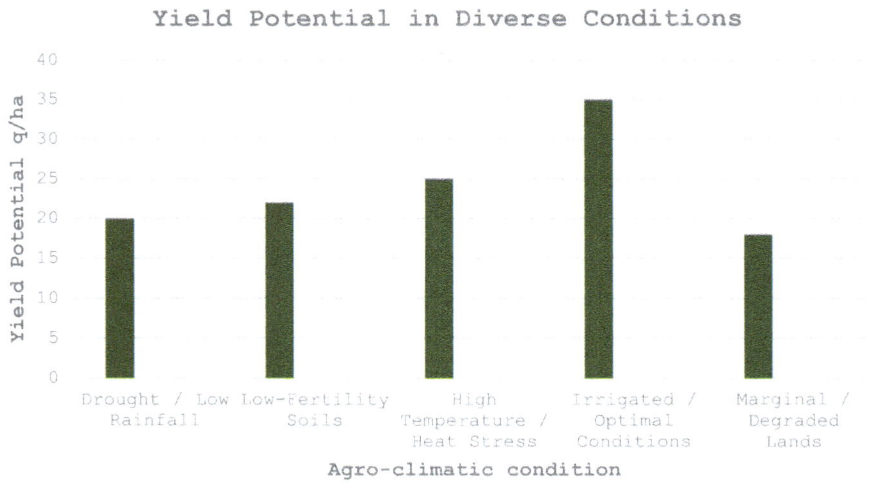

Fig. 9.8 Yield potential comparison of Indian barnyard millet under diverse conditions (Mukherjee et al., 2025)

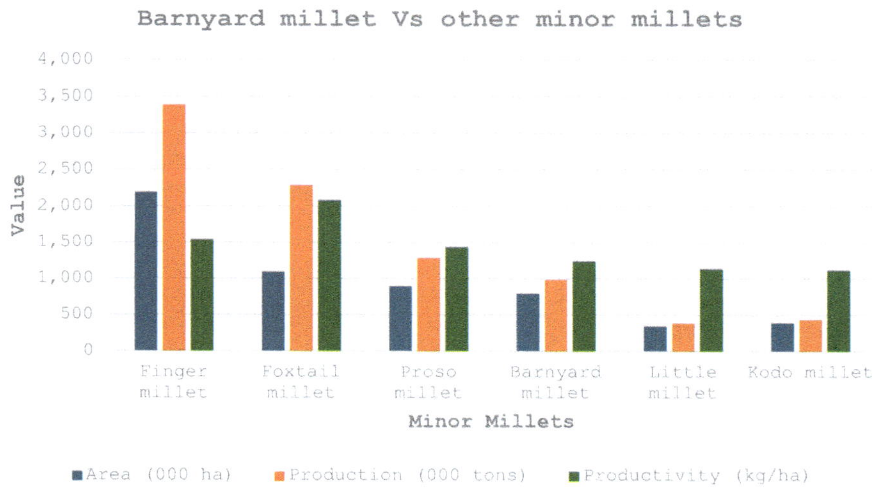

Fig. 9.9 Comparison of barnyard millet with other minor millets in area of cultivation (ha), production (tonns) and productivity (kg/ha) (Renganathan et al., 2020)

Table 9.5 Environmental and soil requirements of Indian barnyard millet (Maitra et al., 2022; Nagaraja et al., 2023)

Parameter	Requirement/Range
Temperature (optimum)	25–30 °C
Temperature (tolerance)	18–38 °C
Altitude range	200–2000 m
Light	Full sunlight
Soil type	Sandy loam to clay loam
Soil pH	5.5–7.5
Soil fertility	Low to moderate
Soil drainage	Good drainage required

9.6 Agronomic Requirements

9.6.1 Soil and Environmental Requirements

Conventionally, the barnyard millet is cultivated in different agro-ecological areas within India, both the plains and the hills, tribal and marginal areas (Table 9.5). The states such as Uttarakhand, Madhya Pradesh, Tamil Nadu, Andhra Pradesh, Karnataka and Maharashtra notable areas of production. Sandy loam to loamy soils are suitable for Indian barnyard millet cultivation typically well drained with 5.5–7.5 pH (Nagaraja et al., 2023).

9.6.2 Water Requirements

Because of high water use efficiency, low rates of irrigation in vegetative growth of Indian barnyard millet can boost the yield as compared to major cereals, especially rice (Table 9.6). This makes barnyard millet a highly marketable crop in those regions where there is water shortage (Singh et al., 2025).

Table 9.6 Water requirements of Indian barnyard millet (Nagaraja et al., 2023)

Parameter	Requirement
Rainfall	400–600 mm (seasonal)
Moisture	Low to medium

9.7 Remote Sensing and Crop Monitoring

9.7.1 Phenological Monitoring Via Satellite (or UAV)

Remote sensors (RS) (satellites or UAVs) provide powerful tools for monitoring and information of spatial/temporal dynamics of crops. In most millet producing regions (rainfed regions, marginal soils, hilly slopes), agronomic surveillance is challenging due to the lack of resources within the field, small-scale plots, and soil and climatic changes. RS avoids intensive ground surveys and enables frequent, non-destructive and spatially explicit monitoring.

Remote sensing helps in the early detection of stress (drought and nutrient deficiency) and regulate timely responses (irrigation and fertilization). It also helps to tracks the barnyard millet phenology which is important because of its short duration and climate resilience.

According to studies, combination of measures derived by RS with statistical/mixed-effect modeling (e.g. NDVI time-series, thermal time, calendar date) is enough to predict crop-progress stages (phenology) of many crops such as Indian barnyard millet (Oikonomidis & Trevezas, 2024).

9.7.2 Vegetation Indices (e.g. NDVI, SAVI, EVI)

A widely used vegetation index is calculated by red and near-infrared reflective bands called as Normalized Difference Vegetation Index (NDVI). It correlates very well with live green vegetation (i.e. canopy cover, biomass and vigour). Other indices are also used depending on the density of the canopy, the soil background, water stress and sensor bands. These indices include Soil Adjusted Vegetation Index (SAVI), Ratio Vegetation Index (RVI), Enhanced Vegetation Index (EVI), red-edge-based indices (NDRE) and multi-sensor/ texture/thermal indices (Arshad et al., 2023).

9.7.3 Yield Estimation, Stress Detection, and Precision Agriculture Applications

Crop development with time (phenology) and stress or moisture deficit can be tracked with such indices which can be calibrated for yield and biomass monitoring.

Current studies on millets demonstrate high potential. A UAV-based multispectral experiment (on millet) utilized multi-source data fusion, i.e. multi-spectral parameter, multi-spectral texture index, multi-spectral canopy temperature, and multi-spectral plant height. These were combined to calculate above-ground biomass (AGB) with high accuracy such as R^2 up to 0.877, RMSE up to 0.207 kg m^{-2} and RPD about 2.847 (Yang et al., 2024).

It implies that remote sensing (satellite or UAV) can be used to assist with the correct monitoring of growth, biomass and even yield in millet crops, such as barnyard millet, particularly when the local calibration procedure is performed. This type of RS-based monitoring can help precision agriculture for variable-rate fertilization or irrigation, early stress identification and efficient management. This makes resource management more efficient and resilient to climatic fluctuations.

9.8 Crop Modelling and Decision-Support Tools

Crop simulation models are computer-aided simulation models that simulate growth and development of crops based on weather, soil, management practices and variety traits. In the case of underutilized cereals (e.g. millets), these models can be used to investigate the potential of yield, the efficiency of water and nutrient use, and climate-resilience in various scenarios.

9.8.1 Rationale of Crop Modeling

Recent changes of climate (erratic rainfall, heat, drought), the smallholder farms that have limited resources find it hard to maximize the planting period, nutrient and water application, and reduction of risks. An example of crop models is the virtual environment where a crop (barnyard millet) behaviour in various environmental and management (sowing date, fertilizer, irrigation) conditions is simulated before the crop was put in the field. This can be particularly useful in marginal and low-input agro-ecologies in which field experimentation is too expensive or unfeasible. Crop models assist in designing climate efficient, resource efficient, climate sensitive, crop designs, planning and decision support to farmers, extension service, and policymakers.

9.8.2 Existing Models for Millets and Underutilized Cereals

According to studies, generic crop models (such as AquaCrop, DSSAT, APSIM and others) can be modified for simulating less commonly used cereals like millets. An evaluation of underutilized cereals was recently done in multi-model (compared AquaCrop, DSSAT, and a simpler model (SIMPLE)). The authors discovered that among the models, AquaCrop had the largest mean simulated yield and reasonable water-use estimates of millet and had rather low input data necessities. This implies that even in the absence of detailed physiological or soil data (usually typical of minor millets), even more simplistic or moderately complicated models can still offer valuable information, provided they are appropriately adjusted to local conditions.

Recently, the researchers created a new and improved version of DSSAT-CERES-Millet which can dynamically predict grain protein and starch accumulation using different irrigation and nitrogen regimes. Though this model was constructed using foxtail millet, the modelling was the integration of carbon/nitrogen dynamics and grain composition modelling. It could be used as a template in modelling barnyard millet particularly quality and nutrition-based research models (Zhou et al., 2025).

9.8.3 Simulating Growth, Yield, Water- and Nutrient-Use Efficiency of Barnyard Millet

Parameters such as phenology (thermal time, maturity of sowing to crop), crop characteristics (growth, water-use, transpiration), biomass proportion, leaf area index (LAI) dynamics, root depth, water/nutrient uptake can be measured by crop modelling. But these might require calibration using experimental data or analogies with other related millets. The yields, water-use efficiency (WUE), biomass and risk can be simulated under the various scenarios such as different management practices (sowing date, fertilizer, irrigation/rainfed vs irrigated), soil types (marginal soils, low fertility, sandy loam, inceptisols) and climates (rainfed, semi-arid, erratic rainfall) conditions. Quality simulation can be done by using refined models (such as the protein/starch module) which not only give yield simulation but also nutritional quality of the grain in different nitrogen and water conditions.

9.8.4 Simulation Under Climate Change and Stress Scenarios

Crop modelling provides an opportunity to evaluate the performance of barnyard millet in the future under diverse climate conditions. Past climate data, projected weather using downscaled GCMs, and local soil/management data can be used to model yield curves, risks of crop failure, water consumption and resource requirement.

This type of simulation can teach us whether barnyard millet is in fact a valid climate-smart solution for future climate versatility. It can be used to design the strategies of adaptation (e.g. changes in sowing date, selection of tolerant genotypes, maximising the use of inputs). In addition, trade-offs can also be analysed using models such as yield against water-use, yield against quality (protein/starch), sustainability (depletion of soil fertility against nutrient-use efficiency), and helps to balance productivity with conservation and resource-use objectives.

9.8.5 Integration of Remote Sensing with Crop Models and Field Data

A new trend in agricultural monitoring is to jointly use crop simulation models with remote sensing (and GIS/Earth Observation) to estimate crop growth and yield in a spatially extensible and data-informed manner. This integration will combine the strengths of both mechanistic knowledge (through models) and timely, spatial information (through RS).

It has been demonstrated in studies that RS data (e.g. LAI, biomass through MODIS) can be used with DSSAT model to estimate canopy state variables and biomass during different seasons and farms (e.g. R 2 of 0.82–0.90 and 0.92–0.99 of LAI and biomass respectively) (Amin et al., 2025).

These hybrid RS-modelling strategies provides an alternative to monitor crop performance on the scale, deciding on climate-smart properties, predicting yield potential, and decision-making of barnyard millet especially in resource-poor or data-limited environments.

9.9 Nutrient Cycling, Soil Health and Sustainability Aspects

Little work is done regarding the study of nutrient cycling in the long term when using barnyard millet as a crop. Agronomic experiments and susceptible suggestions shed more light on its possible role in low-industrial harvesting agricultural practices.

9.9.1 Nutrient Use Efficiency and Fertility Management

A recent experiment indicated that there was a lot of variation in grain yield with the level of fertilizers. The average yields varied between ~1.9 and 2.4 t/ha (between control and NPK-treated plots). This means that although this crop tolerates low-fertility, its yield and nutrient uptake positively respond to balanced fertilization i.e. it can be more nutrient-use efficient when well managed (Selvam et al., 2022).

Further, experimental organic agriculture conducted on the varieties of barnyard millet in different amounts of farm-yard manure (FYM) revealed that amendments on organic farms had an enormous impact on the production of grain and straw. In a research trial, the use of 7.5 t FYM/ha resulted in greater yield of grains than the use of 5 t FYM/ha or control. These organic fertilization practices can help to sustain the sustainability, minimizing the use of chemical fertilizers and can enhance the soil health in the long run.

9.9.2 Role in Sustainable Agriculture and Soil Carbon Sequestration

The ability of barnyard millet to grow in marginal and low-fertility soils as well as its low fertilizer needs makes it an appropriate crop to use in resource-poor farms, soils with low fertility, and rainfed agro-ecologies. Its planting on marginal areas lowers competition on high-quality fertile areas, promotes the diversification of crops and may enhance land use resilience.

The Barnyard millet has been popular in the organic/low-input farming since it gives not only grain but also straw -grain and fodder. It is valuable in mixed small-holder farms that rely on crop-livestock integration. The straw (remainder) can be recycled to soil as an organic matter and nutrients. It also improves soil structure and soil fertility. Decreased dependence on synthetic fertilizers and pesticides, particularly when it's cultivated in an organic or low-input system will decrease energy requirement, production of fertilizer-associated greenhouse gases, and pollution of the environment. Therefore, the introduction of barnyard millet into the cropping systems can relate to the sustainable agriculture targets and climate-smart agriculture models.

The low-input, organic farming and return of residual biomass in the form of straw suggests accumulation of soil carbon, improved soil structure as well as microbial community. These contributes to long-term soil health and ecological stability.

9.9.3 Nutrient Cycling

Indian barnyard millet absorbs nitrogen from soil in the form of ammonium (NH_4^+) and nitrates (NO_3^-). These forms of readily soluble nitrogen are present because of ammonification and nitrification of organic nitrogen by microorganisms. Farmers also use urea and NPK fertilizers to provide direct soluble nitrogen to crop. Crop rotations with legumes can also improve nitrogen availability in soil by biological nitrogen fixation.

Phosphorus exists in soil as organic residues and inorganic mineral forms. Barnyard millet absorbs this phosphorus in form of $H_2PO_4^-$ ions for ATP production. Farmer increases available phosphorus in soil by adding fertilizer such as DAP, SSP and farmyard manure.

Potassium in soil is present as exchangeable, non-exchangeable and mineral forms. Indian barnyard millet takes it as K^+ ions to supporting osmoregulation and enzyme activation. Farmers also add potash fertilizers (KCl, K_2SO_4) in potassium deficient soils.

Some micronutrients such as Fe, Zn, Mn, Cu, B and Mo are also absorbed by barnyard millet in ionic forms, e.g., Zn^{2+}, Fe^{2+}/Fe^{3+}.

The crop uses these nutrients for it vegetative and reproductive growth. These nutrients are then returned to soil as organic matter and residue recycling in form of straw and roots after harvest. The organic matter is then decomposed by microbes which releases nutrients slowly into soil and enhances soil fertility.

9.10 Case Studies

9.10.1 Case Study 1: Current Status and Future Research Direction of Barnyard Millet for Food and Nutritional Security

Renganathan et al. (2020) have discussed the potential of barnyard millet as a nutritious and climate resilient crop. The authors also analysed the statistics of production, the data of the nutritional composition, and collections of the germplasm to explain the role of barnyard millet in food security and sustainable agriculture. The research is based on the macronutrients and micronutrients composition of barnyard millet and agronomic attributes. The above characteristics mentioned included drought tolerance, tolerance to salinity and lack of adaptability to poor soils and this enables it to be grown in the marginal and rainfed regions. It was concluded that it is significant to recommend strategic interventions to achieve conservation of the germplasm by incorporating genomics-aided breeding and advertising value-added merchandise of barnyard millet. It was further concluded that scientific evidence can offer policy level repositioning of the neglected millets as part of the sustainable agriculture system in India.

9.10.2 Case Study 2: Genetics, Breeding and Genomics of Indian Barnyard Millet

Bhinda et al. (2023) investigated history of breeding and genetic enhancement of barnyard millet in India. The performance of different varieties, pedigree tracks and trait-selection methods were analysed on agro-climatic conditions of the Indian regions. Better Indian varieties were also developed by selection and crossbreeding. In contrast to major cereals, the breeding of barnyard millet has depended so much on phenotypic selection with very little integration of molecular technologies. In this experiment, special consideration was made on characteristics like suitability to mechanized harvesting, drought resistance, nutritional increase and salinity endurance. A roadmap was used to conclude the study, which involved genomics-assisted selection, doubled haploids, and multi-location trial to modernize the breeding of Indian barnyard millet in marginal farmers.

9.10.3 *Case Study 3: Value Addition Through Modified Barnyard Millet Starch*

Bhatt et al. (2023) illustrated how Indian barnyard millet was transformed by starch modification to a valuable industrial component. Grains of barnyard millet were ground into starch which was absorbed chemically to be phosphorylated. It was then tested as per its functional and tableting characteristics. The study promotes the idea of a Nutri-industrial millets in connecting the new-age processing technologies with grains. These findings demonstrate that barnyard millet starch has a significant role to play as pharmaceutical ingredients and functional ingredients in processed foods. The research serves as an example of a relevant value-chain intervention. New industrial markets can be open by extraction of starch and by alteration and not confining the barnyard millet to subsistence use.

9.10.4 *Case Study 4: Identifying Sodicity Tolerant Barnyard Millet Genotypes in India*

Sampath et al. (2023) evaluated 92 genotypes of barnyard millet under sodic soil conditions in Tamil Nadu, India. Morphological, yield-associated characteristics and genetic divergence were evaluated to determine promising lines in the poor soils. The genetic variation was found to be significant in case of plant height, panicle length, grain yield and days to maturity. Genotypes were grouped together to form specific clusters in the multivariate analysis whereby it was possible to identify the better genetic lines with high adaptability and yield performance in sodic conditions. It presents the fact that sodic soils occupy huge hectares of Indian land under farming and constrain productive efficiency of the cereals. Barnyard millet has a role in the reclamation of degraded lands and enhancement of livelihood security of the smallholder farmers to abiotic stress.

9.11 Limitations, Challenges and Research Gaps

Although there are Indian barnyard millet germplasm resources and phenotyping but there are still major gaps. In research gaps there is absence of reference genomes, small number of molecular markers sets, absence of studies of QTL mapping, and few breeding programmes.

Simulation accuracy is compromised without good parameterization, and this is mostly experienced in the stress or marginal conditions. Although nowadays more sophisticated models are simulated to experiment with grain starch and protein under different water/nitrogen ratios, they are yet to be conducted on barnyard millet. Modelling of a quality- and nutrition-focused is still having a research gap.

The hybrid methods (RS + models + ML) require the availability of adequately developed data infrastructure, repetitive measurements, accuracy and local expertise. It is usually missing in smallholder, marginal settings in the case of Indian barnyard millet. Different agro-ecologies have different fertility, planting density, and spacing requirements which cannot be adopted on a large scale due to the absence of agro-regional specifications.

9.12 Future Prospects and Research Priorities

9.12.1 Genomic and Breeding Innovations

Future study of the millet in the barnyard should be done on exploiting the genetic variance and come up with climate resistant and high yielding cultivars. Recent genome-editing techniques, such as CRISPR/Cas9, have the potential to enhance anti-nutritional aspects, seed size, and photosynthetic performance, which offer prospects of fast varietal enhancement. The genetic resources preservation and reduction in time of breeding can be done by establishing comprehensive gene banks and molecular databases. With both genomic and phenotypic analysis, the time spent in development of climate- smart varieties in multi-environment tests can be reduced tremendously (Mukherjee et al., 2025).

9.12.2 Digital Agriculture and Precision Farming

Digital technologies have become important to maximize the production of barnyard millet. Multispectral and AI algorithms can be combined with drones to monitor crop development, nutrient and stress in real-time. Precision irrigation can be enhanced by soil moisture sensors with weather forecasts. This will increase the efficiency of water-use in semi-arid and rainfed areas. Mobile apps are assisting farmers to integrate climate-smart practices by allowing farmers in remote areas to receive site-specific information regarding the optimal time to plant, when to apply fertilizer, and pests. Optimization of resources, minimization of the cost of inputs and increased productivity can be achieved by data-driven decision-making especially when the weather is subjected to irregular rainfall and extreme heat levels (Mukherjee et al., 2025).

9.12.3 Market Integration and Development of Value-Chain

The barnyard millet value chains should be developed to guarantee that the farmers are profitable and that there is food security. Innovative techniques for milling, dehulling and packaging will be able to enhance efficiency and market appeal. The increasing demand of gluten-free, low-glycaemic foods in the urban centres will promote the opportunities of a high value product, including fortified breakfast cereals, noodles, vermicelli and snack bars. The access to seeds and prevention of loss of varietal diversity can be facilitated by Farmer Producer Organizations (FPOs) and community seed banks. Stable demand may be achieved using government-sponsored procurement systems, including access of schools feeding programs. The consumer awareness and adoption could be boosted by improving logistics, cold storage and branding process (Mukherjee et al., 2025).

9.12.4 Climate Resilience Research

Barnyard millet is more resistant to heat, drought, and low soil fertility making it a strategic crop during the climate change conditions. Multi-location and long-term tests facilitate the identification of genotypes that are more economical to stresses and are also stable regarding variability in rainfalls. Studies of root development, osmotic regulation and early-maturity phenotype can improve the capacity of the crop to endure water-stressful conditions. With crop modeling and climate forecasting, it is possible to predict the yielding results under variations in temperature and precipitation to support climate wise-smart agriculture policy and planning. Absorption of carbon and minimization of the emission of greenhouse gases by millet cropping systems also add more weight to the contribution of the crop to sustainable agriculture (Mukherjee et al., 2025).

9.12.5 Ecological and Social Research

In addition to agronomy, the research on ecological and social impacts of barnyard millet should be conducted. Intercropping systems and agroforestry with millet integration boost soil fertility and reduce soil erosion along with the enhancement of biodiversity. Traditional knowledge, especially that kept by women and the tribal farmers who save various landraces is of great importance in conservation of the genetic resources and its culture heritage. The role of growing millet to achieve livelihood security, nutritional outcome and empowerment can be evaluated by the socio-economic studies. Sustainable adoption and sustainability in crops are being derived by active involvement of communities in the conservation of seeds, exchange of knowledge and market linkages (Mukherjee et al., 2025).

9.12.6 *Correlation Heatmap of Nutritional and Functional Components*

The Indian Barnyard Millet in Fig. 9.10 is illustrated in a correlation heatmap that gives a clear representation of the connection between its nutritional, biochemical and bioactive components. The variables have been subdivided into four broad categories, namely, macronutrients, amino acids, minerals, and bioactive compounds, and have been colour-coded to make them easier to understand. The heatmap reveals that there are high positive correlations among macronutrients including protein and energy meaning that high protein content is related to high caloric value. The amino acids, Lysine and Methionine display a medium positive relation with protein indicating the contribution of the two amino acids to the entire protein profile. The minerals display strong positive relationships between phosphorus, magnesium, and calcium, which highlights their occurrence together in the grain. Total phenolics, flavonoids, tannins and carotenoids are some of the bioactive compounds that exhibit different levels of correlation demonstrating a possible interconnected antioxidant potential. The colour labelling and the representation of the lower triangle allows clear interpretation of the functional category and increasing differentiation between them respectively. Overall, the given heatmap is an effective way of demonstrating the interdependence between some of the major nutritional and functional constituents of Indian barnyard millet, which can be used in nutritional assessment and crop enhancement policies.

9.13 Conclusion

Indian barnyard millet (Echinochloa frumentacea) is a short cycle and nutritionally versatile cereal with potential to be a successful sustainable and climate resilient crop. Its distinctive characteristics allow it to achieve constant yields and to thrive and remain stable in the face of adverse environmental factors. These characteristics are its tolerance towards low-fertility and poor-quality soils, drought, heat as well as high water use efficiency. Its short life cycle and occurrence of flowering and grain filling at the same time makes it perfect for unpredictable moisture and rainfall areas. It can combine with modern agronomic technologies, including precision agriculture, remote sensing (Sentinel-2, MODIS), UAV surveillance and crop simulation models (DSSAT, APSIM, AquaCrop). This opens the opportunity in water and nutrient management, yield prediction and greater application in smallholder farming and large-scale farming systems. Case studies in India have also revealed that with the introduction of better improvements to the crop varieties and best practices in agricultural management, the crop can be utilized to enhance food security (Zero Hunger), rural livelihoods and environmental sustainability of the country. The average yields varied between ~1.9 and 2.4 t/ha (between control and NPK-treated plots). This means that although this crop tolerates low-fertility, its yield and nutrient uptake positively respond to balanced fertilization i.e. it can be more nutrient-use

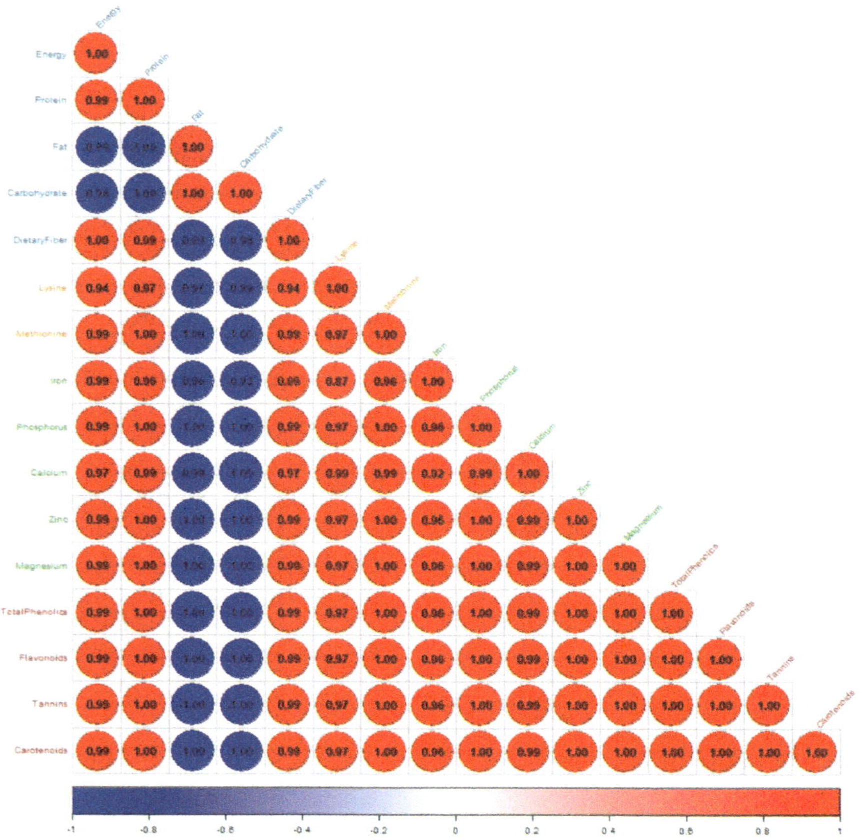

Fig. 9.10 Heatmap of nutritional and functional components in Indian barnyard millet grain

efficient when well managed. It is necessary to integrate the efforts of research, policy support, and community involvement to get maximum output from barnyard millet in adverse climatic circumstances. Research shows that the average barnyard dry fodder yield under Indian conditions is about 5 tons/ha. It has also been determined that barnyard millet generates 2–3 times more biomass than traditional cereals with the same amount of water. It is concluded that Indian barnyard millet is a multi-purpose, resilient and low-carbon grain that is environmentally sustainable in terms of productivity. Increasing its production and ensuring its adoption in the climate smart agriculture will also make a great contribution to the attainment of sustainable development goals.

References

Amin, N. U., Islam, F., Umar, M., Muhammad, W., Rahman, S. U., Gaafar, A.-R. Z., Shah, T. A., Dauelbait, M., & Bourhia, M. (2025). Evaluation of crop phenology using remote sensing and decision support system for agrotechnology transfer. *Scientific Reports, 15*(1), 11582. https://doi.org/10.1038/s41598-025-95109-4

Arshad, S., Kazmi, J. H., Javed, M. G., & Mohammed, S. (2023). Applicability of machine learning techniques in predicting wheat yield based on remote sensing and climate data in Pakistan, South Asia. *European Journal of Agronomy, 147*, 126837. https://doi.org/10.1016/j.eja.2023.126837

Bhatt, D., Rasane, P., Singh, J., Kaur, S., Fairos, M., Kaur, J., Gunjal, M., Mahato, D. K., Mehta, C., Avinashe, H., & Sharma, N. (2023). Nutritional advantages of barnyard millet and opportunities for its processing as value-added foods. *Journal of Food Science and Technology, 60*(11), 2748–2760. https://doi.org/10.1007/s13197-022-05602-1

Bhinda, M. S., Joshi, D. C., Parihar, M., Meena, R. P., Joshi, P., Gupta, A., Kant, L., & Sood, S. (2023). Genetics, breeding, and genomics of Indian barnyard millet (Echinochloa frumentacea). In *Neglected and Underutilized Crops* (pp. 115–135). Elsevier. https://doi.org/10.1016/B978-0-323-90537-4.00017-X

Dhanalakshmi, R., Subramanian, A., Thirumurugan, T., Elangovan, M., & Kalaimagal, T. (2019). Genetic variability and association studies in barnyard millet (*Echinochloa frumentacea* (Roxb.) link) germplasm under sodic soil condition. *Electronic Journal of Plant Breeding, 10*(2), 430. https://doi.org/10.5958/0975-928X.2019.00055.3

Hasan, N., Budakoti, M., Meena, R. P., Bhinda, M. S., Rani, V., Nandini, C., & Joshi, D. C. (2025). Barnyard millet. In S. A. Ceasar, S. Penna, C. W. P. Carvalho, & S. M. Jain (Eds.), *Millets: Crops for climate resilience and for food and nutritional security* (pp. 67–91). Springer Nature Singapore. https://doi.org/10.1007/978-981-95-1256-0_3

Ishaq, T., Masood, A., Saad, A., Singh, L., Mehdi, S., Rashid, S., Sharma, A., Aziz, M. A., Rai, U., Qasim, N., Mir, M. S., Palmo, T., & Rafiq, A. (2025). Climate resilience of millets: A sustainable solution for future agriculture. *International Journal of Research in Agronomy, 8*(5), 744–754. https://doi.org/10.33545/2618060X.2025.v8.i5j.2979

Jayakodi, M., Madheswaran, M., Adhimoolam, K., Perumal, S., Manickam, D., Kandasamy, T., Yang, T.-J., & Natesan, S. (2019). Transcriptomes of Indian barnyard millet and barnyardgrass reveal putative genes involved in drought adaptation and micronutrient accumulation. *Acta Physiologiae Plantarum, 41*(5), 66. https://doi.org/10.1007/s11738-019-2855-4

Kuraloviya, M., Vanniarajan, C., Sudhagar, R., & Vetriventhan, M. (2022). Phenotypic diversity and stability of early maturing barnyard millet (Echinochloa sp.) germplasm for grain yield and its contributing traits. *Indian Journal of Experimental Biology, 60*(12), 918–924. https://doi.org/10.56042/ijeb.v60i12.36064

Maithani, D., Sharma, A., Gangola, S., Bhatt, P., Bhandari, G., & Dasila, H. (2022). Barnyard millet (Echinochloa spp.): A climate resilient multipurpose crop. *Vegetos, 36*(2), 294–308. https://doi.org/10.1007/s42535-022-00420-4

Maitra, S., Praharaj, S., Hossain, A., Patro, T. S. S. K., Pramanick, B., Shankar, T., Pudake, R. N., Gitari, H. I., Palai, J. B., Sairam, M., Sagar, L., & Sahoo, U. (2022). Small millets: The next-generation smart crops in the modern era of climate change. In R. N. Pudake, A. U. Solanke, A. M. Sevanthi, & P. Rajendrakumar (Eds.), *Omics of climate resilient small millets* (pp. 1–25). Springer Nature Singapore. https://doi.org/10.1007/978-981-19-3907-5_1

Mishra, N., Tiwar, S., Sapre, S., & Nema, S. (2022). Qualitative characterization and categorization of barnyard millet (Echinochloa spp.) accession, collected from different regions of India. *Research Trend*. https://www.researchtrend.net/

Mukherjee, B., Jha, R. K., Sattar, A., Dutta, S., Bhattacharya, U., Samanta, S., Huirem, B., Singh, S. K., Das, S., & Bal, S. K. (2025). Harnessing the potential of millets for climate-resilient and sustainable agriculture. *Frontiers in Plant Science, 16*, 1574699. https://doi.org/10.3389/fpls.2025.1574699

Muskan, F., Redhu, M., Redhu, S., & Rahimi, M. (2025). Millets in the global market: A critical review of challenges and opportunities. *Food Production, Processing and Nutrition, 7*(1), 37. https://doi.org/10.1186/s43014-025-00316-1

Nagaraja, T. E., Nandini, C., Bhat, S., Parveen, S., & G. (2023). Artificial hybridization techniques in small millets—A review. *Frontiers in Plant Science, 14*, 1112117. https://doi.org/10.3389/fpls.2023.1112117

Oikonomidis, I., & Trevezas, S. (2024). Cumulative link mixed-effects models in the service of remote sensing crop progress monitoring. *Biometrics, 80*(4), ujae137. https://doi.org/10.1093/biomtc/ujae137

Renganathan, V. G., Vanniarajan, C., Karthikeyan, A., & Ramalingam, J. (2020). Barnyard millet for food and nutritional security: Current status and future research direction. *Frontiers in Genetics, 11*, 500. https://doi.org/10.3389/fgene.2020.00500

Sampath, L., Dhanalakshmi, R., Kalaimagal, T., Susmitha, D., & Antony, B. J. (2023). Evaluation of genetic diversity in sodic soil-grown barnyard millet (Echinochloa frumentacea (Roxb.) Link) germplasm. *International Journal of Plant & Soil Science, 35*(18), 387–392. https://doi.org/10.9734/ijpss/2023/v35i183301

Selvam, R., Santhi, R., Maragatham, S., Gopalakrishnan, M., & Rejeswari, R. (2022). Targeted yield model-based balanced nutrient recommendation for barnyard millet on Inceptisol of Tamil Nadu. *Indian Journal of Agricultural Research*. https://doi.org/10.18805/IJARe.A-5891

Singh, B., Pandey, S., Nivedhitha, S., Shekhawat, N., Singh, M., Jat, B., Pandey, C. D., Semwal, D. P., Arya, L., Gautam, R. K., & Singh, G. P. (2025). Comprehensive phenotyping of 1,807 Indian barnyard millet (Echinochloa frumentacea link) accessions from Indian national genebank: Unlocking diversity for core set development. *Frontiers in Plant Science, 16*, 1644491. https://doi.org/10.3389/fpls.2025.1644491

Sood, S., Khulbe, R. K., Gupta, A. K., Agrawal, P. K., Upadhyaya, H. D., & Bhatt, J. C. (2015). Barnyard millet—A potential food and feed crop of future. *Plant Breeding, 134*(2), 135–147. https://doi.org/10.1111/pbr.12243

Sukanya, T. S., Chaithra, C., Nagaraja, T. E., Latha, H. S., Hadli, C., & D. (2022). Sustainable production of barnyard millet through improved production technologies. *International journal of environment and climate change*, 16–23. https://doi.org/10.9734/ijecc/2022/v12i230625

Yang, Z., Yu, Z., Wang, X., Yan, W., Sun, S., Feng, M., Sun, J., Su, P., Sun, X., Wang, Z., Yang, C., Wang, C., Zhao, Y., Xiao, L., Song, X., Zhang, M., & Yang, W. (2024). Estimation of millet aboveground biomass utilizing multi-source UAV image feature fusion. *Agronomy, 14*(4), 701. https://doi.org/10.3390/agronomy14040701

Zhou, S., Liu, Z., & Chen, F. (2025). Developing the DSSAT-CERES-millet model for dynamic simulation of grain protein and starch accumulation in foxtail millet (Setaria italica) under varying irrigation and nitrogen regimes. *Plants, 14*(6), 910. https://doi.org/10.3390/plants14060910

Chapter 10
Finger Millet (*Eleusine coracana*): Conservation Agriculture and Soil Erosion Control

Umar Farooq and Muhammad Imtiaz Rashid

Abstract Finger millet (*Eleusine coracana*) is a climate-tolerant cereal highly required in food security in semi-arid and marginalized conditions but the production is dramatically limited by soil degradation and erosion. In this chapter, the author discusses the importance of conservation agriculture (CA) as a sustainable measure to improve soil health, productivity and climate resiliency of finger millet systems. These aimed to test CA principles, including minimum soil disturbance, permanent soil cover, and crop diversification, and to test their performance in the reduction of erosion, enhancement of water-use efficiency and maintenance of yield in dryland conditions. Both field-based experiments, long-term CA trials, and modelling strategies were addressed using RUSLE (Revised Universal Soil Loss Equation), APSIM (Agricultural Production Systems simulator), and DNDC (Denitrification–Decomposition) model) with the assistance of remote sensing instruments, i.e., Sentinel-2, Landsat, and SMAP to monitor soils and crops. Case studies in Zimbabwe, India, and the East African highlands show that CA practice potential can decrease soil loss by 40–55% and raise soil moisture by 15–20% and raise finger millet grain output in the conventional systems of 1.5–1.8 t ha^{-1} to 2.2–2.4 t ha^{-1} with CA practices. The retention of residues, intercropping, and stress-tolerant varieties enhance the fertility of soil, the biomass accumulation, and the stability of yield. The directions in the future are CA-adapted breeding, digital agriculture, AI/ML-based decision support, and long-term monitoring in order to optimize the adoption and sustainability. All these plans facilitate the production of climate-resilient finger millet, conservation of soil and livelihood of smallholders.

Keywords Finger millet · Conservation agriculture · Soil erosion · Climate resilience · Remote sensing

U. Farooq (✉)
Department of Agronomy, Pir Mehr Ali Shah Arid Agriculture University, Rawalpindi, Pakistan
e-mail: umerfaroooq42@gmail.com

M. I. Rashid
Center of Excellence in Environmental Studies, Department of Environmental Sciences, King Abdulaziz University, Jeddah, Saudi Arabia

M. Ahmed (ed.), *Climate Change and Neglected Underutilized Crops*,
https://doi.org/10.1007/978-3-032-21005-0_10

10.1 Introduction

Finger millet (*Eleusine coracana*), is a small-grained cereal that was domesticated in the African highlands and has since attracted International attention because of incredible ability to withstand climate variations and ability to survive in marginal areas where major cereals are usually killed. With rising frequency of droughts, soil deterioration, and production unpredictability in the tropics due to the rising climate change, finger millet has become more important as a strategic crop that can be used to provide food and nutrition security in affected areas. International evaluations have noted its appropriateness to dryland agriculture because it has C4 photosynthetic system, good water-use efficiency, and high tolerance to temperatures and can deliver constant yields in low-input conditions. These qualities make finger millet one of the priority crops towards enhancing climate-resilient food systems in sub-Saharan Africa and South Asia (Dzudzor & Gerber, 2023).

Finger millet at international level serves the purpose of giving food security to millions of small hold farmers, specifically in East Africa (Ethiopia, Uganda, Kenya, and Tanzania) and South Asia (India, Nepal, Sri Lanka). It is a good source of essential amino acids, calcium, iron, dietary fiber and antioxidants and therefore a useful cereal in the prevention of micronutrient deficiencies and enhance the nutritional diversity of rural populations. Nutrition aside, finger millet also has agronomic benefits including poor-soil tolerance, not only in shallow or sloping fields, but also in mixed and inter-crop programs common to farming systems of Africa and Asia. However, the crop is not fully used at the international scale because little research has been done on it, it has not gained popularity in the market, and it lacks full integration into modern farming technologies (Fikadu, 2022).

Finger millet is a key component at the national and regional scales and especially in India and East African states, which is important in maintaining rural livelihoods, traditional food systems and resilience in low-input and rainfed agricultural systems. Nevertheless, increasing threats to production areas have been caused by soil degradation and erosion, which destroys the soil fertility, the water retention capacity, and the ecological stability in the end. The steep slopes, unpredictable rainfall patterns, and intensive cultivation activities that lack proper soil conservation measures increase the level of nutrient depletion and speed up the topsoil erosion in most regions where millet thrives. Such challenges have adverse impacts on crop production, susceptibility to climate shocks, and farm-level profitability of smallholders who mostly rely on rainfed production (Ahmadi Mirghaed & Souri, 2022).

Conservation agriculture (CA) has come into the limelight as a viable initiative to overcome these limitations to enhance the soil health, reduce soil erosion, improve moisture-conservation, and stabilize yields during unpredictable climatic conditions. The CA principles of minimum soil disturbance, permanent soil cover, and diversified cropping are in agreement with the ecological needs and the production features of the finger millet. The crop is also sensitive to residue retention, decreased tillage and legume based diversification that brings about better soil structure, stabilized soil temperature, lower runoff as well as increased cycling of nutrients. The experience in Africa and India proves that CA systems can significantly enhance the soil moisture content, decrease soil erosion on slanting land, and enhance a grain yield and yield stability of finger millet in comparison with the traditional tilling practices (Mbanyele et al., 2022).

This chapter discusses the ecological principles, agronomic potential and the benefits of soil conservation of finger millet under the conservation agriculture systems. It examines production dynamics in the world and the region, erosion dilemma in millet systems, and CA effects on soil health, water dynamics, and crop productivity. Recent developments in the areas of modelling and remote sensing to monitor soil erosion, determine the performance of CA and assist in decision-making in millet-growing landscapes are also incorporated in the chapter. This chapter combines the evidence available in the present day science, and it will give us an excellent insight into the way in which conservation agriculture will enhance the climate resilience and sustainability of the finger millet production systems.

10.2 Ecology, Taxonomy, Phenology, Distribution and Production Systems

10.2.1 Taxonomy and Botanical Classification

Finger millet (*Eleusine coracana L.* Gaertn.) is a grass species that is in the Poaceae family, subfamily Chloridoideae, and tribe Chlorideae, and this includes other drought-tolerant grasses adapted to dry and semi-arid environments. Eleusine is a genus that has nine species, out of which only E. coracana subsp. coracana is grown with E. coracana subsp. Africana being considered as its wild ancestor. The East African highlands observed domestication with the selection process being done by the early farmers of the area and the selection was based on compact panicles, non-shattering grains and better thresh ability in the plants (Fig. 10.1). The crop is typified by its very specific inflorescence consisting of the spikes shaped in fingers radiating around the central rachis, tiny grains of the caryopsis, strong root zone, and the high adaptability to low or eroded soils. Such botanical properties make it resilient in areas with low precipitation and frequent cases of soil erosion (Steed et al., 2022) (Fig. 10.2).

10.2.2 Phenology and Growth Stages

Finger millet has a regular growth cycle, which depends on the temperature, moisture content of the soil as well as the varietal properties. Immaturity and early germination when warm and moist, germination takes between one and two weeks. The vegetative stage entails the growth of the plant in terms of leaf development and tillering,

Fig. 10.1 Flower morphology of different landraces of finger millet. **a** Coracana; **b** Plana; **c** Elongata; **d** Compacta; and **e** Vulgaris (Ceasar et al., 2023)

Fig. 10.2 Morphology of finger millet (Ceasar et al., 2023). **a** Entire plant; **b** stem and leaf arrangement; **c** entire inflorescence; **d** spikelets; **e** floret [1) Lemma, 2) Palea and 3) outer glume]; and **f** seeds (Ceasar et al., 2023).

which when combined dictate the density of the plants and the possible formation of the yield. The transition to the reproductive stage is the panicle initiation after which flowering occurs with temperature and moisture stress potentially having a great influence on formation of grains. The grain filling process is fast when the soils are moist and when there is a long period of dryness, the grain size and ultimate yield decreases. It normally takes 90–120 days until they attain physiological maturity; this varies depending on the genotype and the environment. The phenological reactions are very delicate to climate and management, and therefore timely sowing, retention of the residues, and preservation of moisture are critical in stable performance (Grzesiak et al., 2023).

10.2.3 Ecological Adaptation and Climatic Requirements

The finger millet is specifically adapted to semi-arid and sub-humid areas whereby the rainfall is between 350 and 900 mm per year. The C4 photosynthetic pathway makes it efficient with respect to water-use and enables it to absorb carbon efficiently even at high temperatures and low humidity. The crop can be planted on a variety of soils such as sandy loams, lateritic and shallow upland soils with an average level of acidity. It grows best between 27 °C to 32 °C and can endure short-term exposure to high temperatures. Its ability to establish a deep-rooted system that is fibrous supports survival during intermittent drought periods, and the ability to grow on slopes, makes it effective on the landscape where there is erosion potential. These ecological characteristics enable finger millet to survive in most of the difficult environmental conditions compared to a variety of cereals (Zhang et al., 2023a, 2023b, 2023c).

10.2.4 Global Distribution and Production Zones

A wide spread of finger millet is found in East Africa and specifically in Ethiopia, Uganda, Kenya and Tanzania where it forms a major source of home food and local market. India is the biggest producer of this in South Asia, with Nepal and Sri Lanka coming next, and the planting is focused on the rainfed uplands. Small pockets of Southeast Asia and highlands of South America are also, where the crop is grown. Its production zones are spread over various agro-ecological zones that span through dry savannah areas up to high altitude plateaus (Fig. 10.3). The current trends in production show that millet is increasingly gaining popularity in terms of agricultural benefits and resistance to the climate, but the total acreage of the crop is not very high compared to the major cereals because of the lack of policy (both in terms of support and integration) and its inability to integrate into the market (Tsunemi et al., 2023).

The global area and production of finger millet is properly collected, analysed and summarised Fig. 10.2. When compared to the entire area (11 million ha) and production (14.6 million tons), finger millet is produced in over 2.1 m ha of land and a production of 3.7 million tons per year. These values are to the 20% and 26% in the total area covered and production respectively of finger millet to the world millets. The main producer of finger millet in the world is India, with total production area of one million ha, which covers 46% of the total production, and then comes Ethiopia (with total production area of 480,511 ha and 22% share of the world) (Fig. 10.4).

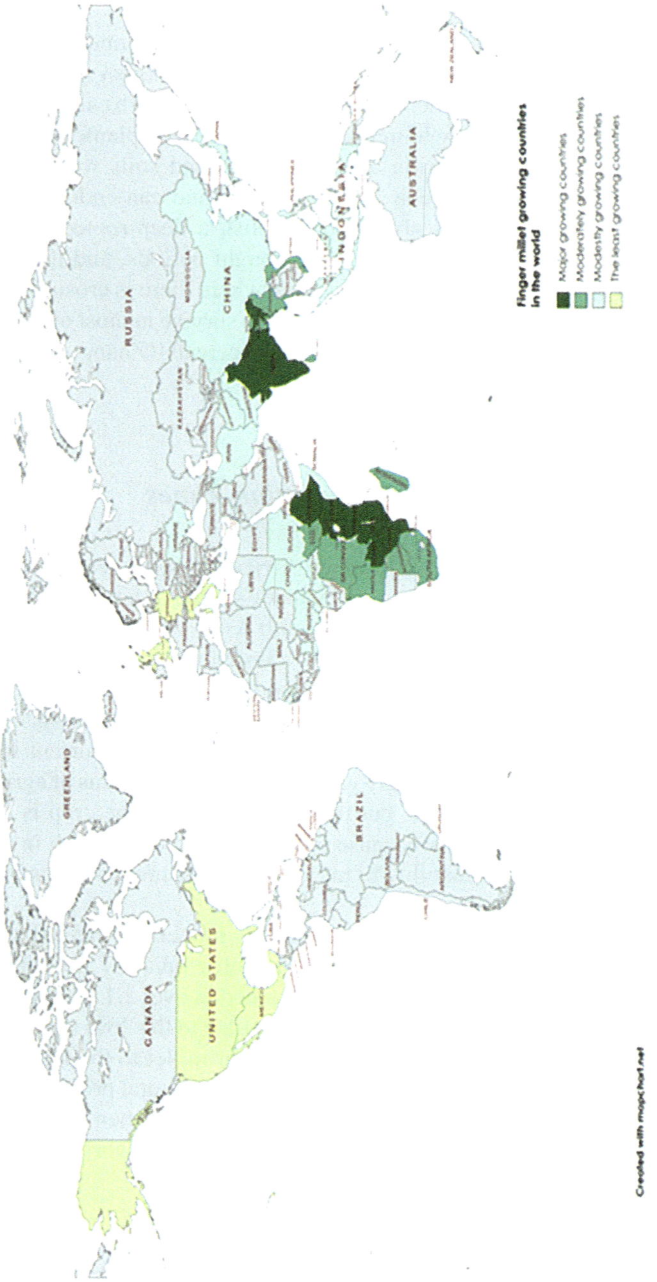

Fig. 10.3 Map of finger millet growing countries in the world (Gebreyohannes et al., 2024)

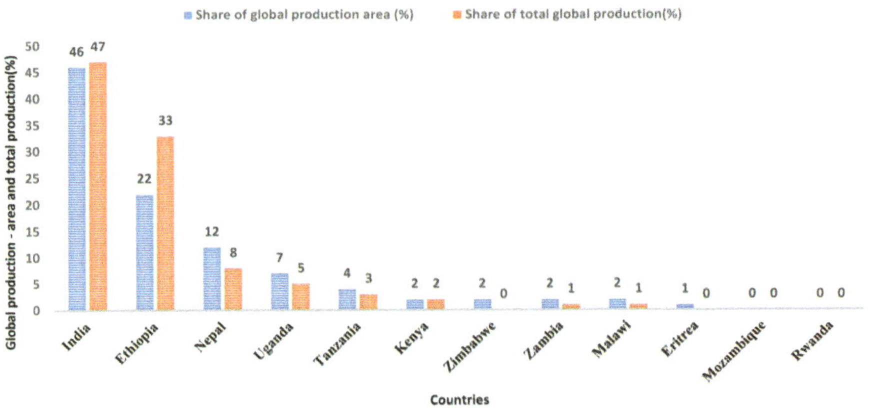

Fig. 10.4 Global area and total production (%) of finger millet in major growing countries (Gebreyohannes et al., 2024)

10.2.5 Cropping Systems

Finger millet is typically planted in intercropping with legumes such as cowpea, pigeon pea and horse gram that enhance fertility of the soil and prevent soil erosion by enhancing the ground cover. Mixed farming, which is sorghum or maize, is also used by small hold farms to distribute the risk of climate. Most parts where moisture conservation is a concern plant the crop on rainfed upland systems. Finger millet-legume rotations are more favorable to the cycle of nitrogen and help in sustenance of long-term soil productivity (Fig. 10.5). It has been shown to be effective when mixed with conservation agriculture practices, in particular residue retention and minimum tillage to improve the soil structure, moisture retention, and yield stability in drylands (Zhang et al., 2023a, 2023b, 2023c) (Table 10.1).

10.2.6 Production Constraints

Even at its toughness, the production of finger millet is limited by the decreasing soil fertility and low organic content with the widespread erosion of soils in upland and hilly areas. Unpredictable rainfall patterns and extended dry periods also decrease the establishment of crops and grain filling. The pressure of weeds is a significant issue, particularly at the initial growth stage, and the yields in low-input systems can be highly restricted by the competition. Diseases like the blast also affect productivity and pests like the stem borers. Inaccessibility to better varieties, inadequate use of fertilizers and low levels of mechanization are the factors that lead to the continued

Fig. 10.5 Red Gram +black gram +Finger millet

Table 10.1 Comparison of finger millet yield and nitrogen use efficiency in legume-based intercropping systems (Panotra et al., 2025)

Intercropping system	Finger millet yield (kg/ha)	Legume yield (kg/ha)	Nitrogen use efficiency (NUE)	Soil nitrogen (mg/kg)
Finger millet + green gram	1500	800	High (25% increase)	50
Finger millet + cowpea	1450	700	Moderate (15% increase)	45
Finger millet + soybean	1400	850	High (22% increase)	52
Sole finger millet	1200	N/A	Low (no increase)	40

low yield in most areas of production. These limitations indicate that conservation-oriented management practices are required to promote productivity on the long-term basis (Grzesiak et al., 2023)

10.3 Soil Degradation and Erosion Challenges in Finger Millet Landscapes

10.3.1 Overview of Soil Degradation in Finger Millet Systems

The process of soil degradation has been a major challenge to the productivity and stability of the finger millet systems especially in the dry lands areas where the crop thrives mostly. Splash erosion occurs primarily in regions that have bare soil surfaces whereby the impact of raindrops carries away fine particles and causes them to move downstream. The intensity of rainfall causes sheet erosion whereby the rainy water in large field areas removes thin layers of the topsoil and with time, the soil becomes poorer in nutrients. Rill erosion occurs on sloping terrain because of the runoff as small channels, which become deeper as time passes and increase the soil erosion. All these types of erosion tend to coexist and to erode the soil, the levels of organic matter and long-term high fertility needed to support sustainable cultivation of finger millet (Feizpoor et al., 2022).

10.3.2 Drivers of Soil Erosion

A number of interrelated forces enhance erosion in areas that grow millet. Deforestation and clearing of the natural plants decrease the protective cover on the ground and this leaves the soil surfaces very susceptible to erosion by raindrops and overland water. Overgrazing also leaves the soil bare, by decreasing biomass, undermining root systems and compacting the surface, preventing infiltration. Slopey terrain (characteristic of the highlands in the East Africa and uplands in South Asia) is prone to speedy run-offs during severe storms particularly in areas where farmers use conventional means of tillage that expose the soil to the elements. The occurrence of unpredictable rainfall, which relates to climate change, contributes to the occurrence of heavy storms, which lead to sudden runoff that mobilizes sediment and increases the risk of erosion (Pineiro-Juncal et al., 2022).

10.3.3 Link Between Erosion and Soil Fertility Decline

Nutrient loss is strongly associated with soil erosion in that the topsoil, which is eroded, has the largest amount of organic carbon, nitrogen, phosphorus and other vital micronutrients. With the deteriorating soil fertility, there has been limited nutrient absorption, root growth, and resistance to drought in the finger millet plants. The decrease in yield is usually realized in a short period of seasons especially in areas that are not covered with residues or those whose soil is not covered with soil-conservative methods. There is also constant erosion that decreases the soil depth and this decreases the water storage and limits the penetration of roots, which compounds the stress of moisture in dry seasons. This compounded loss of nutrients, inadequate water supply, and poor soil structure have a dire effect on the yield capacity of finger millet in affected areas (Xiong et al., 2022).

10.3.4 Vulnerability of Smallholder Dryland Systems

Dryland systems of smallholder farmers are particularly susceptible to soil erosion caused by the insufficient access to the fertilizers, organic amendments, and the soil conservation technology. The vast majority of farmers have to do with rainfed production, and they have little means to rehabilitate damaged soils after they have been eroded. The extensive application of conventional tillage methods, and low residue cover, exposes fields during the beginning of the growing cycle when the amount of rainfall is most erosive. In most societies, the land fragmentation and population pressure compel farming in marginalized and steep slopes leading to high erosion vulnerabilities. Unless conservation agriculture and other interventions are applied, the degradation of soil in finger millet landscapes will persist and hinder food security and low sustainability of these agricultural systems over the long term (Li et al., 2023).

10.4 Conservation Agriculture Principles and their Relevance to Finger Millet

10.4.1 Minimum Soil Disturbance

Minimal soil disruption is one of the fundamental concepts of conservation agriculture that focuses on less tillage activities including ripping, basin planting and minimal soil inversion. The practices preserve the soil structure; minimize soil erosion, and increase water infiltration and retention, most of which are important to the production of finger millet in drylands. Farmers will be able to conserve soil aggregates, enhance porosity, and preserve a conducive microenvironment of microbial activity by reducing soil disturbance, promoting nutrient cycling and improving soil health (Sartori et al., 2022).

10.4.2 Permanent Soil Cover

Permanent soil cover is essential in preventing soil erosion by mulching and retention of crop residues to maintain moisture in the soil. The residence cover keeps the soils safe against precipitation influence, lessens runoff, and evenly distributes the soil temperature. This practice is also associated with the boosting of soil organic matter content in finger millet systems and it increases soil fertility and improves microclimate of the soil to seed germination and plant growth. Permanent cover therefore helps in the short-term crop yield and long-term soil sustainability (Araya et al., 2022).

10.4.3 Crop Diversification

Crop diversification, including legume intercropping and crop rotation, strengthens soil fertility and mitigates erosion in finger millet landscapes. Intercropping with legumes adds biologically fixed nitrogen to the soil, while rotations interrupt pest and disease cycles and maintain nutrient balance. Diversified cropping systems enhance soil cover throughout the year, reducing vulnerability to erosion and improving resilience against climatic variability. For finger millet, integrating crop diversification strategies contributes to sustainable intensification and soil conservation simultaneously (Porwollik et al., 2022).

10.4.3.1 Adaptation of Conservation Agriculture to Smallholder Finger Millet Farmers

The adoption of conservation agriculture by smallholder finger millet farmers through adaptation would not always be a total adoption, as resources and field conditions would limit it. Such obstacles as inadequate equipment access, labor scarcity, and broken land tenures require feasible adjustments of CA principles. Even partial adoption like lower tillage with retention of the residue or selective intercropping can still result in meaningful erosion reduction, enhanced soil fertility and maintenance of yields. Individualized strategies putting into consideration local situations and the capacity of farmers are thus needed to scaffold conservation farming in finger millet systems (Teng et al., 2024).

10.5 Impacts of Conservation Agriculture on Finger Millet Productivity

10.5.1 Soil Moisture and Water-Use Efficiency

The application of conservation agriculture especially in residue retention and mulching contributes a great deal to soil moisture and water-use efficiency in the finger millet systems. Mulch lowers the amount of evaporation at the soil surface, increases the rate of infiltration and alters the fluctuations in soil temperatures, thus providing a more stable moisture regime at times of important growth. The field research in semi-arid and dryland areas has proved that CA practices can enrich the supply of soil water and enhance the capacity of the crop to endure the water stress, which is especially critical in the rainfed cultivation of finger millet (Sartori et al., 2022).

10.5.2 Soil Structure and Erosion Reduction

Conservation agriculture has a great effect on soil structure, and it diminishes the risk of soil erosion in the finger millet landscapes. Minimal tillage and retention of residues are some of the practices that enhance aggregate stability and the development of stable macro aggregates in the soil. The improvements increase the retention of water, decrease overland runoff, and consequently decrease the loss of sediments and nutrients. Long-term CA trials have shown that the soils that are managed through conservation techniques have superior porosity and infiltration ability as compared to the traditionally plowed systems and thus this leads to the protection of soil and increases the sustainability of finger millet production (Teng et al., 2024).

10.5.3 Soil Biological and Nutrient Cycling Improvements

The main advantage of conservation agriculture is the resulting enhanced soil biological activity, which enhances nutrient cycling in the finger millet systems (Table 10.2). The resultant retention and minimized disturbance lead to a positive environment of microbial communities, which enhances the accelerated breakdown of organic matter and release of nutrient essentials. Bigger microbial biomass and activity enhance soil fertility and promote sustainable provision of nutrients, and dynamics of organic matter under CA promote improvement of soil structure and water-holding capacity. The benefits of this type of improvement of soil biology are reflected in the growth of crops and their resistance to the fluctuating environmental conditions (Singh et al., 2023).

10.5.4 Crop Growth and Yield Response

Finger millet is sensitive to conservation agriculture practice, as shown by the enhancement in the production of the tillers, formation of fingers, biomass and stability in the yield. CA does not only increase productivity when conditions are

Table 10.2 Impact of conservation tillage and organic fertilization on soil health and microbial activity (Panotra et al., 2025)

Tillage system	Fertilization type	Soil organic matter (%)	Soil pH	Microbial activity (CFU/g soil)	Water retention (%)
Conventional tillage	Inorganic fertilizers (NPK)	1.5	6.0	5×10^6	18
Conservation tillage (no-till)	Organic fertilizers (FYM)	3.0	6.4	9×10^6	25
Conservation tillage (strip-till)	Organic fertilizers (vermicomposting)	3.5	6.5	12×10^6	27
Conventional tillage	Organic fertilizers (poultry manure)	2.8	6.2	7×10^6	22

favorable but also helps in protecting the crop against effects of drought and unpredictable rainfall. Comparative analysis indicates that the finger millet that is cultivated under CA records a higher and more consistent yield than that cultivated using conventional tillage, which indicates the composite advantages of increased soil moisture, nutrient sustainability, and less soil erosion (Table 10.3). These benefits cause CA to be a significant approach in maintaining the ability of finger millet to grow and thrive in marginal and drylands (Porwollik et al., 2022).

Table 10.3 Yield improvement in finger millet and legume intercropping systems with conservation tillage and organic fertilization (Panotra et al., 2025)

Intercropping system	Tillage system	Fertilization type	Finger millet yield (kg/ha)	Legume yield (kg/ha)	Total yield (kg/ha)
Finger millet + green gram	Conservation tillage (no-till)	Organic fertilizers (FYM)	1600	900	2500
Finger millet + cowpea	Conservation tillage (strip-till)	Organic fertilizers (Vermicompost)	1700	850	2550
Finger millet + soybean	Conservation tillage	Organic fertilizers (NPK)	1400	800	2200
Sole finger millet	Conventional tillage	Organic fertilizers (poultry manure)	1300	N/A	1300

10.6 Conservation Agriculture as a Soil Erosion Control Strategy

10.6.1 Mechanisms of Erosion Reduction under Conservation Agriculture

Conservation agriculture is an effective measure in reducing soil erosion by a combination of various ways. The decreased tillage and the removal of crop residues slow down runoff velocity and increase the water infiltration, thus reducing the detachment and transportation of soil particles. The presence of residues and the lack of disturbance on the soil surface protect the soil surface against the impact of raindrops, consequently preventing the splash erosion of the soil, as well as preventing the development of rills. All these mechanisms enhance the aggregate stability of the soil and hence the topsoil depth was also important in ensuring the nutrient availability and crop yields of finger millet systems. The conservation agriculture also helps slow down the movement of sediments to lower streams by maintaining soil structure and facilitating infiltration as well as fostering wider watershed protection (Zhang et al., 2023a, 2023b, 2023c).

10.6.2 Role of Vegetative and Residue Cover

The finger millet canopy being vegetated cover is a key in the safeguarding of the soil surface. Lush plant covers absorb rain, they absorb and dissipate energy and reduce soil erosion, and roots increase soil cohesion and surface stability. Intercropping of legumes or cover crops further enhances ground cover, which has the effect of offering year round cover when the main crop canopy is weak. Cover on the surface of the soil enhances vegetative cover by ensuring that there is a physical obstacle against erosion processes, as well as, balancing the microclimate of the soil. Such mixture of both living and dead biomass will make sure that the soil erosion is reduced but at the same time increase the moisture retention and soil fertility (Farias et al., 2021).

10.6.3 Quantitative Evidence from Millet-Growing Regions

Quantitative evidence of millet-producing regions Field studies on the effectiveness of conservation agriculture on soil loss reduction provide solid quantitative evidence. Comparative experiments in Zimbabwe, India and East African highlands reveal that plots in CA practices have much reduced surface runoff and sediment yield as compared to traditional tilled fields. The changes have been recorded to reduce soil loss by 30–60% based on the residual management, slope, and the intensity of rainfalls. The extensive trials conducted over long periods prove that better infiltration and aggregate stability and the cover of the soil with continuous covers lead to the erosion control and to the permanent yields of finger millet. These results affirm that conservation agriculture is a feasible and viable approach that ensures that the health of the soil is maintained and that any degradation of the soil is reduced in the case of smallholder millet systems (Mallu et al., 2026).

10.7 Genetic and Agronomic Advances Supporting Conservation Agriculture Adoption in Finger Millet

10.7.1 Genetic Diversity and Breeding Progress

Finger millet has a high genetic diversity and this is the basis of breeding programs to increase the adaptability of the crop in conservation agriculture. Samples of different agro ecological areas have shown a high level of variation with respect to characteristics including tillering capacity, root architecture and biomass production which are important in soil cover and reduction of erosion. This diversity has been used by modern breeding practices to come up with breeds that breed well under low-input and minimal-tillage practices. This is because the incorporation of locally adapted germplasm along with better breeding lines will enable smallholder farmers to practice CA without yield potential being affected (Gebreyohannes et al., 2024).

10.7.2 Stress-Tolerant Cultivars

The use of conservation agriculture in marginal environments has been enhanced by the production of stress-tolerant finger millet varieties. Resistance to blast disease, drought and heat stress resistance has been the basis of breeding. The cultivars have the characteristics of constant growth with unpredictable rainfall, decreased soil disruption and consequent trash retention as the main characteristics of CA systems (Table 10.4). With merged biotic and abiotic stress tolerance, these enhanced varieties contribute to long-term productivity and minimized the requirement of intensive management inputs, which increases the viability of CA adoption by resource-constrained farmers (Mbanyele et al., 2022).

10.7.3 Traits Supporting Conservation Agriculture

Certain characteristics of plants have a great impact on the success of conservation agriculture. Biomass production and early vigor helps in the improvement of soil

Table 10.4 Drought-tolerant finger millet varieties reported globally (Gebreyohannes et al., 2024)

Variety name	Trait description	Breeding method	Days to maturity	Year of release	Yield (t/ha)	Country	References
Mereb-1 (KNE#622)	Drought tolerance	PR	83–121	2016	2.5–3.2	Ethiopia	MOA (2022)
Kako-1 (ACC#203351)	Drought tolerance	PR	139	2016	2.6–3.0	Ethiopia	MOA (2022)
Axum (ACC#229355)	Drought tolerance	PR	147	2016	2.2–3.6	Ethiopia	MOA (2022)
Dibya Sinha	Drought tolerance	MB	90–100	1971	2.0–2.5	India	Hatakeyama et al. (2018), Aralikatti & Chaturvedi (2020)
K6	Drought tolerance	–	100–105	1982	2.3–2.5	India	–
K1 (Kolar Gidda Ragi)	Drought tolerance	PR	130–135	1939	0.9–1.0	India	–
CO2	Drought tolerance	–	110–115	1942	2.0–2.5	India	–
CO1	Drought tolerance	–	120–125	1942	2.5–3.0	India	–
Aruna	Drought tolerance	–	95–100	1956	1.5–1.8	India	–
Vzm1	Drought tolerance	–	105–110	1958	2.0–2.5	India	–
CO8	Drought tolerance	–	90–95	1963	1.5–2.0	India	–
CO10	Drought tolerance	–	110–120	1976	3.0–3.5	India	–
PR202 (Godavari)	Drought tolerance	–	110–115	1978	3.0–3.2	India	–
CO11 (EC4849)	Drought tolerance	–	106–110	1982	3.0–3.5	India	–
VL204	Drought tolerance	–	90–100	1985	1.8–2.0	India	–

cover to minimize the erosion and loss of moisture. Improved root systems enhance aggregation of the soil and infiltration of water that supplements the physical advantages of residue retention and minimum tillage (Table 10.5). The finger millet lines that display the characteristics have two advantages, namely, the boosted soil health and the competitiveness of crops in CA systems. Such agronomic traits should be selected as such in breeding programs as a strategic focus in adopting conservation agriculture (Brhane et al., 2025).

Table 10.5 Genetic gains for agronomic traits and nutritional compositions in finger millet (Gebreyohannes et al., 2024)

Traits	Country	Years of cultivar release	Mean increase/ decrease ($unit^{-1}$ $year^{-1}$)	Genetic gain (% $year^{-1}$)	References
Grain yield (kg ha^{-1})	Ethiopia	1999–2019	27.64	1.35	Marefia et al. (2022)
Grain yield (kg ha^{-1})	Ethiopia	1998–2022	43	1.87	MOA (Ministry of Agriculture) (2022)
Ear weight $plant^{-1}$ (g)	India	1959–2016	0.026	0.0005	Aralikatti and Chaturvedi (2020)
1000 grain mass (g)	India	1959–2016	0.003	0.0001	Aralikatti and Chaturvedi (2020)
1000 grain mass (g)	Kenya	2000–2015	0.43	0.028	Manyasa et al. (2016)
Straw yield $plant^{-1}$ (kg)	India	1959–2016	−0.001	−0.00001	Aralikatti and Chaturvedi (2020)
Number of tillers $plant^{-1}$	Ethiopia	1999–2019	0.11	0.005	Marefia et al. (2022)
Number of tillers $plant^{-1}$	India	1959–2016	−0.005	− 0.0001	Aralikatti and Chaturvedi (2020)
Number of tillers $plant^{-1}$	Kenya	2000–2015	3.02	0.201	Manyasa et al. (2016)
Number of ears $plant^{-1}$	Ethiopia	1999–2019	0.1	0.005	Marefia et al. (2022)
Number of fingers ear^{-1}	India	1959–2016	−0.012	−0.0002	Aralikatti and Chaturvedi (2020)
Number of fingers ear^{-1}	Uganda	1969–2017	0.83	0.017	Owere et al. (2015)
Number of fingers ear^{-1}	Kenya	2000–2015	3.39	0.236	Manyasa et al. (2016)
Plant height (cm)	India	1959–2016	−0.006	−0.0001	Aralikatti and Chaturvedi (2020)
Days to maturity	India	1959–2016	0.027	0.001	Aralikatti and Chaturvedi (2020)
Days to maturity	Ethiopia	2002–2022	−0.58	−0.029	MOA (2022)

(continued)

Table 10.5 (continued)

Traits	Country	Years of cultivar release	Mean increase/ decrease (unit^{-1} year^{-1})	Genetic gain (% year^{-1})	References
Days to 50% flowering	India	1959–2016	−0.033	−0.001	Aralikatti and Chaturvedi (2020)
Days to 50% flowering	Uganda	1969–2017	−0.03	0.002	Owere et al. (2015)
Ear head length (cm)	India	1959–2016	0.009	0.0002	Aralikatti and Chaturvedi (2020)
Flag leaf length (cm)	India	1959–2016	0.019	0.0005	Aralikatti and Chaturvedi (2020)
Finger length (cm)	India	1959–2016	0.017	0.0003	Aralikatti and Chaturvedi (2020)
Finger length (cm)	Kenya	2000–2015	5.15	0.344	Manyasa et al. (2016)
Panicle exertion (cm)	India	1959–2016	0.11	0.002	Aralikatti and Chaturvedi (2020)
Protein content (mg/ 100 g)	India	1976–2019	−0.048	−0.001	Udama et al. (2020)
Protein content (g/ 100 g)	India	1958–1938	0.036	0.007	Vadivoo et al. (1998)
Calcium content (mg/ 100 g)	India	1953–1989	0.13	0.005	–
Iron content (mg/100 g)	India	1976–2019	4.11	0.026	Udama et al. (2020)
Zinc content (mg/100 g)	India	1976–2019	−0.001	−0.00002	Udama et al. (2020)

10.7.4 Genomic Approaches

Genome-wide association studies (GWAS), marker-assisted selection (MAS), and genome editing are modern genomic tools that have transformed the finger millet by revolutionizing the process of its improvement. GWAS has been used to find loci relating to major traits CA-relevant including early vigor, high biomass and root architecture. MAS makes them introgress quickly into elite lines, whereas genome editing has potential of making specific changes in order to increase stress tolerance and yield residue. The practices help in accelerating the cultivar breeding process to create low-input and sustainable production systems that are aligned to CA principles (Choudhary et al., 2023).

10.7.5 Integration of Improved Varieties with Conservation Agriculture

Genetic enhanced varieties of finger millet combined with conservation agriculture gives the best option to high productivity and sustainability. Cultivars that have good agronomic characteristics maximize the cover of the soil surface, erosion, and nutrient cycling and CA practices boost the crop growth in the soil. Through this combined strategy, smallholder systems become better resilient to climatic variability and is associated with sustainable intensification. Field trials have shown that finger millet lands with improved variety together with CA practices result in increased and more sustainable yield, better soil health, and sustainability long-term (Panotra et al., 2025).

10.8 Modelling and Remote Sensing Applications

10.8.1 Soil Erosion Modelling in Millet Systems

The soil erosion modelling offers quantitative data on the effectiveness of the conservation agriculture of the finger millet landscapes. RUSLE, RUSLE2, SWAT and WEPP are empirical and process-based models that have been extensively applied to model soil loss in a variety of management conditions. By using these models, it is possible to evaluate CA practices in terms of the intensity of rainfall, slope, the type of soil, and the extent of its residue cover. Finger millet parameterization such as crop height, canopy cover and residue biomass allows proper predictions of runoffs and sediment transport. As shown in the scenario simulations, minimum tillage, retention of residue, and intercropping help to cut the soil loss by a significant margin relative to regular tillage, which is vital information on sustainable land management (Yan et al., 2022).

10.8.2 Crop Growth and Climate Modelling

Process based crop models such as APSIM and DNDC are needed to examine the interaction of soil water, nitrogen cycling and yield responses of finger millet in response to various CA practices. These models are used to model the interaction between crop growth, soil process, and environmental factors to be able to predict the CA effects under different climatic conditions. By incorporating the projections of the climate change into such models, it is possible to assess both the stability of yield, the water-use efficiency, and the nutrient management of finger millet. These predictive modelling methods help to design site-specific interventions to maximize the CA uptake at the expense of productivity of smallholder rainfed systems.

10.8.3 Remote Sensing for Monitoring Crop and Soil Health

Remote sensing technologies offer a cost-effective method of tracking the performance of crops and soil conditions in large scale regions. The vegetation indices, including the NDVI and SAVI, make it possible to gauge the development of the canopy and biomass of finger millet. In CA systems, high-resolution satellite images of Sentinel-2 and Landsat can be used to estimate surface residue cover, which is essential to protect soil. By taking soil moisture measurements using a SMAP and Sentinel-1, it is possible to perform real-time analysis of the water availability, which can be used to implement changes in management. These tools offer spatially explicit data that is complementary to those collected by the field and builds the accuracy of conservation agricultural activities (Liu et al., 2022).

10.8.4 Integrating Modelling and Remote Sensing into Decision Support

The combination of both the modelling results and the remote sensing data into the GIS-based decision support systems allows focused soil and crop control. The risk erosion maps created based on integrated model and satellite data are useful in detecting the vulnerable regions, which need to be addressed. Digital advisory services that integrate CA-specific advice to smallholder farmers offer advice on how to use land in a sustainable way, as well as increase productivity. The combined strategy arches the gap between research and practice and enables informed choices to be made in soil conservation and climate-resistant production of finger millet (Chimdi et al., 2024).

10.8.5 Modelling the Soil Erosion in Millet Systems

The soil erosion modelling offers quantitative data on the effectiveness of the conservation agriculture of the finger millet landscapes. RUSLE, RUSLE2, SWAT and WEPP are empirical and process-based models that have been extensively applied to model soil loss in a variety of management conditions. By using these models, it is possible to evaluate CA practices in terms of the intensity of rainfall, slope, the type of soil, and the extent of its residue cover. Finger millet parameterization such as crop height, canopy cover and residue biomass allows proper predictions of runoffs and sediment transport. As shown in the scenario simulations, minimum tillage, retention of residue, and intercropping help to cut the soil loss by a significant margin relative to regular tillage, which is vital information on sustainable land management (Yan et al., 2022).

10.9 Case Studies

10.9.1 Case Study1: Zimbabwe Conservation Agriculture Experiments

The study by Nyagumbo et al. (2019) is entitled Effects of Conservation Agriculture on Runoff and Soil Loss in Smallholder Maize and Millet Fields in Zimbabwe that was carried out in the semi-arid regions of Zimbabwe. The experiment was based on the comparison of the conventional tillage and reduced tillage and retaining residues on the finger millet plots across three growing seasons. The moisture of the soil, biomass, tillering and yield were measured under the two systems. Findings indicated that compared to standard tillage, CA plots had 18 percent more soil moisture and were 25% more increased in the above Ground biomass. The number of tiller rose by 12% and the grain yield raised to 2.4 t ha^{-1} as compared to 1.8 t ha^{-1} when using CA practices. The paper suggested the increase in residue retention and use of less tilling to increase soil water conservation and stabilize the millet yield under semi-arid environments.

10.9.2 Case Study2: India Millet-Based Conservation Agriculture Systems

Sharma et al. (2022) evaluated the impact of CA practices in millet-based systems in the Deccan Plateau, India, in their study "*Impact of Zero Tillage and Residue Management on Soil Erosion and Finger Millet Productivity*". Field trials were conducted using residue retention, zero tillage, and conventional tillage treatments. Soil erosion was measured using sediment traps, while crop performance parameters, including biomass, yield, and nutrient status, were recorded. The results indicated a 40% reduction in soil loss under CA compared with conventional fields. Finger millet grain yield increased from 1.5 t ha^{-1} to 2.2 t ha^{-1}, while total aboveground biomass improved by 30%. Residue retention also enhanced soil organic carbon by 0.8% over three seasons. The authors recommended integrating zero tillage and residue management in smallholder systems to reduce soil degradation and sustain productivity.

10.9.3 Case Study3: East African Highlands

The study conducted by Mowe and Kihara (2024) focuses on the adoption of CA in the sloping millet fields of the East African highlands in the paper titled Soil Conservation through Crop Residue Retention and Intercropping in East African Highlands. Smallholder plots were controlled by less tillage, intercrop of finger millet-legumes and retention of residue. Crops performance, soil loss as well as soil runoff were assessed after two consecutive cropping seasons. The plots that were CA had a 50% surface runoff and 55% smaller sediment loss than the conventionally tilled plots. The grain production in finger millet became 2.3 t ha^{-1} as compared to 1.6 t ha^{-1} and the biomass accumulation was increased by 28%. The research paper derived that the interaction of residue retention and intercropping is effective in eroding the soil and enhancing the productivity. It was recommended that CA practices should be encouraged in sloping lands and training be given to the farmers on how to manage the residue and crop diversification.

10.10 Opportunities and Challenges for Scaling Conservation Agriculture in Finger Millet Systems

10.10.1 Socioeconomic Opportunities

There are enhanced socioeconomic advantages of conservation agriculture to the smallholder finger millet farmers. The production cost decreases because of the minimum tillage, which reduces labor and fuel requirements, and the residue retention and return to soil fertility increase the crop production and stability, thus, increasing food security and income of the household. Legume intercrops with CA systems offer more sources of protein and possible commercial products, which are diversified sources of income. In addition, CA adoption can enhance the resilience to climate variability in the long-term to mitigate crop failure risk in marginal drylands (Asmara et al., 2024).

10.10.2 Barriers to Adoption

Although there are evident advantages, there are a number of obstacles to the adoption of CA in finger millet systems. A restricted supply of machinery to manage minimal tilling, inadequate understanding of residue management as well as inadequate technical education all restrain its adoption. The initial investments in CA equipment and labor-intensive CA practices, including residue handling and mulching are often prohibitive to farmers. Cultural biases towards traditional plowing and distrust in long-lasting changes in the fertility of soils further reduce the adoption rates (Qian et al., 2022).

10.10.3 Policy and Extension Needs

Good policy frameworks and extensions services are very important to support the CA scaling. Adoption can be facilitated by subsidies of CA-compatible equipment, quality seeds and incentives to retain residue. The extension programs should provide real-life examples, peer-to-peer education, and region-specific instructions to enhance awareness on CA benefits and methods. The incorporation of CA practices in national development plans of agriculture can establish enabling conditions of the sustainable intensification of finger millet systems (T. Araya et al., 2024).

10.10.4 Residue Competition and Labor Constraints

There is a high competition in the livestock feed and fuel in the area where finger millet straw is widely consumed in various forms. The process of maintaining enough residues to be used as soil cover and at the same time fulfilling other domestic demands would demand a careful planning and possibly trade-offs. In addition, the collection, handling, and distribution of the residue can raise the workload, particularly where the manpower is limited in the smallholder household. To deal with these limitations, there is need to offer innovative ideas like dual-purpose crops, mechanized control of residues, and labor sharing schemes within the community in order to realize sustainable CA implementation (Sithole & Olorunfemi, 2024).

10.11 Future Directions for Conservation Agriculture in Finger Millet Systems

10.11.1 Integration of Conservation Agriculture and Breeding

The strategies in the future ought to be based on the synergistic incorporation of conservation agriculture and the advanced breeding programs. The residue retention and minimal tillage can be maximized by development of finger millet cultivars that are optimized to CA system, and have properties that include high biomass, early vigor, and superior root architecture. This will enhance resistance to climate variability and marginal soils by incorporation biotic and abiotic stress tolerance in CA-adapted variety into varieties. The breeders and agronomists should work together to make sure that the new varieties fulfill the practical needs of the CA smallholder systems and maintain the productivity (Fuentes-Penailillo et al., 2024).

10.11.2 Digital Agriculture and AI/ML Applications

The new opportunities of site-specific CA interventions are represented by digital technologies of agriculture, such as artificial intelligence (AI) and machine learning (ML). The models based on AI may be used to fill in multi-source data such as weather conditions, soil characteristics, and crops performance to optimize tillage, residue treatment, and planting routines. Machine learning models are capable of forecasting erosion risk, deficit of soil moisture, and yield response during CA, which can be used to provide evidence-based recommendations in time. The inclusion of these technologies in mobile advisory systems improves the decision-making of farmers, the efficiency of their decisions, and the risks linked to climatic variations (Hu et al., 2025).

10.11.3 Long-Term Conservation Agriculture Trials

The design and observation of long-term CA trials is essential in the learning of the accumulative impacts of soil health, nutrient cycling, and productivity. These experiments offer information about multi-season relationships involving soil, crop and climate, the advantages of retaining residues, having low levels of soil disturbance and crop turns. Sustainability indicators such as carbon sequestration, carbon buildup in the soil, and reduction of soil erosion can also be determined using longitudinal studies. Results of such trials will inform situational context-based guidelines and optimal practices on how to scale CA in finger millet landscapes (Ingram & Maye, 2023).

10.11.4 Remote Sensing–Model Integration

Remote sensing combined with the models of the process provides strong instruments to control and optimize the use of CA. Spatially explicit data on soil erosion, moisture availability and crop growth can be offered by high-resolution satellite data which is then used together with APSIM, DNDC or RUSLE-based models. These combined platforms provide a way to monitor and analyze scenarios and predictive evaluation of CA interventions on landscape scales. Using simulations in a combination with observational data, farmers and policy makers will be able to apply adaptive management techniques to improve productivity, soil protection and climate stability (Ceasar et al., 2023).

10.11.5 Correlation Heatmap of Finger Millet

The correlation heatmap shows the interaction of soil, nutritional, and yield-related characteristics within the system. Correlation coefficients among soil organic matter, microbial activity, soil pH, and water retention were found to be strong and positive, and thus the enhancement of the components of soil health was found to be supportive of each other and lead to the overall increase of finger millet deliver and nitrogen use efficiency. Finger millet yield also demonstrated there were good positive correlations between microbial activity and soil pH indicating that soils that may be biologically active and chemically balanced are important in terms of their productivity. Negative relationships like that between fiber and NUE or SOM to protein content, on the other hand, point to the existence of significant trade-offs between nutritional characteristics and soil or efficiency attributes. The heat map, in general, naturally provides a brief visualization of the interactions of traits that can inform the soil management strategies, crop improvement strategies, and an improved planning of nutrient utilization (Fig. 10.6).

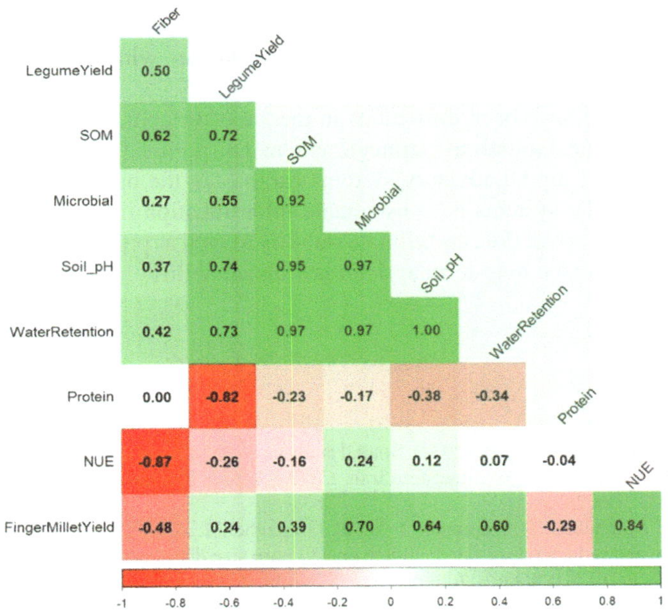

Fig. 10.6 Correlation heatmap among different traits

10.12 Conclusions

Conservation agriculture (CA) provides a radical solution to the production of finger millet given the process of soil degradation in drylands and slopes. Through the focus on minimum disturbance of the soil, constant soil cover, and crop diversification, CA improves soil structure, erosion eradication, and water-use efficiency. Residue retention and intercropping helps in keeping the soil organic matter strong, and assisting in the nutrient cycling, and stabilizing the microclimates, which are all positive to enhancing the stability of crops in the face of climate variability. It is well adapted to marginal soils and its natural drought resistance, the short growing period and its adherence to the CA systems, as a smallholder farmer, finger millet is the best option to be incorporated in the CA systems as it gives them a stable climate-resistant staple crop.

Data in Zimbabwe, India and highlands of East Africa prove that CA practices greatly decrease the runoff and sediment loss and increase the biomass accumulation, tillering and grain yield. Quantitative understanding of the soil, water, and crop responses insights is offered by long-term experiments and modelling, such as RUSLE, APSIM, and DNDC, and remote sensing tools, such as Sentinel-2 and SMAP, to aid in making informed decisions and landscape-scale planning. Genetic and agronomic advances such as stress-tolerant cultivars with high residue output make additional gains of CA.

Although there have been difficulties in the areas of labor, residue competition, and socioeconomic limitations, strategic extension, political encouragement, and implementation of digital advisory systems can hasten the uptake. In general, CA within finger millet systems is a sustainable intensification route, which improves food security, soil protection, and climate stability as well as provides the opportunity to innovate and pursue long-term agricultural sustainability.

References

Ahmadi Mirghaed, F., & Souri, B. (2022). Spatial analysis of soil quality through landscape patterns in the Shoor River basin, Southwestern Iran. *CATENA, 211*, 106028. https://doi.org/10.1016/j.catena.2022.106028

Araya, S. N., Mitchell, J. P., Hopmans, J. W., & Ghezzehei, T. A. (2022). Long-term impact of cover crop and reduced disturbance tillage on soil pore size distribution and soil water storage. *The Soil, 8*(1), 177–198. https://doi.org/10.5194/soil-8-177-2022

Araya, T., Ochsner, T. E., Mnkeni, P. N. S., Hounkpatin, K. O. L., & Amelung, W. (2024). Challenges and constraints of conservation agriculture adoption in smallholder farms in sub-Saharan Africa: A review. *International Soil and Water Conservation Research, 12*(4), 828–843. https://doi.org/10.1016/j.iswcr.2024.03.001

Asmara, D. H., Rala, A., Batas, M. A. A., Laborte, A., Flor, R. J., Khumairoh, U., Sembiring, H., & Yadav, S. (2024). Spatial identification of lost-and-found carbon hotspots at Javanese rice systems, Indonesia. *Agricultural Systems, 214*, 103834. https://doi.org/10.1016/j.agsy.2023.103834

Brhane, H., Haileselassie, T., Tesfaye, K., Lule, D., Abreha, K. B., Hammenhag, C., Ortiz, R., & Geleta, M. (2025). Genome-wide association analysis revealed genetic markers linked to grain yield and yield related traits in finger millet grown in acidic soils. *Plant and Soil, 511*(1–2), 447–462. https://doi.org/10.1007/s11104-024-07000-2

Ceasar, S. A., Maharajan, T., Krishna, T. P. A., & Ignacimuthu, S. (2023). Finger millet (*Eleusine coracana* (L.) *Gaertn*). In *Neglected and underutilized crops* (pp. 137–149). Elsevier. https://doi.org/10.1016/B978-0-323-90537-4.00031-4

Chimdi, A., Tesso, B., & Daba, C. (2024). Genetic variability analysis for yield and yield-associated traits of finger millet [*Eleusine coracana* (L.) Gaertn.] Accessions at Mechara, Eastern Ethiopia. *International Journal of Biomedical Science and Engineering, 12*(1), 10–18. https://doi.org/10.11648/j.ijbse.20241201.12

Choudhary, P., Shukla, P., & Muthamilarasan, M. (2023). Genetic enhancement of climate-resilient traits in small millets: A review. *Heliyon, 9*(4), e14502. https://doi.org/10.1016/j.heliyon.2023.e14502

Dzudzor, M. I., & Gerber, N. (2023). Urban households' food safety knowledge and behaviour: Choice of food markets and cooking practices. *Journal of Agriculture and Food Research, 14*, 100728. https://doi.org/10.1016/j.jafr.2023.100728

Farias, T. R. L., Medeiros, P. H. A., Araújo, J. C. D., & Navarro-Hevia, J. (2021). The role of unpaved roads in the sediment budget of a semiarid mesoscale catchment. *Land Degradation & Development, 32*(18), 5443–5454. https://doi.org/10.1002/ldr.4120

Feizpoor, S., Habibi-Yangjeh, A., & Luque, R. (2022). Design of TiO_2/Ag_3BiO_3 n-n heterojunction for enhanced degradation of tetracycline hydrochloride under visible-light irradiation. *Environmental Research, 215*, 114315. https://doi.org/10.1016/j.envres.2022.114315

Fikadu, G. (2022). Determination of selected physicochemical water quality parameters of the upper stream of Amerti watershed of Western Ethiopia. *Environmental Challenges, 8*, 100558. https://doi.org/10.1016/j.envc.2022.100558

Fuentes-Peñailillo, F., Gutter, K., Vega, R., & Silva, G. C. (2024). Transformative technologies in digital agriculture: Leveraging internet of things, remote sensing, and artificial intelligence for smart crop management. *Journal of Sensor and Actuator Networks, 13*(4), 39. https://doi.org/10.3390/jsan13040039

Gebreyohannes, A., Shimelis, H., Mashilo, J., Odeny, D. A., Tadesse, T., & Ojiewo, C. O. (2024). Finger millet (*ELEUSINE CORACANA*) improvement: Challenges and prospects—A review. *Plant Breeding, 143*(3), 350–374. https://doi.org/10.1111/pbr.13169

Grzesiak, M. T., Hura, T., Maksymowicz, A., Dziurka, K., Rut, G., Rzepka, A., & Grzesiak, S. (2023). Effect of soil drought on growth and yield of maize (*Zea mays* L.) hybrids grown under compacted soil. *Journal of Agronomy and Crop Science, 209*(2), 300–315. https://doi.org/10.1111/jac.12626

Hu, X., Chen, H., Duan, Q., & Zhang, D. (2025). A comprehensive review of diffusion models in smart agriculture: Progress, applications, and challenges (version 6). arXiv. https://doi.org/10.48550/ARXIV.2507.18376

Ingram, J., & Maye, D. (2023). "How can we?" The need to direct research in digital agriculture towards capacities. Journal of Rural Studies, 100, 103003. doi:https://doi.org/10.1016/j.jrurstud.2023.03.011.

Li, C. H., Lee, T. T., & Lau, S. S. Y. (2023). Enhancement of municipal solid waste management in Hong Kong through innovative solutions: A review. *Sustainability, 15*(4), 3310. https://doi.org/10.3390/su15043310

Liu, Y., Shahbaz, M., Fang, Y., Li, B., Wei, X., Zhu, Z., Lynn, T. M., Lu, S., Shibistova, O., Wu, J., Guggenberger, G., & Ge, T. (2022). Featured front cover. *Land Degradation & Development, 33*(2). https://doi.org/10.1002/ldr.4200

Mallu, T. S., Okbagabir, S. G., Negassi, B. T., Tesfay, T., Negash, T. T., Menghis, T. B., Medhn, T. A., Teklay, S. G., Gebremeskal, Y. H., Tewolde, R. H., Gebremariam, G. K., Tecleab, A. A., Tewelde, K. G., Brhane, D., Asghede, K. M., Adem, A. H., & Solomon, M. T. (2026). A comprehensive review on finger millet (*Eleusine coracana* (L.) Gaertn): A gluten-free superfood and feed. *Journal of Agriculture and Food Research, 25*, 102528. https://doi.org/10.1016/j.jafr.2025.102528

Mbanyele, V., Mtambanengwe, F., Nezomba, H., Rurinda, J., & Mapfumo, P. (2022). Conservation agriculture in semi-arid Zimbabwe: A promising practice to improve finger millet (*Eleusine coracana* Gaertn.) productivity and soil water availability in the short term. *Agriculture, 12*(5), 622. https://doi.org/10.3390/agriculture12050622

Mowe, K., & Kihara, J. (2024). Soil conservation through crop residue retention and intercropping in sloping finger millet systems of the east African highlands. *Soil and Tillage Research, 236*, 105756.

Nyagumbo, I., Nyamadzawo, G., & Madembo, C. (2019). Effects of three in-field water harvesting technologies on soil water content and maize yields in a semi-arid region of Zimbabwe. *Agricultural Water Management, 216*, 206–213.

Panotra, N., Yadav, S. S., Zimik, L., Kumar, V., Mohanty, L. K., Tiwari, D. K., Ramya, M., Gopal, R., Upadhyay, L., & Sharma, A. (2025). Impact of conservation tillage and organic nutrient sources on finger millet performance in legume-based cropping systems. *Archives of Current Research International, 25*(8), 635–652. https://doi.org/10.9734/acri/2025/v25i81446

Piñeiro-Juncal, N., Serrano, O., Mateo, M. Á., Diaz-Almela, E., Leiva-Dueñas, C., & Martinez-Cortizas, A. (2022). Review of the physical and chemical properties of seagrass soils. *Geoderma, 428*, 116219. https://doi.org/10.1016/j.geoderma.2022.116219

Porwollik, V., Rolinski, S., Heinke, J., Von Bloh, W., Schaphoff, S., & Müller, C. (2022). The role of cover crops for cropland soil carbon, nitrogen leaching, and agricultural yields—A global simulation study with LPJmL (V. 5.0-tillage-cc). *Biogeosciences, 19*(3), 957–977. https://doi.org/10.5194/bg-19-957-2022

Qian, C., Antonides, G., Heerink, N., Zhu, X., & Ma, X. (2022). An economic-psychological perspective on perceived land tenure security: Evidence from rural eastern China. *Land Use Policy, 120*, 106294. https://doi.org/10.1016/j.landusepol.2022.106294

Sartori, F., Piccoli, I., Polese, R., & Berti, A. (2022). Transition to conservation agriculture: How tillage intensity and covering affect soil physical parameters. *The Soil, 8*(1), 213–222. https://doi.org/10.5194/soil-8-213-2022

Singh, V., Gupta, R. K., Kahlon, M. S., Toor, A. S., Singh, K. B., Al-Ansari, N., & Mattar, M. A. (2023). Effect of different tillage and residue management options on soil water transmission and mechanical behavior. *Land, 12*(10), 1895. https://doi.org/10.3390/land12101895

Sithole, A., & Olorunfemi, O. D. (2024). Sustainable agricultural practices in sub-Saharan Africa: A review of adoption trends, impacts, and challenges among smallholder farmers. *Sustainability, 16*(22), 9766. https://doi.org/10.3390/su16229766

Steed, A., King, J., Grewal, S., Yang, C., Clarke, M., Devi, U., King, I. P., & Nicholson, P. (2022). Identification of fusarium head blight resistance in *Triticum timopheevii* accessions and characterization of Wheat-*T. Timopheevii* Introgression lines for enhanced resistance. *Frontiers in Plant Science, 13*, 943211. https://doi.org/10.3389/fpls.2022.943211

Teng, J., Hou, R., Dungait, J. A. J., Zhou, G., Kuzyakov, Y., Zhang, J., Tian, J., Cui, Z., Zhang, F., & Delgado-Baquerizo, M. (2024). Conservation agriculture improves soil health and sustains crop yields after long-term warming. *Nature Communications, 15*(1), 8785. https://doi.org/10.1038/s41467-024-53169-6

Tsunemi, K., Kawamoto, T., & Matsumoto, H. (2023). Estimation of the potential global nitrogen flow in a nitrogen recycling system with industrial countermeasures. *Sustainability, 15*(7), 6042. https://doi.org/10.3390/su15076042

Xiong, H., Peng, H., Ye, X., Kong, Y., Wang, N., Yang, F., Meni, B.-H., & Lei, Z. (2022). High salt tolerance hydrogel prepared of hydroxyethyl starch and its ability to increase soil water holding capacity and decrease water evaporation. *Soil and Tillage Research, 222*, 105427. https://doi.org/10.1016/j.still.2022.105427

Yan, J., Ren, T., Wang, K., Li, H., Li, X., Cong, R., & Lu, J. (2022). Improved crop yield and phosphorus uptake through the optimization of phosphorus fertilizer rates in an oilseed rape-rice cropping system. *Field Crops Research, 286*, 108614. https://doi.org/10.1016/j.fcr.2022.108614

Zhang, C., Qi, Z., Jiao, P., Jiang, Z., Liu, S., Fan, S., Zhao, C., Ma, Y., & Guan, S. (2023). Cloning and disease resistance analysis of the maize ZmBON3 gene. *Agronomy, 13*(1), 152. https://doi.org/10.3390/agronomy13010152

Zhang, X.-F., Zhong, N.-H., Li, R., Shi, F.-X., & Mao, R. (2023). Nitrogen addition mediates monospecific and mixed litter decomposition in a boreal peatland. *Catena, 225*, 107051. https://doi.org/10.1016/j.catena.2023.107051

Zhang, Y., Li, Z., Wu, T., Chen, F., Xu, M., Wang, Y., He, S., Tan, B., Hou, G., Luo, Z., Zheng, J., & Hu, X. (2023). Soil erosion rates on sloping cropland fragment underlain by contrasting lithologies. *Catena, 220*, 106622. https://doi.org/10.1016/j.catena.2022.106622

Chapter 11
Teff (*Eragrostis tef*): Water Resource Management and Climate Adaptation

Umar Farooq, Wajahat Hussain, Ahmed M. S. Kheir, Mukhtar Ahmed, and Shakeel Ahmad

Abstract Teff (*Eragrostis tef*) is now being considered as an excellent climate-resilient crop much better suited to drought-prone, low-input, and rainfed agroecosystems that are under increasing climatic variability. The impact of global warming, unpredictable precipitation, and escalating water shortages pose a risk to food systems that rely on cereals, especially smallholder farming. This necessitates combined approach should be undertaken for climate change adaptation and water management. The chapter purpose is to summarize the existing information about the physiological, agronomic, and technological adaptation of teff to water and climate stress with the focus on water resource management and resilience-farming strategies. The chapter uses a synthesis type approach, which incorporates experimental data, crop modelling (Food and Agriculture Organization Aqua Crop model, the Decision Support System for Agro technology Transfer and Agricultural Production Systems Simulator) and application of a remote sensing, and case studies of various agro ecological locations. Major points have been examined as the soil-plant-atmosphere interactions, phenological sensitivity, water-use efficiency, yield stability and digital decision-support tools. Field and case studies indicated that near optimum water supply and water management enhanced teff yield and water productivity. Trained farmers were able to gain productivity by a maximum of 22%. Seasonal coefficients of crop were 0.8–1.0 (initial), 0.95–1.1 (mid-season), and 0.4–0.5 (late season), and yield response factor (KY) was 1.04 which was very sensitive to water stress. The simulated FAO Aqua Crop model showed that the additional irrigation of up to 95 mm

U. Farooq (✉) · W. Hussain · M. Ahmed
Institute of Geo-Information and Earth Observation, Pir Mehr Ali Shah Arid Agriculture University, Rawalpindi, Pakistan
e-mail: umerfaroooq42@gmail.com

A. M. S. Kheir
Julius Kühn Institute (JKI)—Federal Research Centre for Cultivated Plants, Institute for Strategies and Technology Assessment, Kleinmachnow, Germany

Soils, Water and Environment Research Institute, Agricultural Research Center, Giza, Egypt

S. Ahmad
Institute of Agronomy Faculty of Agricultural Sciences and Technology, Bahauddin Zakariya University, Multan, Pakistan

M. Ahmed (ed.), *Climate Change and Neglected Underutilized Crops*,
https://doi.org/10.1007/978-3-032-21005-0_11

post-flowering enhanced the grain-water use efficiency but reduces the biomass-water use efficiency. The results show that the scheduling of irrigation and optimization of water management play important roles in improving the productivity and water use efficiency of teff.

Keywords Teff · Climate resilience · Water-use efficiency · Crop modeling

11.1 Introduction

Teff (*Eragrostis tef*) is a native Ethiopian cereal crop that is becoming widely known as being climate resilient as well as nutritionally valuable. Rainfall variability due to climate change, rising temperatures, and rising water scarcity are putting extreme strain on cereal-based production systems in the world, posing a threat to the agricultural sustainability and world food security. Globally, teff is estimated to produce about 6.5 million tonnes annually, with Ethiopia producing a majority of them amounting to over 95% of the global output, which is mostly produced in the Oromia, Amhara and Tigray regions. Smaller-scale production takes place in Kenya, the United States, and Europe, and in Pakistan, teff is very experimental and only takes up less than 500 ha. These difficulties are specifically acute in rainfed zones where climate extremes start to weaken the productivity of crops and the efficiency of their use of resources. The rainfed agriculture of smallholders serving millions of people in the Sub-Saharan Africa and other parts of the developing world is extremely sensitive to erratic rains, frequent droughts and heat stress. The smallholders also have a limited access to irrigation systems and adaptive technologies because these limits the ability to tolerate climatic uncertainty (Woldeyohannes et al., 2022).

Teff (*Eragrostis tef*) has drawn much interest because of its extraordinary adaptability to various agro ecological factors as well as intrinsic adaptability to climatic stresses. Teff has short growing cycle, phenological plasticity, and ability to grow in moisture-limited conditions, which makes it a climate resilient crop with strategic value. In addition to agronomic strength, teff is majorly important to maintain food security, nutrition quality and income of small hold farmers. Teff being a staple food commodity with increased world demand gives a significant contribution to rural livelihoods in addition to ensuring culturally and economically sustainable farming systems. Adequacy in managing water resources is the essence of climate adaptive mechanisms to teff-based systems, because water controls the establishment, development and stability of the yield of crop during changing climatic conditions. Use of agronomic practices, modelling tools and climate-informed decision-making plans are then crucial in enhancing the water-use efficiency to sustain the production of teff (Kordbacheh et al., 2023).

The chapter provides a synthesis of the existing information on the climatic adaptability of the teff with special emphasis on the water resource management, physiological and agronomic adaptation to stress levels, modelling and adaptation strategies. The chapter will focus on identifying ways to boost the resilience and sustainability of teff-based farming systems to evolving climatic conditions through the integration of evidence and information in the fields of experimental studies, modelling frameworks, and climate projections.

11.2 Botanical and Agronomic Profile of *Eragrostis tef*

11.2.1 Taxonomy, Origin, and Domestication

Eragrostis tef (Zucc.) Trotter is a member of the family Poaceae and genus Eragrostis that has over 350 species worldwide. Teff is an auto flowered and self-pollinating cereal that was domesticated in the Ethiopian highlands, which can be regarded as the main centre of origin and diversity of teff (Fig. 11.1). The years of domestication in heterogeneous conditions and farmer-based selection have led to the high level of genetic and phenotypic diversity, making it possible to adapt to a large variety of agro ecological factors (Woldeyohannes et al., 2020).

11.2.2 Growth Habit and C_4 Photosynthesis

Teff has a straight cut growth and has the thin culms and a small canopy structure. The growth period of the crop is relatively short; it usually takes 90–140 days depending on the cultivar and the condition of the environment. Teff being a C_4 plant has high photosynthetic efficiency, especially in intense temperature and light conditions, which increases carbon assimilation, water-use efficiency and biomass production during heat and moisture stressed conditions (Jones et al., 2025).

11.2.3 Rooting Behavior, Lodging Susceptibility, and Harvest Index

Teff has a light yet thick fibrous root system that enables it to take in moisture soil faster under the low water conditions. The crop is however very susceptible to lodging due to its tall stature and slim stems; in a situation where there is excess rainfall or high fertility. Lodging has a great impact on the stability of yield and productivity of harvest (Fig. 11.2). The teff harvest index is usually poor to major cereals, representing higher investments in straw, yet recent breeding purposes are aimed at enhanced stem strength and development of semi-dwarf varieties (Gebru et al., 2023).

Fig. 11.1 Diversity of panicle morphology and grain colour in tef. **a** Comparison of a bread wheat spike (cv. 'Paragon', far left) with tef accessions exemplifying four categories of panicle morphology (from left to right; very lax, lax, semi-compact, and compact). **b** Comparison of bread wheat grains (cv. Paragon, bottom) with grains from brown and white-grained tef varieties (Jones et al., 2025)

11.2.4 Nutritional and Socio-Economic Importance

Teff grain is healthy with high concentrations of dietary fiber, essential amino acids, and minerals of iron, calcium, and zinc. It is gluten-free and this aspect has enhanced its relevance as a health food crop in the world market (Table 11.1). Teff is one of the sources of food security and agricultural livelihoods in Ethiopia, both as staple food and a key source of farm income. An increasing international need has also contributed to increase its market worth and export opportunities (Thers et al., 2024).

11.2.5 Agronomic Traits Enabling Low-Input and Climate-Resilient Production

Teff has several agronomic traits that enable it to be highly adapted to low input and climate-tolerant systems of production. These involve drought, waterlogging,

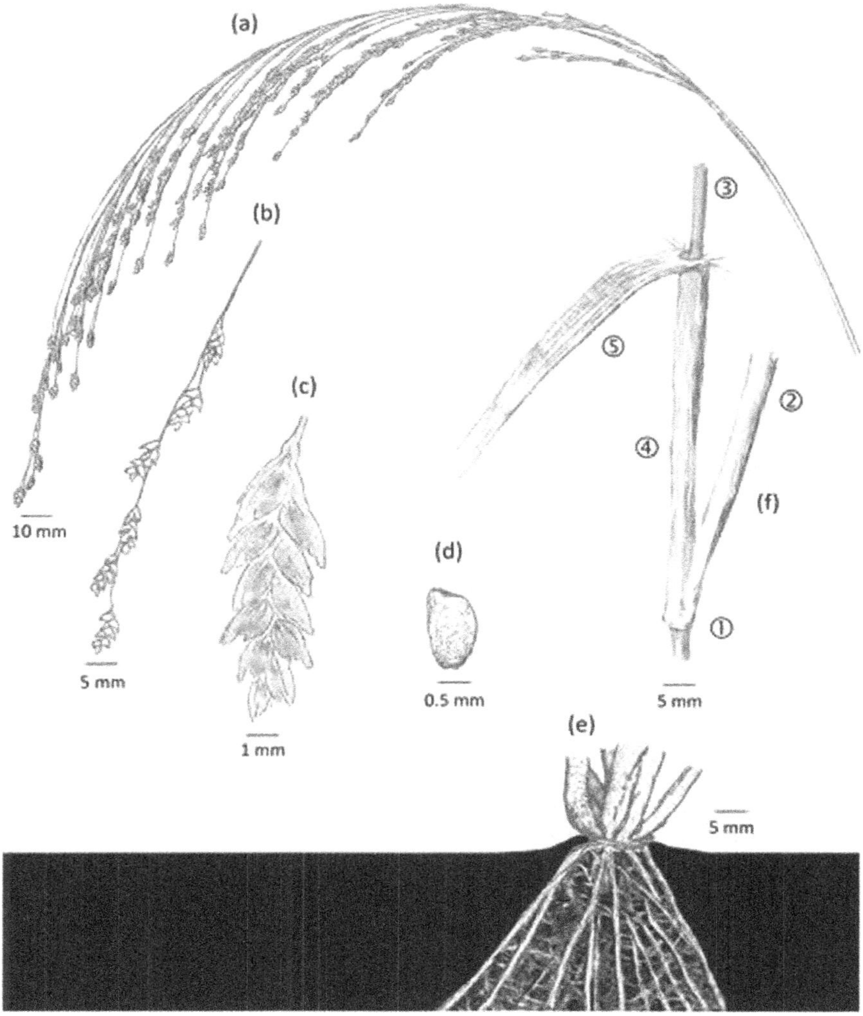

Fig. 11.2 Morphology of teff, **a** ripening inflorescence of teff (drooping panicle); **b** branches of the panicle, containing spikelets; **c** individual spikelet, containing teff grains; **d** individual teff grain; **e** teff stubble and root system; **f** teff phytomer: (1) node, (2) axillary branches, (3) internode, (4) leaf sheath and (5) leaf blad (Seyfu Ketema, 1997)

and acidity of the soils, low nutrient demands, and ability to flex in the planting of crops. Teff is an effective crop in suboptimal conditions because other cereals tend to perform poorly, thus making it a viable crop in rainfed agricultural practices and climate change in the resource-limited setting (Ramirez Gonzales et al., 2024a, 2024b).

Table 11.1 Comparative proximate and mineral contents of teff with quinoa and rice (Gebremedhin & Abraha, 2025)

Parameter	Quinoa* (*Chenopodium quinoa*)	Teff (Kuncho) (*Eragrostis tef*)	Teff (Key-teff) (*Eragrostis tef*)	Rice* (*Oryza sativa*)
Proximate (g/100 g)				
Moisture	11.3	12.9	11.5	13.1
Ash	2	1.9	1.9	0.4
Protein	12.1	12.8	12.8	7.1
Fat	6.3	3.2	3	0.6
Fiber	10.4	7.6	8.3	1.5
Starch	57.2	61.6	62.5	76.8
Amylose	19.7	29.2	–	29.2
Minerals (mg/100 g)				
Copper	0.6	1.01	1.11	0.1
Manganese	2	15.8	18	0.8
Iron	5.5	1		

11.3 Phenology, Distribution, and Production Zones

11.3.1 Phenology

The important regulators of the teff phenological development are temperature and the availability of soil moisture. Germination and seedling establishment normally require a period of 7–10 days post-sowing in the presence of adequate moisture in the soil and are extremely sensitive to early-season drought stress. The phase of vegetative growth, active tillering, and canopy building is the active process that typically takes 40–60 days and defines the ultimate biomass accumulation to a considerable extent (Table 11.2). The most vulnerable phases to both drought and heat stress are the reproductive phases and especially the flowering (anthesis) and grain filling stages; when the stress is experienced during an thesis phase, the result could be huge losses in the grain set and final yield (Fig. 11.3). The effective teff production systems require that teff production be synchronized with the rainfall patterns to have an adequate supply of soil moisture during the most critical period of reproductive development and to reduce yield losses because of climatic variability (Nagangoudar et al., 2024).

Table 11.2 The standardized BBCH scale for teff (Seyfu Ketema, 1997)

Principal stage	BBCH code	Description	Visual characteristics
0: Germination	00–09	Dry seed to emergence	Tiny, needle-like spikes emerging from soil (4–11 days after sowing)
1: Leaf development	10–19	Main shoot leaf growth	First leaf is very narrow; looks like fine grass (3–6 days post-emergence)
2: Tillering	20–29	Side shoot formation	Plant forms a dense tuft or "bunch". Can produce 10–40 tillers
3: Elongation	30–39	Stem elongation	Stems reach 30–100 cm but remain extremely slender (hair-like)
4: Booting	40–49	Swelling of flag leaf	In teff, this stage is subtle; the flag leaf sheath swells slightly before the "lace-like" head appears
5: Heading	50–59	Inflorescence emergence	The delicate, branching panicle emerges fully from the sheath
6: Flowering	60–69	Anthesis (pollination)	Mostly self-pollinating (cleistogamous); the panicle looks like a fine "mist"
7: Milk stage	70–79	Early grain filling	Grains are watery/milky. Panicle is still green and heavy
8: Maturity	8		

Fig. 11.3 Phenology of teff (*Eragrostis tef*) (Mengistu et al., 2025)

Table 11.3 Country-wise teff production and comparison with Ethiopia (FAOSTAT 2023)

Country	Teff production (Thousand Tons, 2022)	Percentage compared to Ethiopia (%)
Ethiopia	5900	100
Uganda	100	1.7
Eritrea	50	0.8
Sudan	40	0.7
United States	30	0.5
Others[a]	80	1.4

[a]Others: Includes Kenya, South Africa, India, Australia, and New Zealand

11.3.2 Geographical Distribution

Teff is native to Ethiopia, where it is grown in a wide geographical condition of agro ecological conditions of highlands to semi-arid agro ecological conditions. Ethiopia is still the major producer and centre of genetic diversity. Outside its native habitat, teff has since been planted in East Africa, North America, Europe and Australia, with its nutritional benefit and resistance to adverse climate conditions being its primary motivators (Table 11.3). The crop has large eco-geographic adaptability and it thrives well within an annual rainfall of 300–1200 mm and moderate to high temperature. Climate suitability modelling suggests that the teff crop can continue to expand in the areas where the climatic variability is rising (Felix et al., 2018).

11.3.3 Production Zones and Farming Practices

The Ethiopian highlands produce major Teff commercial belts, which are mostly rainfed. Most farmers who cultivate their crops are smallholder farmers whose production methods are traditional where they broadcast and use minimal external inputs (Fig. 11.4). These systems are well adjusted to local conditions but in most cases, lead to poor productivity (Table 11.4). Such practices as the row planting, more rational use of the seed, and the combination of balanced nutrient rates have been proved beneficial in increasing the yield stability and resource-use efficiency especially when they are integrated with additional irrigation in unpredictable rainfall environments (Lee, 2018).

Table 11.4 shows that teff production in Ethiopia saw a steady increase from 2013 to 2022. Production volumes, expressed in thousand tons, are compared year by year. The simple index uses 2013 as the base year, showing a 48% increase in teff production by 2022. The chain index calculates yearly growth, demonstrating a consistent annual increase from 2013 to 2022.

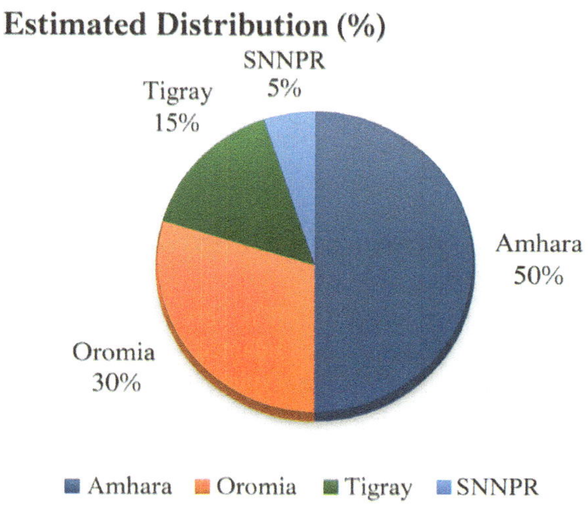

Fig. 11.4 Teff production areas and estimated distributions in Ethiopia (2023). *Source* Ethiopian Central Statistical Agency (CSA) data

Table 11.4 Teff production in Ethiopia (2013–2022)

Year	Teff production (Thousand Tons)	Simple index (2013 = 100)	Chain index
2013	4000	100	–
2014	4200	105	105
2015	4400	110	105
2016	4650	116	105
2017	4850	121	104
2018	5000	125	103
2019	5200	130	104
2020	5400	135	104
2021	5650	141	104
2022	5900	148	104

Source Ethiopian Central Statistical Agency (CSA) data

11.4 Climate Adaptation Strategies in Teff-Based Systems

11.4.1 Climate Risks Affecting Teff Production

The teff production has been increasingly challenged by the effects of climate stress (especially high temperature and heat waves with growing frequency during sensitive development phases). The variability of rainfall such as the late start and early end of the rains has usually led to the reduction of growing seasons and the crop has been prone to moisture stress in flowering and filling of the grains. Yield instability is further aggravated by frequent occurrence of drought and extreme weather occurrences. The interaction between heat and water stress, which accelerates the level of phenological development and decreases the accumulation of grain set and biomass (Emeru, 2022).

11.4.2 Genetic and Phenotypic Adaptation

The genetic diversity of teff is very high, and much of it has been preserved by farmer-managed landraces adapted to a variety of environmental eco-geography. This diversity gives phenotypic plasticity, which helps to increase resilience in changing climates. Teff can survive terminal drought and heat stress by relying on adaptative traits like early maturity, plastic tillering ability and mechanisms of escaping stress. Landraces are always important even in the future climatic conditions, since they provide useful characteristics in breeding programs for climatic resilience enhancement (Wilson, n.d.).

11.4.3 Agronomic Adaptation Strategies

Teff adaptation is essential in stabilizing yields of teff in climatic situations that are varied. The strategy of sowing dates to match the rainfall distribution has been used to reduce the effects of terminal stress. Practices such as reduced tillage, retention of residues and proper land preparation are soil moisture conservation practices that enhance water availability during dry spells. Crop rotation and diversification will increase the resilience of systems by promoting healthier soils, exposure to risk of climate, as well as farm income stability (Mallu et al., 2025).

11.4.4 Water-Centered Climate Adaptation

Climate adaptation is based on water management to teff-based systems. Additional irrigation at the vital growth periods especially flowering and grain filling has been found to be very effective in minimizing yield losses during unpredictable spells of rainfall. The optimal use of water by using efficient irrigation technologies and better scheduling can boost the resilience to mounting climatic uncertainty (Fig. 11.5). The combination of water management and the wider agronomic and genetic solutions enhances the ability to adapt the tiff-based production systems (Research Institute (Ifpri) et al., 2018).

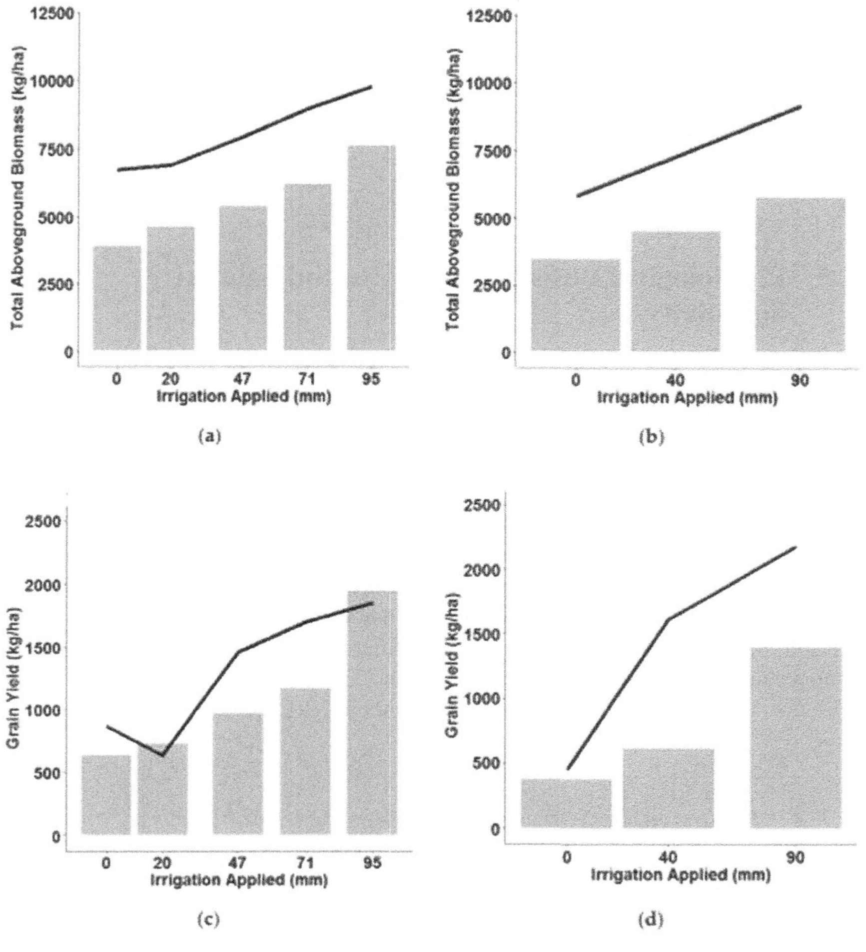

Fig. 11.5 Effects of irrigation on (**a**, **b**) total aboveground biomass and (**c**, **d**) on grain yield for (**a**, **c**) Mekelle, Ethiopia, in 2008 and (**b**, **d**) Mekelle, Ethiopia, in 2009. Bars represent the observed data, while the lines represent the simulated data (Paff & Asseng, 2019)

11.4.5 Socioeconomic and Policy Dimensions

Access to financial resources, technical knowledge and institutional support by the smallholder farmers play a significant role in the success of the climate adaptation strategies implemented in the teff-based systems. Resource scarcity/lack of adaptive capacity—The scarcity of resources and insufficient availability of credit and lack of information is still a significant impediment to the implementation of climate resilient practices. To improve the farm-level decision-making, climate services, early warnings system and targeted extension program should be integrated to make climate information more practical and provide management guidance. Along with these, enabling institutional and policy environments that favour climate-smart water management technologies, the preservation and utilization of locally adapted teff landraces, and investment in the smallholder building of capacity are also necessary to increase the adaptation effort and enhance resilience in teff-based production systems. (Wosene et al., 2025).

11.5 Physiological Basis of Water Use and Climate Sensitivity

11.5.1 Plant Water Relations and Transpiration Control

Teff demonstrates good physiological processes of controlling plant water status in changeable moisture situations. The strategies in dehydration avoidance stabilize the leaf water potential such as a slow leaf growth rate and regulated transpiration rate during water scarcity. Teff stomatal conductance is a rapid response to deteriorating soil moisture and increased vapour pressure deficit, which respond to situation by partial stomatal closing to reduce water loss without halting photosynthesis. In addition, the fibrous root system of the crop increases its ability to take in water at an upper area of the soil, and thus teff can take advantage of intermittent rainfall incidents that happen in rainfed areas (Alemu et al., 2024).

11.5.2 Photosynthesis and Gas Exchange under Stress

Being a C_4 species, teff has a physiological advantage in conditions characterized by high temperatures and water limitation as it has efficient carbon-concentrating mechanism. Photosynthetic carbon assimilation does not change significantly during moderate drought and heat stress as C_4 metabolism lowers photorespiration, and increases CO_2 fixation efficiency. Stomatal conductance and mesophilic activity decrease under extreme stress circumstances to limit gas exchange, but teff has a relatively high intrinsic water use efficiency through stabilizing the ratio of carbon accession and transpiration loss. This physiological characteristic adds to its strength in the face of climatic changes and low water supply (Farquhar et al., 1989).

11.6 Biochemical and Molecular Responses to Water and Climate Stress

11.6.1 Oxidative Stress and ROS Generation

Teff is affected by water deficit and heat stress, which causes the excessive generation of reactive oxygen species (ROS), such as hydrogen peroxide, superoxide radicals, and hydroxyl ions. These oxidative species cause cellular homeostasis disruption by affecting the lipids, proteins, and nucleic acid especially during the long period of the drought or a mixture of the heat-water stress. The strength of oxidative stress is firmly connected with the severity of stress and the development stage, and the reproductive ones are more sensitive. (Ramirez-Gonzales et al., 2024a, 2024b).

11.6.2 Antioxidant Defense Mechanisms

Teff triggers enzymatic and non-enzymatic defence against oxidative damage. Enzyme based major antioxidants that include the examples of superoxide dismutase, catalase and peroxidases are central to the scavenging of ROS. Additional antioxidant activity is provided by non-enzymatic antioxidants such as ascorbate, glutathione and phenolic compounds, which protect cells. These antioxidant systems differ in their genotypic efficiency, indicating a dissimilarity in the ability to endure stress and adaptive ability (Cotter et al., 2023).

11.6.3 Osmotic Adjustment and Metabolic Protection

Osmotic adjustment: is a key biochemical response mechanism, which makes teff to retain cell turgor and metabolic activity in water-stress conditions. Complementary solutes of proline, soluble sugars, and other osmolytes stabilize membranes, as well as protect and stabilized enzymatic functions. They are also signalling molecules affecting downstream stress-responsive pathways like the recognition of osmotic stress; this and other indicators improved the chances of survival during dehydration and thermal stress (Assefa et al., 2024).

11.6.4 Genetic and Molecular Regulation

At the molecular level, water and climatic stress induce teff to react to stressful situations by activating stress-sensitive genes under the control of hormonal signalling cascades, especially abscisic acid (ABA). ABA-mediated signalling controls stomata behaviour, antioxidants action, and osmolytes production. Transcriptomic studies have found the genes linked with the ROS detoxification process, osmotic balance, and heat shock reactions as core elements of the stressful resistance. These molecular processes give breeding programs a good understanding of the development of climate-resilient teff cultivars having enhanced water-use efficiency and stress tolerance (Bekele-Alemu & Ligaba-Osena, 2023).

11.7 Morphological and Anatomical Adaptations

11.7.1 Root System Architecture

Teff exhibits considerable plasticity in its root structure architecture by which the root structure has become efficient in extracting soil moisture in response to changing water availability. The crop produces extensive fibrous root system and large proportional growth in the number of lateral roots thereby increasing access to the shallow soil moisture after the intermittent rainfall. When water is scarce, root length density and root-to-shoot allocation enhance, and this enhances the efficiency of water uptake and physiological functioning during drought. This intrinsic plasticity is important to maintain the growth and stability of yields under rainfed conditions that are vulnerable to drought and fluctuating rainfall (Bayable et al., 2020).

11.7.2 Shoot and Leaf Structural Adaptations

Morphological adaptation of the shoot and leaves also helps teff to survive climatic changes. Water is minimized by regulating the amount of leaf area by inhibiting the growth of the leaf and enhancing senescence during stress. Teff leaves have narrow blades and adaptive cuticle traits, which lowers the evaporative demand of the leaves without interfering with photosynthesis. These morphological characteristics allow finding a compromise between carbon uptake and water-retention and made survival and productivity more resilient to combined stress (Vlajkov et al., 2023).

11.8 Agronomic and Yield Responses under Water and Climate Stress

11.8.1 Establishment and Biomass Accumulation

The moisture of soil is very sensitive to teff establishment and initial development. Through uniform germination, the tillers grow and the canopy is developed optimally and these factors are resourceful in the absorption of light and in carbon assimilation. Water limitations at the initial stages of vegetative growth may decrease biomass and leaf area together with weakening the overall hydrostaticity of the plants (Araya et al., 2011).

11.8.2 Yield Response to Drought and Heat

Losses in yields are specific to stages. Flowering and grain filling under heat stress or drought lead to sterility of the spikelet, and low grain filling weight, but did not lead to overall depressions in tiller development. Heat stress increases the rate of phenological development, decreasing the length of the grain filling, further dropping the ultimate yield. These adverse impacts are worsened by a combination of water and heat stress (Mihretie et al., 2021).

11.8.3 Water Productivity and Yield Stability

Teff has adaptive capability in low input systems. Irrigation supplement or moderate levels in stages of sensitivity enhance the water-use efficiency without reducing the yield. Flexible tillering, early maturity, and efficient root systems are some of the traits that make it stable in the fluctuating rainfall and temperature regimes, which increases resilience in semi-arid environments (Araya et al., 2010).

11.9 Crop Modelling for Water Resource Management in Teff

11.9.1 Role of Crop Models

Quantitative Crop models give a framework through which one can make sense of the soil-plant-atmosphere interactions in teff-based systems. They allow simulating growth, phenology and yield in conditions of different water availability, and help to note yield gaps. Crop models can measure the consequences of the water constraints and make judicious management decisions by combining climatic, soil and management information, to allocate resources efficiently to smallholder and low-input systems (Yihun et al., 2013).

11.9.2 Models Applied to Teff

Some models are calibrated and tested with regard to teff production. The use of aqua Crop has been successful in simulating yield and water-use efficiency of rainfed and supplemental irrigation. One can use DSSAT models and APSIM models to assess teff responses in terms of phenology, water stress impact, and management responses in current and future climate conditions (Table 11.5). Model outputs offer information regarding stage levels of sensitivity to drought and heat (Dinauer et al., 2022).

11.9.3 Modelling Water Productivity

Crop simulations can be used to obtain water productivity measures, such as crop coefficients (Kc), evapotranspiration and efficiency of transpiration. The use of deficit/supplemental irrigation methods is virtually tested to maximize water utilization, determine the key growth phases, and maximize water-use efficiency. These simulations help with evidence-based guidelines of maximizing yield per unit of water in semi-arid situations (Yihun, 2015).

11.9.4 Model-Based Applications

Crop models have been used as the decision-support model in climate adaptation planning. They advise the best dates to be used in sowing, the timing of irrigation, and the handling of fertilizers, besides allowing scenarios of climate analysis to predict the risks ahead. The outputs of models provide the smallholders with an opportunity to adopt adaptive practices, hence effective and resilient systems of teff production under different water regimes (Mulatu & Tamir, 2023).

Table 11.5 The HI, Grain-WUE, BM-WUE and Green-WUE under various water level treatments at Mekelle and Ilala sites for the cropping season 2008 and 2009 (Araya et al., 2010)

Site and year	Treatment	T (m^3 ha^{-1})	Rain + Irr. (m^3 ha^{-1})	GY (kg ha^{-1})	BY (kg ha^{-1})	HI	Grain-WUE (kg m^{-3})	BM-WUE (kg m^{-3})	Green-WUE (%)
Mekelle (2008)	R + 8	1399	2795	1950	7620	0.26	1.39	5.45	50
Mekelle (2008)	R + 6	1185	2556	1170	6220	0.19	0.99	5.25	46
Mekelle (2008)	R + 4	985	2316	970	5400	0.18	0.98	5.48	43
Mekelle (2008)	R + 2	759	2046	730	4630	0.16	0.96	6.10	37
Mekelle (2008)	R + 0	663	1846	640	3930	0.16	0.98	5.93	36
Ilala (2008)	R + 0	690	1879	350	3700	0.09	0.51	5.36	37
Mekelle (2009)	R + 9	1280	3670	1400	5800	0.24	1.09	4.53	35
Mekelle (2009)	R + 4	945	3176	610	4530	0.13	0.65	4.79	30
Mekelle (2009)	R + 0	696	2776	370	3480	0.11	0.53	5.00	25

Note that transpiration (T) is determined by Aqua Crop model. R, rainfall; Gy, grain yield (measured); HI, harvest index; BM, final aboveground biomass (measured); WUE, water use efficiency

11.10 Remote Sensing and Digital Tools for Climate and Water Monitoring

11.10.1 Stress Detection Principles

Remote sensing can offer a non-invasive method by which the health of crops can be studied to identify stress development in teff systems caused by climate conditions. Through the spectral reflectance pattern, it is possible to determine early water deficit, heat stress and nutrient limitations. The methods increase the accuracy and promptness of detecting stress, which can then be used to proactively implement management interventions (Kebede et al., 2025).

11.10.2 Vegetation and Water Indices

The indices of vegetation like NDVI (Normalized Difference Vegetation Index) and EVI (Enhanced Vegetation Index) measure the canopy vigor, vegetation, and biomass growth. Water-sensitive indices, such as NDWI (Normalized Difference Water Index) and WDI (Water Deficit Index), evaluate the level of moisture of the soil and the amount of water in plants. The spatiotemporal variability of teff growth can be tracked by combining these indices when the conditions are rainfed and irrigated (Tasew et al., 2025).

11.10.3 Platforms and Technologies

Data on remote sensing is gathered on various platforms: satellites (e.g., Sentinel-2, Landsat 9), UAVs (drones), and ground based sensors. The development of high-resolution imaging and multispectral sensors allows monitoring of stress and water demand in the agricultural field on the field scale. The combination of these tools and geographic information systems (GIS) helps in mapping water constraints and hotspots of water stress (Kazemi Garajeh et al., 2024).

11.10.4 Model–Remote Sensing Integration

Connections between models and remote sensing would be useful in increasing the predictability of a water-management and climate-adaptation. The growth simulations can be calibrated, the yield predictions can be verified and the water productivity can be evaluated under changing conditions using this integration. Digital tools are therefore used to support decision-making by smallholder farmers and therefore help them to make interventions and optimize resources in teff-based systems in time (Huang et al., 2025).

11.10.5 Pearson Correlation Heat map between Growth and Yield Traits

A Pearson correlation heat map refers to a graphical representation that is employed to visualize the strength and direction of correlation between a set of multiple growth and yield traits. The heat map allows describing the correlation coefficients with a color gradient, thus, allowing quick recognition of positively or negative associated traits and those with weak or independent associations. This visualization can be used to comprehend the interdependence of traits, determine the main characteristics that impact the formation of the toggle yield, and be used as the direction in the selection of crop improvement programs (Mengistu et al., 2025) (Fig. 11.6).

11.10.6 Case Study 1: Crop Coefficient and Water Productivity of Teff in Northern Ethiopia

A field experiment was carried out in Tigray, Northern Ethiopia (2008–2009) by Araya, Stroosnijder, Girmay & Keesstra to assess Crop coefficient, response of yield to water stress and water productivity of teff using simple field experiments and locally prepared lysimetres. The experiment was used to determine seasonal evapotranspiration (ETc), crop coefficient (Kc), and yield to water stress and water productivity at varying rainfall conditions. The averages of potential evapotranspiration were 260–317 mm, and Kc was 0.8–1.0 (initial), 0.95–1.1 (mid-season), and 0.4–0.5 (late season). The response factor of seasonal yield (KY) was 1.04, which showed that it was a sensitive response to water stress. Water productivity in the economies and crop industries was maximized during very wet periods and 22% when the skill of a farmer was high in comparison to moderately skilled farmers, mainly owing to the fact that there was less water loss. The research hypothesized that the teff yield would increase greatly with almost optimum amount of water and proposed that they adopt better water management practices coupled with training of farmers to increase their productivity.

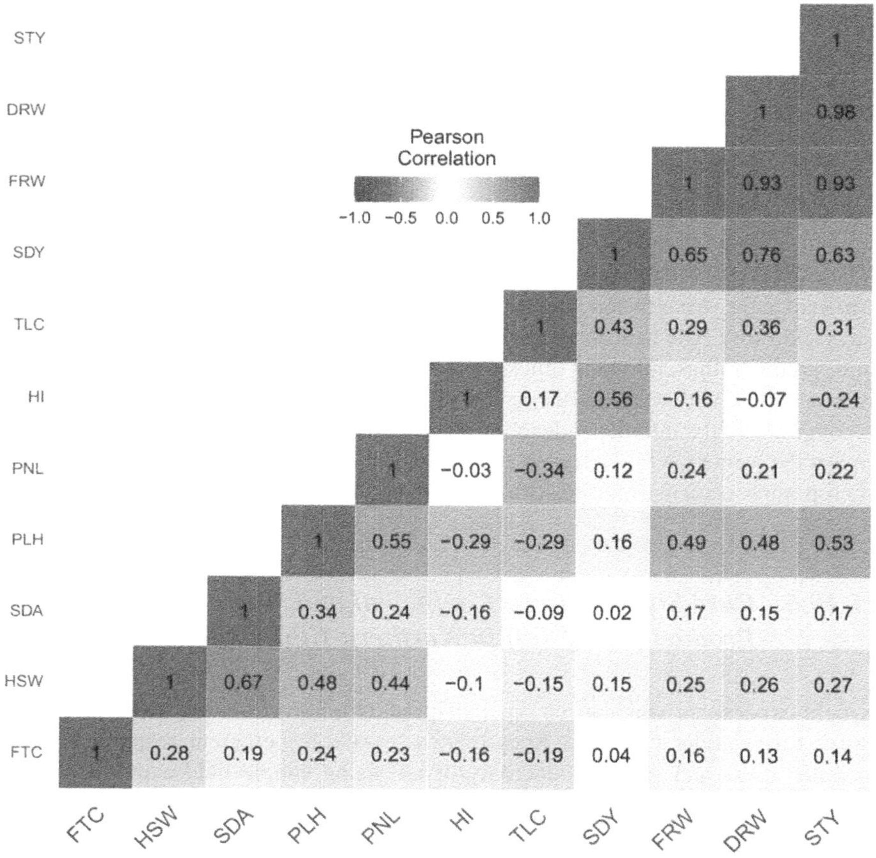

Fig. 11.6 Correlation heatmap and Pearson correlation coefficients analysis of 368 E. tef accessions based upon 11 quantitative agronomic traits. Floret count per spikelet (FTC), hundred-seed weight (HSW), seed area (SDA), plant height (PLH), panicle length (PNL), harvest index (HI), tiller count (TLC), seed yield (SDY), fresh weight (FRW), dry weight (DRW), and straw yield (STY). Qualitative traits for panicle morphology and seed color were not included. Heat map scale indicates the degree of negative or positive correlations observed (Mengistu et al.,2025)

11.10.7 Case Study 2: Aqua Crop Model Simulation of Teff Water Response in Semi-Arid Ethiopia

Araya et al. (2010) used FAO Aqua Crop model in Mekelle and Ilala, Northern Ethiopia to predict the response of teff to water and assess irrigation management in semi entire dry conditions. They used measured data to calibrate and validate Aqua Crop on teff using measured data of 2008–2009 on canopy cover, biomass and soil water. This model showed that the efficiency of grain-water use improved with the addition of supplementary irrigation of up to 95 mm of irrigation following flowering and biomass-water use efficiency declined. Aboveground biomass and grain yield were simulated well, and the models were valid. The findings revealed that optimal irrigation practices have the highest grain production and deficit irrigation is inclined towards biomass production when land is in abundance. In the case study, it was concluded that Aqua Crop is a useful tool to consider options of teff water management and advised its application in the extension planning to inform irrigation scheduling to enhance water productivity.

11.10.8 Case Study 3: Eco-geographic Adaptation and Phenotypic Diversity of Teff across Ethiopia

Woldeyohannes et al. (2020) have compared 3850-teff accessions sampled at various Ethiopian agro-ecological regions with the aim of determining eco-geographic adaptation and phenotypic diversity. The research incorporated the phenotype characterization of a large ex situ germplasm collection and climatic and geographic information to determine patterns of adaptation and climate-fitness. Findings showed a great deal of diversity in major characteristics, plant height had a range of 75–130 cm, days to maturity of 90–130 days, and a range of 0.9–2.4 t ha^{-1} in various habitats. High correlations were also found between the phenotypic traits and the environmental gradients especially the annual rainfall (400–1200 mm), mean temperature (12–24 °C), and elevation (1200–2800 m above sea level). Under future climate conditions, climate suitability modelling predicted a decline in the highly suitable areas of teff growth of between 15–25%. The paper suggested the use of locally adapted accessions in breeding initiative to produce climatically resistant varieties of teff and fortifying both in situ and ex situ conservation to preserve adaptive genetic resources.

11.10.9 Case Study 4: Seed Priming with Gas Plasma-Activated Water to Enhance Teff Establishment

A study conducted in 2024 in Ethiopia examined the impacts of gas plasma-activated water (GPAW) seed priming on the germination of teff, seed vigour, and stress tolerance. The experiment involved the comparative germination tests and accelerated ageing of conventional hydro priming, GPAW priming and non-primed seeds in controlled conditions. Final germination percentage (92–96%) using GPAW priming was significantly higher than that of the hydro-primed (82–86%) and non-primed seeds (70–75%). The rate of germination was elevated by 12–18%, and seedling vigour index was enhanced by 20–30% among the different genotypes that were tested. GPAW-treated seeds had 15–22% better viability when compared to controls following accelerated ageing. Such findings point to the fact that GPAW priming increases early seedling establishment and stress resilience. The study suggested the validation and incorporation of GPAW priming at the field level and integration in the practice of smallholder seed management to enhance the establishment of uniform crop in the climate variability.

11.11 Future Directions

The future of the teff-based modifying rests in its incorporation of climate-resilient breeding, increased genomics, and accuracy cultivation to protect efficiency under circumstances that are going to vary greater relative to current fluctuations. In breeding to climate-resilience, emphasis must be on the identification and deployment of tolerant genotypes that are more drought, heat, and waterlogging tolerant, based on the abundance of landrace diversity of teff. With the help of modern genomics and phenomics tools, alongside crop modelling, the identification of the stress-adaptive traits can be made faster and the prediction of the performance under the conditions of the future climate observed. Moreover, internet based agricultural systems, remote sensing, and decision aids facilitate real-time controls and responsible table of water, nutrient and planting procedures. The innovations of the research should be scaled to the smallholder systems to give the maximum impact as they should combine the local knowledge with the sustainable agronomic practices that can make it economically viable and also environmental sustainable to ensure that the productivity gains are sustainable. Such coordination will be essential in creating resilient teff-based food production systems that will be able to sustain food security in semi-arid areas (Huang et al., 2025).

11.12 Conclusion

Teff (*Eragrostis tef*) is discovered as a very climate independent cereal that can be used to produce in fluctuating and harsh conditions. Its natural physiological, morphological and biochemical characteristics such as C4 photosynthesis, water-use efficiency, plasticity in roots and osmotic adjustment allow teff to endure the heat and drought stress, and preserve yield constancy. Its capacity to be adapted to suit modern agronomic and genetic interventions and combining it with traditional landrace diversity enables it to have greater adaptive capacity, permitting it to have a flexible sowing date, in addition to supplemental irrigation, soil moisture preservation, and crop diversification strategies that vary depending on rainfed and semi-arid systems.

Complex methods, such as crop modelling (Aqua Crop, DSSAT, and APSIM), remote sensing, and digital monitoring instruments, offer accurate information on the interaction between soil-plant-atmosphere, water productivity and stress identification to aid evidence-based decision-making among smallholder farmers. Low-input and context-specific management practices prove to be both effective and yield enhancing and health promoting of the soil, as shown in case studies, which supports the value of integrating indigenous and modern technologies.

Further studies are needed to support and scale up teff-based systems as climate-resilient breeding, molecular and genomics-assisted selection, and optimization of water management at projected climate scenarios should be the focus of future research efforts. Adaptive agronomic practices and digital agriculture solutions, which help in building food and nutritional security, should be adopted with the aid of policy frameworks and institutional support. In general, teff can serve as a model crop example of climate-smart agriculture, where physiological, agronomic, and technological methods can be integrated to mitigate climate change effects and enhance viable, sustainable, and high-productive cereal production.

References

Alemu, M. D., Barak, V., Shenhar, I., Batat, D., & Saranga, Y. (2024). Dynamic physiological response of tef to contrasting water availabilities. *Frontiers in Plant Science, 15*, 1406173. https://doi.org/10.3389/fpls.2024.1406173

Araya, A., Keesstra, S. D., & Stroosnijder, L. (2010). Simulating yield response to water of teff (*Eragrostis tef*) with FAO's AquaCrop model. *Field Crops Research, 116*(1–2), 196–204. https://doi.org/10.1016/j.fcr.2009.12.010

Araya, A., Stroosnijder, L., Girmay, G., & Keesstra, S. D. (2011). Crop coefficient, yield response to water stress and water productivity of teff Eragrostis tef (Zucc.). *Agricultural Water Management, 98*(5), 775–783. https://doi.org/10.1016/j.agwat.2010.12.001

Assefa, M., Mekonnen, T., Mohammed, H., Umer, Z., & Song, Y. (2024). Characterization of genetic diversity in Eragrostis tef (zucc.) Trotter through microsatellite and morphological markers. *Ecological Genetics and Genomics, 33*, 100292. https://doi.org/10.1016/j.egg.2024.100292

Bayable, M., Tsunekawa, A., Haregeweyn, N., Ishii, T., Alemayehu, G., Tsubo, M., Adgo, E., Tassew, A., Tsuji, W., Asaregew, F., & Masunaga, T. (2020). Biomechanical Properties and Agro-Morphological Traits for Improved Lodging Resistance in Ethiopian Teff (Eragrostis tef (Zucc.) Trottor) Accessions. *Agronomy, 10*(7), 1012. https://doi.org/10.3390/agronomy10071012

Bekele-Alemu, A., & Ligaba-Osena, A. (2023). Comprehensive in silico analysis of the underutilized crop tef (Eragrostis tef (Zucc.) Trotter) genome reveals drought tolerance signatures. *BMC Plant Biology, 23*(1), 506. https://doi.org/10.1186/s12870-023-04515-1

Cotter, C. J., Wright, A. J., Romanov, A. V., Graf, T. N., Whisnant, E. D., Flores-Bocanegra, L., Doldron, M. S., Oberlies, N. H., Jia, Z., & Ligaba-Osena, A. (2023). Evaluating the Antioxidant Properties of the Ancient-Crop Tef (Eragrostis tef) Grain Extracts in THP-1 Monocytes. *Antioxidants, 12*(8), 1561. https://doi.org/10.3390/antiox12081561

Dinauer, A., Laufkötter, C., Doney, S. C., & Joos, F. (2022). What Controls the Large-Scale Efficiency of Carbon Transfer Through the Ocean's Mesopelagic Zone? Insights From a New, Mechanistic Model (MSPACMAM). *Global Biogeochemical Cycles, 36*(10), e2021GB007131. https://doi.org/10.1029/2021GB007131

Emeru, G. M. (2022). The perception and determinants of agricultural technology adaptation of teff producers to climate change in North Shewa zone, Amhara Regional State, Ethiopia. *Cogent Economics & Finance, 10*(1), 2095766. https://doi.org/10.1080/23322039.2022.2095766

Farquhar, G. D., Wong, S. C., Evans, J. R., & Hubick, K. T. (1989). Photosynthesis and gas exchange. In H. G. Jones, T. J. Flowers, & M. B. Jones (Eds.), *Plants under Stress* (1st ed., pp. 47–70). Cambridge University Press. https://doi.org/10.1017/CBO9780511661587.005

Felix, M. T., Tewodros, T., & Wales, S. (2018). Perceptions and choices of adaptation measures for climate change among teff (Eragrostis tef) farmers of Southeast Tigray, Ethiopia. *Journal of Agricultural Extension and Rural Development, 10*(1), 11–19. https://doi.org/10.5897/jaerd2017.0929

Food and Agriculture Organization of the United Nations (FAO). (2013). Teff production in Ethiopia. Retrieved from http://www.fao.org/faostat.

Gebremedhin, H., & Abraha, A. (2025). Teff: a healthy crop of the century–challenges and opportunities for enhancing productivity under climate change. *Discover Agriculture, 3*(1). https://doi.org/10.1007/s44279-025-00179-7

Gebru, M., Alemayehu, G., & Bitew, Y. (2023). Yield and lodging response of tef [Eragrostis tef (Zucc) trotter] varieties to nitrogen and silicon application rates. *Heliyon, 9*(12), e22576. https://doi.org/10.1016/j.heliyon.2023.e22576

Huang, Z., Yan, H., Zhan, Q., Yang, S., Zhang, M., Zhang, C., Lei, Y., Liu, Z., Liu, Q., & Wang, Y. (2025). A survey on remote sensing foundation models: from vision to multimodality (Version 1). arXiv. https://doi.org/10.48550/ARXIV.2503.22081

Jones, M. R. W., Kebede, W., Teshome, A., et al. (2025). Population genomics uncovers loci for trait improvement in the indigenous African cereal tef (*Eragrostis tef*). *Communications Biology, 8*, 807. https://doi.org/10.1038/s42003-025-08206-5

Kazemi Garajeh, M., Haji, F., Tohidfar, M., Sadeqi, A., Ahmadi, R., & Kariminejad, N. (2024). Spatiotemporal monitoring of climate change impacts on water resources using an integrated approach of remote sensing and Google Earth Engine. *Scientific Reports, 14*(1), 5469. https://doi.org/10.1038/s41598-024-56160-9

Kebede, A. S., Muluneh, T. W., & Adege, A. B. (2025). Detection of weeds in teff crops using deep learning and UAV imagery for precision herbicide application. *Scientific Reports, 15*(1), 30708. https://doi.org/10.1038/s41598-025-15380-3

Ketema, S. (1997). *Tef: Eragrostis tef (Zucc.) Trotter*. IPK.

Kordbacheh, F., Flaten, D. N., & Gulden, R. H. (2023). Weed community dynamics under repeated fertilization with different nutrient sources over 5 years. *Agriculture, Ecosystems & Environment, 346*, 108328. https://doi.org/10.1016/j.agee.2022.108328

Lee, H. (2018). Teff, A Rising Global Crop: Current Status of Teff Production and Value Chain. *The Open Agriculture Journal, 12*(1), 185–193. https://doi.org/10.2174/1874331501812010185

Mallu, T. S., Okbagabir, S. G., Negassi, B. T., Tesfay, T., Negash, T. T., Menghis, T. B., Medhn, T. A., Teklay, S. G., Gebremeskal, Y. H., Tewolde, R. H., Gebremariam, G. K., Tecleab, A. A., Tewelde, K. G., Brhane, D., Asghede, K. M., Adem, A. H., & Solomon, M. T. (2025). The agronomic, nutritional, environmental, and health benefits of Teff (*Eragrostis tef*) as an underutilized gluten-free superfood and feed source. *International Journal of Agronomy, 2025*(1), 6128837. https://doi.org/10.1155/ioa/6128837

Mengistu, M. A., Yim, W. C., Solomon, J. K. Q., & Cushman, J. C. (2025). Exploring the phenotypic diversity of Eragrostis tef for biomass and grain production under optimum growth conditions. *Frontiers in Plant Science, 16*, 1538510. https://doi.org/10.3389/fpls.2025.1538510

Mihretie, F., Tsunekawa, A., Bitew, Y., Chakelie, G., Derebe, B., Getahun, W., Beshir, O., Tadesse, Z., & Asfaw, M. (2021). Teff [*Eragrostis tef* (Zucc.)] rainfed yield response to planting method, seeding density, and row spacing. *Agronomy Journal, 113*(1), 111–122. https://doi.org/10.1002/agj2.20462

Mulatu, A. N., & Tamir, E. (2023). Prediction of Teff Yield Using a Machine Learning Approach. In B. H. Woldegiorgis, K. Mequanint, M. A. Bitew, T. B. Beza, & A. M. Yibre (Eds.), *Artificial Intelligence and Digitalization for Sustainable Development* (Vol. 455, pp. 159–176). Springer Nature Switzerland. https://doi.org/10.1007/978-3-031-28725-1_10

Nagangoudar, M. B., Ashoka, P., Km, L., Jadipujari, J., Thimmareddy, H., Jyothsna, K., Ma, S., Dj, K., & Mopagar, M. M. (2024). The lexicon on agro techniques in Teff- A super millet.

International Journal of Research in Agronomy, 7(9S), 190–196. https://doi.org/10.33545/2618060X.2024.v7.i9Sc.1459

Paff, K., & Asseng, S. (2019). A crop simulation model for Tef (Eragrostis tef (Zucc.) Trotter). *Agronomy, 9*(12), 817. https://doi.org/10.3390/agronomy9120817

Ramírez Gonzales, L. Y., Cannarozzi, G., Jäggi, L., Assefa, K., Chanyalew, S., Dell'Acqua, M., & Tadele, Z. (2024a). The role of omics in improving the orphan crop tef. *Trends in Genetics, 40*(5), 449–461. https://doi.org/10.1016/j.tig.2024.03.003

Ramirez-Gonzales, L., Cannarozzi, G., Rindisbacher, A., Jäggi, L., Schneider, R., Weichert, A., Plaza-Wüthrich, S., Chanyalew, S., Assefa, K., & Tadele, Z. (2024b). Transcriptomic Profile of Tef (Eragrostis tef) in Response to Drought. *Plants, 13*(21), 3086. https://doi.org/10.3390/plants13213086

Research Institute (Ifpri), I. F. P.(2018). Agronomics of teff (0 ed). International Food Policy Research Institute. doi:https://doi.org/10.2499/9780896292833_03.

Tasew, E., Dessie, M., Walraevens, K., Nigate, F., & Asmamaw, D. K. (2025). A comparative study of conventional irrigation and CropWat model-based irrigation scheduling: Effects on water productivity under onion and teff cultivation in the upper Blue Nile basin. *Environmental Research Communications, 7*(5), 055010. https://doi.org/10.1088/2515-7620/add46c

Thers, H., Jensen, J. L., Rasmussen, J., & Eriksen, J. (2024). Increasing cattle slurry application to grass-clover leys of different ages did not affect nitrate leaching but increased legacy effect in mixed organic crop rotations. *Field Crops Research, 306*, 109233. https://doi.org/10.1016/j.fcr.2023.109233

Vlajkov, V., Pajčin, I., Vučetić, S., Anđelić, S., Loc, M., Grahovac, M., & Grahovac, J. (2023). Bacillus-loaded biochar as soil amendment for improved germination of maize seeds. *Plants, 12*(5), 1024. https://doi.org/10.3390/plants12051024

Wilson, M. K. L. (n.d.). *Discovery of domestication, resilience, and agronomic traits in the underutilized cereal teff*. Michigan State University. https://doi.org/10.25335/5925-HF12

Woldeyohannes, A. B., Accotto, C., Desta, E. A., Kidane, Y. G., Fadda, C., Pè, M. E., & Dell'Acqua, M. (2020). Current and projected eco-geographic adaptation and phenotypic diversity of Ethiopian teff (Eragrostis teff) across its cultivation range. *Agriculture, Ecosystems & Environment, 300*, 107020. https://doi.org/10.1016/j.agee.2020.107020

Woldeyohannes, A. B., Desta, E. A., Fadda, C., Pè, M. E., & Dell'Acqua, M. (2022). Value of teff (Eragrostis tef) genetic resources to support breeding for conventional and smallholder farming: A review. *CABI Agriculture and Bioscience, 3*(1), 27. https://doi.org/10.1186/s43170-022-00076-9

Wosene, G., Pang, Y., Cao, J., Aynalem, M., & Jadoon, A. U. (2025). Socioeconomic Drivers of Adoption and Use Intensity of Improved Teff Seed in Northwest Ethiopia. *Food Science & Nutrition, 13*(10), e71030. https://doi.org/10.1002/fsn3.71030

Yihun, Y. M. (2015). *Agricultural water productivity optimization for irrigated teff (Eragrostic Tef) in a water scarce semi-arid region of Ethiopia*. CRC Press/Balkema.

Yihun, Y. M., Haile, A. M., Schultz, B., & Erkossa, T. (2013). Crop Water Productivity of Irrigated Teff in a Water Stressed Region. *Water Resources Management, 27*(8), 3115–3125. https://doi.org/10.1007/s11269-013-0336-x

Chapter 12
Buckwheat (*Fagopyrum esculentum*): Agro Ecological Benefits and Pollinator Support

Muhammad Asif, Umar Farooq, Ahmed M. S. Kheir, Farid Asif Shaheen, Mukhtar Ahmed, and Shakeel Ahmad

Abstract Buckwheat (*Fagopyrum esculentum*) playing an important component in promoting agro ecological stability and pollinator communities, which play a crucial role in ensuring food availability in the world. But with growing habitat degeneration, exposure to pesticides, climatic pressure and agricultural intensification, the populations of pollinators have reduced drastically. In this chapter, the authors discuss the ecological and economic significance of pollinators highlighting the necessity to adopt sustainable policies ensuring pollination services that are essential to crop productivity. It brings together the existing body of scientific knowledge, surveys successful methods of conservation, and shows the quantitative items of the latest empirical researches and case studies. The strategies that have been emphasized are the restoration of habitats, improvement of floral resources, combined methods of managing pests, diversified farming systems and community-based programs of monitoring. The existing quantitative measures demonstrate that the creation of pollinator friendly habitats is able to boost the frequency of pollination to a 30–60% range and enhance yields of crops that rely on pollinators by 20–45%. Case studies also show that diversified landscapes retain at least 50% of pollinator richness, and lowers (25–40%) exposure of pesticides in diversified landscapes reduces colony mortality. The buckwheat, oilseed and fruit production systems demonstrate that the lack of

M. Asif (✉) · U. Farooq · M. Ahmed
Department of Agronomy, Pir Mehr Ali Shah Arid Agriculture University, Rawalpindi, Pakistan
e-mail: asiffacho19@gmail.com

A. M. S. Kheir
Julius Kühn Institute (JKI)—Federal Research Centre for Cultivated Plants, Institute for Strategies and Technology Assessment, Kleinmachnow, Germany

Soils, Water and Environment Research Institute, Agricultural Research Center, Giza, Egypt

F. A. Shaheen
Department of Entomology, Pir Mehr Ali Shah Arid Agriculture University, Rawalpindi, Pakistan

S. Ahmad
Institute of Agronomy Faculty of Agricultural Sciences and Technology, Bahauddin Zakariya University, Multan, Pakistan

M. Ahmed (ed.), *Climate Change and Neglected Underutilized Crops*,
https://doi.org/10.1007/978-3-032-21005-0_12

pollination can decrease yields by 15–35%. Altogether, the results highlight the need to consider ecological principles in agricultural planning, as well as to increase the awareness of farmers and use nature-based approaches to guarantee the resilience of pollinators in the long term and sustainability in crop production.

Keywords Buckwheat · Pollination services · biodiversity conservation · crop productivity · ecological management

12.1 Introduction

The modern trend in strategic use of pseudocereals in sustainable agriculture has brought buckwheat (*Fagopyrum esculentum*) back as an important crop due to its adaptability, nutrition profile and the multiple ecosystem functions. This crop, is treasured globally due to its ability to grow in less productive soils, short growing periods and minimum input requirements. Hence, it is more appropriate in organic and low-input systems. Studies in the last ten years have highlighted its agroecological value: e.g. buckwheat has Argo-benefits due to the biomass and competitive growth. It can be useful in suppressing weeds, and can be used with other crops as a green manure (Xu et al., 2023a, 2023b). Moreover, its nutrient-dense grains, including balanced amino acids and bioactive compounds like rutin, allow it to be a potential functional food, which contributes to improving its potential in food security and human health. Buckwheat main benefit across the globe has been observed as an attractive crop of the pollinators. It has open flowers with high concentration of nectar. Insect are main drivers of pollination in this crop. Thus, in order to achieve optimal seed, set insect pollination is compulsory. Furthermore, this also determines the crop yield. Latest field and experimental evidence have shown that environmental stressors like drought and increased temperatures have the potential to significantly decrease floral rewards (e.g. nectar and pollen volumes). This causes a sharp decrease in pollinator visitation and, consequently, yield losses (Defalque et al., 2025). Furthermore, various groups of pollinators react differently to the change in weather conditions. A study has found that the visitation of Hymenoptera group decreases under specific weather conditions and the visitation behavior of Diptera remains relatively stable in response to environmental change (Miyashita et al., 2023). This suggests that diversity of response by pollinators can be used to counteract some effects of environmental change. However, this crop-pollinator relationship is going down the drain. Climate variability such as increase in frequency of heatwaves and water stress is changing floral characteristics and decreasing the availability of resources. Both of which have the potential to decouple key plant-pollinator interactions. In addition, habitat fragmentation, intensification of land-use, and agricultural homogenization leads to the further decline of the pollinators, increasing the threat of the pollination deficit. This risk can be concretely determined based on empirical studies in Nepal: in other agroecosystems (semi-natural, organic, and intensively farmed), unrestricted pollination on plots increased yields more than 200–300% than controls without pollinators (Rijal et al., 2018). These results underline the idea that even with being ecologically resilient, buckwheat productivity is closely connected with the effective pollination.

Modern methods like computational modelling, remote sensing, and geographic information systems (GIS) provide possible solutions to these problems and therefore present some potential opportunities to learn, as well as improve, buckwheat-pollinator dynamics. The flowering phenology, pollinator visitation behavior and yield outcomes can be simulated using mechanistic and statistical models in the context of different stresses. In the meantime, remote sensing (satellites or drone images) can be used to track flowering intensity, field health and stress cues, and GIS analyses can be used to map landscape connectivity of habitats and create pollinator friendly landscapes. Through the combination of these technologies, researchers and practitioners can develop a system of decision support systems that maximize the yield of buckwheat production in a manner that does not undermine or weaken ecosystem services.

These different yet connected themes are what this chapter tries to synthesize. First, we discuss the botanical and agronomic properties of buckwheat. The next step is discussing its agro ecological advantages, such as soil health, support of biodiversity, and potential as a functional food. We then move on to its pollination ecology, recording the effect of environmental stressors on floral characteristics and pollinator behavior, and their effects on yield. After which, we analyze empirical data of the pollination shortage particularly in small-scale agricultural systems and then explore the use of modelling and geospatial tools to predict and mitigate these risks. Lastly, we suggest agro ecological solutions, including habitat improvements, the conservation of pollinators, and the selection of genotypes, to increase the resilience of buckwheat and its utility to pollinators and indicate areas of future research that could help sustainability in agriculture.

12.2 Botanical, Physiological, and Agronomic Overview

12.2.1 Taxonomy and Morphological Traits

Fagopyrum esculentum (Polygonaceae) is a short-growth herbaceous pseudo-cereal that stands upright and has triangular leaves. This crop has a fibrous root system that can act as a soil stabilizer. It has small pentamerous insect-pollen-adapted flowers in its racemose inflorescences (Fig. 12.1). Recent reports have underlined morphological plasticity of the crop, which allows its cultivation in different agro ecological niches including temperate zones and high-altitude soils or soils that have low nutrient values. The traits of the rapid canopy closing, early flowering, and soil-enhancing, proves the multi-dimensionality of the use of buckwheat in a climate-resilient systems (Stefanova et al., 2023) (Wu et al., 2024).

Fig. 12.1 Parts of the buckwheat plant. Adapted from Farooq and Siddique (2023). Diversity analysis of tartary buckwheat (*Fagopyrum tataricum* Gaertn) in Nepal (Ph.D. dissertation). University of Tsukuba, Japan.

12.2.2 Floral Biology, Distyly, and Pollination Mechanisms

Buckwheat flowers display distyly and they possess pin and thrum morphs that favor obligate cross-pollination, and inhibit self-fertilization, which improves outcrossing. The insects that are pollinated are bees and dipterans through floral accessibility and nectar characteristics. The environmental conditions, such as temperature, humidity, water supply, have a powerful influence on the nectar secretion, the pollen viability, and the floral duration, which affects the pollinator behavior and the reproductive success. The latest evidence has shed light on the fact that the lack of pollination activity or floral desynchronization due to stress results in the cases of pollination deficiency and low grain set (Cheng et al., 2022; Ren et al., 2023)

12.2.3 Agronomic Requirements and Adaptability to Marginal Environments

Buckwheat would grow well in the well-drained soils of moderate fertility, under low-input, rainfed, or marginal conditions. Despite being sensitive to balanced nutrition, there is excessive nitrogen, which facilitates vegetative growth at the detriment of reproductive growth. It can be included in rotations, relay cropping and intercropping because of its brief growth period. According to recent agronomic analysis, it is a high yield of degraded soils and environmentally advantageous to regenerate soil, accumulate carbon, and support pollinators, which makes it a strategic crop in the face of climate uncertainty (Zhang et al., 2022).

12.2.4 Phenology and Growth Stages: Implications for Pollination and Management

The vegetative development all the way through anthesis up to seed maturation as the keys of phenological stages directly control the success of pollination and seed formation. The phenology progresses from germination, growth, flowering, and yield of Tartary buckwheat are shown in (Fig. 12.2) (Table 12.1). The fast growth of canopy promotes weed inhibition, and the flowering period needs to be in pace with the pollinators to achieve good seed set. The shortening of the anthesis period, decreasing nectar supply, and changing the patterns of pollinator visits have been demonstrated to be caused by heat stress, drought, and changing weather patterns. Recent studies highlight the need to use phenology-based management to maximize the sowing dates, irrigation, and nutrient supply to maintain reproductive stability in the fluctuating climatic regimes (Kamble et al., 2022).

Fig. 12.2 Phenology growth of buckwheat (Ren et al., 2018), International Journal of Food Properties

Table 12.1 Phenological growth stages of common buckwheat *(Fagopyrum esculentum Moench):* An extended BBCH scale (Grimes et al., 2021). University of Hohenheim

Principal growth stage	BBCH code	Description
Principal growth stage 0: Germination	00	Dry seed
	01	Beginning of seed imbibition
	03	Seed imbibition complete
	05	Radicle (root) emerged from seed
	06	Elongation of radicle; formation of root hairs and/or lateral roots
Principal growth stage 6: Flowering	68	Conclusion of flowering: only individual flowers still open
	69	End of flowering: no more flowers visible
Principal growth stage 7: Fruit development	71	First fruits visible
	72	First fruits reach variety-specific size; main phase of fruit set
	76	50% or more fruits reach variety-specific size
	79	Nearly all fruits reach variety-specific size
Principal growth stage 8: Fruit maturation	81	First ripe fruits
	82	All plants have at least one ripe fruit; main phase of seed ripening
	86	More than 50% of fruits are ripe
	89	Nearly all fruits ripe, terminal coloration, hard and dry
Principal growth stage 9: Senescence	91	Leaves still green
	93	Leaves begin changing color or falling
	96	50% leaves discolored or fallen
	97	End of leaf fall; plants or above-ground parts dead/dormant
	99	Plant fully senescent

12.2.5 Global Production Trends, Cropping Patterns, and Ecological Niches

Production of buckwheat has been growing in ecologically limited areas because of its versatility, limited requirements, and its role in the agroecological differentiation. It is also being applied more in rotations and intercropping systems to promote the health of the soil, the availability of floral resources, and the population of pollinators. The buckwheat is also of significant value in the high-altitude agro-ecosystems (Fig. 12.3) in South and East Asia, particularly the Himalayan regions, as it is important in terms of food security and biodiversity protection. It is an ecologically diverse crop, which

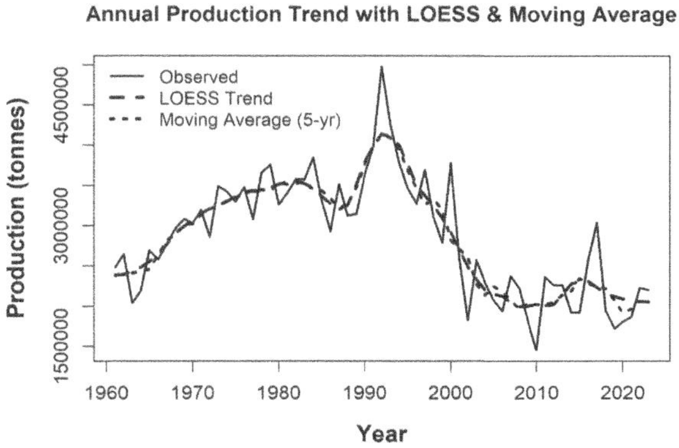

Fig. 12.3 Globally production of Buckwheat (FAOSTAT 2023) (LOESS (Locally Estimated Scatterplot Smoothing) is a non-parametric smoothing method that reveals the underlying long-term trend in your data by reducing short-term fluctuations)

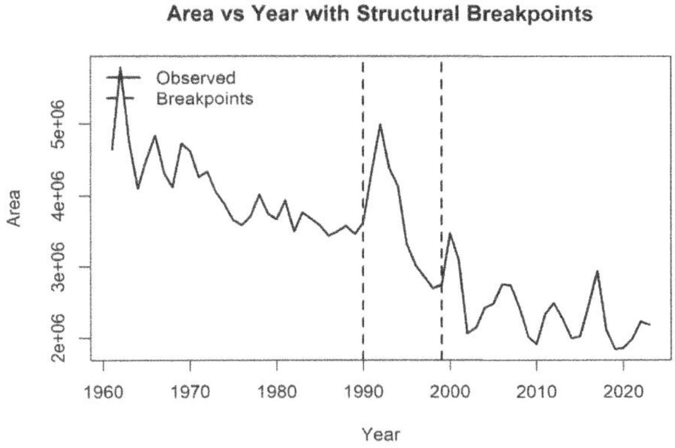

Fig. 12.4 Area versus year structural breakpoints in yield

makes it a priority crop in sustainable and pollinator-friendly and climate-healthy agriculture (Figs. 12.4 and 12.5) (FAOSTAT, 2025).

12.3 Agro ecological Benefits of Buckwheat

12.3.1 Soil Health Enhancement

Buckwheat is a significant addition to the soil as it builds biomass quickly enhancing the soil organic carbon and activating microorganisms. Its residues break down fast enhancing nutrient cycling and maintaining biological processes in the soil. Being a green manure, buckwheat increases the turnover of nitrogen and helps in the further productivity of crops. The fibrous root system increases the aggregation of the soil, the infiltration and the minimization of the erosion as well as the fast cover of the canopy helps in the prevention of the degradation of the soil. All these qualities

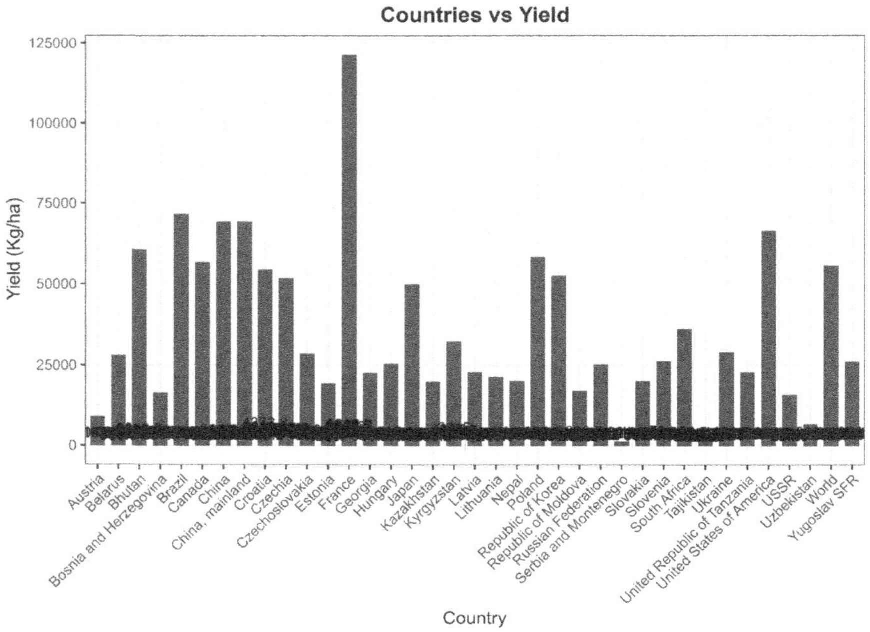

Fig. 12.5 Yield trends of buckwheat in different countries (FAOSTAT, 2025)

make buckwheat a good option in restoring poor and unfertile soils in sustainable agroecosystems (Wieneke et al., 2022).

12.3.2 Weed Suppression and Allelopathy

Buckwheat is very effective in controlling the growth of weeds as it grows very fast and provides a tight cover and thus less light reaches the soil surface (Fig. 12.6). The release of allelopathic compounds into the crop through root exudates and decomposing residues also prevents germination of weed seeds and premature development of seedlings. The natural suppression strategies reduce the use of herbicides and contribute to the shift towards low-input and organic food production systems (Wen et al., 2021).

12.3.3 Nutrient Management and Phosphorus Mobilization

Buckwheat is a component in the nutrient management especially in the mobilization of soil phosphorus. It releases organic acids that dissolve insoluble phosphates so

Fig. 12.6 Weed suppression (%) in buckwheat monocultures and two- and three-way mixtures throughout the growing season, measured by GDD5. Worthing (2025)

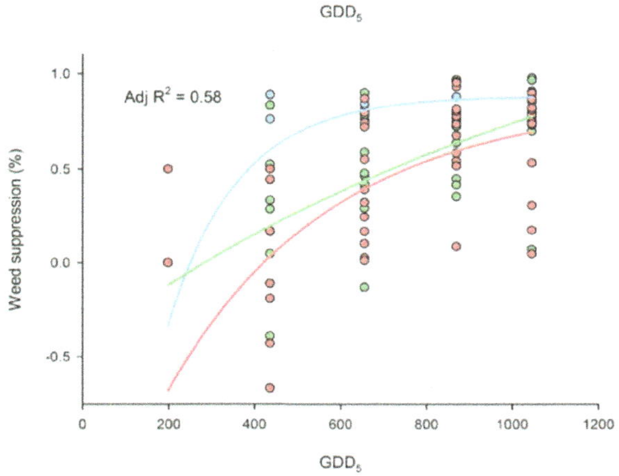

that they are better accessible to the buckwheat and other crops. This ability to mobilize nutrients enables the buckwheat to grow in low nutrient content soils and minimizes the reliance of external sources of phosphorus fertilizer. It contributes to the nutrient cycling of the soil, thereby supporting sustainable management of soils and the increase in productivity of the diversified cropping systems (Buck et al., 2023).

12.3.4 Enhancement of Functional Biodiversity

The functional biodiversity of buckwheat fields is favorable because functional floral resources are readily available and abundant. The crop has a high diversity of pollinating arthropods and other helpful arthropods such as predatory insects and parasitoids that help in the regulation of pests in nature. Predatory beetles, lacewings, and syrphid flies are common users of buckwheat habitats which increases their pollination and biological control. Buckwheat enhances the resilience of agroecosystems and increases the decrease in the reliance on pesticides by developing various positive communities (Mele et al., 2025).

12.3.5 Contribution to Resilient, Low-Input, and Climate-Smart Agriculture

The fact that it has the shortest growth cycle, good utilization of resources and their ability to adapt to marginal environments makes buckwheat well suited to climate-smart farming. It has been shown to be reliable to drought, temperature change, and low soil fertility conditions and thus production can be maintained well with minimal inputs. This stability is used to counteract the effects of climate variability and be able to sustain an intensification in areas under environmental stress. Its low-input characteristic also makes it have lower environmental footprints compared to other crops of conventional varieties (Sun et al., 2023).

12.3.6 Integration in Agroecological Frameworks

Buckwheat is a highly compatible crop with various agroecological models that have system-level advantageous effects. It also enhances the soil fertility, lowers pest and disease accumulation and improves the yield potential of subsequent crops in crop rotations. In intercrops, buckwheat enhances the efficiency of resource use and continuity of floral resources which boosts the pollinator population. Being used as a cover crop, it suppresses any type of weeds, discourages soil erosion, enhances the amount of organic matter, and provides a home to useful species. Such multifunctional functions indicate the ability of the crop to improve the ecological operations of diversified farm landscapes (Wang et al., 2024).

12.4 Floral Traits, Nectar Dynamics, and Pollination Ecology

12.4.1 Floral Architecture and Anthesis Behavior

Buckwheat is a plant with a unique floral structure with small, pentamerous, distylous flora in small racemes as shown in (Fig. 12.7). Pin and thrum morphs forcefully stimulate cross-pollination, which increases genetic diversity and the reproductive effectiveness (Fig. 12.8). Anthesis normally takes place early in the morning when the weather is favorable in terms of temperature and humidity in stigma receptivity and presentation of pollen. The floral characters, such as corolla size, exposition of nectar glands, and petal reflective value, have a strong effect on pollinator attraction and the following pollen transfer efficiency. The traits are important in seed set and yield stability especially when there are changes in agroecological conditions (Quinet et al., 2004).

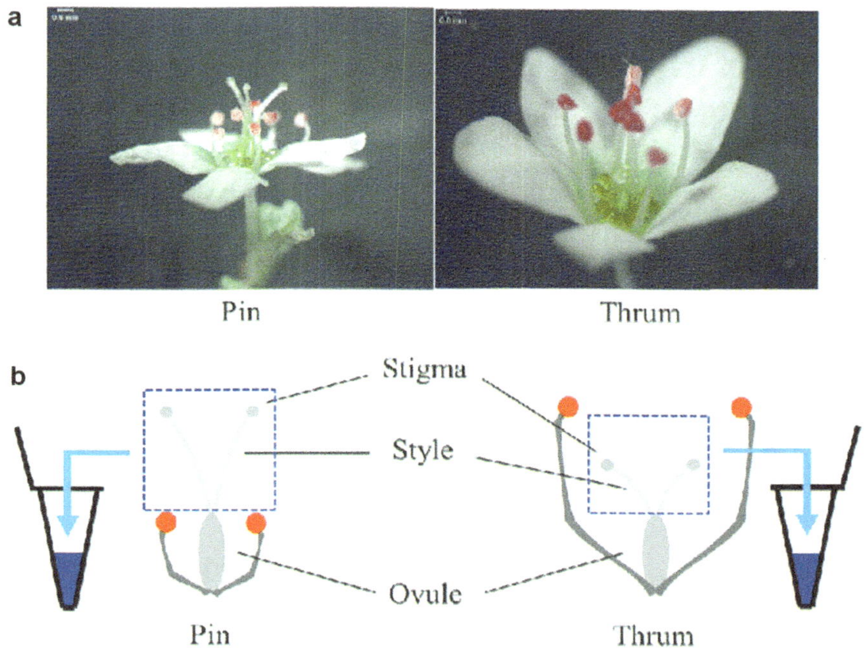

Fig. 12.7 Floral morphotypes of common buckwheat (*Fagopyrum esculentum*) and method of sampling for 2D-PAGE analysis. A Pin flower has long styles and low anthers (left panel); thrum flower has short styles and high anthers (right panel) (Takeshima et al., 2019)

12.4.2 Nectar Quantity, Composition, and Environmental Sensitivity

Buckwheat flowers contain nectar with high simple sugars like glucose, fructose and sucrose, which is one of the major attractants to honeybees, bumblebees as well as syrphid flies. The amount, viscosity and sugar composition (Table 12.2) of nectar are extremely sensitive to both environmental and physiological conditions, such as nutrient status of soil, temperature and water. Higher temperatures or quick changes in humidity can cause a significant decrease in nectar volume whereas drought stress can cause higher sugar concentration and lower nectar secretion rates. The pollinator visitation patterns are also affected by such shifts, and the results have a direct impact on the pollen deposition and reproductive success (Zhang et al., 2022; Li et al., 2024).

Fig. 12.8 Flowers and seeds of (**a**) common and (**b**) Tartary buckwheat along with (**c**) pin flower and (**d**) thrum flower. Adapted from Farooq and Siddique (2023). Phaper. Sthaniya/raithane baali kheti prabidhi, utpadan bibidhikaran taalim srot pustika. Baali Bikas tatha Krishi Jaiwik Bibidhata Samraxan Kendra, Harihar Bhawan, pp. 27e57 (In Nepali).

12.4.3 Effects of Abiotic Stress on Reproductive Biology

Buckwheat is under a significant impact of abiotic stressors such as heat, drought, and excessive humidity in the reproductive processes. Flowering heat stress decreases pollen viability, enhances floral senescence, and decreases the anthesis period. Water shortage inhibits flowering and nectar release and excessive humidity inhibits dispersal of pollen and encourages fungal attack on floral structures. Such stress-related disturbances frequently result in the decreased pollinator visitation and lack of correspondence between pollinator activity and the floral receptivity, which eventually causes pollination deficits and lower seed production (Duan et al., 2023; Qu et al., 2022)

Table 12.2 Sugars and minerals in buckwheat and wild plant honey from three locations in Virginia during 2015 (Corey et al., 2022)

Component	Buckwheat	Wild plants
Fructose (%)	41.69	51.57
Glucose (%)	1.72	1.96
Sucrose (%)	1.26	1.19
Maltose (%)	0.26	0.18*
Melesios (%)	0.54	0.69
Sulfur (%)	0.02	0.01
Phosphorus (%)	0.02	0.01
Potassium (%)	0.17	0.04**
Magnesium (%)	0.01	0.01
Calcium (%)	0.05	0.02
Boron (ppm)	6.67	4.67
Zinc (ppm)	1.33	1
Manganese (ppm)	11.33	9.33
Iron (ppm)	1	1
Copper (ppm)	5	3.33*
Aluminum (ppm)	8	1.67

Note *, **, ***: Means were statistically different at 10, 5, and 1 percent level, respectively.

12.4.4 Behavioral Ecology of Pollinators in Buckwheat Fields

The foraging behavior of pollinators in buckwheat fields depends on species-species, species-floral and species-landscape structure. Honeybees (*Apis mellifera*) tend to monopolize visitation as a result of efficient foraging behaviors with bumblebees (*Bombus* spp.) making an unequal contribution to the successful pollen transfer through buzz-pollination like vibrations and burrowing deep into the flowers. Pollination is additionally obtained by the syrphid flies and small solitary bees, especially in unfavorable weather conditions. The shape of patches of flowering, the presence of neighboring floral resources, and microclimatic conditions are all the factors that determine the movement of pollinators and foraging levels. Floral continuity and habitat complexity can be controlled to increase stability of the ecosystem and improve the results of pollination (Devkota et al., 2026).

12.4.5 Interplay Between Floral Traits, Pollinator Foraging, and Reproductive Success

Buckwheat floral traits in combination with pollinator foraging behavior play an important role in the successful reproduction. The efficiency of pollen transfer between morph-distinct distylous morphs is determined by characteristics of the flower density, nectar provisioning, and morph-specific compatibility. The fluctuations in the environment that change these characteristics will interfere with the processes of plant-pollinator synchrony and decrease cross-pollination efficiency. Floral rewards and density respond to foraging with preference-dependent foraging, which results in preference-dependent reproductive outcomes. Ecologically friendly strategies, including the continuity of flowering, improvement of pollinator habitats, and minimization of stressors can be used to stabilize the process of pollination and reproductive success of various agricultural landscapes (Akter et al., 2023).

12.5 Pollinator Diversity, Function, and Ecosystem Services

12.5.1 Composition of Pollinator Guilds

Buckwheat is the most popular pollinator attracting various types of beetles, honeybees (managed and wild), hoverflies, lepidopterans, and several types of bees. Honeybees (*Apis mellifera*) are superior in that they are high visitation insects which maintain stable foraging frequencies. The wild bees like *Bombus* spp solitary *Andrena* spp can play a crucial role due to their capacity to forage in more cool and changeable conditions and also to forage complicated architectures. The non-bee insects, such as syrphid flies and nocturnal moths, have functional redundancy, and this provides a stable level of pollination in case the primary pollinators are scarce. This multispecies guild enhances the resilience of the ecosystem and yields stability in the landscapes that are different in terms of agroecology (Table 12.3) (Kovacs-Hostyanszki et al., 2022; Li et al., 2024)

12.5.2 Pollinator Visitation Patterns and Foraging Efficiency

The frequency, duration and constancy of visitation by pollinators have a strong impact on the reproductive performance of buckwheat (Table 12.4). Honeybees show high levels of morph fidelity, which enhances the efficiency of pollen transfer between floral types of compatibility. Wild bees show increased foraging areas and are able to pollute even under extreme temperatures when honeybees forage less. The variability in the weather such as a wind change, humidity increase, or heat stress in the middle of the day is used to modulate foraging windows and may influence the synchrony

Table 12.3 Taxonomic composition of large and small insects observed visiting buckwheat flowers combined at all sites (Miyashita et al., 2023)

Large insects		Small insects	
Taxon	Abundance	Taxon	Abundance
Scarabaeid beetle	1596	Ant	2502
Butterfly	259	Hoverfly	538
Syrphid fly	159	Non-syrphid fly	463
Non-syrphid fly	146	Others	313
Others	113	Wasp	115
Wasp	104	Bee	43
Bee	19		
Total	2396	Total	3974

between anthesis and pollinator activity. The analysis of these foraging patterns allows creating the optimized interventions in the form of pollinator management (Fig. 12.9) (Devkota et al., 2026).

12.5.3 Functional Contributions to Seed Set, Fruit Quality, and Yield

Pollination by insects has a significant positive effect on seed set, seed weight and grain uniformity in buckwheat. Empirical evidence indicates that flower yield is significantly greater in open-pollinated flowers than in controls with pollinators excluded referring to the reliance of the crop on biotic pollination. Enhanced fill and concentration of nutrients and uniformity of maturation of kernels-attributes of great value in commerce and downstream processing are also improved with improved pollination. Their contributions are especially important in low-input systems, in which dependence on pollination by nature is the greatest (Xiang et al., 2022).

12.5.4 Climate Change Impacts on Pollinator Behavior and Plant–Pollinator Synchrony

Climate variability changes the phenology of flora, as well as pollination activity, and affects the synchrony that should be present to conduct successful pollination. Heatwaves have the potential to inhibit nectar secretion, decrease pollen viability, drought decreases flower production and may reduce the length of anthesis. At the same time, pollinators can change the time of their emergence or daily behavioral patterns in relation to the changes in temperature, which is associated with the risk of unsuitable interactions. This is a set of adaptive strategies that must be put into practice in order to protect the pollination services in the future climate conditions

Table 12.4. The number of compatible pollen grains per single visit for the thrum and pin flowers, visitation rate and per flower handling time (mean ± SEM values in brackets refer to number of observations or 95% CI where indicated), and estimated individual pollination efficiency (compatible pollen grains transferred per minute) for different species. A dash (–) indicates that the data was not available (Liu et al., 2020)

Group	Species	Per flower handling time (log of second)	Visitation rate (number of flowers per minute)	Number of compatible pollen grains per visit Pin flower	Thrum flower	Pollination efficiency (per individual, 95% CI
Honey bee	*Apis mellifera*	0.65 ± 0.07 (38)	31.8 ± 1.5 (38)	8.1 ± 1.2 (77)	3.7 ± 0.7 (54)	167 (135–203)
Honey bee	*Apis cerana*	0.68 ± 0.05 (58)	31.4 ± 1.1 (58)	5.1 ± 0.8 (120)	4.4 ± 0.6 (99)	151 (126–177)
Hoverfly	*Eristalis tenax*	0.81 ± 0.04 (122)	27.6 ± 0.9 (122)	2.7 ± 0.6 (44)	5.5 ± 1.6 (41)	113 (76–154)
Wasp	*Tiphia vernalis*	1.99 ± 0.11 (9)	8.2 ± 2.2 (9)	9.2 ± 4.0 (8)	11.8 ± 4.1 (16)	97 (57–145)
Wasp	*Vespula germanica*	1.31 ± 0.10 (11)	16.6 ± 2.1 (11)	4.4 ± 1.5 (20)	4.7 ±1.6 (18)	75 (48–106)
Fly	*Parasarcophaga sericea*	1.21 ± 0.05 (66)	19.8 ± 1.1 (66)	2.6 ± 0.7 (49)	5.4 ± 2.2 (39)	73 (44–114)
Wasp	*Dolichovespula saxonica*	1.35 ± 0.10 (10)	15.5 ± 2.1 (10)	3.0 ± 0.9 (14)	6.1 ± 1.6 (14)	73 (51–96)
Wasp	*Polistes gallicus*	1.70 ± 0.11 (10)	10.4 ± 2.1 (10)	3.3 ± 1.0 (35)	9.5 ± 2.5 (23)	73 (47–99)
Hoverfly	*Eristalis cerealis*	1.20 ± 0.12 (8)	19.2 ± 2.4 (8)	3.1 ± 1.2 (9)	4.2 ± 1.1 (20)	71 (45–101)
Fly	*Lucilia sericata*	1.18 ± 0.05 (74)	19.1 ± 1.0 (74)	2.9 ± 0.6 (34)	4.0 ± 1.1 (27)	64 (45–84)
Fly	*Linnaemya comta*	1.03 ± 0.09 (14)	22.3 ± 1.8 (14)	1.2 ± 0.4 (39)	4.1 ± 1.0 (33)	59 (40–82)
Fly	*Lucilia illustris*	1.22 ± 0.05 (82)	18.6 ± 1.0 (82)	2.3 ± 0.5 (44)	3.7 ± 1.4 (16)	56 (35–82)
Hoverfly	*Eristalinus aeneus*	1.25 ± 0.06 (52)	18.6 ± 1.1 (52)	3.2 ± 0.7 (15)	3.1 ± 1.4 (8)	54 (33–76)
Wasp	*Polistes snelleni*	1.57 ± 0.19 (3)	13.5 ± 3.8 (3)	1.0 ± 0.6 (4)	6.7 ± 2.7 (3)	48 (24–80)
Ant	*Formica japonica*	2.36 ± 0.11 (9)	7.4 ± 2.2 (9)	3.0 ±0.8 (29)	6.6 ± 1.7 (44)	28 (17–40)
Hoverfly	*Episyrphus balteatus*	1.48 ± 0.05 (63)	15.0 ± 1.1 (63)	1.8 ± 0.3 (43)	1.4 ± 0.5 (34)	22 (16–29)
Hoverfly	*Sphaerophoria scripta*	1.97 ± 0.15 (5)	8.7 ± 3.0 (5)	0.2±0.2 (4)	1.6 ± 1.0 (11)	8 (2–15)
Fly	*Chrysomya pinguis*	1.09 ± 0.12 (7)	22.0 ± 2.5 (7)	5.5 (2)	0.0 (1)	–

(continued)

Table 12.4. (continued)

Group	Species	Per flower handling time (log of second)	Visitation rate (number of flowers per minute)	Number of compatible pollen grains per visit Pin flower Thrum flower		Pollination efficiency (per individual, 95% CI
Hoverfly	*Eristalis himalayensis*	0.93 ± 0.12 (8)	23.9 ± 2.4 (8)	9.0 (1)	8.5 (2)	–
Hoverfly	*Eupeodes corollae*	1.31 (1)	15.9 (1)	10.0 (1)	–	–
Hoverfly	*Eupeodes nitens*	1.58 (2)	12.7 (2)	1.3 ± 1.3 (3)	0.01 (1)	–
Hoverfly	*Phytomia zonata-*	0.77 ± 0.16 (4)	28.2 ± 3.3 (4)	3.0 (1)	–	–
Wasp	*Vespa bicolor*	–	–	3.0 (1)	–	–
Hoverfly	*Orthonevra elegans*	–	–	0 (1)	3.0 ± 0.6 (3)	–
Bumblebee	*Bombus hypocrita*	–	–	0 (1)	–	–
Fly	*Tachinidae sp1*	–	–	0 (1)	–	–
Fly	*Tachinidae sp2*	–	–	0 (1)	–	–

Fig. 12.9 Examples of flower-visiting species that were evaluated in this study (Liu et al., 2020)

such as the use of climate-resistant cultivar, habitat diversification, and the adjustment of sowing phenology (Fattorini et al., 2023)

12.6 Pollination Services, Deficit Assessment, and Yield Responses

12.6.1 Measuring Pollination Dependence in Buckwheat

The distylous nature of the reproductive system of buckwheat causes high dependence on biotic pollination, and pollination dependence needs to be quantitatively measured. There are consistent comparative assessments of the yield comparing open-pollinated flowers, hand-pollinated flowers, and pollinator-excluded flowers yielding much better with insect involvement in the transfer of pollen. Such evaluations can be used to identify the ecological significance of pollinators and the degree to which reproductive achievement depends on pollinators in different production systems, such as organic, low-input, and commercial enterprises (Farooq & Siddique, 2023; Sun et al., 2023).

12.6.2 Experimental Evidence from Exclusion/Inclusion Studies

Experiments based on pollinator exclusion (mesh bags or cages) are also a clear indicator that insect visitation does greatly increase the buckwheat yield components. Plants that are open-pollinated exhibit better seed set, grain weight and kernel development than the excluded flowers. Inclusion experiments also support the benefits of inclusion of honeybees with diverse wild pollinator communities being synergistic, and in most cases, wild bees maintain pollination during unfavorable weather conditions (Akter et al., 2023).

12.6.3 Regional Case Studies

Regional evaluations indicate that agroecosystems are different in their effectiveness of pollination. In mid-hills and valley systems, buckwheat produces in South Asia show a loss of yield of more than 30% in the case of pollinator exclusion. According to European field research, seed uniformity will improve significantly when honeybee colonies are added to the natural assemblages of pollinators. Wild bee guilds such as Bombus, Halictus, and Andrena are used as the main pollinators in the North

American habitats and, in particular, in large monocultural landscapes where the density of honeybees is uneven (Fijen et al., 2022).

12.6.4 Characterizing Pollination Deficit: Biological and Environmental Causes

The problem of pollination lack is a situation that occurs when the level of pollen deposition is below the level required to maximize seed production. Some of the biological aspects that contribute to this are the lack of pollinator abundance, low flower attractiveness, morphological differences between the floral structure and the visitor characteristics. Unpredictable rainfall, heat spikes, and habitat fragmentation are other environmental stressors which decrease effective visitation. The discovery of such interacting drivers will allow the provision of a specific intervention to stabilize pollination and improve the efficiency of yield (Oliveira et al., 2025).

12.6.5 Effects of Pollination on Yield Parameters

Pollination is very effective in increasing the important yield traits in buckwheat, such as seed number, thousand grains weight, total biomass partitioning. Proper cross-pollination enhances fill of kernels and uniformity of grains, which determine commercial as well as processing quality. This reduces seed set, grain mass and abnormal maturation, which is a highly important point to note, as pollination services are central in determining economic returns in buckwheat production (Rijal et al., 2018).

12.6.6 Socioeconomic Relevance of Pollination Services in Smallholder Systems

Pollination services are especially important at the smallholder level production landscapes where reliance on natural populations of pollinators is large and access to the managed bees is scarce. Improved pollination leads to increased yields, household income and also food security (Fig. 12.10). Wild pollinator communities can be sustained by practices like the conservation of field margins, creation of floral resources, and minimization of the pesticides and offer long-term socio-ecological returns. Enhancement of pollinator-friendly management is, therefore, one option of enhancing rural livelihoods and maintaining agroecosystem functioning (Razze et al., 2016).

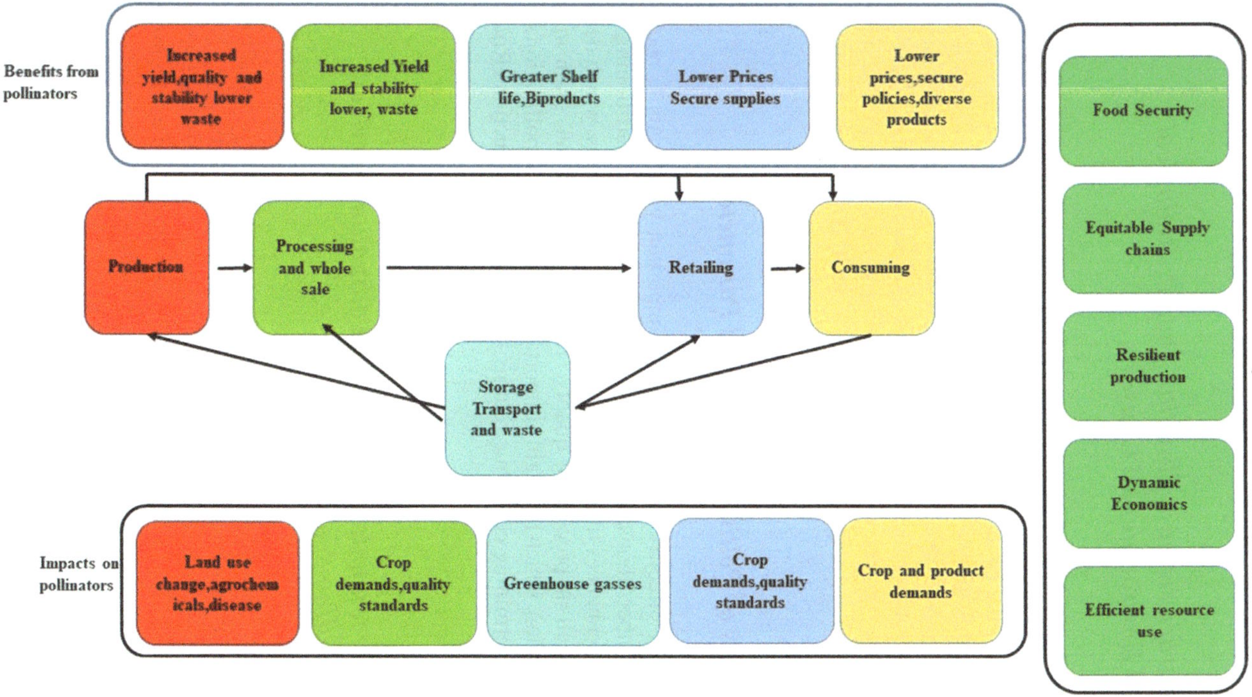

Fig. 12.10 Representation of a food system, including the benefits of and impacts on pollinators for each Factor

12.7 Environmental, Agronomic, and Climatic Drivers of Pollination Success

12.7.1 Microclimatic Influences

Pollination efficacy in buckwheat is severely controlled by microclimatic conditions such as temperature, relative humidity, wind movement, etc. The average temperatures at anthesis favor the pollen viability and stigma receptivity but heat stress enhances the speed of floral senescence and lowers the reproductive success. Humidity has an influence on the release of pollen and the release of nectar; high humidity may interfere with the pollen release and low humidity may interfere with the nectar production and thus the visitation may be decreased by the pollinators. Wind also affects insect foraging behavior such that at low intensity it can facilitate or impede the transfer of pollen. The study of the interaction between these variables allows forecasting the results of pollination in the field setting, as well as creating climate-receptive management plans (Xu et al., 2023a, 2023b).

12.7.2 Water Stress, Nutrient Regimes, and Floral Physiology

The availability of water and its nutrients have a direct effect on the floral biology and pollination potential. Drought stress decreases flower initiation, anthesis period and decreases the amount of nectar, reducing attractiveness of buckwheat flowers to pollinators. Nutrient constraints especially those of nitrogen and phosphorus are disadvantageous to the floral morphology, pollen viability and composition of nectar. Even balanced fertilization and sufficient moisture are thus both necessary to keep the floral resources strong and to sustain successful pollination with the aim of achieving a high-level of reproductive success (Antala et al., 2025).

12.7.3 Pest Management and Pesticide Impacts on Pollinators

The health of the pollinators and the reliability of the pollination services heavily depend on the pest control practices. Both wild and under management pollinators have a high likelihood of having a reduced abundance, diversity, and foraging behavior due to the use of broad-spectrum insecticides. The sublethal exposures can reduce the navigation, learning, and handling efficacy of flowers. Integrated pest management (with a focus on biological control agents, pest survey, habitat diversification, and selective or less risky pesticides) will support in protecting the pollinator communities and provide crop protection. This is necessary to maintain the ecosystem services in buckwheat production through such strategies (Matsuzawa et al., 2023).

12.7.4 *Genotype-Specific Variation in Floral Attractiveness and Resilience*

The genotypes of buckwheat vary in the most important floral characters with the size of corolla, pigmentation, quantity of nectar and time of anthesis, which affect pollinator attractiveness. Some genotypes are considered more resilient to stressors such as heat waves, drought and nutrient imbalance and hence maintain reproductive fitness. These are tough lines that in most cases show a stable nectar secretion and a long-lasting floral life cycle in a changing environment. The efficient pollination can be enhanced by breeding and selection of the genotypes with increased floral attractiveness and stress resistance in response to the pressure of climate-related challenges (Gao et al., 2023).

12.7.5 *Field Layout, Landscape Configuration, and Pollinator Movement*

The field design and landscape structure are critical towards determining the movement of pollinators and the efficiency of cross-pollination (Fig. 12.11). The creation of large monoculture fields can limit the accessibility of some wild pollinators to heterogeneous landscapes, which are rich in hedges, flowering strips, fallow areas, and semi-natural environment, would allow increased pollinator diversity and functional redundancy to become common. The spatial layout which coincides with the foraging areas of the pollinator enhances the rate of visitation and seed set. Ecological infrastructure-based landscape large scale planning improves connection, contributes to the existence of pollinators, and pollination services of buckwheat systems (Yang et al., 2023).

12.8 Modelling Pollination, Yield Formation, and Ecosystem Services

12.8.1 *Conceptual and Mechanistic Models of Plant–Pollinator Interactions*

The superior modelling methods offer insights that are important to understand the relationship between plants and pollinators in buckwheat agroecosystem. The connection between flower density, visitation rates and pollen deposition is modeled in conceptual and mechanistic models to measure pollination efficiency. Flower density models assess the impact of the spatial distribution of inflorescences in influencing pollinator foraging behavior, whereas visitation rate models reflect the

Fig. 12.11 Field layout, landscape configuration

temporal dynamics of the pollinator behavior in a condition of variable environmental conditions. Pollen deposition models determine the effectiveness of cross-pollination of various pollinator groups. The combination of these strategies permits forecasting the results of reproduction in a variety of conditions and guides measures directed to increase pollination services (Peralta et al., 2024).

12.8.2 Crop Growth and Yield Modelling for Buckwheat

Crop growth models are based on the phenology, biomass accumulation and yield formation, to model performance across diverse management and environmental variations. Phonological prediction models can be used to predict flowering and a thesis time, which is necessary to align the action of pollinators with the most productive periods. The effects of drought, heat or nutrient restrictions in the production and supply of flowers and nectar are measured by stress-response models. The simulation of pollination-dependent yield incorporates those with pollination visitation and deposition of pollen to predict the seed set, thousand-grain weight, and total yield. These types of models offer a solid structure of optimization of management activities to promote productivity and resilience in buckwheat agroecosystems.

12.8.3 Ecosystem Service Modelling

The ecological contribution of pollinators is more than yield and the ecological contribution is modeled through ecosystem service modelling. The population dynamics models of pollinators determine abundance of species and activity patterns over time whereas the habitat suitability models determine areas that facilitate the optimal pollinator communities. The pollination services of buckwheat are measured at the landscape scale to quantify spatial diversity of pollination activity and its influence on the buckwheat productivity. With the help of these strategies, conservation planning, habitat management, and the introduction of pollinator-friendly practices into the landscapes of the struggle against cropping are directed (Dyer et al., 2022).

12.8.4 Integrating Climate Models to Predict Future Pollination Scenarios

It is possible to predict the way in which changes in temperature, precipitation regimes, and exceptionally adverse weather changes can modify interactions between plants and their pollinators using climate-integrated models. By connecting the climate predictions with the phenology, nectar production and pollinator foraging models, temporal mismatches during flowering and pollinator foraging can be predicted. These lessons are used to implement adaptive management measures, such as the choice of resilient genotypes, changing the sowing times, and interventions in the habitat to preserve pollination services in the future climatic conditions (Krzyzanska et al., 2022).

12.8.5 Modelling Pollination Deficit and Identifying Intervention Thresholds

Predictive modelling frameworks are used to quantify the deficit of pollination through the simulation of situations at different levels of pollinator abundance, visitation rate and floral resource supply. Critical levels where reproductive achievement and yield are impaired may be found, which can be used to cause specific interventions, including the introduction of supplemental pollinators, habitat improvements, or flowering synchrony management. Such strategies facilitate the process of decision-making in order to maximize the pollination, minimize the yield losses, and sustain ecosystem services in different buckwheat production systems (Satterthwaite et al., 2002).

12.9 Applications of Remote Sensing and GIS in Buckwheat Agroecosystems

12.9.1 Remote Sensing for Crop Monitoring

The remote sensing technologies are needed to monitor the buckwheat growth and reproductive development both on field and landscape level. Vegetation indices including NDVI (Normalized Difference Vegetation Index) and EVI (Enhanced Vegetation Index) facilitate the continuous evaluation of the crop vitality, biomass build-up and canopy coverage. The accurate observation of flowering phenology can be done with high-resolution satellite or UAV images, which is imperative to coordinate pollinator activity and predict the possible pollination period. These instruments aid in making decisions in a timely manner and improving the knowledge on crop performance under changing environmental conditions (Sun et al., 2023).

12.9.2 Using Drones and UAVs for Pollinator Activity Monitoring

The use of drones and UAVs will offer new possibilities of mapping floral resources and pollinator activity in buckwheat fields. Flower mapping measures the amount of floral and their spatial distribution whilst pollinator hotspots determine the pattern of visitation and behavioral specificity to the species. Thermal imaging can be used to measure stress in plants including heat or water shortage that may cause a reduction in nectar production and pollinator habits. Combining the UAV-based observations with ground data can also lead to more accurate pollination estimates, as well as precision management to enhance a better reproductive success.

12.9.3 Satellite-Based Environmental Monitoring

The rising method of satellite remote sensing can facilitate monitoring of environmental factors that affect the development of buckwheat and pollination at large scale. The changes in soil moisture, temperature, and vegetation health are monitored with time and give information regarding the abiotic stresses that affect flower production, nectar secretion, and pollinator activity. Moreover, satellite images can be used to determine pollinator abundance and movement by detecting the areas of habitat fragmentation and heterogeneity in the landscape, which are important predictors of pollination. Such data are used to make adaptive measures and maximize pollination services to ensure the functioning of the ecosystems (Wang et al., 2023).

12.9.4 GIS-Based Spatial Analysis

Using Geographic Information Systems (GIS), buckwheat agroecosystems can be spatially analyzed to determine the pollinator habitats, connectivity and configuration of the landscape. GIS mapping can be used to determine the appropriate habitats of the wild and managed pollinators and to assess connectivity involving the floral resources, which is a vital requirement in maintaining the movement and diversity of pollinators. With GIS landscape planning, it is possible to plan the location of semi-natural habitats, hedges, or flowering strips to ensure the maximum access of pollinators and enhance cross-pollination. Spatial information can be utilized to uphold agroecological interventions based on evidence to increase the efficiency of pollination and yield stability (Houtman et al., 2025).

12.9.5 Integration of Remote Sensing Outputs into Modelling and Decision-Support Systems

The products provided by the remote sensing and GIS can be integrated into crop growth and pollination models to enhance the prediction accuracy and guide decision-making. Flowering time, vegetation index, and connectivity of habitat data are used to simulate yield and ecosystem services that are dependent on pollination. The decision-support systems based on these datasets allow optimized schedules of sowing, controlled pollinators, and reduction of the effects of environmental stress. These combined systems enhance accuracy farming and enhance tough and climate-wise production systems of buckwheat (Sun et al., 2023).

12.10 Enhancing Pollination and Agroecological Performance

12.10.1 Habitat Diversification for Pollinator Conservation

Habitat diversity should be encouraged as a basic measure of increasing the pollination in buckwheat agroecosystems. Creation of wildflower strips, hedges, and buffer strips is a constant source of flowers and nesting places of wild and controlled pollinators. These habitat attributes alleviate the bad consequences of monoculture, decrease the isolation of pollinators, and maintain the levels of visitation during flowering. By establishing refuges in and around buckwheat fields, the pollinators also get stronger and more successful during reproductive processes (Costa et al., 2025).

12.10.2 Management of Domesticated Pollinators

Honeybees and other domesticated pollinators are strategic to manage and improve the efficiency of pollination. Migration of the best hives, introductions of colonies at the right time when flowers are in full bloom and healthiness of the colonies are the important practices. Managed pollinators have an adjuvant effect on the wild species, enhancing cross-pollination and overcoming temporal or spatial pollinator ecological deficits. Wild and managed pollinator services together make sure that the seed set and yield can be reliable in all types of agroecological environments (Zhang et al., 2022).

12.10.3 Breeding and Selecting Varieties for Improved Pollinator Attraction

One of the main reproductive success strategies that can be used to improve pollination includes genetic enhancement of buckwheat that will attract more pollinators. Selection of varieties which have bigger floral displays, increased production of nectar, and available floral morphology will enhance the rates of visitation and pollen transfer. Resistant to stress varieties which retain flower characteristics with heat or drought conditions also improve the reliability of pollination. Integrating breeding with the contemporary phenotyping instrumentation enables the formation of cultivars that will facilitate the sound pollinator activity and yield (Luthar et al., 2020).

12.10.4 Soil and Water Management for Stable Floral Resources

The ability of the soil to sit, the water control has a direct impact on the production of flowers, on the availability of nectar and on the quality of pollen. Balanced nutrient supply and adequate irrigation ensure a constant supply of floral resources which are important in keeping the pollinators active during flowering. Intensive irrigation, organic additions, and mulching decrease floral attractiveness losses caused by stress and thus provide sufficient pollination and better reproduction (Tamburini et al., 2023).

12.10.5 Designing Pollinator-Friendly Landscapes

It is possible to optimize the movement of pollinators and improve pollination services only through landscape-level planning. Connectivity between fields is provided by incorporating semi-natural habitats, corridors and mixed cropping system systems that enable the efficient foraging of the pollinators in the fields. This type of spatial planning will contribute to the protection of biodiversity of pollinators, reduce environmental uncertainty, and help to increase the resilience of buckwheat production systems (Lecegui et al., 2022).

12.11 Future Directions and Research Priorities

12.11.1 Integrating Genomics, Ecology, and Modelling for Precision Breeding

Genomic breeding and ecological analyses using mechanistic modelling processes of buckwheat are likely to result in a precise breeding of buckwheat varieties with improved floral characteristics, stress resistance and pollinator preferences. Such an integration helps the cultivation of cultivars with the capacity of continuing with high reproductive performance and yielding stability under varied environmental conditions (Zhang et al., 2022; Peralta et al., 2024).

12.11.2 Digital Agriculture and Sensor-Based Management

The adoption of digital agriculture tools, including high-resolution sensors, UAVs, and IoT-based monitoring, allows real-time assessment of flowering phenology, nectar availability, soil moisture, and overall crop health. Data-driven interventions enhance pollination efficiency, synchronize flowering with pollinator activity, and optimize yields across varying agroecological zones (Luthar et al., 2020). Phenological data of flowering, nectar availability, soil moisture and general crop health can be assessed in real time with the use of digital tools of agriculture (high-resolution sensors, UAVs, and IoT-based monitoring). Using data-driven interventions would help maximize the efficiency of pollination and match flowering and pollinator activity, as well as maximize yields in different agroecological zones (Luthar et al., 2020).

12.11.3 Multi-location, Multi-year Pollination Trials

Spatially separated and long-term pollination trials give important information regarding genotype x environmental interactions, abundance of pollinators and fluctuations in reproductive performance. These experiments result in the development of strong data sets that are used to shape adaptive management and determine the varietal choice in the production of buckwheat in a sustainable manner (Houtman et al., 2025).

12.11.4 Predictive Modelling Under Climate Change Scenarios

Predictive models that combine climate predictions (Fig. 12.12), flowering phenology, and pollinator behavior can be used to predict temporal mismatches and possible pollination deficits on the vegetation in future climatic conditions. The models guide modification of sowing schedules, crop genotype and habitat manipulation to ensure successful pollination and achieve harvest in the face of climate variability (Sun et al., 2023; Powlen et al., 2022).

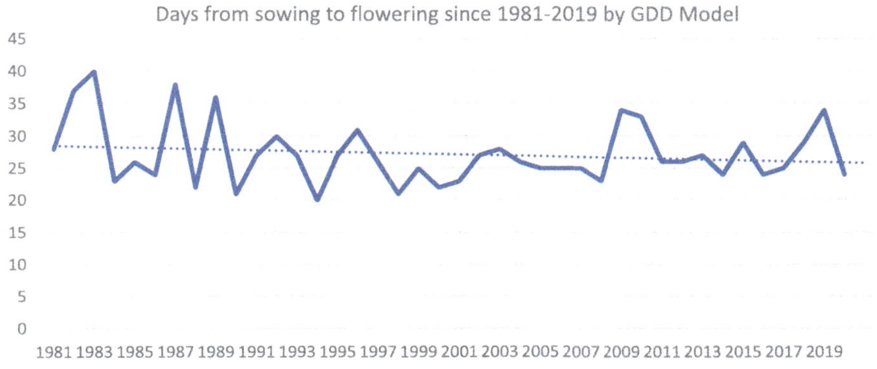

Fig. 12.12 Trend in days from sowing to flowering in buckwheat (1981–2019) based on the GDD model. The gradual decrease in days to flowering reflects rising temperatures over the years, leading to a shorter growth cycle. This indicates that warmer conditions may accelerate buckwheat development, enabling faster maturity in traditionally colder cultivation regions.

12.11.5 Policy Frameworks for Pollinator-Safe and Agroecological Buckwheat Farming

Sustainable production of buckwheat should also establish policy frameworks that will encourage the conservation of pollinators, low-chemical input practices and agro ecological practices. The supportive policies will boost an ecosystem service, safeguard the smallholder livelihoods and promote climate-sensitive, pollinator-friendly farming.

12.11.6 Opportunities for International Collaboration

Through international cooperation, there is exchange of germplasm, collaborative research, and transfer of knowledge, which enhances the ability of countries across the globe to come up with climate-resistance and pollination-friendly buckwheat agroecosystems. The dissemination of research, technology, and best practices will allow consistent methods in different countries and other agroecological regions with the promotion of sustainable buckwheat production at the global level (Gao et al., 2024).

12.12 Case Studies

12.12.1 Case Study 1: Climate-Resilient Buckwheat Production in Eastern Europe

The adaptability of buckwheat (*Fagopyrum esculentum*) under low temperature continental climate conditions in terms of varying water levels and periodical heat stress was assessed in 2023, in Poland and Ukraine. The study was entitled Evaluation of Genotypic and Agronomic Performance of Buckwheat under Climate variability in Eastern Europe where 18 farmer cooperatives were surveyed and field trials of 20 local and improved buckwheat varieties in various sites were done. The data collected were Agronomic, such as flowering phenology, accumulation of biomass, yield of seeds, and drought tolerance. Statistical analysis through ANOVA, PCA, and Pearson correlation found that yields were 1.2–4.5 t/ha with various genotypes, different irrigation and the management of soil fertility. Cultivars with early flowering and deep roots did well in a situation that was characterized by water scarcity. According to socioeconomic surveys, there was moderate level of competitiveness among the smallholder cooperatives because of inaccessibility to quality seed and mechanized harvesting machines. The paper concluded that better participatory breeding initiatives, better extension services and better genotype resistant to climatic changes are important in stabilizing yields and sustaining smallholder livelihood in Eastern European buckwheat systems.

12.12.2 Case Study 2: Integrating Buckwheat in Agroforestry Systems in Sub-Saharan Africa

The paper Evaluation of Buckwheat Intercropping in smallholder agroforestry systems to support soil fertility and pollination by insects (2024) is a field experiment carried out in Kenya and Uganda. The aim of the study was to determine agronomic, improvement in soil health, and pollinator diversity in case of intercropping buckwheat with legumes and shade-tolerant trees. A randomized complete block design was used to set up 25 experimental plots, in which data on canopy development, biomass, nitrogen fixation, flowering phenology and pollinator visitation rates were measured. Findings showed that there were immense changes concerning soil organic matter (10–15% enhanced) and nitrogen supply in intercropped plots relative to monoculture. When using pollinator monitoring, it was found that 40–50% more of the systems were visited by intercropping, and that the honeybees and native solitary bees preferred buckwheat flowers. Analysis of yield showed that buckwheat yielded 1.5–3.2 t/ha with respect to shade levels and soil fertility. The paper indicated that the strategic incorporation of buckwheat into agroforestry can improve ecosystem services, improve the resilience to climate variability, and biodiversity and keep smallholders profitable.

12.12.3 Case Study 3: Enhancing Pollination Efficiency in Organic Buckwheat Systems in Japan

In 2023, a study in the name of Optimizing Pollination Services and Yield in Organic Buckwheat Production in Japan found out the impacts of habitat diversification and pollinator management that influence the production of organic buckwheat. The analysis was performed in four prefectures where 30 organically managed farms were located and through participatory workshops, structured surveys and experimental trials. Plots of buckwheat were planted with different degrees of floral diversity i.e. hedges, strips of wild flowers and buffer strips to increase the presence of pollinators. During the flowering period, observations of the pollinators and data of the seed yield were taken. ANOVA and regression analyses provided showed that farms that had a wide range of floral habitats had 25–40% more seed set than control plots, with the benefit of the managed honeybees supplementing the role of the wild pollinators in poorly naturally-represented areas. Production was between 1.8 and 3.7 t/ha with the interventions of habitats and the presence of pollinators. The research made the following conclusion: the development of pollinator-friendly landscapes and the implementation of reduced-pesticide management practices can be critical to maximizing yield and ensuring the sustainability of organic buckwheat systems, providing useful implications of the policy and adoption at the level of farmers in the temperate regions of Asia.

12.13 Conclusion

Buckwheat (*Fagopyrum esculentum*) plays a central role in the sustainable agroecosystems by promoting soil health, nutrient cycling, weeds, and biodiversity in general. This is a multifunctional and climate-resistant crop due to its adaptability to marginal soils, compatibility with crop rotations, intercropping, and cover cropping systems and provision of ecosystem services. Buckwheat cannot be successful in its reproductive processes without pollination activity (wild and managed), and effective pollen transfer by both pollinators has been found to increase seed set, grain quality, and grain size. It is important to protect and enhance pollinator diversity by management of the habitat and low-pesticide farming and flowering synchronization to achieve stable yields. Recent advances like the incorporation of modern technologies like modelling, remote sensing and GIS have provided new possibilities in the monitoring of flowering phenology, pollinators activity, environmental stress and landscape connectivity. These strategies enable accuracy interventions, which enable farmers and researchers to maximize pollination services and commercial crop functions in unpredictable climatic conditions. Buckwheat can be regarded as a keystone crop in the sustainable and pollinator-friendly agriculture because of its ecological advantages, stress resistance and ability to serve pollinators. Ecological, technological innovations and conducive policies all incorporated strategically make buckwheat a model crop in climate-smart, biodiversity-enhancing, and sustainable production of crops globally.

References

Akter, M., Hossain, M., Rahman, S., Banu, M., & Akhtar, M. (2023). Insect diversity and pollination effect on buckwheat (*Fagopyrum esculentum*) yield. *Journal of Biodiversity Conservation and Bioresource Management, 9*(1), 109–118. https://doi.org/10.3329/jbcbm.v9i1.66636

Antala, M., Kovar, M., Sporinová, L., Filacek, A., Juszczak, R., Zivcak, M., Shomali, A., Prasad, R., Brestic, M., & Rastogi, A. (2025). High-throughput phenotyping of buckwheat (*Fagopyrum esculentum Moench.*) genotypes under water stress: Exploring drought resistance for sustainable agriculture. *BMC Plant Biology, 25*(1), 444. https://doi.org/10.1186/s12870-025-06429-6

Buck, N., Fountain, M. T., Potts, S. G., Bishop, J., & Garratt, M. P. D. (2023). The effects of non-crop habitat on spotted wing drosophila (*Drosophila suzukii*) abundance in fruit systems: A meta-analysis. *Agricultural and Forest Entomology, 25*(1), 66–76. https://doi.org/10.1111/afe.12531

Cheng, Y., Zhang, T., Qiao, R., Yan, S., Luo, M., Wang, C., Zhang, T., Dong, Q., & Feng, H. (2022). Photosynthesis and salt cations adsorption response of spring maize (*Zea mays* L.) to salinity stress caused by different drip irrigation regimes in an arid saline area, Northwest China. *Journal of Agronomy and Crop Science, 208*(5), 693–708. https://doi.org/10.1111/jac.12601

Corey, K., Hamama, A. A., Li, H., Siddiqui, R. A., Kim, C., & Bhardwaj, H. L. (2022). Composition of Buckwheat honey. *Journal of Agricultural Science, 14*(9), 59. https://doi.org/10.5539/jas.v14n9p59

Costa, L. O., Lira, R., Silva-Torres, C. S. A., & Torres, J. B. (2025). Quality traits of the lady beetle *Eriopis connexa* (Germar) resistant to pyrethroids over 130 generations of rearing. *Agricultural and Forest Entomology, 27*(3), 452–462. https://doi.org/10.1111/afe.12678

Defalque, C., Laeremans, J., Drugmand, J., Tcheutchoua, C. F., Meng, Y., Zhou, M., Zhang, K., & Quinet, M. (2025). Drought and high temperatures impact the plant–pollinator interactions in *Fagopyrum esculentum. Plants, 14*(1), 131. doi:https://doi.org/10.3390/plants14010131

Devkota, K., Rijal, P., & Dos Santos, C. F. (2026). Flower-visiting insect diversity within buckwheat crops: An underutilized crop for sustainable economic livelihoods. *Insects, 17*(2), 200. https://doi.org/10.3390/insects17020200

Duan, C., Li, X., Li, C., Yang, P., Chai, Y., & Xu, W. (2023). Positive effects of fungal β diversity on soil multifunctionality mediated by pH in the natural restoration succession stages of alpine meadow patches. *Ecological Indicators, 148*, 110122. https://doi.org/10.1016/j.ecolind.2023.110122

Dyer, J. E., Talamas, E. J., Leskey, T. C., & Bergh, J. C. (2022). Evaluating chemical cues associated with *Halyomorpha halys* toward enhanced sensitivity of surveillance for *Trissolcus japonicus. Environmental Entomology, 51*(4), 679–687. https://doi.org/10.1093/ee/nvac045

Farooq, M., & Siddique, K. H. M. (Eds.). (2023). *Neglected and underutilized crops: Future smart food*. Academic Press, an imprint of Elsevier.

Fattorini, N., Costanzo, A., Romano, A., Rubolini, D., Baillie, S., Bairlein, F., Spina, F., & Ambrosini, R. (2023). Eco-evolutionary drivers of avian migratory connectivity. *Ecology Letters, 26*(7), 1095–1107. https://doi.org/10.1111/ele.14223

Fijen, T. P. M., Bodegraven, V. V., & Lucassen, F. (2022). Limited honeybee hive placement balances the trade-off between biodiversity conservation and crop yield of buckwheat cultivation. *Basic and Applied Ecology, 65*, 28–38. https://doi.org/10.1016/j.baae.2022.09.003

Food and Agriculture Organization of the United Nations. (2025). *FAOSTAT statistical database: Crops and livestock products – Buckwheat production*. https://www.fao.org/faostat/

Gao, J., Yang, B., Mura, C., Boucher, Y., & Rossi, S. (2023). Increased inter-annual variability in budburst dates towards the northern range edge of black spruce. *Agricultural and Forest Meteorology, 333*, 109410. https://doi.org/10.1016/j.agrformet.2023.109410

Gao, Y., Hou, H., Cang, Q., Sangdan, Z., Zhao, G., Yuan, Y., & Feng, B. (2024). Global research trends and future directions for buckwheat as a smart crop: A bibliometric and content analysis. *Foods, 13*(24), 4068. https://doi.org/10.3390/foods13244068

Houtman, L., Robinson, A. C., McLaughlin, D., & Grozinger, C. M. (2025). Evaluating the usability and utility of a spatial decision support system for pollinator ecology. *Ecological Informatics, 89*, 103182. https://doi.org/10.1016/j.ecoinf.2025.103182

Kamble, N. U., Petla, B. P., Ghosh, S., Achary, R. K., & Majee, M. (2022). Oryza coarctata PROTEIN L-ISOASPARTYL METHYLTRANSFERASE (PIMT) repairs isoaspartyl modification to antioxidative enzymes and is implicated in seed traits in rice. *Environmental and Experimental Botany, 202*, 105027. https://doi.org/10.1016/j.envexpbot.2022.105027

Kovács-Hostyánszki, A., Piross, I. S., & Shebl, M. A. (2022). Non-native plant species integrate well into plant-pollinator networks in a diverse man-made flowering plant community. *Urban Ecosystems, 25*(5), 1491–1502. https://doi.org/10.1007/s11252-022-01242-7

Krzyzanska, M., Hunt, H. V., Crema, E. R., & Jones, M. K. (2022). Modelling the potential ecological niche of domesticated buckwheat in China: Archaeological evidence, environmental constraints and climate change. *Vegetation History and Archaeobotany, 31*(4), 331–345. https://doi.org/10.1007/s00334-021-00856-9

Lecegui, A., Olaizola, A. M., López-i-Gelats, F., & Varela, E. (2022). Implementing the livelihood resilience framework: An indicator-based model for assessing mountain pastoral farming systems. *Agricultural Systems, 199*, 103405. https://doi.org/10.1016/j.agsy.2022.103405

Li, D., Clements, C. F., & Memmott, J. (2024). Isolation limits spring pollination in a UK fragmented landscape. *PLOS ONE, 19*(9), e0310679. https://doi.org/10.1371/journal.pone.0310679

Liu, R., Chen, D., Luo, S., Xu, S., Xu, H., Shi, X., & Zou, Y. (2020). Quantifying pollination efficiency of flower-visiting insects and its application in estimating pollination services for common buckwheat. *Agriculture, Ecosystems & Environment, 301*, 107011. https://doi.org/10.1016/j.agee.2020.107011

Luthar, Z., Zhou, M., Golob, A., & Germ, M. (2020). Breeding buckwheat for increased levels and improved quality of protein. *Plants, 10*(1), 14. https://doi.org/10.3390/plants10010014

Matsuzawa, Y., Fukuda, S., Ohira, M., & De Baets, B. (2023). Modelling fish co-occurrence patterns in a small spring-fed river using a machine learning approach. *Ecological Indicators, 151*, 110234. https://doi.org/10.1016/j.ecolind.2023.110234

Mele, A., Canella, J., Mirandola, E., Ceccato, E., Tirello, P., Scaccini, D., & Pozzebon, A. (2025). Flowering buckwheat supports *Trissolcus japonicus* and *Trissolcus mitsukurii* egg parasitism on *Halyomorpha halys*. *Biological Control, 208*, 105851. https://doi.org/10.1016/j.biocontrol.2025.105851

Miyashita, T., Hayashi, S., Natsume, K., & Taki, H. (2023). Diverse flower-visiting responses among pollinators to multiple weather variables in buckwheat pollination. *Scientific Reports, 13*(1), 3099. https://doi.org/10.1038/s41598-023-29977-z

Oliveira, J. R. D., Chambó, E. D., Santos, N. B., Cerqueira, L. D. C., Santos, É. B. D., Silva, D. G., & Carvalho, C. A. L. D. (2025). Impact of environmental factors and sowing date on flowering phenology, pollinator behavior, and buckwheat yield. *Ciência e Agrotecnologia, 49*, e010725. https://doi.org/10.1590/1413-7054202549010725

Peralta, G., CaraDonna, P. J., Rakosy, D., Fründ, J., Pascual Tudanca, M. P., Dormann, C. F., Burkle, L. A., Kaiser-Bunbury, C. N., Knight, T. M., Resasco, J., Winfree, R., Blüthgen, N., Castillo, W. J., & Vázquez, D. P. (2024). Predicting plant–pollinator interactions: Concepts, methods, and challenges. *Trends in Ecology & Evolution, 39*(5), 494–505. https://doi.org/10.1016/j.tree.2023.12.005

Powlen, J. S., Patton, A. J., Jiang, Y., Fraser, M. L., & Bigelow, C. A. (2022). Response of drought susceptible and resistant Kentucky bluegrass and tall fescue cultivars and mixtures to limited irrigation. *Crop Science, 62*(5), 2024–2034. https://doi.org/10.1002/csc2.20789

Qu, Q., Zhang, J., Hai, X., Wu, J., Fan, J., Wang, D., Li, J., Shangguan, Z., & Deng, L. (2022). Long-term fencing alters the vertical distribution of soil δ13C and SOC turnover rate: Revealed by MBC-δ13C. *Agriculture, Ecosystems & Environment, 339*, 108119. https://doi.org/10.1016/j.agee.2022.108119

Quinet, M., Cawoy, V., Lefevre, I., Van Miegroet, F., Jacquemart, A.-L., & Kinet, J.-M. (2004). Inflorescence structure and control of flowering time and duration by light in buckwheat (*Fagopyrum esculentum* Moench). *Journal of Experimental Botany, 55*(402), 1509–1517. https://doi.org/10.1093/jxb/erh164

Razze, J. M., Liburd, O. E., & Webb, S. E. (2016). Intercropping buckwheat with squash to reduce insect pests and disease incidence and increase yield. *Agroecology and Sustainable Food Systems, 40*(8), 863–891. https://doi.org/10.1080/21683565.2016.1205541

Ren, Q., Liu, W., Zhao, M., Sai, C., & Wang, J. (2018). Changes in α-glucosidase inhibition, antioxidant, and phytochemical profiles during the growth of Tartary buckwheat (*Fagopyrum tataricum* Gaertn). *International Journal of Food Properties, 21*(1), 2689–2699. https://doi.org/10.1080/10942912.2018.1560314

Ren, L., Liu, P., Xu, F., Gong, Y., Zhai, X., Zhou, M., Wang, J., & Wang, Z. (2023). Rice–fish coculture system enhances paddy soil fertility, bacterial network stability and keystone taxa diversity. *Agriculture, Ecosystems & Environment, 348*, 108399. https://doi.org/10.1016/j.agee.2023.108399

Rijal, S. P., Thapa, R. B., Sharma, M. D., Sah, S. K., & Gc, Y. D. (2018). Assessment of pollination deficit in buckwheat (*Fagopyrum esculentum* moench) and its impact on production in different agroecosystems of chitwan, Nepal. *International Journal of Research-Granthaalayah* 6(11)

Satterthwaite, W. H., Menges, E. S., & Quintana-Ascencio, P. F. (2002). Assessing scrub buckwheat population viability in relation to fire using multiple modeling techniques. *Ecological Applications, 12*(6), 1672–1687. https://doi.org/10.1890/1051-0761(2002)012%5B1672:ASBPVI%5D2.0.CO;2

Stefanova, K. T., Brown, J., Grose, A., Cao, Z., Chen, K., Gibberd, M., & Rakshit, S. (2023). Statistical analysis of comparative experiments based on large strip on-farm trials. *Field Crops Research, 297*, 108945. https://doi.org/10.1016/j.fcr.2023.108945

Sun, Y., Zhu, X., Sui, L., Zhu, Z., Ren, B., Pan, D., Wan, H. Y., & Zhu, H. (2023). Effects of water stress on plant volatile emission and insect oviposition preference in an agroecosystem. *Agricultural and Forest Entomology, 25*(3), 458–467. https://doi.org/10.1111/afe.12567

Takeshima, R., Nishio, T., Komatsu, S., Kurauchi, N., & Matsui, K. (2019). Identification of a gene encoding polygalacturonase expressed specifically in short styles in distylous common buckwheat (*Fagopyrum esculentum*). *Heredity, 123*(4), 492–502. https://doi.org/10.1038/s41437-019-0227-x

Tamburini, G., Laterza, I., Nardi, D., Mele, A., Mori, N., Pasini, M., Scaccini, D., Pozzebon, A., & Marini, L. (2023). Effect of landscape composition on the invasive pest Halyomorpha halys in fruit orchards. *Agriculture, Ecosystems & Environment, 353*, 108530. https://doi.org/10.1016/j.agee.2023.108530

Wang, F., Men, R., Lai, H., Feng, K., Yan, S., Gao, S., Wang, Z., Tian, Q., Guo, W., & Yang, H. (2024). The response of vegetation dynamics to drought and its driving factors identification in Inner Mongolia of China. *Ecological Indicators, 164*, 112125. https://doi.org/10.1016/j.ecolind.2024.112125

Wang, L., Zhao, Z., Walter, G. H., & Furlong, M. J. (2023). Predicting the impacts of climate change on the biological control of *Plutella xylostella* by *Diadegma semiclausum*. *Agricultural and Forest Entomology, 25*(2), 251–260. https://doi.org/10.1111/afe.12548

Wen, W., Li, Z., Shao, J., Tang, Y., Zhao, Z., Yang, J., Ding, M., Zhu, X., & Zhou, M. (2021). The distribution and sustainable utilization of buckwheat resources under climate change in China. *Plants, 10*(10), 2081. https://doi.org/10.3390/plants10102081

Wieneke, S., Balzarolo, M., Asard, H., Abd Elgawad, H., Peñuelas, J., Rascher, U., Ven, A., Verlinden, M. S., Janssens, I. A., & Vicca, S. (2022). Fluorescence ratio and photochemical reflectance index as a proxy for photosynthetic quantum efficiency of photosystem II along a phosphorus gradient. *Agricultural and Forest Meteorology, 322*, 109019. https://doi.org/10.1016/j.agrformet.2022.109019

Wu, H., Luo, Z., Li, H., Feng, J., Wang, C., Zhang, J., Xu, J., & He, P. (2024). Thermal fluctuations and CO_2 enrichment synergistically accelerate biomass yield of Neopyropia yezoensis. *Environmental and Experimental Botany, 219*, 105624. https://doi.org/10.1016/j.envexpbot.2023.105624

Xiang, D., Song, Y., Song, C., Wan, Y., Ye, X., Liu, C., Liang, C., & Zhao, G. (2022). Seed setting and its spatial characteristics in tartary buckwheat (*Fagopyrum tataricum*). *Phyton, 91*(8), 1659–1669. https://doi.org/10.32604/phyton.2022.020338

Xu, C., Zhang, Q., Yu, Q., Wang, J., Wang, F., Qiu, S., Ai, M., & Zhao, J. (2023a). Effects of land use/cover change on carbon storage between 2000 and 2040 in the Yellow River Basin, China. *Ecological Indicators, 151*, 110345. https://doi.org/10.1016/j.ecolind.2023.110345

Xu, S., Zhang, X., & Zhu, Y. (2023b). Evidence for the anaerobic oxidation of methane coupled to nitrous oxide reduction in landfill cover soils: Promotor and inhibitor. *Science of The Total Environment, 900*, 166752. https://doi.org/10.1016/j.scitotenv.2023.166752

Yang, F., Sun, C., Wang, H., Hu, X., Wang, S., Zhang, M., Zhang, L., & Zhong, J. (2023). Significant spatiotemporal variability of nitrous oxide emissions from a temperate reservoir experiencing intensive aquaculture disturbance. *Agriculture, Ecosystems & Environment, 348*, 108427. https://doi.org/10.1016/j.agee.2023.108427

Zhang, Z., Wang, Y., Wang, Z., Ashraf, U., Mo, Z., Tian, H., Duan, M., Li, Y., Tang, X., & Pan, S. (2022). Precise delivery of nitrogen at tillering stage enhances grain yield and nitrogen use efficiency in double rice cropping systems of South China. *Field Crops Research, 289*, 108736. https://doi.org/10.1016/j.fcr.2022.108736

Chapter 13
Proso Millet (*Panicum miliaceum*): Climate Adaptation and Carbon Sequestration Potential

Zohaa Fatima, Umar Farooq, Wajahat Hussain, Mukhtar Ahmed, Shakeel Ahmad, and Muhammad Imtiaz Rashid

Abstract Proso millet (*Panicum miliaceum*) is emerging as a climate-resilient, nutrient-rich cereal with significant potential for carbon sequestration. This chapter examines its role in global food security, especially in drylands, by analyzing its C_4 physiology, stress tolerance, and adaptive traits. The objectives are to evaluate proso millet's agronomic and physiological responses to climate stress, assess its capacity for soil carbon storage, and explore modeling and remote-sensing tools to optimize its production. The study analyses field tests and computer simulation experiments, such as climate change impact modeling using APSIM, yield prediction using remote sensing and multiple site genotype analysis. Intercropping with legumes, precision farming (e.g. Sentinel-2-based nitrogen control), and maintaining residue to develop soil organic carbon are some of the key mitigation methods. Results shows that there is water use efficiency of 4.8–5.4 kg grain/mm in dryland experiments in the U.S., yield is 0.75 t/ha under combined heat, salinity and drought in UAE accessions and soil carbon gain of 0.12–0.34 t C/ha/year in rotation systems. Genetic analysis of the genetic diversity in Kazakhstan revealed accessions that had the lowest GDD of 850 and frost tolerance to − 4 °C. The proso millet versatile and the low input requirements suggests that it can be used in the future as a solution to sustainable low carbon farming in the future climate conditions.

Keywords Proso millet · Nutritional security · Landrace variation · Climate resilience · Sustainable agriculture

Z. Fatima (✉) · U. Farooq · W. Hussain · M. Ahmed
Department of Agronomy, Pir Mehr Ali Shah Arid Agriculture University, Rawalpindi, Pakistan
e-mail: zohaafatimaniazi@gmail.com

S. Ahmad
Institute of Agronomy Faculty of Agricultural Sciences and Technology, Bahauddin Zakariya University, Multan, Pakistan

M. I. Rashid
Center of Excellence in Environmental Studies, King Abdulaziz University, Jeddah, Saudi Arabia

M. Ahmed (ed.), *Climate Change and Neglected Underutilized Crops*,
https://doi.org/10.1007/978-3-032-21005-0_13

13.1 Introduction

New interest has developed in proso millet (*Panicum miliaceum* L.) as a resistant crop to climate particularly under marginal and semi-arid rain-fed systems. It can thrive in unfavourable environments because of the short length of growth that is typically 60–90 days and low input needs. The recent studies have reported high tolerance level to drought, heat and low soil fertility. Thus proso millet is viable in the environment with limited resources (Wimalasiri et al., 2023). High water-use efficiency of proso millet is due to its C_4 plant status. This implies that the CO_2 in the biological pathway is concentrated around the Rubisco, reducing photorespiration and increasing the rate of photosynthesis even in times of water stress (Ghatak et al., 2025).

However, proso millet remains underutilized in most agricultural systems throughout the world despite its adaptation. It is often known as a "minor" or "orphan" crop, but it is a very significant food security crop to the smallholder farmers in Asia, Africa, and the eastern Europe. Proso millet could improves crop system diversification and buffers the community against climate changes (Muskan et al., 2025). The policy movements like the move by the UN to declare 2023 as the "International Year of Millets" are helping to draw such grains back to the agendas of the world. At the physiological level, the proso millet possesses a complicated stress-response regulation: under drought, a recent transcriptomic analysis identified over 16,000 differentially expressed genes in roots. They are mostly linked to hormone signalling, the reactive oxygen species (ROS) scavenging and transcriptional regulation by transcription factors (Zhang et al., 2024). Such molecular responses are likely to be useful in facilitating the crop to survive and give yields in varying conditions of moisture. This chapter attempts to provide an overview of the potential of proso millet under climate-smart agriculture. It covers botanical traits, its distribution across the globe, agronomic performance, nutritional value, physiological response to the different stresses, possibilities of carbon sequestration as well as the modeling and remote sensing for its assessment. The chapter integrates the existing empirical evidence with the progressive strategies to underline the importance of proso millet's scale up and contribution to sustainable food systems and climate resilience.

13.2 Botanical Description and Taxonomy

13.2.1 Taxonomic Classification

Proso millet (*Panicum miliaceum* L.) is a member of the family Poaceae which also contains other global significant cereals as wheat, rice, maize and sorghum. Proso millet has many traits in common to grass family crops in terms of anatomy and reproduction. Recent genomic and phylogenetic research has shown that *P. miliaceum* belongs to the subfamily Panicoideae which includes certain C_4 grasses that

Fig. 13.1 Proso millet (*Panicum miliaceum* L.): **a** plant morphology, **b** panicle, **c** grains, and **d** root (Rajasekaran & Francis, 2021)

can withstand warm and dry conditions (Hunt et al., 2014). Its position in the Panicum complex has been determined using updated molecular studies of 2022–2025 such as whole-genome sequencing and plastid phylogenomic. This also clarified its evolutionary divergence with other millets using chloroplast and nuclear DNA markers (Zhang et al., 2024). These updated taxonomic confirmations highlight its ecological adaptation as well as its association with other C_4 cereals capable of surviving in low moisture and high temperature environment (Ishaq et al., 2025).

13.2.2 Morphological Characteristics

Proso millet is morphologically an annual grass. It possesses fibrous root system that is not particularly deep but spreads at a lateral location therefore, can grow in dry and compacted soil. The culm (stem) is straight, slim and most often branched while leaf blades on both surfaces and edges are hairy with overlapping margins. Its panicle is generally opened, compact and clustered but can vary depending on the genotype (Fig. 13.1). The accessions differ in the grain size and colour. The width of the seeds is 1.8–2.6 mm and the colour is white to dark brown (Boukail et al., 2021). Grain is extremely fine, hulled and in general round to elongated. Its thousand-grain weight and other morphological characteristics varied widely among landraces showing genetic diversity (M. Shalim Uddin et al., 2021).

13.2.3 Growth Habit and Phenology

Proso millet has relatively short phenological life cycle (approximately 60–90 days). Most genotypes are photoperiod insensitive and may vary in growth duration by

Table 13.1 Phenological stages of Proso millet *Panicum miliaceum* L. (Ventura et al., 2020)

BBCH Code	Growth Stage	Description
00	Germination	Seed imbibition and radicle emergence
10–19	Leaf development	Leaves unfolding
20–29	Tillering	Formation of tillers (secondary shoots)
30–39	Stem elongation	Main stem and internodes elongating
50–59	Inflorescence emergence	Panicle (head) visible/emerging
60–69	Flowering	Anthesis and flowering stage
70–79	Milk/Soft dough	Grain filling (soft dough phase)
80–89	Maturity/Ripening	Grain maturation and ripening
90–99	Senescence	Plant aging, drying, and final ripeness

germplasm (Table 13.1). Vegetative growth can be the suspended in the inappropriate conditions (e.g. high temperature), and even the plant can be kept in low stature, this will conserve the water (Fig. 13.2) (Habiyaremye et al., 2017b).

Phenological data of proso millet was collected in 2019 and 2020 and compared to test the consistency of Cumulative Growing Degree Days (CGDD) thresholds necessary to reach each stage as in (Fig. 13.3). Single triangle method was used to calculate the Growing Degree Days (GDD), which gave good accuracy with an acceptable error level as compared to hourly air temperature values (Ventura et al., 2022).

13.2.4 Genetic Diversity and Landraces

Proso millet has high genetic diversity, which is largely preserved for centuries within traditional cultivated landraces across Eastern Europe, Asia and the Himalayan region. This can be because of variations in plant height, grain colour, tillering, seed size and time of maturation. These variations enable the crop to survive under drought stress, short-season, and low-fertility environments and offer genes for high biomass, short maturity and efficient water use traits. Proso millet consists of multiple genetic clusters, deep phylogeographic structure and broad allelic richness playing important role in independent domestication pathways. Integration of landrace diversity into breeding programs can develop climate-adaptable cultivars having stability and nutritional quality as shown in (Fig. 13.4).

Landraces are agro-ecological which exhibit a considerable diversity in the length of panicle, maturity, plant height, drought resilience and yield traits (Ghimire et al., 2018). These landraces are important reservoirs of adaptive traits for stress situations. Heat tolerance, grain quality attributes and water-stress resilience come with these landraces that are absent in modern varieties. Based on panicle shape, there are five

Fig. 13.2 Phenological stages of Proso millet *Panicum miliaceum* L. (Ventura et al., 2020)

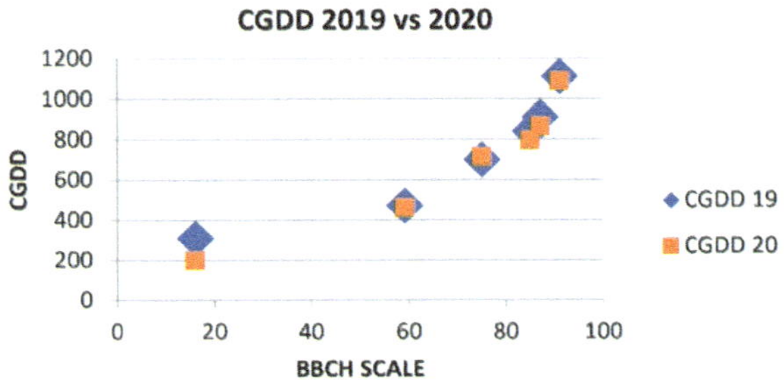

Fig. 13.3 The CGDD necessary for Proso millet to reach the same BBCH stage in the 2019 and 2020 seasons (Ventura et al., 2022)

Fig. 13.4 Land races of Proso millet based on panicle shape (Narciso & Nyström, 2023)

races of cultivated proso millet: *miliaceum*, *patentissimum*, *contractum*, *compactum* and *ovatum* as shown in Fig. 13.4 (Narciso & Nyström, 2023).

13.3 Global Distribution and Production Status

13.3.1 Historical Domestication and Geographic Origin

Proso millet was apparently domesticated in East and Central Asia at the start of the Holocene (about 10,000 years ago) (Zargar et al., 2023). It was then spread to other countries for its resilience and adaptation under different climatic conditions. In 1875, it was introduced in North America and planted along the eastern Atlantic coast by the German and Russian immigrants. Until the start of the twentieth century, proso millet was a highly significant crop in China, but recently higher-yielding varieties of other crops (such as rice, maize, wheat) have replaced proso millet and led to a significant decline in its cultivation (Narciso & Nyström, 2023). In Nepal, Himalayan landraces collections also indicate its existence in high altitude traditional agricultural systems (Ghimire et al., 2018).

13.3.2 Current Global Production Patterns

In Asia and Africa, pearl millet accounts for about 94% of global output. This percentage shows production of average 30 million tons (2023/2024) with proso millet accounting for 4 million tons. The statistics of the proso millet production in the world show that it takes a relatively small area in comparison to the major cereals but is significant among the minor millets. Millet proso is grown on approximately 757,000 hectares over the world (Ramadas et al., 2023). The proso millet trade is not very widespread. Approximately 100,000–150,000 tonnes proso millet are traded each year, primarily as bird food. The U.S., Australia and Argentina are the largest exporters and the EU, Japan and Canada are the most common importers of it as bird food (FAO, 2023).

13.3.3 Importance in Asia, Eastern Europe, and Semi-Arid Regions

Proso millet is still used in some areas of Asia (China, India, Nepal) and Eastern Europe (e.g. Russia, the Danube region) and is grown in semi-arid regions. It is one of the several traditional millets cultivated in a low-input situation in India (FAO, 2023). It is best suited for the rainfed agricultural system because annual rainfall is $\leq$ 100 mm. The shallow root system, high-water use efficiency and short growing cycle make it an ideal for dryland. Such as in the USA, it is grown on an average of 204,366 ha as a dryland crop. Proso millet is used in crop rotation with sunflower and corn in the Central Great Plains (CGP), mainly concentrated in the states of South Dakota, Nebraska and Colorado (Das et al., 2019).

13.3.4 Trends in Trade, Consumption, and Market Value

Proso millet's level of production is now small but its role in the niche markets (health foods, gluten-free diets) is growing in developed countries. Its small trade with the international world demonstrates its small size as well as its specialization. Demand of health-conscious people along with increasing interests in millets has increased the global millet seed market. Proso millet is widely used in as bird feed, snacks and cereals as well as by vegetarian and gluten allergic. With increasing awareness among consumers, there is still supply chain challenges such as low value-add facilities and processing infrastructure. The volatility in prices has been high in the past; past records (1990s) indicated that prices in the U.S. were between USD 88 and USD 550 per tonne due to shocks in supplies. Future market insights project that the global millet market will grow from USD 16.1 billion in 2025 to USD 56.7 billion by 2035 (CAGR ~ 13.4%). It is also estimated because of climate-resilient demand and functional food uses, that millet market will grow at ~ 4.95% CAGR (2025–2030) (FAO, 2023).

13.4 Agronomic Characteristics

13.4.1 Climate Adaptation and Environmental Requirements

Proso millet is a plant that can withstand high temperatures and low precipitation hence thriving in semi-arid regions. It is thermophilic because it is C_4 and cannot thrive in low temperature (< 10–13 °C). Optimal temperature for seed germination is 20–30 °C. The climate sensitivity of proso millet has been simulated by modeling the studies with APSIM (Agricultural Production Systems sIMulator) in the context of varying climate conditions. Proso millet can grow in northern side than the other millets (up to 54° N latitude) as it is well adapted to high-elevation conditions and plateau. It can also grow in arid lands under non-irrigated conditions and can produce grain with only 330–350 mm of annual rainfall. Under drought, limited nitrogen and CO_2 and high temperature conditions, it can efficiently fix carbon because of being C_4 crop and low transpiration rate (Table 13.2). Proso millet reaches maturity rapidly to avoid drought sensitivity and stops vegetative growth at temperatures above 30 °C. At this temperature, it ceases to flower and maintains short primary stem to survive drought conditions (Habiyaremye et al., 2017b).

Table 13.2 The optimal environmental conditions for proso millet (Das et al., 2019)

Crop	Scientific name	Altitude range (m)	Temp. (°C)	Rainfall required (cm)	Maturity time (days)
Proso millet	*Panicum miliaceum*	1200–3500	20–30	20–50	60–90

Table 13.3 The optimal soil conditions for proso millet (Das et al., 2019)

Optimum Soil Type	pH Soil	Salinity (dS/m)
Sandy loam and slightly acidic	5.5–6.5	1.5–9.5

13.4.2 Soil Requirements and Tolerance to Poor/Marginal Lands

Proso millet will also grow on light-medium-heavy soils, such as sandy loam, low-fertility, slightly acidic and saline soils. Soil pH of 5.5–6.5 for most of its well seeds germination and pH above 7.8 can cause iron chlorosis while pH ≤ 5 can constrain millet cultivation seriously (Table 13.3) (Habiyaremye et al., 2017b). Waterlogging, compaction and coarse sandy soils can also affect it as its root system is shallow (Hanna et al., 2004).

13.4.3 Water Use Efficiency and Short Growing Cycle

Proso millet has a C_4 pathway, which grows rapidly and is very efficient in the use of water. Comparison of irrigated and rainfed conditions in the field revealed that irrigation gives a high yield, but most accessions can continue to give a decent yield in water stress (Table 13.4). As an example, in a multi-year experiment with 20 different accessions, grain yield in irrigated plots was much greater whereas flowering was not much affected by water stress indicating that phenology is resilient (Habiyaremye et al., 2017a).

13.4.4 Pest and Disease Resistance

Proso millet is moderate resistant to *Exserohilum turcicum* (leaf blight) and *Sclerospora graminicola* (downy mildew) which are considered common fungal pathogens. It is also less affected by rusts and smuts as compared to wheat and maize crop. Due to its short growing cycle and adaptation to low humidity, it is less affected by bacterial and viral diseases. The exposure to pathogens is reduced because of early maturity. It also has some inherent resistance due to genetic diversity among landraces. Pests such as *Atherigona soccata* (shoot fly) and *Ostrinia nubilalis*, it is relatively less affected by aphids and stem borers as compared to maize or sorghum. The hard seed coat and small size of its seed prevent larval survival of pest and early maturation also prevents the long exposure to insect infestation of late season. It is highly predated by birds so crop rotation and proper netting should be used (Habiyaremye et al., 2017a). Proso millet is tolerant to some levels of herbicides: a test trial examined its behaviour to sulfonylurea herbicides (CGA-152005, metsulfuron, triasulfuron). It was found that even though weed-suppressing herbicides led to early

Table 13.4 Proso millet versus other major cereals in terms of water use efficiency and growing cycle (Habiyaremye et al., 2017a)

Crop	Growing season (days)	Water requirement	Tolerance to poor/ marginal soils	Drought tolerance
Proso millet (*Panicum miliaceum*)	60–90	Low (~ 30–40% less than maize)	High: Tolerates sandy, low fertility, eroded soils	High
Maize (*Zea mays*)	90–120	High; requires consistent rainfall or irrigation	Medium: Does not perform well on degraded soils	Medium
Sorghum (*Sorghum bicolor*)	90–120	Low to medium	High: Tolerates drylands	High
Pearl millet (*Pennisetum glaucum*)	70–110	Low	High: Grows in sandy and infertile soils	High
Wheat (*Triticum aestivum*)	120–150	Medium	Medium: Needs fertile soil	Low to medium

stunting but grain yield in the end was not significantly affected even with 2–4 times of recommended doses (Uludag et al., 1997). This indicates that proso millet has the resilience in rotation systems in which herbicides were applied, but the impact depends on the pH and organic matter of the soil.

13.4.5 Yield Potential under Diverse Conditions

Management strategies and environmental conditions have significant impact on the yield of crop. Proso millet intercropping trials with mung bean under moderate nitrogen (120 kg N/ha) produced 6300–6500 kg/ha of grain in two growing periods on the Loess Plateau of China with significant improvements in terms of water use efficiency and nitrogen uptake (Table 13.5) (Gong et al., 2021).

Table 13.5 Yield potential of Proso millet under diverse conditions (Makuchete et al., 2024)

Condition	Observed yield (t/ha)
Drought/Low rainfall	0.5–1.5
Low-fertility soils	0.8–1.5
High temperature/heat stress	1–2
Irrigated/Optimal conditions	2–3
Marginal/Degraded lands	0.5–1

Table 13.6 Macronutrient composition (per 100 g dry weight) of Proso millet (Balli et al., 2023)

Component	Range/Average
Energy	350–360 kcal
Carbohydrates	70–75 g
Starch	60–65 g
Protein	11–13 g
Fat	3–4 g
Dietary fiber	8–10 g

13.5 Nutritional and Functional Profile

Grains are important in human diet as they meet approximately half of an individual's need for energy and protein intake. Proso millet is rich in macronutrients like protein, dietary fiber and carbohydrates and important minerals such as phosphorus, potassium, magnesium and zinc. It also contains B vitamins (like thiamin, niacin, and B6) and vitamin E.

13.5.1 Macronutrient Composition: Protein, Starch, Fat

The carbohydrates of 25 global proso millet accessions according to compositional study had a total of 65.1–75.3 g/100 g of grain, with the variability based on the accession. The same study gave protein contents of the same accessions between 8.70 and 14.00 g per 100 g (Table 13.6) The content of fat in that study was rather low, ranging between 2.6 and 5.1 g/kg/100 g, though there were a few genotypes that had a little higher level (Balli et al., 2023). Proso millet contains starch content according to feed-tables from INRA/CIRAD is of approximately 58.2%.

13.5.2 Essential Amino Acids (Methionine-Rich)

In a protein-quality study of eight varieties of proso millet, crude protein (~ 11.6%) was equivalent to wheat (Fig. 13.5). This showed that proso millet contained higher concentrations of essential amino acids, especially methionine, leucine and isoleucine resulting in a higher Essential Amino Acid Index (~ 51%) than wheat. In that experiment, the lysine was not high (limiting amino acid), approximately 3.3 g per kg of millet protein (Kalinova & Moudry, 2006).

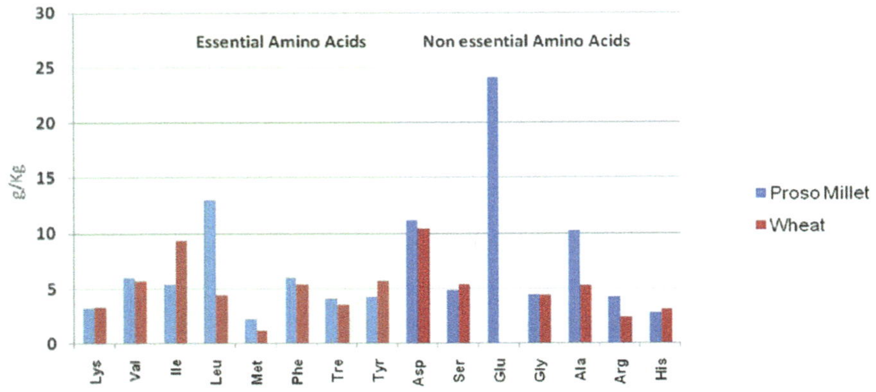

Fig. 13.5 Essential and non-essential amino acids comparison between wheat and proso millet (Rajasekaran & Francis, 2021). Essential and non-essential amino acid content: proso millet versus wheat, expressed in g/kg (Lys lysine, Val valine, Ile isoleucine, Leu leucine, Met methionine, Phe phenylalanine, Tre threonine, Tyr tyrosine, Asp aspartic acid, Ser serine, Glu glutamic acid, Gly glycine, Ala alanine, Arg arginine, His histidine)

13.5.3 Micronutrients and Phytochemicals

In 25-accession study, mineral analysis revealed that proso millet contains good amount of iron and zinc (Table 13.7). The same paper suggests that ash content (that is related to mineral content) differed considerably between coloured and white varieties, which suggests that grain colour can also influence mineral and micronutrient content (Balli et al., 2023). Other sources (FAO/nutritional databases) also reported the proso millet to have calcium, magnesium, phosphorus, and other macro and micro minerals.

13.5.4 Antioxidants and Bioactive Components

In the 25 accessions study, cinnamic acids were the highest in the phenolic compounds with 0.69–1.35 mg/g dry weight as the concentration range. The same study had results that the flavonoids were detected having trace amounts (0.01–0.10 mg/g DW) in most of the accessions (Balli et al., 2023).

A germination experiment indicated that total phenolic content (TPC) of proso millet increased by 6.30–8.66 times (free phenolics) and 77.65–116.18% (bound phenolics) after 6 days of germination, which exhibited significant enhancement of antioxidant activity (Table 13.8). That identical germination paper found new phenolic compounds (e.g. feruloylquinic acid and N,N′+bis(p-coumaroyl) putrescine) that were produced during germination in proso millet (Xiang et al., 2023).

Table 13.7 Phytochemicals and micronutrients (per 100 g dry weight) of Proso millet (Balli et al., 2023)

Content	Range	Functional role/benefit
Iron (Fe)	6–9 mg	Hemoglobin synthesis, oxygen transport
Zinc (Zn)	2–3 mg	Immune function, enzyme cofactor
Magnesium (Mg)	100–120 mg	Muscle and nerve function, bone health
Phosphorus (P)	280–300 mg	Bone health, energy metabolism
Potassium (K)	250–300 mg	Electrolyte balance, heart health
Calcium (ca)	8–12 mg	Bone health
Phenolic acids (ferulic, p-coumaric, vanillic)	100–300 mg GAE/kg	Antioxidant, anti-inflammatory, anti-cancer
Flavonoids (apigenin, quercetin, luteolin)	20–50 mg QE/kg	Antioxidant, anti-inflammatory, cardioprotective
Saponins	0.5–2%	Cholesterol-lowering, potential anticancer
Phytosterols	200–400 mg/100 g	Reduce LDL cholesterol, cardiovascular health
Tocopherols and Tocotrienols	1–3 mg	Antioxidant, protects lipids from oxidative damage
Carotenoids (lutein, zeaxanthin)	0.5–2 mg	Eye health, antioxidant

Table 13.8 Phytochemicals and antioxidants (per 100 g dry weight) of Proso Millet (Rajasekaran et al., 2023)

Phytochemicals and antioxidants	Functional role/benefit
Phenolic acids (ferulic, p-coumaric, vanillic)	Antioxidant, anti-inflammatory, anti-carcinogenic
Flavonoids (apigenin, quercetin, luteolin)	Anti-inflammatory, cardio-protective
Tocopherols and tocotrienols (vitamin E)	Antioxidant, protects cells from oxidative damage
Saponins	Cholesterol-lowering, potential anti-cancer effects
Phytosterols	Reduce LDL cholesterol, support heart health
Tannins	Antioxidant, moderate levels beneficial, high levels may reduce protein digestibility
Carotenoids (lutein, zeaxanthin)	Eye health, antioxidant activity

13.5.5 Gluten-Free Properties and Health Benefits

It is gluten-free, potentially safe and fitting of those with celiac disease or gluten sensitivity. It also prevents intestinal damage and has high dietary fibre which improves the gut health (Habiyaremye et al., 2017a). The high levels of resistant starch (13.41–26.07 g/100 g) could be useful as prebiotic and might produce a smaller glycemic response, which is a healthy factor in metabolism and blood sugar management (Balli et al., 2023). It also provides support to bones and their mineral composition.

The improved antioxidant activity of the germination study lends to the premise that proso millet (particularly germinated) might be converted to functional foods having potential health effects (anti-oxidative, anti-inflammatory). Oxidative stress and inflammation can be reduced by Phenolic acids, flavonoids and saponins which lowers the risk of chronic diseases. (Xiang et al., 2023).

13.6 Climate Adaptation and Stress Tolerance

13.6.1 Physiological Mechanisms of Stress Tolerance

Proso millet at the physiological level has various mechanisms to deal with abiotic stress. During drought, the plant controls osmolytes such as proline, relative water content and conductance of stomata. The transcriptomics of roots of proso millet in drought stress have found out that there are gene networks that are linked to the root growth, hormonal signal (ABA), and the Osmo protectants biosynthesis (Zhang et al., 2024).

13.6.2 Drought Tolerance and Osmoregulation

In water stress, proso millet roots increase genes that regulate osmolyte production, antioxidant defence and root structure, which increases drought resilience. (Zhang et al., 2024).

13.6.3 Heat Stress Adaptation

Proso millet has a unique reaction to high temperature: when the temperature rises above about 30 °C, the vegetative growth is decelerated or ceased, the reproductive development is altered, and the plant is usually kept short to save water (Habiyaremye et al., 2017a).

13.6.4 Cold and Frost Tolerance

Proso millet is not as cold-tolerant as some other millets. It does not grow well at low temperatures (below 10–13 °C).

13.6.5 Salinity and Alkalinity Tolerance

Salinity tolerance is also exhibited in proso millet. In saline and water-stress environment several genotypes retained a biomass yield, particularly when inoculated with arbuscular mycorrhizal fungi in a Mediterranean marginal habitat (Caruso et al., 2018).

13.6.6 Performance in Climate-Smart Agriculture Systems

Nitrogen utilization, water utilization and output yield of proso millet is enhanced when intercropped with legumes (e.g., mung bean) with optimal nitrogen. It also cut down the level of input. These systems that result in greater resource efficiency and resilience come under climate-smart agriculture (Gong et al., 2021).

13.7 Carbon Sequestration Potential

13.7.1 Role of the C_4 Pathway in High Carbon Fixation

Proso millet has a high-efficiency C_4 photosynthetic pathway, which concentrates CO_2 around the Rubisco enzyme, causing the reduction of photorespiration and maximization of carbon assimilation. This physiological capability especially in the absence of high temperature and water-limited conditions favors high net photosynthetic rate and enhanced carbon fixation. There are a very small number of studies that report direct CO_2 assimilation rates of proso millet, but distant C_4 millets (e.g., foxtail millet) have demonstrated continued or increased photosynthesis at elevated CO_2 concentrations, indicating a healthy capability to assimilate carbon (Li et al., 2019).

13.7.2 Root Biomass, Soil Aggregation, and Carbon Storage

Another source of soil organic carbon (SOC) is underground biomass. Although there are few literature-specific measurements of proso millet, the research on biochar-amended proso millet systems indicates that soil aggregates can stabilize more organic carbon, which suggests the contribution of root-derived residues (Tian et al., 2025). In proso millet fields, biochar enhances the stability of soil aggregates and related OC sequestration suggesting that even the simple, branched root structures can effectively contribute to SOC when residue management is used with biochar.

13.7.3 Carbon Sequestration in Dryland Farming Systems

Proso millet will be a low-input and short-season C_4 crop, which covers the soil, produces residues and slows down the SOC loss in drylands. Although the field-based SOC accumulation assessment of millet are not common, catch crop experimentations indicate that antibiosis biomass (both aboveground and belowground) contributes to carbon storage of the soil. Proso millet when grown in rotations or as an intercrop might also build-up SOC over time. There will be better availability of soil nitrogen (N) and microbial communities in response to organic matter processes. These microbial activities may increase the fixation of freshly deposited carbon in soil (Dang et al., 2020).

13.7.4 Comparison with Other Millets and C_4 Crops

Proso millet does not produce large amounts of biomass (above and belowground) as compared to perennial or high-biomass C_4 grasses (including sorghum or switch grass). Nevertheless, it has a short growing period and allows cultivating in marginal soils, so it can be more effective in carbon sequestered per unit input (i.e., water, fertilizer) (Table 13.9). SOC accumulation is also flexible to perennial C_4 (due to their long-lived root systems), but annual proso millet cycle might offer flexibility in the crop cycles and less trade-offs with food production.

13.7.5 Potential Role in Regenerative and Low-Carbon Agriculture

The carbon-reduction potential of proso millet can be improved by incorporating proso millet in regenerative agricultural systems e.g. through intercropping, rotations or cover cropping. The incorporation of proso millet with proso milling and

Table 13.9 Comparison of Proso millet with other millets and C_4 crops (Ventura et al., 2022)

Crop type	Crop name	Biomass (t/ha)	Characteristics
Proso millet (C_4)	*Panicum miliaceum*	2–8	Short duration, drought-tolerant, low input, suitable for marginal lands
Other millets	Pearl millet	4–10	Higher biomass than proso; drought-tolerant; dual-purpose for grain and fodder.
	Finger millet	3–7	Slightly higher or similar biomass; mainly grown for grain
	Foxtail millet	2–5	Comparable to proso; short duration, low input
	Kodo millet	2–5	Like proso; mainly grain-focused
Other major C_4 crops	Maize	15–25	Very high biomass; widely used for silage/fodder; high input
	Sorghum	10–20	High biomass; drought-tolerant; dual-purpose crop
	Sugarcane	60–100	Extremely high biomass; industrial crop
	Switchgrass	10–20	Perennial bioenergy crop; high biomass potential

soil additive was observed to increase the overall stability as well as the carbon sequestration of the soils. Simultaneously, legumes can be used to intercrop with other crops (mung bean) to enhance the cycling of nitrogen, microbial activity, and indirectly promote the SOC build-up (Dang et al., 2020).

Overall, direct, long-term SOC quantification of proso millet is quite limited, but mechanistically and proxy evidence indicates that this crop does have considerable potential of soil carbon sequestration when applied in smart crop management systems. Field research of root biomass, time-varying change of SOC and microbial decomposition will be crucial in future to legitimize its contribution to low-carbon agriculture (Gong et al., 2021).

13.8 Modelling Approaches for Proso Millet

13.8.1 Crop Simulation Models (DSSAT, APSIM, AquaCrop)

The simulation models of crop growth and yield are very effective in predicting the performance of proso millet to be used in different environmental conditions and different management practices. The proso millet has been calibrated in APSIM (Agricultural Production Systems Simulator), based on field data on 35 plots of proso millet farmers in Sri Lanka, where measured yields are 1187 $\pm$ 336 kg/ha and simulated yield is 1208 $\pm$ 255 kg/ha, and the difference is not statistically significant (Wimalasiri et al., 2023). Proso millet can also be scaled to the DSSAT (Decision Support System to Agro technology Transfer) platform, but calibrations are not made public. Aqua Crop, developed to operate in water-limiting environments, has the capacity to simulate water use efficiency, biomass growth and yield during a drought situation, and thus provides a model of the assessment of irrigation and water-conserving interventions.

13.8.2 Modelling Growth, Yield, and Water Use Efficiency

Proso millet phenology, leaf area index (LAI), biomass accumulation and yield have been modelled using APSIM simulations in a variety of agronomic practices (Wimalasiri et al., 2023). As an example, in silico, calibration of the sowing dates and nitrogen fertilizer rates was able to predict a variation in yield of 1.0–1.5 t/ha, as well as estimate the water productivity of crops, which was 1.2–1.4 kg/m^3, under dryland conditions. These models help to optimize the sowing windows, the fertilization dates, and the water usage to enhance the efficiency of resource use.

13.8.3 Climate Change Impact Modelling (CMIP6, RCP/SSP Scenarios)

The sensitivity of proso millet yield to climate change has been evaluated under the scenario of future climate. Average yield differences at the mid-century (2040–2069) based on CMIP6 downscaled simulations on RCP4.5 and RCP8.5 took place to create average yield reductions of about 7.3 and 16.6%, respectively, in the same time interval. Such simulations are important to the breeding programs and policy planning, as they confirm the necessity to work on climate-resistant varieties and change the approaches to management in response to the future (Ruiz-Aracil et al., 2023).

13.8.4 Integrating Soil Carbon Models (CENTURY, RothC)

An alternative approach to measuring carbon sequestration is to combine crop model outputs (e.g. root biomass) with soil carbon models (e.g. CENTURY or RothC). Although no direct proso millet, studies have been conducted using this method, it has been demonstrated to be effective to predict SOC accumulation of 0.12–0.34 t C ha^{-1} yr^{-1} in residue-retention and low-tillage systems. This kind of integration allows projecting the potential of long-term carbon storage, at the same time projection of crop productivity is simulated (Ren et al., 2022).

13.8.5 Uncertainty Analysis and Model Validation Techniques

The most significant sources of uncertainty in proso millet simulations are parameter uncertainty (e.g. genotype-specific coefficients), variability in climate projections, and assumptions in the model structure. The APSIM validation model in Sri Lanka used field level observations to reduce parameters uncertainty. Future studies must incorporate Monte Carlo simulations, multi-model ensembles and cross-validation to gain more precision in the predictions of yield and water-use efficiency in different agroecology (Asia, Europe, Africa) (Wimalasiri et al., 2023).

13.9 Remote Sensing Applications

13.9.1 Remote Sensing for Crop Monitoring

Remote sensing offers a scalable methodology, which helps to measure proso millet growth, health and yield in heterogeneous or marginal landscapes. Because of the distribution of millet on small fields with a high rate of growth, intense and geographically detailed monitoring based on satellite or UAVs can be used to detect the stress at the earliest stages, estimate biomass, and monitor phenology (Xue & Su, 2017).

13.9.2 Vegetation Indices (NDVI, EVI, SAVI) in Millet Assessment

The most common vegetation indices that are used in the measurement of plant vigor, biomass and stress are NDVI, EVI and SAVI. The ideal growth of the canopies is usually represented in the NDVI of 0.65–0.85 in the case of small cereals like millet (Ortega-Farias et al., 2021). These indices are highly correlated with leaf area index (LAI) and above ground biomass while indirect yield prediction (Poorazimy et al., 2023).

13.9.3 Satellite Platforms (Sentinel-2, Landsat-8/9, MODIS)

Sentinel 2 with images 10–20 m resolutions and red-edge strips and are appropriate with small millet plots and the determination of stress indications. Landsat 8/9 with medium resolution (30 m) with long-term temporal resolution, which is good in historical monitoring. MODIS is daily high-temporal data, which is appropriate to study phenology, but not to small plots (Liu et al., 2022b).

13.9.4 Estimating Biomass and Yield Using Remote Sensing

Indices such as spectral, canopy temperature, texture, and plant-height features can be used to predict aboveground biomass of proso millet by combining multi-sensor UAV platform (spectral, textural, thermal, and point cloud) with machine-learning (random forest) with high accuracy. $R^2 = 0.78–0.89$ in the prediction of millet and small cereal yield have been obtained with random Forest regression and Partial Least Squares (PLS) models (Yang et al., 2024).

13.9.5 Monitoring Drought Stress and Soil Moisture

Water stress can be detected by remote sensing indices that are sensitive to the water content e.g. NDWI or Thermal Vegetation Index (TVI). With the integration of optical (Sentinel 2) and radar (Sentinel 1), indirect soil moisture can be estimated (Ortega-Farias et al., 2021). Field calibration is also critical to small millet plots.

13.9.6 UAV/Drone-Based Precision Agriculture for Millet

UAVs have multispectral/hyperspectral sensors that provide resolution of centimetre scale to detect canopy health, early stress and phenotyping. Research claims that nitrogen stress can be identified through UAVs with more than 85% precision in small cereals. UAV data can be used to guide precision fertilization and irrigation of proso millet improvement trials (Liu et al., 2022a).

13.10 Industrial, Nutraceutical, and Value-Added Uses

13.10.1 Food Products (Flour, Porridge, Bakery, Beverages)

Proso millet flour is becoming increasingly functional and nutritional. Its starch content is between 58 and 75%, depending on the genotype, and is one reason it is used in porridge, flatbreads and gluten-free bakery goods its protein content ranges between 11–14, and this may be used in the nutritional fortification of composite flours when it is mixed with maize or wheat. Fermented proso millet drinks maintain high antioxidant activity and enhanced bioavailability of phenolic compounds showing potential of value-added nutraceutical product (Xiang et al., 2023). Resistant starch varies in range of 13 to 26%, which is positively impacting glucose as it slower its release and is also applicable in diabetic-friendly diets (Balli et al., 2023). Studies have also reported that germination/malting can also enhance total phenolic content (6–8 times), antioxidant capacity and functional food potential moreover, sensory analyses revealed that bakery goods using millet have acceptable taste and texture, and thus, they can be used as alternatives to populations that are gluten intolerant. (Johnston et al., 2023).

13.10.2 Feed and Fodder Utilization

Straw and Proso millet grains are useful as feeds to poultry and small ruminants as well as birds. It has proven that lambs and goats fed on millet grain diets increase by an average of 7–9% faster than those in control groups that reside in similar (Persson et al., 2022). Small-scale livestock systems are more likely to be using straw and husks since options of forage are few in semi-arid areas. Due to the low inputs of proso millet, it can be adopted as a dual-purpose crop, where grain and fodder are collected without affecting the fertility or sustainability of the soil (Hassan et al., 2021).

13.10.3 Industrial Uses (Bioethanol, Starch Industries)

The starch component of proso millet can be applied in the industry, such as production of bioethanol and bioplastics. Millet starch can also be utilized in biodegradable adhesives, packaging materials and bioplastics, and these can be used as a substitute of petroleum-based products. The cultivars of millet starch differ in their physicochemical characteristics, which can be used in customized industrial operations. Moreover, the pilot studies are managed by small-scale, indicating that millet-based bioethanol has a potential to cut down to 25% of the CO_2 emissions as compared to corn ethanol, which is a high indicator of its possible input to low-carbon industrial systems (Kumar et al., 2023).

13.10.4 Nutraceutical and Functional Food Applications

Polyphenols, flavonoids and antioxidants are found in high concentration in proso millet. The overall phenolic concentration is 6–8 times increased during germination to improve radical-scavenging action and possible effects of cardio protective, anti-diabetic, and anti-inflammatory benefits. The presence of cinnamic acids and other phenolic acids explain the benefits of the antioxidants (Balli et al., 2023). Proso millet is also a prime product to be developed into a functional food to address metabolic health due to its high fiber content, insoluble starch, and being gluten free (Xiang et al., 2023).

13.11 Constraints and Challenges

13.11.1 Limited Farmer Awareness and Research Gaps

Proso millet has low levels of adoption because of low levels of awareness of farmers, particularly in non-traditional areas despite its resilience and nutritional qualities. It has been surveyed that in the world, less than 10% of the arable land is now farmed to grow millet (FAO, 2023). Genotype-environment research gaps, physiological processes of stress, water management, as well as the optimization of post-harvest operations are limitations to the utilization of data-driven agronomic recommendations (Wimalasiri et al., 2023).

13.11.2 Weak Seed Systems and Variety Availability

The farmers still use the traditional landraces since the access to better or certified seeds is limited. High performance varieties can only be utilized in about 15–20% of semi-arid region farmers (Zargar et al., 2023). There is no better channel of distributing seeds, which limits the possibility of yield increase and climate resilience. Seed production and certified seed production in communities is thus essential in increasing the uptake.

13.11.3 Post-harvest Challenges and Processing Limitations

The low grain size and delicate husks make the mechanism of harvesting and milling a problem, which frequently leads to 12–15% after-harvest losses (FAO, 2023). There is a lack of infrastructure related to storage, cleaning, and processing, which further restricts commercial use on a large scale particularly in the case of functional food or industrial use.

13.11.4 Market and Policy Constraints

Proso millet is not in high demand on the markets in comparison with such staples as rice and wheat. This reduces profitability by the fluctuating prices, the low value chains, and the absence of appropriate processing plants (FAO, 2023). The modest policy incentives to promote the cultivation, processing, and marketing of millet are still not adequate to facilitate the use of the practice on a large scale, even in the appropriate agro ecology.

13.11.5 Climate Vulnerabilities in Non-traditional Regions

Even though proso millet is drought-tolerant, its growth to new agro-ecological regions is limited due to the unavailability of locally adapted germplasm. Breeding outside the boundaries of the natural habitat may result in lower performance and exposure to severe weather conditions, and it is the reason why regional breeding programs are essential (Wimalasiri et al., 2023).

13.12 Opportunities for Scaling Up

13.12.1 Role in Sustainable and Low-Input Farming Systems

Proso millet suits low-input systems best since it does not need much fertilizer and irrigation, yet the process yields between 1–2 t/ha under rainfed conditions. Its rapid growing period enables to be dual cropped and better land use efficiency, especially to the smallholder farmers of the semi-arid areas (Ishaq et al., 2025).

13.12.2 Potential in Organic and Regenerative Agriculture

The crop brings organic deposits and favours soil microbes, which is beneficial to soil health and biodiversity. Its rotational cultivation or as a cover crop also reduces erosion and enhances soil organic carbon, and therefore it meets the objectives of regenerative agriculture (Ishaq et al., 2025).

13.12.3 Use in Crop Rotation and Diversification

Proso millet crop rotation helps to keep ground cover, minimizes fallow and increases nutrient cycling. The system of Millet-based rotational systems would result in the decrease of disease pressure and climate variability resilience of smallholder farmers (FAO, 2023).

13.12.4 Climate-Resilient Livelihood Opportunities

The drought, heat, and poor soils adaptability of proso millet allows consistent production in the marginal lands. The advantage is that farmers will get a stable crop in times of climate strain, which will improve food security and increase household income (Chaturvedi et al., 2022).

13.12.5 Value Chain Development and Market Expansion

The markets of gluten-free as well as functional foods are expanding around the world, which opens chances to the millet-based products. The government activities promote the awareness and investment of millet value chains, such as the UN International Year of Millets 2023, which assists farmers in earning more and industries in developing (FAO, 2023).

13.13 Breeding and Genomic Improvement

13.13.1 Genetic Resources and Germplasm Repositories

Breeding programs have a variety of germplasm collections such as the 88-accession GWAS panel (Fig. 13.6). These panels record variation in drought tolerance, early maturity and nutrient content, which can be used as a basis of marker-assisted selection (Zargar et al., 2023).

There are more than 29,000 germplasm collections of proso millet conserved worldwide. The major gene banks of proso millet germplasms are listed in Table 13.10.

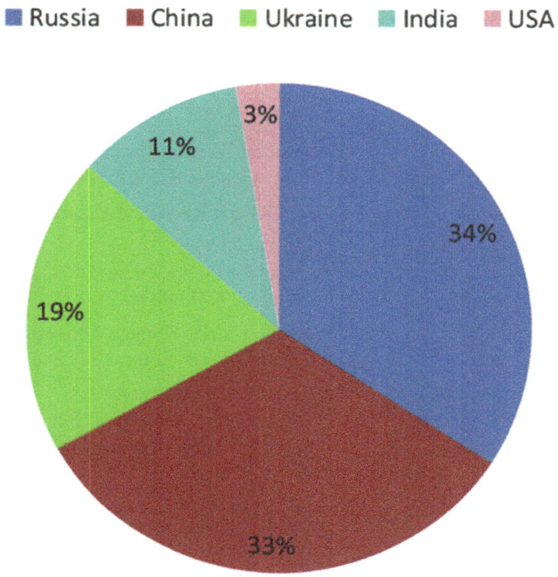

Fig. 13.6 % Share of major countries to Proso millet germplasm collection (Rajasekaran & Francis, 2021).

13.13.2 Breeding Goals: Stress Tolerance, Nutrition, Early Maturity

The breeding activities focus on drought and heat tolerance, high protein and micronutrient features and early flowering to reduce growing seasons. Such characteristics guarantee the ability to be adapted to different agro-ecologies and the improvement of the nutritional value of foods based on millet (Ghatak et al., 2025).

13.13.3 Molecular Breeding and Genomics (SNPs, QTL Mapping)

GWAS and SNP analysis identified loci of seed size, flowering time, and stress resilience. Marker-traits help in faster breeding and allows better choices of better varieties in marginal habitat (Fig. 13.7) (Ghatak et al., 2025).

Table 13.10 Major gene banks conserving germplasm of Proso millet worldwide (Narciso & Nyström, 2023)

Country	Institute	Germplasm accessions	
		Cultivated	Wild
Australia	Australian Tropical Crops and Forages Genetic Resources Centre	228	
Bangladesh	Plant Genetic Resources Centre, BARI	198	
Bulgaria	Institute for Plant Genetic Resources "K. Malkov"	489	
China	Institute of Crop Science, Chinese Academy of Agricultural Sciences	8451	
Czech Republic	Gene Bank Department, DIVISION of Genetics and Plant Breeding, Research Institute of Crop Production	171	
Germany	Gene bank, Leibniz Institute of Plant Genetics and Crop Plant Research	165	1
Hungary	Institute for Agro-Botany	243	1
India	AICRP on Small Millets	920	4
	International Crop Research Institute for the Semi-Arid Tropics (ICRISAT)	849	
	National Bureau of Plant Genetic Resources	994	
Japan	Department of Genetic Resources I, National Institute of Agro-Biological Sciences	516	
Mexico	Estación de Iguala, Instituto Nacional de Investigaciones Agrícolas	400	
Poland	Botanical Garden of Plant Breeding and Acclimatization Institute	354	

13.13.4 Genome Editing (CRISPR, Targeted Traits)

CRISPR/Cas9 has possibilities of achieving targeted enhancement in root architecture, photoperiod sensitivity, and metabolic pathway. The editing of these characteristics can elevate the stress tolerance, efficiency of the resource use and provide stability in the climate-stressed conditions (Ghatak et al., 2025).

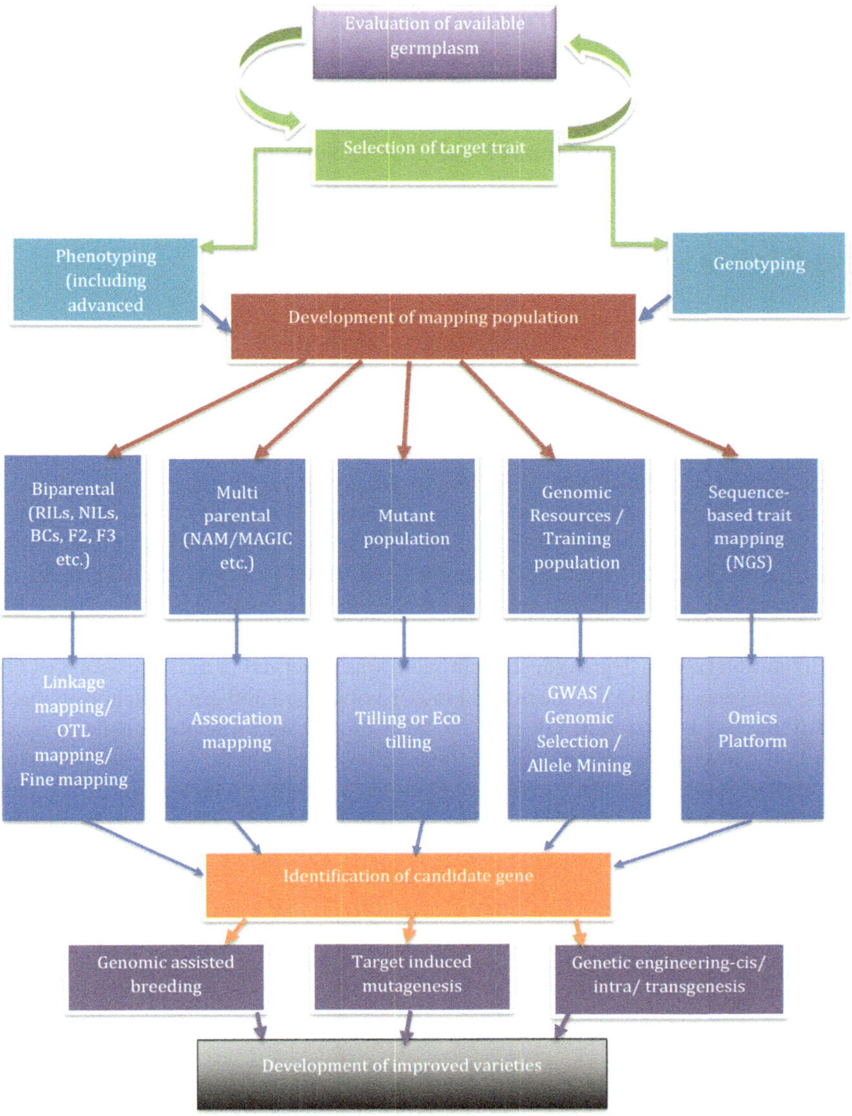

Fig. 13.7 Roadmap for genetic and genomic improvement in Proso millet

13.13.5 Seed System Strengthening for Wider Adoption

Certified seed production and community-based network of distribution are essential in the delivery of better varieties to the farmers. Good seed systems make available cultivars of genetically better and climatically stable cultivars (Zargar et al., 2023).

13.14 Case Studies

13.14.1 Case Study: United States—Irrigation Response in the Pacific Northwest

Habiyaremye et al. (2017a) used a three-year (2012–2014) period to assess the phenotypic reaction of 20 different proso millet (*Panicum miliaceum* L.) accessions to irrigation in organic farming conditions, in Pullman, Washington, USA. Experiment was based on a randomised complete block design using two irrigation conditions fully irrigated and non-irrigated to recreate optimal and water-restricted conditions. The researchers used ANOVA and Pearson correlation statistical models. The results also showed that mean emergence was also higher with irrigation (52%) than with non-irrigation (24%). The highest mean yield was observed in the accession GR 665 whose yield was found to be the highest in 2013 of around 365 g/plot. In non-irrigated conditions, the accession of GR 658 and Minsum showed low but consistent yield. Correlation tests showed that in 2012, the height of plants at the later extension stem was highly correlated with yield ($r = 0.75$, $p \leq 0.0001$), whereas in 2014, yield was statistically correlated with days to heading ($r = 0.55$) and days to maturity ($r = 0.47$), which indicates that the duration of growth has an impact on yield when water is variable. The authors ended up concluding that breeding programs must be oriented towards the production of drought tolerant genotypes that can be used in rain-fed systems as well as lodging resistant types that can be used in irrigated systems.

13.14.2 Case Study: United Arab Emirates—Abiotic Stress Resilience under Heat, Salinity, and Water Deficit

Samineni et al. (2024) tested the resilience of 24 genotypes of proso millet in extreme desert environments in UAE where the heat, water, and salinity levels were high and severe. Field trials were taken across two seasons which were normal and summer with three irrigation regimes: complete freshwater, 50% irrigation and saline irrigation at 10 dS/m. The experiment was used to measure grain yield, chaffy grain ratio, weight of 1000 grains, height of the plant, forage yield. Environmental data in terms of temperature peaks per day, which was higher than 40 °C at the reproductive period, and it reached 49 °C in the summer. ANOVA and GGE biplot were used to determine stable and high-performing lines. Results show that some genotypes such as G7, G10 and G12 continued to produce grain greater than 0.75 t/ha as compared which were inefficient in developing reproductively. Correlation results indicated that the grain yield was positively linked with the plant height, forage yield, and 1000-grain weight in all the environments, indicating them as important breeding goals when the environment is arid.

13.14.3 Case Study: Nepal—On-Farm Evaluation of Landraces in Hilly Terrains

Bhatta et al. (2025) in Bajura District (Nepal) conducted an on-farm evaluation on 8 proso millet genotypes (two local landraces (Mal Chino and Dudhe Chino) and 6 accessions of National Gene bank (NGRC)). The Himalayan hill condition performance was tested with the use of a randomized complete block design. Morphological traits such as number of tillers, flag leaf length, number of leaves and grain yield were determined using regression and correlation analysis. The results are domestic landrace Mal Chino performed better with 2.06 t/ha and the best-performing introduced accession "NGRC07345" had a yield of 1.70 t/ha. These findings indicate that certain morphological characteristics can be used as selection factors during breeding initiatives against hilly conditions. The research suggested long and multi-site trials to ensure stability of successful genotypes, inclusion of the locally adapted landraces in formal seed systems and application of trait-based selection to improve productiveness in marginal and remote areas.

13.14.4 Case Study: Kazakhstan—Genetic Diversity and Population Structure

In a, Zargar et al. (2023) concluded the genetic diversity and population structure of 100 proso millet accessions by 20 SSR selections of SSR markers in Akmola, Northern Kazakhstan. It was used to identify allele richness, polymorphic information content (PIC) and population structure. They have found 20 loci containing 47 alleles with a mean average of 2.35 alleles/locus. The PIC values were 0.424, but markers SSR-85 and SSR-86 displayed the greatest polymorphism (0.795 and 0.758). STRUCTURE analysis provided accessions in three large genetic clusters having accessions of Central Asian origin in all of them. Authors highlighted that these markers may be implemented in breeding programs to speed up the process of achieving yield-related traits and that good strategic crossing should be undertaken between different clusters to obtain maximum cross heterosis.

13.14.5 Case Study: Iran—Drought Stress and Nitrogen Supply Interaction

Nematpour et al. (2019) at Isfahan University of Technology in Iran, explored how drought stress, nitrogen nutritional status, and sowing date interactively affect proso and foxtail millet. The field experiment on factorial field involved two levels of nitrogen (low and optimal) and two sowing dates having drought induced. Biomass, photosynthetic pigments, final grain yield, relative water content (RWC), leaf water potential and proline content were measured. In the conditions of drought and low nitrogen status, the grain yield was reduced by approximately 30% as compared to non-stressed controls. RWC was also kept at the optimal level of nitrogen at approximately 68% but dropped to approximately 53% in low-N drought. The osmotic adjustment was demonstrated by an increase in Proline, an osmolyte, by about 250% in stressed low-N plants. Low nitrogen early sowing incurred a 10% decrease in yield loss than the late sowing, which emphasizes on the role of sowing date in stress mitigation. The authors suggested the combination of nitrogen management and optimized sowing to increase drought resilience, the breeding lines with greater osmolyte accumulation capacity.

13.15 Policy, Socioeconomic, and Market Perspectives

13.15.1 Government and FAO Initiatives

The international agencies and national governments have come to recognize the potential of the millets such as proso millet in terms of food security, adaptation to climate and nutrition. In India and China and some regions of Africa, programs have involved subsidies of millet farming, capacity-building workshops provided to farmers and funding of research of better varieties (FAO, 2023). It has also contributed to productivity and resilience as well as pilot schemes have experienced a 25–30% increase in yield with the introduction of proposed varieties and the introduction of soil management measures (Ghatak et al., 2025).

13.15.2 UN International Year of Millets 2023

On January 15, 2023, to create awareness in the world, the United Nations proclaimed 2023 as International Year of Millets on advantages of millets (FAO, 2023). Its campaign stimulated the integration of policies, such as the promotion of millet-based products in public distribution systems and school meals and to encourage the involvement in the value-chain development by the private sector. Countries participating have already cited that the area under millet cultivation has increased and India alone has reported an increase of 15% per year on the area under millet cultivation in 2022–2023. Another area of priority of the initiative was the digital extension services to inform farmers and consumers about the benefits of millet, which connects policy advocacy with market development.

13.15.3 Farmer Cooperatives and Community-Based Approaches

Farmer cooperatives are important for the adoption of millet and market access. The low risks to the smallholders are also decreasing as community seed banks, common storage sites and processing plants are expanding access to high quality proso millet seed. Coupled resources and expertise make cooperatives more resilient to market fluctuations, minimize transaction expenses, and facilitate the adoption of technology to support the practice of climate smartness.

13.15.4 Economic Viability and Market Opportunity

Proso millet has economic potential in the niche of health and gluten-free markets because it has low glycemic index, high fiber, and high levels of minerals (Xiang et al., 2023). It has been discovered in the research of the Indian and Eastern European cost-benefits that the net returns received by the smallholders by growing proso millet and sold in the organic and gluten free or health-conscious value chain are comparable to those of maize or sorghum (Ghatak et al., 2025).

13.15.5 Social Inclusion and Gender Equity

Smallholder and women farmers are some of the frequent farmers in millet cultivation especially in semi-arid areas. The post-harvest processing, seed selection, and influence on the household nutrition are carried out by women, and involvement in it is essential in food security, empowerment, and empowerment (FAO, 2023). Gender policies to favor female farmers, fair access to seeds, and inclusion of small farmers have enhanced the nutrition outcomes in households and economic empowerment. By incorporating gender-sensitive extension and seed sovereignty, resilience is enhanced and the marginalized communities will enjoy the millet commercialization.

13.16 Future Prospects and Research Needs

13.16.1 Integrated Climate-Carbon Modeling

The research on proso millet should combine the crop growth models with soil carbon models so that it is ready to achieve its full potential in climate smart agriculture. Though simulating the yield of milk operating under different conditions of the environments has been achieved with the help of APSIM, DSSAT, and AquaCrop, the relationship between the models and the carbon processes in the soils could be made based on such models like CENTURY or RothC, with the assistance of which the sequestration of the soil organic carbon could be clearly predicted. It is a kind of integrated modeling that can be applied to make recommendations of cover cropping, residue retention and optimum rotations and enhance carbon storage without yield loss (Wimalasiri et al., 2023).

13.16.2 Digital Agriculture and Remote Sensing Applications

The proso millet potential can be improved by digital agriculture. Phenotyping, sensing drought stress, and precise irrigation with the assistance of satellites, UAVs, or drones may be performed in real-time by means of the high-resolution imaging. With machine learning algorithms, breeders can screen large sets of germplasm to the degree that they are able to identify accessions that are resistant to climate changes to improve planting schedules. Precision agriculture approaches may also be applied to make the water usage more efficient, and pilot results demonstrate that the water input may be decreased by 20% without influencing the yields (Xiang et al., 2023).

13.16.3 Multi-environment Trials for Climate-Resilient Varieties

Multi-location, multi-season trials are significant to establish the stability and adaptability of the proso millet types to the altering climatic conditions. The trials organized in a large variety of agroecology, including the semi-arid, hilly and steppe areas. These will help to discover the genotypes with high water use efficiency, heat and drought tolerance and positive nutritional properties. Some of the agronomic traits that can be used are number of tillers, length of flag-leaf, weight of a thousand grains and osmolytes accumulation to make a drought-resistant selection.

13.16.4 Enhancing Value Chains and Consumer Products

Processing, value addition and market development should be researched which are not in agronomy. Not only can Farmer earn more, but also consumers can use gluten-free flours, functional foods, bioethanol, and starch-based bioplastics. By making investments in the milling infrastructure, cold stores and the smallholder cooperative networks, the losses that occur after harvest will be reduced to increase the economic viability. Indian and Eastern European studies also suggest that milling of products using millet and selling them as branded products could be able to generate 20–40% more net returns per hectare than that of raw grains.

13.16.5 Aligning Proso Millet Promotion with Sustainable Development Goals (SDGs)

To support SDGs, including Zero Hunger (SDG 2), Climate Action (SDG 13), and Land Restoration (SDG 15), promotion of the proso millet planting and use can be done. By integrating climate-resistant varieties, carbon-binding activities, and creation of value-chains, the stakeholders will be able to enhance food security, reduce greenhouse gas emission as well as rehabilitate degraded soils in the dryland ecosystems. The proso millet production systems need to be implemented to coordinate the policy, create extension services, enhance awareness among consumers (FAO, 2023).

13.17 Conclusion

Proso millet (*Panicum miliaceum*) is identified as a climate resilient, nutritionally enriched cereal. It has tremendous capabilities of improving food security, carbon sequestration and sustainable agriculture. The C4 physiology of its species allows it to have high water use efficiency and resistance to abiotic stresses like drought, heat, salinity, and frost. This makes it uniquely adapted to dryland and semi-arid agroecosystems where conventional crops typically fail. In a wide range of geographies, such as the USA, China, India, Kazakhstan and African drylands, proso millet has proved to be a reliable performer. With grain yield of 0.75–2.3 t/ha in stressful environments and the rate of soil carbon sequestration of 0.12–0.34 t C ha^{-1} yr^{-1} are evidence of its dual attributes in food production as well as climate mitigation. It can Combine with modern agronomic technologies, including precision agriculture, remote sensing (Sentinel-2, MODIS), UAV surveillance, and crop simulation models (DSSAT, APSIM, AquaCrop). This opens the opportunity in water and nutrient management, yield prediction and greater application in smallholder farming and large-scale farming systems. Socio-economic opportunities to the resource-constrained farmers in the form of facilitating diversified rotations, low-input systems and value-chain development also exist. Enhancing seed systems and supporting landraces that are resistant to climate can speed up the improvement of the high-performing varieties by using genomic and molecular breeding technologies (QTL mapping, SNPs, genome editing). The increase in adoption and maximization of its potentials should be done by policy interventions, carbon-credit programs, and awareness efforts in farmers. In summary, proso millet is a multi-purpose, resilient and low-carbon grain that is environmentally sustainable in terms of productivity. Increasing its production and ensuring its adoption in the world agri-food strategies will also make a great contribution to the attainment of sustainable development goals, especially Zero Hunger, Climate Action, and Land Restoration.

References

Balli, D., Bellumori, M., Masoni, A., Moretta, M., Palchetti, E., Bertaccini, B., Mulinacci, N., & Innocenti, M. (2023). Proso millet (*Panicum miliaceum* L.) as alternative source of starch and phenolic compounds: A study on twenty-five worldwide accessions. *Molecules, 28*(17), 6339. https://doi.org/10.3390/molecules28176339

Bhatta, K., Sipkhan, P. R., Bhusal, B., Khatri, L., & Dhakal, R. (2025). Performance Evaluation and Trait Association Analysis of Proso Millet Genotypes in Bajura District, Nepal. *Agriculture Development Journal, 18*, 82–94. https://doi.org/10.3126/adj.v18i1.82090

Boukail, S., Macharia, M., Miculan, M., Masoni, A., Calamai, A., Palchetti, E., & Dell'Acqua, M. (2021). Genome wide association study of agronomic and seed traits in a world collection of proso millet (*Panicum miliaceum* L.). *BMC Plant Biology, 21*(1), 330. https://doi.org/10.1186/s12870-021-03111-5

Caruso, C., Maucieri, C., Berruti, A., Borin, M., & Barbera, A. (2018). Responses of different *Panicum miliaceum* L. Genotypes to Saline and Water Stress in a Marginal Mediterranean Environment. *Agronomy, 8*(1), 8. https://doi.org/10.3390/agronomy8010008

Chaturvedi, P., Govindaraj, M., Govindan, V., & Weckwerth, W. (2022). Editorial: Sorghum and Pearl Millet as Climate Resilient Crops for Food and Nutrition Security. *Frontiers in Plant Science, 13*, 851970. https://doi.org/10.3389/fpls.2022.851970

Dang, K., Gong, X., Zhao, G., Wang, H., Ivanistau, A., & Feng, B. (2020). Intercropping alters the soil microbial diversity and community to facilitate nitrogen assimilation: A potential mechanism for increasing Proso millet grain yield. *Frontiers in Microbiology, 11*, 601054. https://doi.org/10.3389/fmicb.2020.601054

Das, S., Khound, R., Santra, M., & Santra, D. K. (2019). Beyond bird feed: Proso millet for human health and environment. *Agriculture, 9*(3), 64. https://doi.org/10.3390/agriculture9030064

FAO. (2023). *Unleashing the potential of millets*. FAO. https://doi.org/10.4060/cc7484en

Ghatak, A., Pierides, I., Singh, R. K., Srivastava, R. K., Varshney, R. K., Prasad, M., Chaturvedi, P., & Weckwerth, W. (2025). Millets for a sustainable future. *Journal of Experimental Botany, 76*(6), 1534–1545. https://doi.org/10.1093/jxb/erae507

Ghimire, K. H., Joshi, B. K., Dhakal, R., & Sthapit, B. R. (2018). Diversity in proso millet (*Panicum miliaceum* L.) landraces collected from Himalayan mountains of Nepal. *Genetic Resources and Crop Evolution, 65*(2), 503–512. https://doi.org/10.1007/s10722-017-0548-7

Gong, X., Dang, K., Liu, L., Zhao, G., Lv, S., Tian, L., Jin, F., Feng, Y., Zhao, Y., & Feng, B. (2021). Intercropping combined with nitrogen input promotes proso millet (*Panicum miliaceum* L.) growth and resource use efficiency to increase grain yield on the loess plateau of China. *Agricultural Water Management, 243*, 106434. https://doi.org/10.1016/j.agwat.2020.106434

Habiyaremye, C., Barth, V., Highet, K., Coffey, T., & Murphy, K. (2017a). Phenotypic responses of twenty diverse Proso millet (*Panicum miliaceum* L.) Accessions to Irrigation. *Sustainability, 9*(3), 389. https://doi.org/10.3390/su9030389

Habiyaremye, C., Matanguihan, J. B., D'Alpoim Guedes, J., Ganjyal, G. M., Whiteman, M. R., Kidwell, K. K., & Murphy, K. M. (2017b). Proso millet (*Panicum miliaceum* L.) and its potential for cultivation in the Pacific northwest, U.S.: A review. *Frontiers in Plant Science, 7*. https://doi.org/10.3389/fpls.2016.01961

Hanna, W. W., Baltensperger, D. D., & Seetharam, A. (2004). Pearl millet and other millets. In L. E. Moser, B. L. Burson, & L. E. Sollenberger (Eds), Agronomy monographs (1st edn, Vol. 45, pp. 537–560). Wiley. doi:https://doi.org/10.2134/agronmonogr45.c15.

Hassan, Z. M., Sebola, N. A., & Mabelebele, M. (2021). The nutritional use of millet grain for food and feed: A review. *Agriculture & Food Security, 10*(1), 16. https://doi.org/10.1186/s40066-020-00282-6

Hunt, H. V., Badakshi, F., Romanova, O., Howe, C. J., Jones, M. K., & Heslop-Harrison, J. S. P. (2014). Reticulate evolution in Panicum (Poaceae): The origin of tetraploid broomcorn millet, P. miliaceum. *Journal of Experimental Botany, 65*(12), 3165–3175. https://doi.org/10.1093/jxb/eru161

Ishaq, T., Masood, A., Saad, A., Singh, L., Mehdi, S., Rashid, S., Sharma, A., Aziz, M. A., Rai, U., Qasim, N., Mir, M. S., Palmo, T., & Rafiq, A. (2025). Climate resilience of millets: A sustainable solution for future agriculture. *International Journal of Research in Agronomy, 8*(5), 744–754. https://doi.org/10.33545/2618060X.2025.v8.i5j.2979

Johnston, M. G., Breakspear, A., Samwald, S., Zhang, D., Papp, D., Faulkner, C., & De Keijzer, J. (2023). Comparative phyloproteomics identifies conserved plasmodesmal proteins. *Journal of Experimental Botany, 74*(6), 1821–1835. https://doi.org/10.1093/jxb/erad022

Kalinova, J., & Moudry, J. (2006). Content and quality of protein in Proso millet (*Panicum miliaceum* L.) varieties. *Plant Foods for Human Nutrition, 61*(1), 43–47. https://doi.org/10.1007/s11130-006-0013-9

Kumar, S. R., Tangsrianugul, N., & Suphantharika, M. (2023). A review on isolation, characterization, modification, and applications of Proso millet starch. Food, 12(12), 2413. doi:https://doi.org/10.3390/foods12122413, Foods.

Li, P., Li, B., Seneweera, S., Zong, Y., Li, F. Y., Han, Y., & Hao, X. (2019). Photosynthesis and yield response to elevated CO2, C4 plant foxtail millet behaves similarly to C3 species. *Plant Science, 285*, 239–247. https://doi.org/10.1016/j.plantsci.2019.05.006

Liu, Y., Zhang, W., Li, F., Zuo, Z., & Huang, Q. (2022a). Real-time Lidar Odometry and mapping with loop closure. *Sensors (Basel), 22*(12), 4373. https://doi.org/10.3390/s22124373

Liu, Z., Zhang, Q., & Li, K. (2022b). A smart noise jamming suppression method based on atomic dictionary parameter optimization decomposition. *Remote Sensing, 14*(8), 1921. https://doi.org/10.3390/rs14081921

Uddin, M. S., Azam, M. G., Bagum, S. A., & Hakim, M. A. (2021). *Genetic diversity analysis of proso millet (Panicum miliaceum L.) in relation to phenotypic characters*. https://doi.org/10.5281/ZENODO.4421795

Makuchete, L., Hove, A., Nezomba, H., Rurinda, J., Mbanyele, V., Mlambo, S., Nyakudya, E., Mtambanengwe, F., & Mapfumo, P. (2024). Productivity of sorghum and millets under different in-field rainwater management options on soils of varying fertility status in Zimbabwe. *Frontiers in Agronomy, 6*, 1378339. https://doi.org/10.3389/fagro.2024.1378339

Muskan, F., Redhu, M., Redhu, S., & Rahimi, M. (2025). Millets in the global market: A critical review of challenges and opportunities. *Food Production, Processing and Nutrition, 7*(1), 37. https://doi.org/10.1186/s43014-025-00316-1

Narciso, J. O., & Nyström, L. (2023). The genetic diversity and nutritional quality of proso millet (*Panicum miliaceum*) and its Philippine ecotype, the ancient grain "kabog millet": A review. *Journal of Agriculture and Food Research, 11*, 100499. https://doi.org/10.1016/j.jafr.2023.100499

Nematpour, A., Eshghizadeh, H. R., & Zahedi, M. (2019). Drought-tolerance mechanisms in foxtail millet (Setaria italica) and proso millet (Panicum miliaceum) under different nitrogen supply and sowing dates. *Crop & Pasture Science, 70*(5), 442–452. https://doi.org/10.1071/CP18501

Ortega-Farias, S., Esteban-Condori, W., Riveros-Burgos, C., Fuentes-Peñailillo, F., & Bardeen, M. (2021). Evaluation of a two-source patch model to estimate vineyard energy balance using high-resolution thermal images acquired by an unmanned aerial vehicle (UAV). Agricultural and Forest Meteorology, 304–305, 108433. https://doi.org/10.1016/j.agrformet.2021.108433.

Persson, Y., Gustafsson, K., Hoffman, M., Nyman, A.-K., Söderquist, L., & Persson Waller, K. (2022). Risk factors for intramammary infection in meat- and pelt producing ewes with clinically healthy udders. *Small Ruminant Research, 206*, 106595. https://doi.org/10.1016/j.smallrumres.2021.106595

Poorazimy, M., Shataee, S., Aghababaei, H., Tomppo, E., & Praks, J. (2023). First demonstration of space-borne polarization coherence tomography for characterizing Hyrcanian Forest structural diversity. *Remote Sensing, 15*(3), 555. https://doi.org/10.3390/rs15030555

Rajasekaran, R., & Francis, N. (2021). Genetic and genomic resources for improving proso millet (*Panicum miliaceum* L.): A potential crop for food and nutritional security. *The Nucleus, 64*(1), 21–32. https://doi.org/10.1007/s13237-020-00331-2

Rajasekaran, R., Francis, N., Mani, V., & Ganesan, J. (2023). Proso millet (*Panicum miliaceum* L.). In *Neglected and underutilized crops* (pp. 247–278). Elsevier. https://doi.org/10.1016/B978-0-323-90537-4.00005-3

Ramadas, S., Joseph, J., & Mohapatra, S. (2023). *Global status of millets ecosystem—Trends.* Research Progress and Way Forward.

Ren, L., Cornelis, W. M., Ruysschaert, G., De Pue, J., Lootens, P., & D'Hose, T. (2022). Quantifying the impact of induced topsoil and historical subsoil compaction as well as the persistence of subsoiling. *Geoderma, 424*, 116024. https://doi.org/10.1016/j.geoderma.2022.116024

Ruiz-Aracil, M. C., Valverde, J. M., Beltrà, A., Carrión-Antolí, A., Lorente-Mento, J. M., Nicolás-Almansa, M., & Guillén, F. (2023). Putrescine increases frost tolerance and effectively mitigates sweet cherry (Prunus avium L.) cracking: A study of four different growing cycles. *Agronomy, 14*(1), 23. https://doi.org/10.3390/agronomy14010023

Samineni, S., Gummadi, S., Thushar, S., Khan, D. N., Gkanogiannis, A., Becerra, L. A., & Singh, R. K. (2024). *Exploring Proso Millet Resilience to Abiotic Stresses: High-Yield Potential in Desert Environments of the Middle East*. Biology and Life Sciences. https://doi.org/10.20944/preprints202412.2296.v1

Tian, L., Wang, Y., Jin, D., Zhou, Y., Mukhamed, B., Liu, D., & Feng, B. (2025). The application of biochar and organic fertilizer substitution regulates the diversities of habitat specialist bacterial communities within soil aggregates in proso millet farmland. *Biochar, 7*(1), 6. https://doi.org/10.1007/s42773-024-00398-z

Uludag, A., Lyon, D. J., Nissen, S. J., & Kachman, S. D. (1997). Proso millet (*Panicum miliaceum*) response to CGA-152005, Metsulfuron, and Triasulfuron. *Weed Technology, 11*(1), 138–143. https://doi.org/10.1017/S0890037X00041476

Ventura, F., Poggi, G. M., Vignudelli, M., Bosi, S., Negri, L., Fakaros, A., & Dinelli, G. (2022). An assessment of Proso millet as an alternative summer cereal crop in the Mediterranean Basin. *Agronomy, 12*(3), 609. https://doi.org/10.3390/agronomy12030609

Ventura, F., Vignudelli, M., Poggi, G. M., Negri, L., & Dinelli, G. (2020). Phenological stages of Proso millet (*Panicum miliaceum* L.) encoded in BBCH scale. *International Journal of Biometeorology, 64*(7), 1167–1181. https://doi.org/10.1007/s00484-020-01891-3

Wimalasiri, E. M., Ashfold, M. J., Jahanshiri, E., Walker, S., Azam-Ali, S. N., & Karunaratne, A. S. (2023). Agro-climatic sensitivity analysis for sustainable crop diversification; the case of Proso millet (*Panicum miliaceum* L.). *PLoS One, 18*(3), e0283298. https://doi.org/10.1371/journal.pone.0283298

Xiang, J., Yuan, Y., Du, L., Zhang, Y., Li, C., & Beta, T. (2023). Modification on phenolic profiles and enhancement of antioxidant activity of proso millets during germination. *Food Chemistry: X, 18*, 100628. https://doi.org/10.1016/j.fochx.2023.100628

Xue, J., & Su, B. (2017). Significant remote sensing vegetation indices: A review of developments and applications. *Journal of Sensors, 2017*, 1–17. https://doi.org/10.1155/2017/1353691

Yang, Z., Yu, Z., Wang, X., Yan, W., Sun, S., Feng, M., Sun, J., Su, P., Sun, X., Wang, Z., Yang, C., Wang, C., Zhao, Y., Xiao, L., Song, X., Zhang, M., & Yang, W. (2024). Estimation of millet aboveground biomass utilizing multi-source UAV image feature fusion. *Agronomy, 14*(4), 701. https://doi.org/10.3390/agronomy14040701

Zargar, M., Dyussibayeva, E., Orazov, A., Zeinullina, A., Zhirnova, I., Yessenbekova, G., & Rysbekova, A. (2023). SSR based genetic diversity analysis and population structure of Proso millet (*Panicum miliaceum* L.) in Kazakhstan. *Biology and Life Sciences*. https://doi.org/10.20944/preprints202309.0826.v1

Zhang, P., Wang, B., Guo, Y., Wang, T., Wei, Q., Luo, Y., Li, H., Wu, H., Wang, X., & Zhang, X. (2024). Identification of drought-resistant response in Proso millet (*Panicum miliaceum* L.) Root through Physiological and Transcriptomic Analysis. *Plants, 13*(12), 1693. https://doi.org/10.3390/plants13121693

Chapter 14
Little Millet (*Panicum miliare*): Enhancing Food Security Through Resilient Cropping Systems

Zohaa Fatima, Muhammad Asif, Umar Farooq, Ahmed M. S. Kheir, Mukhtar Ahmed, and Shakeel Ahmad

Abstract Little millet (*Panicum miliare*) is one of small millets with climate resilience, high nutritional value and tolerance to drought. It has great potential to improve food security especially in marginal, rain-fed and climate stressed agro-ecosystems. Little millet is a yield stabilizer crop because of its short growth cycle (70–90 days), adaptability to a variety of soils and C_4 photosynthetic efficiency. Research on genetic resources, breeding and genomics have created better cultivars that have a better yield, stress resistance and nutritional value. The main objective of this chapter is to evaluate role of little millet in sustainable food systems by using different climate resilient cropping systems. For this purpose, an analytical approach with integrating research studies on little millet incorporation in different farming regimes was used. These studies used variety of cropping methods such as crop rotation, intercropping and conservation agriculture which increased efficiency of land-use, ecological sustainability and climate resilience. Field studies indicate that little millet has water use efficiency values ranging from 3.5–4.8 kg grain/mm in drylands. Multi-stress trials also reported that yields of 0.8–1.2 t/ha under combined drought and heat conditions. Measurable improvements are also noticed in soil carbon (0.10–0.28 t C/ha/year) in rotation systems. Overall, strategic intervention in breeding, value chain development, policy support and consumer awareness can open the full potential of little millet and make it a central part of resilient, nutrient-rich and sustainable food systems7.

Z. Fatima (✉) · M. Asif · U. Farooq · M. Ahmed
Department of Agronomy, Pir Mehr Ali Shah Arid Agriculture University, Rawalpindi, Pakistan
e-mail: zohaafatimaniazi@gmail.com

A. M. S. Kheir
Julius Kühn Institute (JKI)—Federal Research Centre for Cultivated Plants, Institute for Strategies and Technology Assessment, Kleinmachnow, Germany

Soils, Water and Environment Research Institute, Agricultural Research Center, Giza, Egypt

S. Ahmad
Institute of Agronomy Faculty of Agricultural Sciences and Technology, Bahauddin Zakariya University, Multan, Pakistan

M. Ahmed (ed.), *Climate Change and Neglected Underutilized Crops*,
https://doi.org/10.1007/978-3-032-21005-0_14

Keywords Little millet · Landrace variation · Food security · Climate resilience · Sustainable agriculture

14.1 Introduction

Little millet (*Panicum sumatrense)* sometimes is also called *Panicum miliare*. It is particularly worth mentioning among the small millets as it has agronomic features such as short growth cycle of 70–90 days, drought tolerance, low input needs. It also has nutritional and health benefits which are desirable to consumers. It can thrive in unfavourable environments especially in marginal, rain-fed and climate stressed agro-ecosystems. Its potential is not being fully exploited due to low awareness, lack of breeding investment, and poor value chains (Shekhar et al., 2023).

The world's food and agriculture have an endemic issue such as to nourish the expanding population and to endure climate change, water shortage, soil erosion and micronutrient malnutrition. Recent studies reconsidered underutilized crops particularly those with resistance to abiotic stresses and those with better nutritional value to achieve food and nutrition security. Little millet is one of those crops and is increasingly being spotted as a possible contributor to more diversified, climate resilient and nutritionally valuable food basket. The policy movements such as United Nations (UN) declaring the year "2023" as the "International Year of Millets", are helping to draw such grains back to the agendas of the world (Muskan et al., 2025).

Due to little millet's C_4 physiology, it has high water-use efficiency because of low stomatal conductance and efficient CO_2 concentration mechanisms. This makes it drought tolerant and has low water and fertility requirements. These tolerance traits are particularly relevant under climate change, as many dryland regions are experiencing shorter rainy seasons, delayed monsoons, and prolonged mid-season droughts (Bandyopadhyay et al., 2017).

This chapter attempts to provide an overview of the potential of little millet under resilient cropping regimes to promote food security particularly in marginal and rainfed agro ecosystems. It covers little millet's botanical traits, its agronomic performance, nutritional value, role in food security using different cropping systems. It also evaluates the studies of modeling and remote sensing for its assessment. The chapter also integrates the existing empirical evidence with the progressive strategies to underline the importance of little millet's scale up and contribution to sustainable food systems and climate resilience.

14.2 Botanical Identity, Taxonomy, and Distribution

14.2.1 Taxonomy and Botanical Description

Panicum miliare (little millet) is also given scientific name as ***Panicum sumatrense*** (Ratnam et al., 2024). The species is tetraploid and belongs to the family Poaceae. The plant is morphologically annual grass with height usually between 30 cm and 1 m depending on the cultivar and the growing conditions. The leaves are linear with hairy laminae, and the inflorescence is a panicle. Grains are very small, round, sub-spherical and smooth. These botanical characteristics of little millet are an indication of adaptation to diverse agroecological environments particularly marginal and semi-arid soils (Shekhar et al., 2023) (Fig. 14.1).

Fig. 14.1 Morphology of little millet (*Panicum miliare*) (Reddy, 2024)

Table 14.1 Phenological stages of Little Millet (**Panicum miliare**) (Vivekavadhani et al., 2024)

BBCH stage no.	Stage name (adapted)
00–09	Germination and emergence
10–19	Seedling
20–29	Tillering
50–59	Inflorescence emergence
60–69	Flowering (anthesis)
70–79	Grain filling/milk-dough stage
80–89	Maturation and harvest readiness

14.2.2 Phenology

A relatively short and flexible life cycle of 70–90 days of little millet compared to most major cereals is a characteristic feature of its life cycle. This helps it to adapt effectively in short or intermittent rainfalls as well as in rainfed and marginal agro ecosystems. Studies based on the germplasm evaluation revealed that across a variety of accessions of little millet, days to flowering varied widely (38 days to 97 days) indicating a high degree of phenological plasticity. Such diversity in terms of developmental duration is a beneficial attribute. Under favourable circumstances, early maturing accessions will be able to complete their process more rapidly, avoiding late-season droughts or heat. Therefore, phenology is an important area of study that is required in management, breeding, crop modeling, and remote sensing-based monitoring (Pahari et al., 2024) (Table 14.1).

Both graphs in Fig. 14.2 show a positive relationship of average cumulative growing degree days (CGDD) with DAS and BBCH. As the average CGDD increases with the increase of DAS and BBCH scale, this shifts phenological stage of little millet from vegetative stage to reproductive stage (Fig. 14.3).

14.2.3 Distribution and Agro-Ecological Spread

Little millet is commonly grown in India (primarily central and southern (including Madhya Pradesh, Chhattisgarh, Odisha, Andhra Pradesh, etc.) and in areas of Sri Lanka, Nepal and other parts of South/Southeast Asia. Little millet can thrive well with the rainfed conditions due to its ability to tolerate low fertility soils, water stress and a relatively short growth cycle. Therefore, it thrives well on the uplands of the hilly marginal lands where major cereals are not readily manageable (Ratnam et al., 2024).

However, area of cultivation reduced over the decades, as farmers were focused to high yield staples (rice, wheat, maize), mechanization and market demands. Recent climate stress, soil degradation, rediscovery of crop diversity and nutritional security

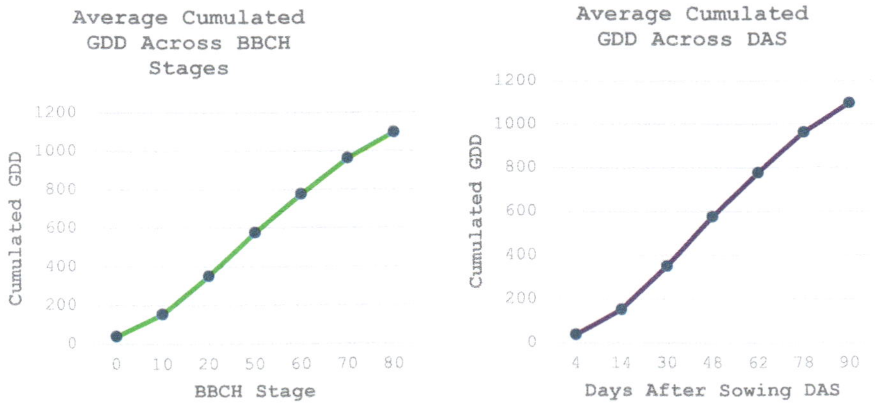

Fig. 14.2 The average CGDD across DAS necessary for little millet to reach the same BBCH stage (Barad et al., 2024)

has brought interest in little millet and other small millets (Pramitha et al., 2023) (Fig. 14.4).

14.3 Agronomic Traits and Climatic Resilience

14.3.1 Short Growth Cycle

The fact that it takes a short time by little millet to mature is one of the most significant agronomic benefits of this crop. Most cultivars take 90 days to mature after sowing, with certain lines taking a shorter time (i.e. 75–80 days) to mature under favourable conditions. Little millet can be used as a catch crop, a relay crop or can be sown within a multi-crop system, particularly in rainfed or haphazard rain systems. It also implies that in case of monsoon being unpredictable, farmers can take advantage of the brief periods of moisture in the soil or the left-over moisture and minimize the chances of ending up with a crop failure (Swamy, 2023).

14.3.2 Environmental Requirements and Water Use Efficiency (WUE)

Little millet (***Panicum miliare***) has been commonly known as a cereal that requires low water hence very favourable in rainfed and dryland farming. This crop can be reliable in dryland regions of India and Africa with 300–600 mm of seasonal rainfall

Fig. 14.3 Phenological stages with BBCH codes for little millet (Reddy, 2024)

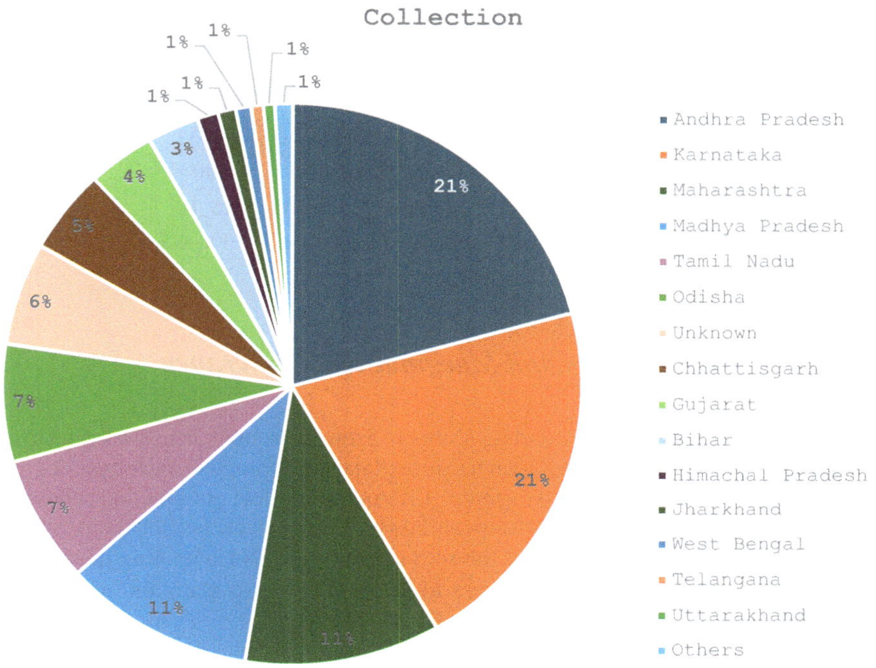

Fig. 14.4 Germplasm distribution of little millet in India (Arya et al., 2025)

that is far lower than rice (1200–1800 mm) or wheat (600–1200 mm) (Kumar et al., 2018).

It's short growth cycle, reduced size of canopy and its C_4 photosynthetic pathway contribute to the water-use efficiency (WUE). This enables the crop to yield moderate even with a restricted supply of moisture. Research on the little millets has shown that the efficiency of transpiration and accumulation of biomass per unit of water used by them is much more than most of the major cereals. These characteristics make little millet productive even when rain fall is inconsistent or unevenly distributed during the season of crop. Even though there is a limited report of specific WUE of little millet because it is an under-studied orphan crop, field observations have regularly reported its potential to achieve its life cycle without consuming one-tenth of the water needed by traditional cereals. This makes little millet an important candidate of climate-adaptive cropping systems especially in semi-arid and sub-humid areas (Ishaq et al., 2025) (Table 14.2).

Table 14.2 Temperature, WUE and water requirement of little millet (Kumar et al., 2018; Vetriventhan et al., 2025)

Parameter	Temperature Range (°C)
Germination optimum	25–35
Vegetative growth optimum	28–34
Grain filling/reproductive optimum	26–32
Functional photosynthesis up to	~ 40
Minimum temperature for growth	15–18
Seasonal water requirement (mm)	300–600
WUE (kg ha^{-1} mm^{-1})	5

14.3.3 Drought, Stress and Marginal Land Tolerance

Little millet has shown an outstanding ability to grow in drought, heat stress and even in low fertility soils. It is very tolerant because of its morphological and physiological features such as high rate of early growth, tillering, wavy leaf cuticles and robust C_4 metabolic pathway. This pathway keeps carbon fixation even in elevated temperatures and dry surroundings. Plasticity in the root system assists in obtaining deeper soil moisture during terminal drought stress to allow the crop to maintain grain filling when the surface is dry (Bandyopadhyay et al., 2017).

The short growth period of little millet (80–100 days) enables it to avoid heat stress in late seasons. The remobilization of stored carbohydrates improves the resistance of little millet to moisture stress in the reproductive phase. It has also been confirmed in genetic research that some transcripts for stress response in little millet help in climate-resilient agriculture to adapt in drought, saline and oxidative stress (Das et al., 2020).

Little millet tends to be a good crop in multi-stress conditions when crops such as maize or rice have completely failed. Little millet can survive in shallow, gravelly, lateritic and upland soils. This has brought into focus its suitability to marginal land areas where the traditional cereals can no longer be profitably grown. Such tolerations are especially applicable in the case of climate changes as most of the dryland areas are facing shorter rainy periods, late monsoons and long mid-season droughts (Sharma, 2022) (Table 14.3).

Table 14.3 Stress tolerance of little millet (Vetriventhan et al., 2025)

Parameter	Temperature Range (°C)
Heat tolerance threshold	Up to 38–40
Critical high temperature	> 42
Functional photosynthesis up to	~ 40
Cold sensitivity threshold	< 12–14
Frost tolerance	Very low
Yield decline per + 1 °C warming	1–2%

14.3.4 Soil Requirements and Low Fertility Tolerance

Little millet displays exceptional adaptability to diverse soil types, thriving in environments where many staple cereals perform poorly. It grows well on shallow, sandy-loam, red, lateritic, gravelly and even acidic soils. This makes it suitable for cultivation in upland and low-input farming systems. It has moderate nutrient requirements that can be grown at acceptable yields without extensive fertilizer application. It often does better than other cereals in low-nitrogen and low-organic-matter conditions. The root system of the crop is fibrous and improves soil structure by minimizing erosion in sloppy and vulnerable areas. The fact that it can grow grain on soils with a pH of between 5.0 and 7.5 also increases its flexibility (Reddy, 2024).

Little millet is productive in nutrient-deficient conditions particularly when cultivated as a component of crop rotations with legumes. The crop is also beneficial to the health of the soil by adding organic matter and the nutrient mining is lower than the high-input cereals. Little millet does not lead to soil carbon accretion and water retention, a key characteristic in long-term sustainability in dryland ecosystems because it leaves behind a large root biomass. Its suitability to intercropping systems (e.g., with pulses such as pigeon pea, cowpea or green gram) increases the efficiency of land use, biological fixation of nitrogen and the performance of the entire system. These benefits associated with soil pose the importance of soil in cultivating crop as it is not only capable of surviving in poor soils, but also it has a beneficial effect on the restoration of soil and ecological sustainability (Vivekavadhani et al., 2024) (Table 14.4).

Table 14.4 Soil property requirements and response of little millet (Himasree et al., 2021)

Soil Property	Requirement
Soil type	Shallow, sandy-loam, red, lateritic, gravelly soils
Soil pH range	5.0–7.5
Organic matter requirement	Low
Nitrogen requirement	20–40 kg/ha (optimal)
Soil property	*Response*
Phosphorus and potassium	Low to moderate
Soil depth tolerance	High
Salinity tolerance	Moderate
Erosion control potential	High

Table 14.5 Macronutrient composition (per 100 g dry weight) of little millet (FAO, 2023)

Component	Per 100 g dry weight
Energy	~ 353 kcal
Carbohydrates (available digestible)	~ 66.2 to 68.1 g
Dietary fiber (total)	~ 7.6 to 15.8 g
Protein	~ 7.7 to 9.4 g
Fat (total lipid)	~ 3.9 to 4.7 g

14.4 Nutritional Value and Health Benefits

The health-promoting value and nutrient value of little millets can be regarded as one of the largest factors leading to re-emergence of interest in it. Recent studies have extensively studied their macronutrient, micronutrient and phytochemical composition (Pramitha et al., 2023).

14.4.1 Macronutrients

Macronutrients in small millet such as carbohydrates, proteins, fats, and dietary fiber are important in the daily energy and physiological needs. Complex starches (mainly carbohydrates) give a slow and sustained energy intake. Therefore, little millet would be an ideal staple food that allows glycaemic stability in people with diabetes or metabolic disorders. Its intermediate level of protein concentrations helps in repairing the tissues, enzyme actions and the sustenance of body mass. Little millet is particularly vital because of its protein content in communities that consume cereal-based diets in large quantities with low levels of pulse consumption (Ratnam et al., 2024).

The lipid fraction is small in little millet, but it supplies the required fatty acids that enhances cell integrity and metabolism. Moreover, the grain is characterized by high levels of dietary fibre that enhances healthy bowel movements, helps to increase gut motility, decrease cholesterol absorption and long-term prevention of cardiovascular diseases. All these macronutrients put little millet in a position to be an energy and nutrition-secure cereal to vulnerable populations (Swamy, 2023) (Table 14.5; Fig. 14.5).

14.4.2 Micronutrients

Little millet has high amounts of vital micronutrients which include iron, zinc, magnesium, phosphorus and potassium which have crucial biochemical and physiological functions. Iron plays a critical role in the synthesis of haemoglobin and immunity.

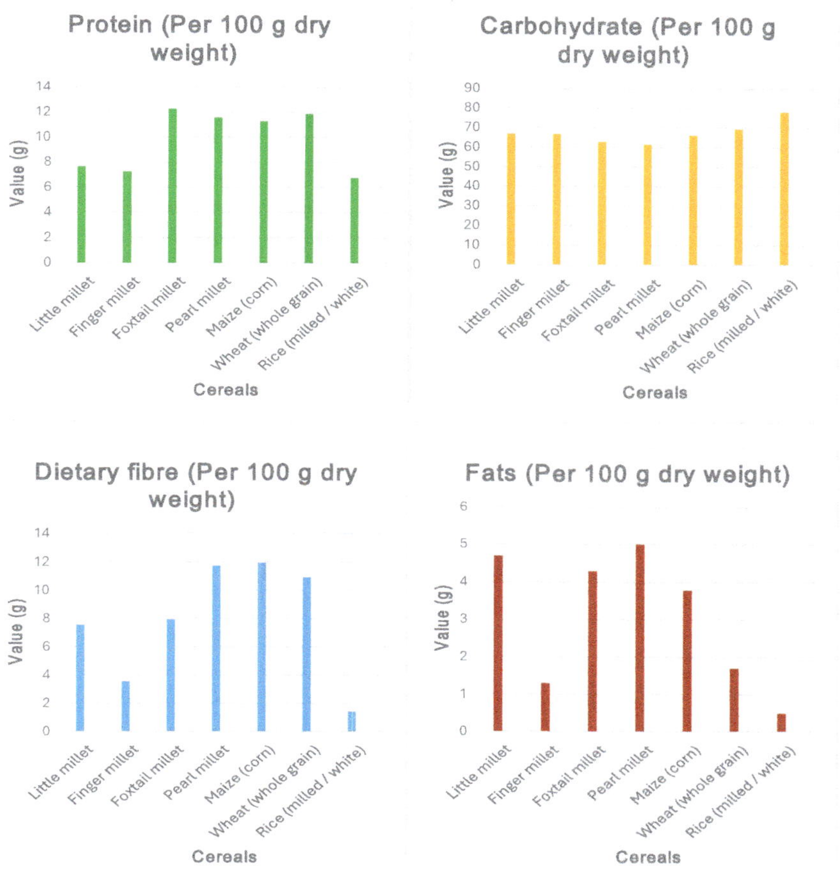

Fig. 14.5 Macronutrient (protein, fats, carbohydrates and crude fibre) composition of little millet versus other cereals (FAO, 2023)

Calcium is beneficial for bones. Zinc helps in boosting immunity, enzymatic activities and heals wounds, which makes the immune system stronger and more balanced to the metabolism. Magnesium is a major substance in the transmission of nerve, muscle contractions, as well as blood pressure regulation. Phosphorus helps in the formation of bones and teeth and in energy metabolism via ATP. Potassium assists in keeping the electrolyte balance and cardiovascular activity. Even though the micronutrient content differs among genotypes, their total sum contributes to little millet being a useful grain to combat hidden hunger (Shekhar et al., 2023) (Table 14.6; Fig. 14.6).

Table 14.6 Micronutrient composition (per 100 g dry weight) of little millet (Jacob et al., 2024)

Component	Per 100 g dry weight
Calcium (Ca)	~ 16 to 20 mg
Iron (Fe)	~ 8 to 9.3 mg
Magnesium (Mg)	~ 114 mg
Phosphorus (P)	~ 207 to 220 mg
Potassium (K)	~ 195 to 250 mg
Zinc (Zn)	~ 1.2 to 1.9 mg
Ash content	~ 5.4 g

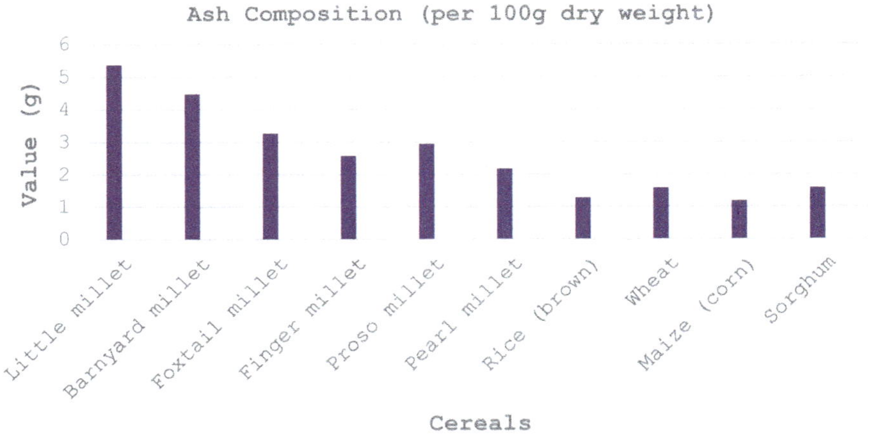

Fig. 14.6 Micronutrient Ash composition of little millet versus other cereals (Meti et al., 2024)

14.4.3 Bioactive Compounds and Phytochemicals

Little millet contains a rich reserve of bioactive substances which play an important role in health protection, including phenolics, flavonoids, tannins and phytates and antioxidants. Phenolic compounds (1021.9 ± 1.4 mg/kg) and flavonoids are natural antioxidants that counteract the harmful free radicals and alleviate oxidative stress. These in turn reduces the prevalence of the chronic diseases, including diabetes, cardiovascular disease, cancer, and neurodegenerative disorders. It also has anti-microbial and anti-inflammatory actions that help the immune system to be more resilient. Although phytates and tannin are regarded as anti-nutritional factors they do also have antioxidant and metal-chelating properties (Pramitha et al., 2023).

A combination of these bioactive compounds can improve the functional food properties of little millet. It is not only a source of calories but also a cereal with a major potential of long-term health benefits and prevention of lifestyle diseases. Due to these characteristics and the fact that all millets are gluten-free, little millet are increasingly being viewed as appropriate to individuals with dietary limitations, diabetics, and health conscious consumers (Pramitha et al., 2023).

14.5 Genetic Resources, Breeding, and Genomic Advances

14.5.1 Germplasm Diversity and Genetic Resources

Little millet has over 3000 accessions in the gene banks of the world. These are landraces, developed lines and the wild relatives. In India, where small millets have been a focus of diversity, much of this diversity is conserved at national gene banks and overseas institutes. The little millet's diversity is a source of genetic reservoir of varieties with traits of yield, nutrient content, abiotic and abiotrophic stress tolerance and adaptation to different agroecology. This genetic diversity is critical to grow little millet on a long-term basis and in fluctuating climates. With broad germplasm, breeders would have more opportunities to combine required qualities (Ratnam et al., 2024) (Fig. 14.7).

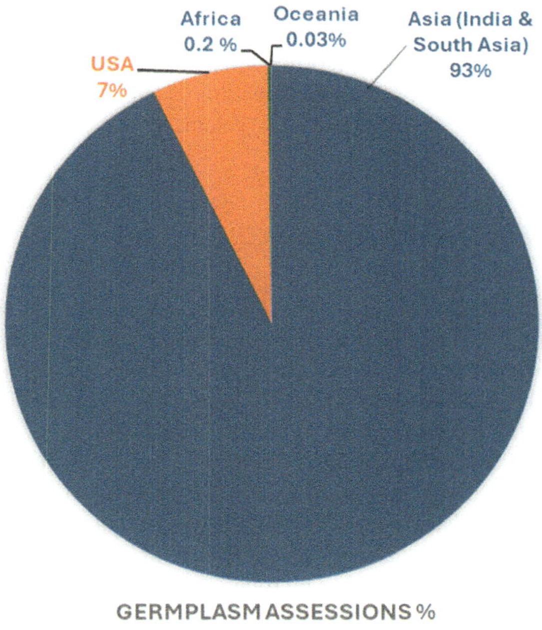

Fig. 14.7 Global little millet's germplasm assessions distribution (Ratnam et al., 2024)

14.5.2 Breeding Approaches

In past, the traditional breeding (selection and crossbreeding between landraces) has been employed to produce improved little millet lines, this process has been slow because of the lack of research focus. The application of genomics, transcriptomics, and molecular marker technologies in recent times has increased the rate of improvement. According to research, an entire transcriptome database of little millet-LMTdb has been created providing an important functional genomics and breeding resource (Shekhar et al., 2023).

Cross-transferability studies have demonstrated that SSR (simple sequence repeat) markers that have been developed in little millet provides a short-cut to improve genetic research and marker-assisted selection. Recent surveys showed that a new generation of better cultivars for current demands can be generated by incorporating the methods of omics with the mainstream breeding technique (Mishra et al., 2024) (Fig. 14.8).

14.6 Resilient Cropping Systems and Sustainable Agriculture

Little millet can be incorporated into diversified cropping regimes and can increase its yield. It may be effective as ecological and nutritional buffer particularly in marginal or climate-stressed agro-ecosystem.

14.6.1 Crop Rotations

Crop rotation is basic principle of sustainable and cropping system. Little millet is suitable for diversified cropping rotations as it has a short growing cycle (70–90 days), minor nutrient requirements and adapted to in marginal areas. Growing of little millet after leguminous crop like pigeon pea, cowpea, or green gram can improves soil fertility. This happens by the biological fixation of nitrogen by the leguminous crop and reduces the need to use external fertilizer (Sharmili & Manoharan, 2018).

Similarly, fungi and diseases can be reduced by planting small plots of little millet before oilseeds crops (sesame or groundnut). Little millets do not host many of the major pathogens and insects as cereals such as rice and wheat. Little millets have also been suggested to be rotated to minimize the weed pressure because they rapid early growth and canopy formation which reduces herbicide dependence. Rotational diversification with little millet also contributes to the stability of the yield in rainfed systems. It also makes sure that the monsoon period is fully used and risks of total crop failure are reduced (Pramitha et al., 2023).

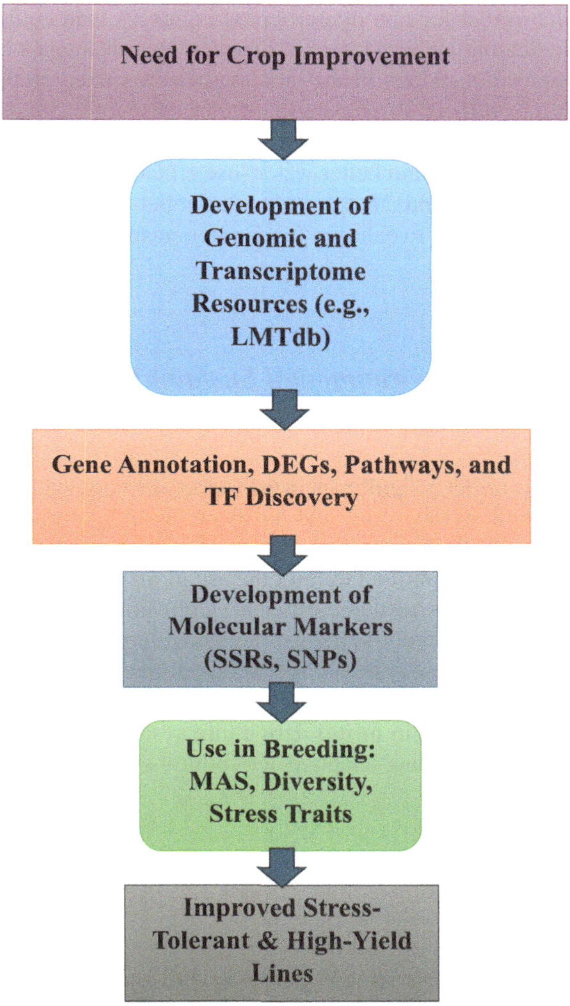

Fig. 14.8 Roadmap of genomic and transcriptomic advances in little millet (Shekhar et al., 2023)

14.6.2 Intercropping and Mixed Cropping

There is a high compatibility of little millet for intercropping with legumes and coarse grains as it has a short period, low requirement of input and a wide range of adaptability. It has a good root system and a small canopy. This allows little millet to be complementary in resources-partitioning and efficiently use space, nutrients and soil moisture. It can be intercropped with crops like horse gram, black gram, red gram, finger millet or sorghum. Legume intercropping is also quite beneficial since the legumes add nitrogen to the soil and enhance the biological activity and little millet offers early ground cover to the soil to prevent weeds. Research has revealed that millet-legume intercropping enhances land equivalent ratio (LER) by 1.2–1.6 indicating evident forms of productivity advantages over monocropping. Likewise, the ecological resilience increases due to mixed cropping of little millet with other minor millets, dispensing the threat of abiotic stress. In research, it was indicated that mixed millet systems tend to reduce soil erosion, improve the storage of organic carbon and attain better water -use efficiency under drylands. These studies support the world trends that promote the millet-based intercropping as a climate-resilient methodology to enhance food system resilience in areas prone to droughts (Sharmili et al., 2022).

14.6.3 *Environmental Sustainability and Resource Use Efficiency*

Little millet is significant in the aspect of ecologically sustainable agriculture because it requires less water and fertilizer. Water scarcity, soil degradation and increased input costs are threatening food systems of the world. Crops such as little millet have been identified to limit the use of irrigation and chemical inputs hence production becomes resource efficient. Cultivation of it also enhances soil structure and organic matter due to regular biomass return particularly in rotation and intercropping systems. These systems also encourage positive microorganisms in soil, reduces pests and disease build-up and improves overall ecosystem functioning. Therefore, the use of little millet in mixed agriculture systems can help in sustainable intensification and more climate-adaptive agricultural systems (Choudhary et al., 2023).

14.6.4 *Food Security and Yield Potential*

Introducing little millet into the resilient agricultural systems helps in ensuring food and nutrition security. Nutrient-rich little millet is helpful in preventing hidden hunger which is widespread in many low-income areas. Little millet can be associated with the development of value chains like processing of flour, multigrain food and small local food businesses. This can help smallholders to provide healthier alternatives to grain to rural and urban markets. This improves yield and at the same time provides nutritious foods which increases dietary diversification and enriches the long-term population health. With these combined functions, little millet enhances food security on the production and nutrition aspects (Ceasar & Baker, 2025).

In the past, little millet was relatively low in productivity as compared to major cereals which somehow shifted attention from its production. Nevertheless, current agronomic and breeding studies show that there is a possibility of increased yield with better management and better varieties. According to a review of genetic improvement (2024), implementation of genomic technologies has resulted in the creation of

Table 14.7 Yield of little millet under diverse cropping systems (Barad et al., 2024)

Condition/Practice	Grain yield (kg/ha)
Rainfed region	1115
Intercropping little millet + black gram	1326
Little millet (mulching/minimal tillage)	2260
Alternative foliar/nutrient treatments	2093–2149 (grain)
Improved cultivar (DHLM-14-1) with multi-location trials	4241 (42.41 q/ha)

opportunities to develop better cultivars with improved yield, stress tolerance, and nutritional value (Mishra et al., 2024).

Additionally, its yield in stress environments is more stable because of its climate resilience. This might result in a lower absolute yield than the high potential cereals in an ideal condition. But the risk of complete crop failure is much less which is more important in marginal and smallholder situations. This stability in yield in unfavourable and fluctuating situations highlights little millet's importance in food security (Baduni et al., 2024) (Table 14.7).

14.7 Crop Modelling for Little Millet

14.7.1 Crop Modelling Approaches

Crop modelling can be used for assessing the response of little millet to diverse conditions in the environment such as temperature, rain distribution, soil limitation and nutrient availability. In comparison with better-investigated cereals, little millet does not have all the parameter sets necessary to operate the models of DSSAT, APSIM, INFOCrop, and AquaCrop. Parameters that are missing are cardinal temperatures that germinate, radiation use efficiency and leaf-area dynamics, root architecture and biomass partitioning coefficients. Scientists have been working to address these shortcomings by synthesizing field level physiological data and calibrating models and optimising them using Bayesian optimization. These developments have increased precision and accuracy of simulation and prediction for the stages of growth, change in yield and water-use patterns (Sharmili et al., 2022).

14.7.2 Modelling Little Millet Under Climate Change Scenarios

Process based models are very important to understand the climate resilience of little millet by examining its performance in future climatic conditions. The results of the simulations with RCP 4.5 and RCP 8.5 scenarios are consistent in terms of little millet showing relatively stable yields despite the rise in temperature up to 2–3 °C. Findings show that cereals like rice and maize experience a dramatic reduction in yield when exposed to the same condition. The ability of little millet to adapt to terminal heat stress is due to its brief growth cycle and effective water-use physiology. Crop models support little millet as a climate-resistant crop in semi-arid climate and rainfed region (Pramitha et al., 2023).

14.7.3 Modelling Cropping System Interactions and Nutrient Requirements

In addition to climate analysis, crop models can be used to investigate the appropriateness of intercropping design, rotation and nutrient design of little millet. The APSIM simulations show that intercropping of little millet with legumes, like pigeon pea improves nitrogen cycling and boosts productivity of the system by 12–25%. Likewise, nutrient-response modelling of little millet with INFOCROP indicates that the crop has moderate fertilizer requirements. It also responds positively to organic type of manures like farmyard manure, green leaf manure and micro-doses of nitrogen. This helps to build low-input and resource efficient crop production in accordance to the ecological demands of marginal farmers (Sharmili & Manoharan, 2018) (Fig. 14.9).

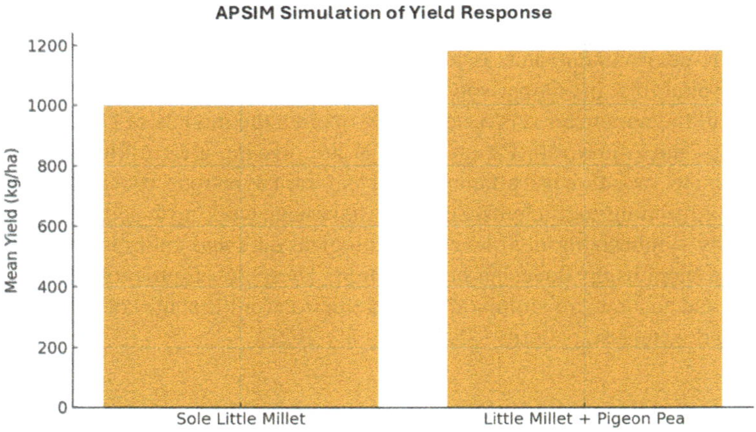

Fig. 14.9 Comparison of simulated APSIM-style yield response of sole little millet and little millet intercropping with pigeon pea (Sharmili & Manoharan, 2018)

14.8 Remote Sensing

14.8.1 Applications in Monitoring Little Millet

The use of remote sensing technologies has revolutionized crop monitoring in agriculture as it can assess large tracts of crops such as little millet on a continuous basis. Sentinel-2, Landsat-8/9, PlanetScope and MODIS are high-resolution multispectral and hyperspectral satellites that offer the spectral reflectance phenomena of the crop. These spectral features can be employed to distinguish little millet and other cereals and to quantify canopy health, biomass levels and photosynthetic activity. This is especially useful since little millet is frequently planted in small fields and mixed cropping systems in which the traditional surveys are time-consuming and inaccurate (Nandini et al., 2025).

14.8.2 Vegetation Indices for Stress and Productivity Assessment

The satellite-generated vegetation indices including NDVI, EVI, SAVI and NDWI provide essential data of crop state, chlorophyll content and water, and drought stress. These indices have been found useful in identifying the signs of water deficiency and nutrient stress in little millet in most cases several weeks before any physical symptoms manifest. This early detection also facilitates timely advisory services to farmers to help them in making decisions that are climate resilient. Moreover, spectral-index-based models of yield prediction also provide useful tools used in predicting production at regional levels. A combination of NDVI time series and soil moisture with topography to predict yields in little millet fields has been demonstrated to be over 80% accurate using machine-learning techniques (Matere et al., 2022).

14.8.3 Integrating Crop Models and Remote Sensing for Climate-Resilient Planning

The methods of data assimilation can include adding satellite-based measurements of leaf area index (LAI) to DSSAT or adding time series of NDVI to APSIM. These integrative methods are found to be far more accurate in growth simulations and seasonal yield forecasts and less predictive uncertainty is observed by 20–40% when using dryland millet-based ecosystem. These methods also support early warning systems, adaptive nutrient management, timing of planting according to climate conditions and drought and heat spatial risks assessment. As a result, the combination of modelling and remote sensing tools improves the scientific foundation to promote little millet as a fundamental part of the resilient cropping systems. This also increases food security and livelihood stability in the susceptible agro-ecosystems (Muthamilarasan & Prasad, 2021).

14.9 Case Studies

14.9.1 Case Study 1: Intercropping Little Millet with Legumes under Rainfed Conditions

Vivekavadhani et al. (2024) conducted field experiment in rainfed conditions to compare the sole little millet with other intercropping treatment of little millet with other pulses. The growth parameters (plant height, and leaf area index), yield parameters (number of tillers per plant, productive tillers, grain yield, stover yields) and nutrient uptake parameters (N, P, K) were measured during the growing season. Out of the treatments, the combination of little millet mixed with black gram in the proportions of 4:1 appeared to be the best. Each plant produced nine tillers on average of which eight of them were productive. The outcome was as grain yield (1326 kg ha^{-1}) and stover yield (4109 kg ha^{-1}). The uptake of nutrients was also improved, with 40.8 kg N/ha, 5.7 kg P/ha and 33.5 kg K/ha. These results prove that intercropping little millet with legumes has a potential to greatly increase the yield and nutrient-use efficiency of crop in rain fed environment. This can provide a smallholder farmer with a stable strategy to diminish the risk of crop failure, increase food grain production and sustain soil fertility.

14.9.2 Case Study 2: Conservation-Agriculture Practices with Little Millet in Rainfed Farming

Sathiya et al. (2024) performed a split-plot experiment comparing various tillage and mulching treatments such as no-tillage, small amount of tillage and mulching to various small millet crops such as little millet. Measurements were grain yield, straw yield, weed population and relative water use efficiency (RWUE). The findings indicated that mulching enhanced the little millet grain production significantly to 1015 kg ha^{-1} which is 24.3–36.1% more than the control test yield. The RWUE increased to 7.03 kg/ha/mm from 5.16 kg/ha/mm in test of mulching. Pressure of the weed was also minimized significantly. The results prove that conservation-agriculture systems in combination with little millet can enhance productivity, water-use efficiency and reduce the cost per the labour and resource requirement.

14.9.3 Case Study 3: Effects of Variety and Row Spacing on Little Millet Performance

Barad et al. (2024) evaluated three little millet varieties (Phule Ekadashi, OLM-203, BL-6) in three row spacing (22.5 × 10 cm, 30 × 10 cm and 45 × 10 cm) using the factorial randomized block design. Growth parameters, biological yield and grain yield were assessed. Results indicate that BL-6 had the best performance with dry matter per plant of 7.85 g, LAI of 4.78 and grain yield of 1308 kg ha^{-1}. When it comes to spacing, the nearest spacing (22.5 × 10 cm) had the highest grain yield (1115 kg ha^{-1}). This study proves that the productivity of little millet under rainfed conditions can be greatly improved by choosing the right variety and spacing of the rows making it more competitive and dependable crop on marginal lands.

14.9.4 Case Study 4: Genetic Variability and Yield-Trait Diversity among Little Millet Genotypes

Sneha et al. (2023) evaluated seventeen landraces and varieties under field conditions to determine the genetic variability (PCV and GCV) and correlation of the traits. The findings showed a moderate-high phenotypic (PCV) and genetic (GCV) variation with low environmental impact, which portrays a high genetic control on the quantitative traits. The qualitative characteristics such as panicle shape, compactness and grain colour varied significantly. Principal component analysis (PCA) revealed that characteristics such as days to 50% flowering, peduncle length and flag leaf length to be important contributors to overall variation hence promising to be targeted in breeding programs. The authors emphasize that little millet landraces contain genetic diversity and trait stability which forms an effective basis to breed superior cultivars. These cultivars can be with increased yield, drought-resistance and adaptation to provide long-term resilience and community-scale food security in unpredictable and marginal environments.

14.10 Challenges and Constraints to Wider Adoption

14.10.1 Limited Research Attention, Genetic Bottlenecks, and Germplasm Underuse

Little millet has long been viewed as an orphan crop with the little financial support, breeding and agronomic research. Relative to other crops such as rice or wheat, little millet has been the focus of fewer omics and functional genomic studies as well as breeding investment. This makes it more difficult to develop improved cultivars to adapt to modern farming and market conditions. This insufficient attention gives a very small range of varietal options to the farmers, inadequate availability of seed and lack of diffusion of better lines (Shekhar et al., 2023).

14.10.2 Agronomic and Post-harvest Constraints

Due to a lack of little millet demand and small-scale production, there is a lack of strong infrastructure in seed production and distribution. Certified seed and mechanization for sowing, harvesting, de-husking and milling of little millet is limited. Moreover, it is traditionally processed manually. Agriculturalists find millets (such as little millet) not to be as productive as rice or wheat, which may not encourage usage, even though they are resilient to stress. Without proper value-addition, storage and processing plant, little millets will always be a subsistence crop, but not a commercial crop.

14.10.3 Institutional, Policy, and Extension Gaps

Millets have never been a prominent national food security program, subsidy program or the focus of an extension campaign unlike staples (rice, wheat, maize). Farmer's outreach and scale-up on little millet is not strong as there is no funding and institutional interest in these activities. Lack of coordinated value chains seed supply, milling, processing, packaging, marketing, etc.—negatively affects the commercial viability of the crop.

14.11 Potential Future Approaches

14.11.1 Strengthening Genomic Research and Breeding

Use of available genetic diversity, germplasm stocks and emerging genomic technologies is important to come up with better little millet varieties. These could have better yield, improved nutritional value and resistance to stress that can enable large-scale farming. According to studies this can be achieved by development of resources like LMTdb (transcriptome database). By promoting marker-assisted selection (MAS), genomic selection and genome-editing to create little millet varieties of the twenty-first century makes it even faster (Kumar et al., 2024).

14.11.2 Efficient Seed Systems and Agro-Value Chains

Community seed banks, decentralized networks of seed production and public–private partnerships for little millet can be established. This can help to generate and supply high quality and certified little millet seed to farmers particularly in areas with marginal land and low input potentials. This promotes marketability, convenience and acceptability particularly to the urban or semi-urban consumers. Encouraging the production, milling and distribution of little millet-based food to markets through farmer producer group (FPO), cooperatives and agro-enterprise.

14.11.3 Raise Awareness for Climate-Resilient Cropping Systems and Improve Consumer Acceptance

Support can be given to planting systems (rotations, intercropping, mixed cropping, etc.) of little millet especially in rainfed, degraded and marginal soils. This can stabilize production, make soil healthy and use resources to the fullest. The extension services are supposed to exhibit proper packages of such systems (sowing time, spacing, pairing crops, residual fertility management, water conservation). Program and policy support e.g. subsidies on millet seeds, value-addition, market access and nutrition programs which incorporate millets should be promoted. Perceptions for little millet can be changed through public awareness and education. By encouraging the production of attractive and convenient food products of little millet (multigrain flour, ready-to-cook, breakfast cereal, snacks) can increase the current interest in little millet production.

14.12 Conclusion

The little millet (***Panicum miliare***) shows great potential as a climate-resistant and nutrient-enriched cereal. It can contribute to food and nutrition security in particularly in marginal, rainfed and stressed agro ecosystems. It is a stable and reliable crop due to its agronomic properties of short growth cycle, high drought tolerance, low requirement of both water and nutrients. Because of its short growth cycle (70–90 days), adaptability to a variety of soils and C_4 photosynthetic efficiency make little millet a yield stabilizer crop. Little millet is a nutritionally rich source of macronutrients, essential minerals, dietary fibre and bioactive compounds. This helps to improve hidden hunger, aiding metabolic health and preventing lifestyle-related chronic illnesses. The little millet grain can be utilized in diversified cropping, which could involve intercropping with legumes, crop rotations, mixed cropping, and conservation agriculture. This boosts productivity, efficiency of use of resources, soil health, and the sustainability of the agro-ecological environment. It can also be proposed that with the usage of crop modelling and remote sensing, accurate tracking, identification of stress, predicting yield and detecting an appropriate cultivation area become available. These could be utilized to maximize production and climate resilience of little millet on a large scale. Little millet has not been adopted because of the lack of research, lack of genetic base exploitation, underdeveloped seed systems, mechanization problems and lack of awareness by consumers. These gaps can be addressed using more powerful breeding, genomic treatments, powerful agro-value chain, policy support and public sensitization programs. By making little millet a commercially viable and healthy food plant, it can take a leading part in creating resilient, diversified and sustainable food systems. This will also benefit vulnerable populations by improving livelihoods, health and nutrition security.

References

Arya, L., Singh, M., Gomashe, S. S., Pandey, C. D., Semwal, D. P., Verma, M., Singh, M., Singh, B., Ganapathy, K. N., Pandey, S., & Bajpai, A. (2025). Little millet. In K. Tripathi, J. Kumari, S. Pandey, B. Singh, S. Archak, & G. P. SIngh (Eds.), *Plant Genebank utilization for trait discovery in millets* (pp. 141–158). Springer Nature Singapore. https://doi.org/10.1007/978-981-96-4043-0_5

Baduni, P., Maikhuri, R. K., Bhatt, G. C., Rawat, H., Singh, R., Semwal, C. P., & Meena, A. K. (2024). Contribution of millets in food and nutritional security to human being: Current status and future perspectives. *Natural Resources Conservation and Research, 7*(1), 5479. https://doi.org/10.24294/nrcr.v7i1.5479

Bandyopadhyay, T., Muthamilarasan, M., & Prasad, M. (2017). Millets for next generation climate-smart agriculture. *Frontiers in Plant Science, 8*, 1266. https://doi.org/10.3389/fpls.2017.01266

Barad, S., Karanjikar, P., Sabale, A., Sb, S., & Gawande, M. (2024). Growth and yield of little millet (*Panicum sumatrense* L.) as influence by varieties and row spacings under rainfed condition. *International Journal of Research in Agronomy, 7*(11S), 29–32. https://doi.org/10.33545/2618060X.2024.v7.i11Sa.1926

Ceasar, S. A., & Baker, A. (2025). Millets for food security and agricultural sustainability. *Planta, 261*(2), 39. https://doi.org/10.1007/s00425-025-04613-4

Choudhary, S., Boruah, A., Ram, N., Gulaiya, S., Choudhary, C. S., Vidhya C. S., & Verma, L. K. (2023). Millet's role in sustainable agriculture: A comprehensive review. *International Journal of Plant & Soil Science, 35*(22), 556–568. https://doi.org/10.9734/ijpss/2023/v35i224165

Das, R. R., Pradhan, S., & Parida, A. (2020). De-novo transcriptome analysis unveils differentially expressed genes regulating drought and salt stress response in Panicum sumatrense. *Scientific Reports, 10*(1), 21251. https://doi.org/10.1038/s41598-020-78118-3

FAO. (2023). About—International year of millets 2023. Food and Agriculture Organization of the United Nations. https://www.fao.org/millets-2023/about/?utm_source=chatgpt.com

Himasree, B., Hemalatha, S., Sumathi, V., Sudhakar, P., & Nagamadhuri, K. V. (2021). Yield improvement and nutrient uptake of little millet (*Panicum sumatrense*) for agronomic interventions. *International Journal of Plant & Soil Science*, 1–9. https://doi.org/10.9734/ijpss/2021/v33i1730544

Ishaq, T., Masood, A., Saad, A., Singh, L., Mehdi, S., Rashid, S., Sharma, A., Aziz, M. A., Rai, U., Qasim, N., Mir, M. S., Palmo, T., & Rafiq, A. (2025). Climate resilience of millets: A sustainable solution for future agriculture. *International Journal of Research in Agronomy, 8*(5), 744–754. https://doi.org/10.33545/2618060X.2025.v8.i5j.2979

Jacob, J., Krishnan, V., Antony, C., Bhavyasri, M., Aruna, C., Mishra, K., Nepolean, T., Satyavathi, C. T., & Visarada, K. B. R. S. (2024). The nutrition and therapeutic potential of millets: An updated narrative review. *Frontiers in Nutrition, 11*, 1346869. https://doi.org/10.3389/fnut.2024.1346869

Kumar, A., Tomer, V., Kaur, A., Kumar, V., & Gupta, K. (2018). Millets: A solution to agrarian and nutritional challenges. *Agriculture & Food Security, 7*(1), 31. https://doi.org/10.1186/s40066-018-0183-3

Kumar, S., Kumar, A., Sen, H., Janeja, H. S., Maity, S., Banerjee, S., Singh, P., & Channapur, A. M. (2024). Small millets: A multifunctional crop for achieving sustainable food security under climate change. *Plant Science Today*. https://doi.org/10.14719/pst.4113

Matere, B. S., Bhavsar, V. V., Mate, P. A., & Girase, V. S. (2022). Genetic variability of yield and its components in little millet [*Panicum sumatrense* L.]. Local Germplasm. *Journal of Agriculture Research and Technology, 47*(03), 258–263. https://doi.org/10.56228/JART.2022.47303

Meti, S. C., Nidoni, U., Hiregoudar, S., Mathad, P. F., Hosamani, A., & Rao, S. N. (2024). Studies on biochemical characteristics of little millet. *International Journal of Advanced Biochemistry Research, 8*(11S), 339–342. https://doi.org/10.33545/26174693.2024.v8.i11Se.2911

Mishra, A., Dash, S., Barpanda, T., Choudhury, S., Mishra, P., Dash, M., & Swain, D. (2024). Improvement of little millet (*Panicum sumatrense*) using novel omics platform and genetic resource integration. *Planta, 260*(3), 60. https://doi.org/10.1007/s00425-024-04493-0

Muskan, F., Redhu, M., Redhu, S., & Rahimi, M. (2025). Millets in the global market: A critical review of challenges and opportunities. *Food Production, Processing and Nutrition, 7*(1), 37. https://doi.org/10.1186/s43014-025-00316-1

Muthamilarasan, M., & Prasad, M. (2021). Small millets for enduring food security amidst pandemics. *Trends in Plant Science, 26*(1), 33–40. https://doi.org/10.1016/j.tplants.2020.08.008

Nandini, C., Mani, V., Joshi, D. C., Maharajan, T., Nandini, B., Bhat, S., Meenakshi, J., Parveen, S. G., Nagaraja, T. E., Divya, M., & Ceasar, S. A. (2025). Little millet. In S. A. Ceasar, S. Penna, C. W. P. Carvalho, & S. M. Jain (Eds.), *Millets: Crops for climate resilience and for food and nutritional security* (pp. 127–156). Springer Nature Singapore. https://doi.org/10.1007/978-981-95-1256-0_5

Pahari, S., Vaid, N., Soolanayakanahally, R., Kagale, S., Pasha, A., Esteban, E., Provart, N., Stobbs, J. A., Vu, M., Meira, D., Karunakaran, C., Boda, P., Prasannakumar, M. K., Nagaraja, A., & Jain, A. K. (2024). Nutri-cereal tissue-specific transcriptome atlas during development: Functional integration of gene expression to identify mineral uptake pathways in little millet (*Panicum sumatrense*). *The Plant Journal, 119*(1), 577–594. https://doi.org/10.1111/tpj.16749

Pramitha, J. L., Ganesan, J., Francis, N., Rajasekharan, R., & Thinakaran, J. (2023). Revitalization of small millets for nutritional and food security by advanced genetics and genomics approaches. *Frontiers in Genetics, 13*, 1007552. https://doi.org/10.3389/fgene.2022.1007552

Ratnam, T. V., Kumar, D. M., & Latha, L. M. (2024). Emerging trends in little millet breeding: From traditional methods to genomic approaches. *Plant Archives, 25*(1). https://doi.org/10.51470/PLANTARCHIVES.2025.v25.supplement-1.004

Reddy, J. (2024). Organic little millet farming: Cultivation practices and production management. *AgriFarming*. https://www.agrifarming.in/organic-little-millet-farming-cultivation-practices-and-production-management

Sathiya, K., Nirmalakumari, A., Sukanya, T. S., Vanitha, C., Karthikeyan, R., & Ayyadurai, P. (2024). Enhancing small millet farming resilience through conservation agriculture in Rainfed regions. *Journal of Experimental Agriculture International, 46*(5), 68–75. https://doi.org/10.9734/jeai/2024/v46i52357

Sharma, S. (2022). Evaluating the performance of little millet (*Panicum flexuosum*) in rainfed agro-ecosystem cropping systems. *International Journal of Agriculture and Nutrition, 4*(1), 91–100. https://doi.org/10.33545/26646064.2022.v4.i1b.157

Sharmili, K., & Manoharan, S. (2018). Studies on intercropping in Rainfed little millet (*Panicum sumatrense*). *International Journal of Current Microbiology and Applied Sciences, 7*(2), 323–327. https://doi.org/10.20546/ijcmas.2018.702.042

Sharmili, K., Parasuraman, P., & Sivagamy, K. (2022). Effect of inter and sequential cropping of pulses in little millet (*Panicum sumatranse* L.) based cropping system. *Indian Journal of Agricultural Research*. https://doi.org/10.18805/IJARe.A-5896

Shekhar, S., Prasad, A. S., Banjare, K., Kaushik, A., Mannade, A. K., Dubey, M., Patil, A., Premi, V., Vishwakarma, A. K., Sao, A., Saxena, R. R., Dubey, A., & Chandel, G. (2023). LMTdb: A comprehensive transcriptome database for climate-resilient, nutritionally rich little millet (*Panicum sumatrense*). *Frontiers in Plant Science, 14*, 1106104. https://doi.org/10.3389/fpls.2023.1106104

Sneha, N. M., Selvaraj, S., Francis, N., & Pramitha, L. (2023). Analyzing the variability parameters of the landraces and varieties of little millet (*Panicum sumatrense* Roth ex Roem. & Schult.). *Electronic Journal of Plant Breeding, 14*(1). https://doi.org/10.37992/2023.1401.040

Swamy, K. R. M. (2023). Origin, domestication, taxonomy, botanical description, genetics and cytogenetics, genetic diversity and breeding of little millet (*Panicum sumatrense* Roth. Ex. Roem. & Schult.). *International Journal of Current Research.*

Vetriventhan, M., Jagadeesh, K., Naresh, D., & Singh, K. (2025). Fodder and grain yield potential of little millet, *Panicum sumatrense* race robusta and trait-specific genetic resources for crop improvement. In Review. https://doi.org/10.21203/rs.3.rs-7373619/v1.

Vivekavadhani, A. R., Mohanapriya, R., Sharmili, K., & Poonguzhali, R. S. (2024). Evaluating Little Millet (Panicum sumatrense L.) Based Intercropping System on Growth, Yield and Nutrient Status of Soil under Rainfed Condition. *International Journal of Environment and Climate Change, 14*(4), 606–613. https://doi.org/10.9734/ijecc/2024/v14i44144

Chapter 15
Kodo Millet (*Paspalum scrobiculatum*): Low-Input Farming and Ecosystem Services

Umar Farooq, Wajahat Hussain, Zohaa Fatima, Mukhtar Ahmed, Shakeel Ahmad, and Muhammad Imtiaz Rashid

Abstract Kodo millet (*Paspalum scrobiculatum* L.) is rapidly gaining recognition as a climate-resistance cereal that could be suitable in low-input and drought prone farming systems and in ecologically vulnerable systems. The increased relevance is motivated by the fact that world requires crops that can support food and nutritional security amidst climatic and resource limitation. In this chapter, the authors summarize the existing knowledge on its taxonomy, phenology, physiological efficiency, biochemical defense mechanisms, ecological benefits and agronomic performance in marginal environments. It further describes the important practices that are oriented to mitigation to include conservation agriculture principles of minimal disturbance soil, retention of residues, and diversified rotations, enhanced nutrient management and stress-reducing mechanisms with osmolytes, antioxidants and molecular networks that control drought resistance and nutrient-use efficiency. The chapter also identifies remote-sensing tools such as Normalized Difference Vegetable Index, Enhanced Vegetable Index, Normalized Difference Water Index, and Photochemical Reflectance Index, which have been used to measure the crop vigor, water status, and stress dynamics. Previous research evidence demonstrates that with refinement, notable productivity and ecological benefits have been obtained. The combination of farmyard manure with seed priming showed a 41.7% grain yield increase, 0.48–0.62% soil organic carbon increase and 32.9% microbial biomass carbon increase. Cowpea intercropping increased the grain yield by 34.5% and improved the soil

U. Farooq (✉) · Z. Fatima · M. Ahmed
Department of Agronomy, Pir Mehr Ali Shah Arid Agriculture University, Rawalpindi, Pakistan
e-mail: umerfaroooq42@gmail.com

W. Hussain
Institute of Geo-Information and Earth Observation, Pir Mehr Ali Shah Arid Agriculture University, Rawalpindi, Pakistan

S. Ahmad
Institute of Agronomy Faculty of Agricultural Sciences and Technology, Bahauddin Zakariya University, Multan, Pakistan

M. I. Rashid
Center of Excellence in Environmental Studies, King Abdulaziz University, Jeddah, Saudi Arabia

M. Ahmed (ed.), *Climate Change and Neglected Underutilized Crops*,
https://doi.org/10.1007/978-3-032-21005-0_15

nitrogen, phosphorus and potassium by 19.8, 14.6 and 11.2%, respectively. The organic Kodo millet systems minimized carbon footprint of 54.2% and nearly doubled the energy-use efficiency. The future directions are further genome-editing methods and digital phenotyping increased to increase cultivar development and resilience.

Keywords Kodo millet · Climate resilience · Low-input agronomy · Drought adaptation · Sustainable systems

15.1 Introduction

Kodo millet (*Paspalum scrobiculatum*) is a small-grained cereal crop whose tradition has been to be grown in semi-arid and marginal land of South and Southeast Asia. It is known to be extremely resilient to poor environmental conditions such as droughts, poor soil fertility, and erratic rainfall. As climate change continues to impact tropical and subtropical areas, Kodo millet has been considered a climate resilient low-input crop that has potential to add to food and nutritional security in vulnerable agricultural systems. It's potential to multiply well with minimal inputs and on small-scale lands makes it a strategic cereal to smallholder farmers who in most instances have poor access to irrigation, fertilizer and other modern farming inputs (Rangaiah et al., 2024).

Kodo millet especially fits in the rainfed and low input farming systems. Its low growing period, drought resistance and water-efficient nature qualify it to fit well in regions where other crop varieties like rice, wheat and maize might fail. Kodo millet is rich in nutrients besides being resilient. It is a viable source of important minerals, dietary fiber and bioactive compounds that help in the prevention of the micronutrient deficiency, the enrichment of the diet of rural populations. This makes Kodo millet fulfill two functions in enhancing food security and human nutrition in marginal areas, respectively (Prashanth et al., 2025).

Kodo millet is cultivated that is closely related to the traditional modes of agriculture and local ecosystems. It has been cultivated mostly through mixed cropping systems with other legumes or other cereals that has enhanced not only the fertility of the soil but also minimizes the pest pressure and productivity of the system. The crop has been largely underused on a mass scale because of a lack of research on it, lack of demand in the market and its inability to fit in the modern agricultural system despite the potential. The awareness about its ecological and nutritional benefits and the creation of sustainable management multiple manners can open all the potential of Kodo millet (Nirubana et al., 2020).

Loss of soil, soil erosion and diminished fertility are among the major problems to the areas where Kodo millet is cultivated. Topsoil may be lost, the water-holding capacity may also be reduced, and the availability of nutrients may be low due to continuous cultivation, absence of soil-conservation practices and unpredictable rainfall. These issues jeopardize crop productivity, level farm profitability and resilience of the smallholder farming systems. Consequently, there is an urgent necessity to implement the methods to keep the soil healthy, stabilize the yields, and increase the ecosystem functions.

The conservation agriculture can be used as a viable method to keep the Kodo millet production. Such principles as minimal soil disturbance, a permanent soil cover, diversification of crops and retention of residues are in line with the ecological needs of Kodo millet. Such practices enhance the soil structure, boost moisture retention, reduce runoff and improve nutrient cycling ultimately leading to an increase in higher and more stable yields. As has been experienced in various parts of the world, conservation agriculture when adopted in millet-based farming in various regions is a method of reducing soil erosion, water conservation, and enhancing the resiliency of the smallholder farmers to climatic changes.

This chapter examines ecological concepts and farming agronomics of the Kodo millet in sustainable farming. It looks at the production patterns on a local, regional and global level, the problems created by the soil degradation and climate variability and analyses the opportunities of conservation agriculture to improve the soil health, water control, and the crop yields. Moreover, the latest innovations in remote sensing and modeling instruments to track the performance of crops and make management decisions in the millet growing landscapes are reported. This chapter offers perspective on how Kodo millet can support resilient and low-input, as well as environmentally sustainable agricultural systems, and the livelihoods of smallholder farmers through the synthesis of existing information on the topic, namely, the role of Kodo millet in infra-red heating influence of amylose and bioactive compounds (Jeeva et al., 2024).

15.2 Botanical and Agronomic Profile of Kodo Millet

15.2.1 Taxonomy, Origin, and Domestication

Kodo millet (*P. scrobiculatum*) is a member of family Poaceae and subfamily Panicoideae and tribe Paniceae. It is among the small types of millet that have co evolved with the South and Southeast Asian human agriculture. It has been proposed (based on archaeobotanical and genetic evidence) that it was domesticated on the Indian subcontinent thousands of years ago, although its wild lineages go back to tropical Africa. During centuries of breeding, a variety of landraces and varieties have developed, which are adjusted to local agro ecological conditions (Vetriventhan & Upadhyaya, 2019) (Fig. 15.1, Table 15.1).

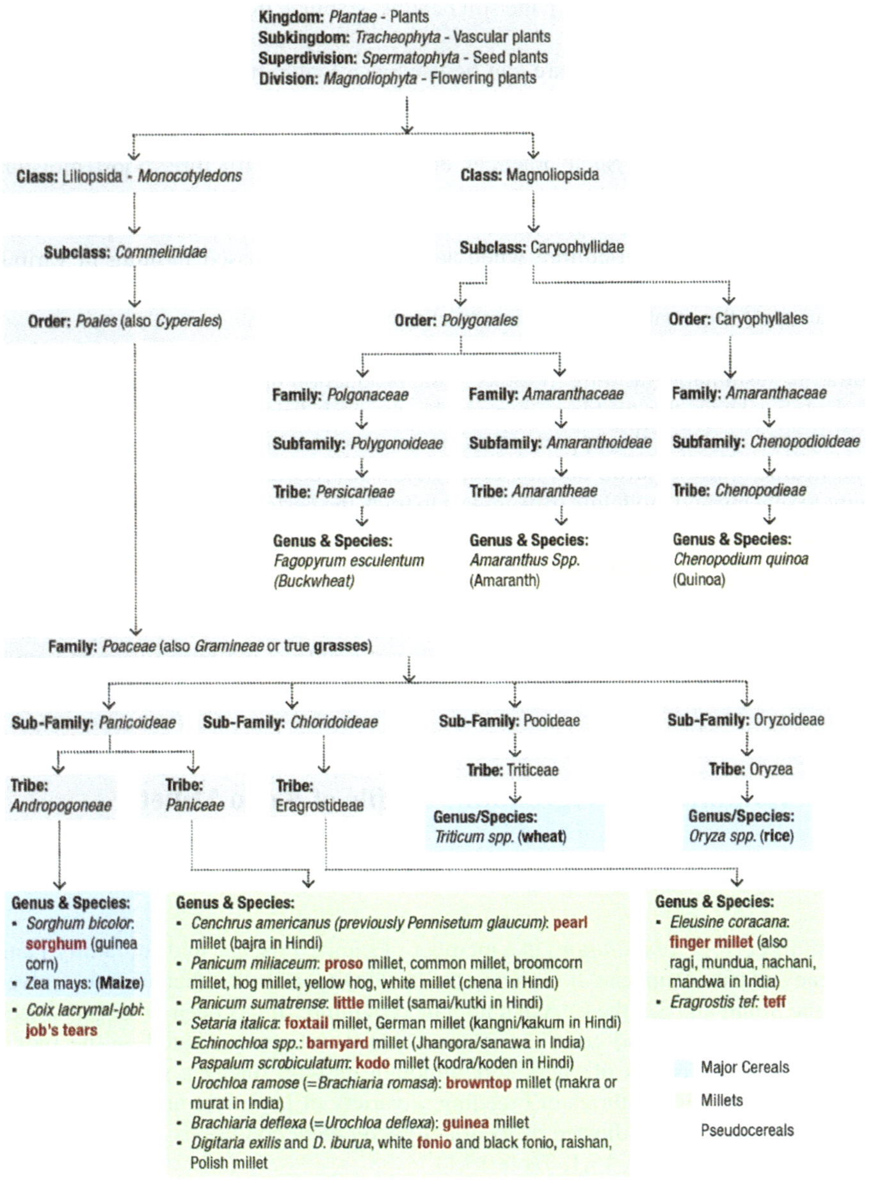

Fig. 15.1 Taxonomical classification of small millets and other major cereals and millets, and pseudo-cereals [*Note* "Millet" is a common term to categorize small-seeded grasses that are often called dryland cereals. The grasses most commonly referred to as millets are: Major millet (pearl millet) and small millets (finger millet, foxtail millet, proso millet, little millet, barnyard millet, kodo millet, browntop millet, fonio, teff and job's tears, and guinea millet; and sometimes sorghum is categorized as a major millet—this is typically included in the common definition of millets in India but less so in other countries]

Table 15.1 The different millet species (information mainly from the USDA germplasm resources information network—GRIN)

Generally used English name	Other common vernacular names	Taxonomy
Finger millet	Ragi, Wimbi	Tribe Eragrostideae *Eleusine coracana* L. Gaertn
Teff	Teff, Teff grass, Abyssinian love grass	Tribe Eragrostideae *Eragrostis tef* (Zuccagni) Trotter
Job's tears	Adlay, Adlay millet	Tribe Andropogoneae *Coix lacryma-jobi* L.
White fonio	Fonio, Acha, Acha millet	Tribe Paniceae *Digitaria exilis* (Kippist) Stapf
Black fonio	Black acha, hungry rice, black acha	Tribe Paniceae *Digitaria iburua* Stapf
Japanese millet	Japanese barnyard millet	Tribe Paniceae *Echinochloa esculenta* (A. Braun) H. Scholz
Sawa millet	Shama millet, awnless barnyard grass, corn panic grass, decan grass, jungle ricegrass, french millet	Tribe Paniceae *Echinochloa colona* (L.) Link
Proso millet	Common millet, broomcorn millet, hog millet	Tribe Paniceae *Panicum miliaceum* L. subsp. *miliaceum*
Little millet	Little millet, panic millet, brown top millet, blue panic	Tribe Paniceae *Panicum sumatrense* Roth
Kodo millet	Kodo millet, creeping paspalum, Indian paspalum, native paspalum	Tribe Paniceae *P. scrobiculatum* L.
Foxtail millet	Foxtail millet, Italian millet, German millet, Hungarian millet, Italian foxtail grass	Tribe Paniceae *Setaria italica* (L.) P. Beauv. subsp. *italica*
Pearl millet	Pearl millet, bulrush millet, cattail millet	Tribe Paniceae *Pennisetum glaucum* (L.) R. Br

15.2.2 Botanical Characteristics and Life Cycle

Kodo millet is an annual grass having tufted culms that may be found erect or decumbent in nature with a height of about 90–100 cm when favorable. Its inflorescence is usually a number of racemes with small and hard-coated grains. The grains are bound up in tenacious husks and the size of the seeds is small (approximately 2–3 mm in diameter) thereby affecting the post-harvest processing demands. Its life cycle between sowing and maturity is comparatively brief—usually between 3.5 and 4 months—and thus is acceptable to practice on rainfed cropping in uncertain rainfall regimes (Mishra et al., 2025) (Figs. 15.2 and 15.3).

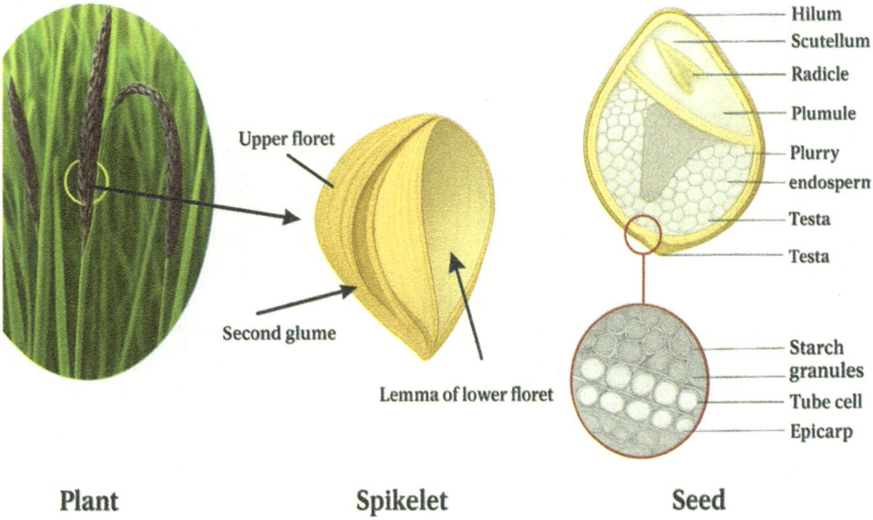

Fig. 15.2 Schematic structure of kodo millet (Shikha et al., 2024)

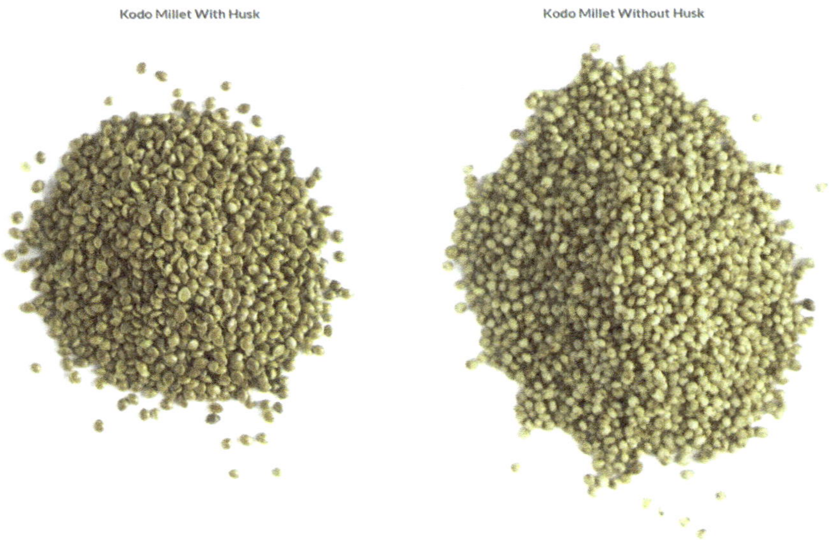

Fig. 15.3 Kodo millet seeds with and without husk (Ceasar & Ignacimuthu, 2010)

15.2.3 Adaptation to Low-Fertility Soils and Harsh Environments

Kodo millet is very tolerant to drought, low soil fertility and low soil structure. It is capable of having a growth on shallow and gravelly soils or light soils where traditional cereals might not develop. It is also well adapted to marginal and semi-arid living conditions because of its ability to survive on limited water and nutrient levels. It has low nutrient requirements and is resilient, and therefore, less external inputs (fertilizers, irrigation) are required, which is why it is a perfect crop in low input farming systems and among farmers with limited resources (Mehta et al., 2024).

15.2.4 Nutritional Composition and Socio-economic Value

The nutritional value of the grain of kodo millet is good: it contains a sufficient portion of proteins, dietary fibers, as well as minerals include iron and calcium. Composition It is a nutritious cereal that can be used to replace major staples in areas that are likely to experience micronutrient deficiencies, being a gluten free option. Other than as a source of human food, kodo millet has socio-economic significance due to its low cost of production, storage flexibility, and application as a backup or subsistence crop in marginal areas. The fact that it has value-added products (e.g., millet based foods) also contributes to its livelihood value to rural communities (Mathur et al., 2025) (Fig. 15.4, Table 15.2).

15.2.5 Key Agronomic Traits Enabling Survival in Low-Input Conditions

The traits of kodo millet that are of great significance in agronomy are; it grows in a short period; it uses water efficiently; it is drought and heat resistant, it yields enough with less fertilizer or less soil fertility; it grows with rainfed, marginal conditions. The genetic variability of its landraces implies that some of the genotypes are more productive, biomass-productive, and resistant to stresses than others. All these characteristics allow using kodo millet as a stable cereal in unfavorable conditions and allow sustaining low input agriculture (Mukherjee et al., 2025).

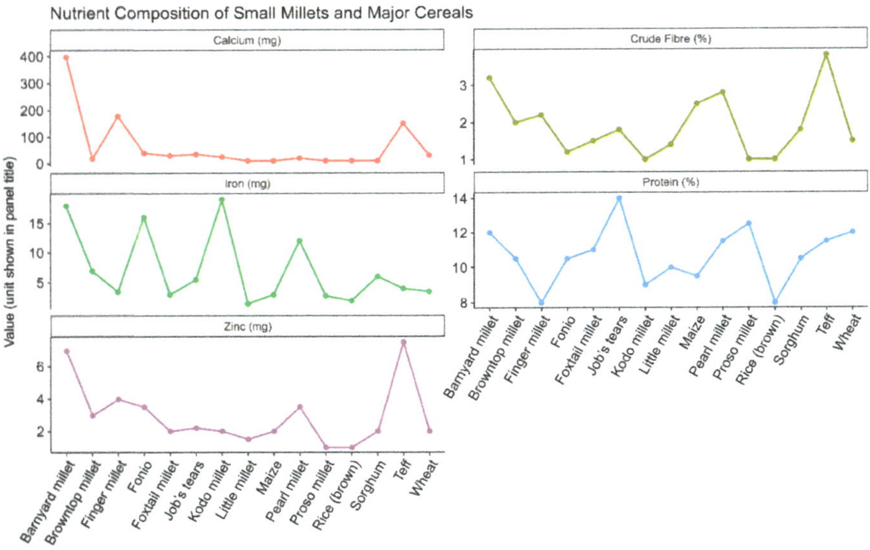

Fig. 15.4 Nutrient composition of small millets in comparison with major cereals (100 g edible part at 12% moisture) (Vetriventhan et al., 2020)

Table 15.2 Nutritional composition of Kodo millet

Component	Mean	SEM	ICMR	FAO
Moisture (%)	7.3	0.18	12.8	–
Ash (g/100 g)	2.6	0.21	2.6	3.3
Crude fiber (g/100 g)	8.5	0.54	9.0	5.5
Fat (g/100 g)	1.29	0.23	1.4	3.6
Protein (g/100 g)	8.03	0.13	8.3	9.8
Carbohydrate (g/100 g)	64.9	1.17	65.9	66.6
Energy (kJ/100 g)	1257	3.18	1282	1402
Iron (mg/100 g)	2.9	0.02	0.5	3.6
Calcium (mg/100 g)	21.7	0.5	27	31

15.3 Phenology, Distribution, and Production Zones

15.3.1 Phenology

The germination of kodo millet occurs easily when the seeds have proper moisture content in the soil and those that germinate grow with relative ease even in low-fertile and marginal soils. High root growth and early vigor assists the crop to withstand the early dry spells and unfavorable soil conditions. As the crop matures, the crop passes into a tillering phase individual crops generate several tillers, which adds to the overall resilience of the biomass and yield under low-input environments. The flowering and the initiation of panicles are initiated when there are good moisture and seasonal conditions. The duration of grain-filling and physiological maturity is dependent on the genotype, as well as the environmental conditions, however most varieties fulfil their life cycle within 3–4 months after sowing. This is a comparatively brief development period, in combination with adaptive phenological alterations, that enables Kodo millet to flourish in rainfall regimes of erratic rainfall, characteristic of semi-arid regions (Mishra et al., 2024a, 2024b) (Figs. 15.5 and 15.6).

15.3.2 Geographical Distribution

Kodo millet is a native of the tropics and subtropics; the records of history and archaeology suggest that it was domesticated in the Indian subcontinent, and has been growing there thousands of years. It has the conventional areas of growth in central, western and southern India particularly in the states of Madhya Pradesh, Chhattisgarh, Karnataka, Tamil Nadu, and Gujarat and so on. Outside India, Kodo millet is grown in subtropical areas of Asia and some parts of Africa but the coverage area in the world is small compared to other major cereals. Its endurance to marginal landscapes is explained by the fact that it grows in low fertility and coarse soils, in rainfed environment (Ghatak et al., 2025).

15.3.3 Production Zones and Practices

The production of kodo millet is largely carried out by small-scale farmers under rainfed low input production within their area of production. The largest producer in the world is India where the crop is concentrated in upland and tribal areas, which have marginal soils. Kodo millet is domesticated in India and cultivated in upland rice regions in India, Indonesia, Philippines, Thailand, Vietnam, Bangladesh, and Myanmar and in western Africa. It has high protein content (11%), low fat (4.2%) and very high fibre content (14.3%). Kodo millet is very easy to digest; it contains a high amount of lecithin. Kodo millet is rich in B vitamins, especially niacin, B6 and folic acid, as well as the minerals such as calcium, iron, potassium, magnesium and zinc In the 2021–2022 season, minor millets that included Kodo occupied approximately 0.44 million hectares with a yield of 0.37 million tons having average productivity of 789 kg/ha. Other states which do a great deal of production are Madhya Pradesh, Chhattisgarh, Karnataka, Andhra Pradesh, Tamil Nadu and Gujarat (Dey et al., 2023).

The yields are moderate because of low soil fertility, low fertilizer application and reliance on rainfall although the performance differs among landraces and lines of improved performance. Studies indicate that the improved agronomic management practices can contribute greatly to increasing the stability of yields and narrowing

a
b
c
d
e
f
g
h

◂**Fig. 15.5** Induction and maturation of somatic embryos in kodo millet cultivar RK98. **a** Initiation of callus from excised shoot apex explants after 1 week incubation in the dark on MS medium supplemented with 9.0 lM 2,4 D. **b** Enlargement of callus after 2 weeks of incubation in the dark on MS medium supplemented with 9.0 lM 2,4 D. **c** Formation of SEs after 7 weeks of incubation in the dark on MS medium supplemented with 9.0 lM 2,4 D and 2.25 lM Kn. **d** Maturation and germination of EC on MS medium containing 4.50 lM TDZ. **e** Shoot and root induction from SEs after 2 weeks of culture in the light on MS medium supplemented with 4.50 lM TDZ and 120 mM maltose. **f** Shoot elongation on MS medium supplemented with 4.50 lM TDZ, 120 mm maltose and 200 lM L-proline. **g** Plants maintained in paper cups covered with polythene bags for hardening. **h** Maturation and seed setting of soil grown plant (Ceasar & Ignacimuthu, 2010)

Fig. 15.6 Phenology of kodo millet (Hariprasanna, 2017)

the disparity between the potential and the actual yield, through the ability to manage spacing, nutrient application, and the use of better varieties. Tradition has the basis of broadcast sowing and low inputs, but better systems have the focus on the establishment of crops, controlled nutrient supply, and soil conservation strategies. All in all, Kodo millet continues to be a good crop in marginal agroecosystems, and it is both resilient and low-input intensive (Kannababu et al., 2023) (Fig. 15.7).

Fig. 15.7 Distribution of Kodo millet in world (Jeeva et al., 2024)

15.4 Physiological Basis of Low-Input Adaptation

Kodo millet is a plant with a mix of physiological characteristics, which enable it to work effectively in low-input, water-controlled, and nutrient-inadequate conditions. It is more resilient due to its good water management, osmotic balance, stable photosynthetic activity, and carbon assimilation, which allows it to survive and perform well in semi-arid areas and on marginal soils where the traditional cereals fail to perform. All these adaptive processes help to establish, grow and develop crops with little irrigation and low fertilizer content (Wang et al., 2024).

15.4.1 Water Relations and Stress Physiology

This crop is incredibly resistant to moisture stress and this is mainly due to the adaptations, which minimize water loss and maximize internal hydration. Evaporative losses are reduced in dry conditions by leaf rolling, cuticular resistance as well as regulative loss through transpiration. It has deep roots, dense roots, this root is highly branched thus able to extract lower layers of soil moisture, and this is very beneficial especially during prolonged dry seasons. Kodo millet has a desirable ratio of root to shoot with a balance of up taking water and less transpiration requirement.

Osmotic adjustment is a major factor involved in stress tolerance because compatible solutes like proline, soluble sugars and organic acids accumulate in cells during the event of dehydration in order to maintain turgor, stabilize membranes and safeguard enzymes. Stomatal control also adds to the water-use efficiency; the stomata can react dynamically to the presence of humidity and water in the soil, and when photosynthesis rates are high enough, the stomata are opened enough, but water

Table 15.3 Water requirement and drought tolerance of millets and major cereals (Tiwari et al., 2022)

Crop	Water requirement (mm)	Drought tolerance
Pearl millet	350–500	High
Foxtail millet	300–400	High
Finger millet	500–1000	Moderate
Proso millet	300–400	High
Barnyard millet	500–700	Moderate
Kodo millet	400–500	High
Little millet	300–400	High
Rice	1000–1500	Low
Wheat	500–800	Moderate
Maize	500–800	Moderate

loss is reduced to the bare minimum by operating the stomata. This accurate regulation enables the crop to continue growing and be physiologically active in either intermittent or extended drought periods (Luo et al., 2024) (Table 15.3).

15.4.2 Photosynthesis and Carbon Assimilation

The photosynthetic function of Kodo millet has been found to be resilient during low levels of fertility and water. The structure and content of the chlorophyll is functional, and the chlorophyll is able to capture light even during environmental stress. To avoid oxidative damage to the species during high temperature or extreme sunlight conditions, the species has photo protective mechanisms, including non-photochemical quenching.

The crop is highly intrinsically water-use efficient with a carbon gain in relation to a low water expenditure. This guarantees further carbohydrate synthesis, biomass accretion and tetanizing grains with low-input conditions. Even in the event of limited nutrient supply, carbon assimilation is facilitated by metabolic flexibility, which is a feature of sustained Calvin cycle activity, stable electron transport and adaptive enzyme activity.

All these attributes provide an innate capacity to survive semi-arid climate, water scarcity, and nutrient deficiencies, and Kodo millet can be a strong candidate as a low-input and resilient agricultural system (Ru et al., 2023).

15.5 Biochemical and Molecular Mechanisms Supporting Low-Input Farming

Kodo millet has an advanced biochemical and molecular defense mechanism that enables this crop to have metabolic stability and productivity under a low-input, nutrient-poor, and limited water conditions. Its resistance is determined by the interaction of oxidative stress response, antioxidant defense, osmolyte build-up, and the expression of stress-sensitive genes. All these mechanisms are useful in ensuring that cellular homeostasis, biochemical integrity, and efficient utilization of resources is maintained; the crop is typically adapted to suit marginal landscapes where conventional cereals usually cannot grow effectively (Zulfiqar et al., 2024).

15.5.1 Oxidative Stress and ROS Management

The plants are usually subjected to high concentration of reactive oxygen species (ROS) including superoxide radicals, hydrogen peroxide and hydroxyl ions when there is a low-input environment especially during drought conditions, nutrient insufficiency and periods of high temperature. The interrupting of the electron flow in chloroplasts and mitochondria, the lack of water, and the irregularities in the metabolism of the nutrient activate the formation of ROS in Kodo millet. Despite the fact that high levels of ROS are known to destroy cellular components, Kodo millet has a physiology that allows it to survive under moderate levels of oxidative stress coupled with defense mechanisms.

Uncontrolled ROS may result in membrane lipid peroxidation resulting in fluidity and integrity loss. Changes these pigments are very sensitive to oxidative stress and they can deteriorate, interfering with photosynthesis. Denaturation or fragmentation can also occur to proteins, especially those that are metabolic enzymes, and those that produce energy. Nevertheless, Kodo millet inhibits this harm by quickly detecting ROS and quickly inoculating protective biochemical processes, stabilizing pigments, membranes and structural proteins (Azar-Kia et al., 1975).

15.5.2 Antioxidant Defense Systems

Kodo millet prevents oxidative stress by a properly coordinated antioxidant defense mechanism, which is a combination of enzymatic and non-enzymatic activity. The enzymatic antioxidants are the superoxide dismutase (SOD) and catalase (CAT) which change the superoxide radicals to hydrogen peroxide and the hydrogen peroxide to water and oxygen respectively. Ascorbate peroxidase (APX), glutathione reductase (GR) also neutralize ROS by the ascorbate glutathione cycle and therefore make sure that any reactive molecule cannot damage the cell.

Table 15.4 Antioxidant profile of different millets (polyphenol and FRAP values) (Almaski et al., 2017)

Millet type	Test	Grain	Flour	Flakes	*P*-value
Pearl millet	Polyphenol	2.15 ± 0.22	1.59 ± 0.23	1.76 ± 0.06	<0.001
	FRAP	4.58 ± 0.64	4.313 ± 0.322	4.737 ± 0.32	0.028
Finger millet	Polyphenol	3.72 ± 0.14	2.55 ± 0.20	1.27 ± 0.14	<0.001
	FRAP	13.98 ± 0.16	9.87 ± 1.02	11.46 ± 1.34	<0.001
Little millet	Polyphenol	1.51 ± 0.21	0.52 ± 0.06	0.75 ± 0.03	<0.001
	FRAP	3.5 ± 0.26	0.64 ± 0.07	1.34 ± 0.06	<0.001
Barnyard millet	Polyphenol	1.41 ± 0.11	0.51 ± 0.03	0.69 ± 0.03	<0.001
	FRAP	2.69 ± 0.37	0.55 ± 0.07	1.08 ± 0.06	<0.001
Foxtail millet	Polyphenol	0.98 ± 0.06	0.29 ± 0.02	0.63 ± 0.05	<0.001
	FRAP	1.89 ± 0.14	0.35 ± 0.05	1.12 ± 0.15	<0.001
Kodo millet	Polyphenol	4.44 ± 0.15	0.22 ± 0.04	N/A	<0.001
	FRAP	20.81 ± 1.31	0.33 ± 0.13	N/A	<0.001
Proso millet	Polyphenol	1.14 ± 0.11	0.54 ± 0.04	N/A	<0.001
	FRAP	4.35 ± 0.62	0.92 ± 0.11	N/A	<0.001

The crop is also dependent on varied number of non-enzymatic antioxidants like carotenoids, phenolic compounds and flavonoids. These molecules are scavenging free radicals, chlorophyll protecting, and membrane stabilizing molecules. Carotenoids are important in protecting photosystems during the presence of photo-oxidative stress whereas phenolic and flavonoid act in the reinforcement of structures and as powerful ROS suppressors. All these antioxidants combine to produce a multi-level defense system, which helps Kodo millet to be metabolically active and able to defend delicate tissues even in cases of low environmental inputs (Bao et al., 2024) (Table 15.4).

15.5.3 Osmolyte Accumulation and Metabolic Adjustments

The other major adaptation in Kodo millet is the osmolytes small organic molecules accumulation which helps keep the cells hydrated and in a state of metabolic stability in the face of drought, salinity and nutrient stress. One of the most significant osmo-protectants, which is accumulated in the plant, plays the role in osmotic balance, membrane protection and enzymes and proteins stabilization is proline. It is also a molecular chaperone, which assists in preserving the structure of proteins in dehydration stresses.

Glucose, fructose, and sucrose are soluble sugars that help in the provision of energy and work as osmotic buffers preventing water loss to the external environment. Such sugars are useful in the regulation of turgor pressure, membrane stability, and carbohydrate metabolism during slowing photosynthesis. There are other osmoprotectants such as glycine betaine and sugar alcohols, which help in maintaining redox balance and conserving cell structures as well. Such metabolic adaptations guarantee Kodo millet that it carries on with essential biochemical activities even in the most unfriendly conditions of growth (Mondal & Jespersen, 2025).

15.5.4 Molecular Regulation and Stress-Responsive Genes

Kodo millet has involved the use of complex regulatory pathways controlled by stress-sensitive genes and signaling pathways at the molecular level. Among the most crucial pathways, there is the abscisic acid (ABA)-dependent signaling pathway, which regulates stomata closure, osmotic, and expression of drought-response genes. ABA signaling induces transcription factors, which cause protective proteins to be expressed, such as late embryogenesis abundant (LEA) proteins, dehydrins, chaperones and osmolyte biosynthetic enzymes. The efficiency of nutrient use of the plant is also supported by the patterns of gene expression in Kodo millet, and this enables the plant to maximize uptake and assimilation of scarce soil nutrients. Nitrogen-transporter genes, stress-regulated enzymes and carbon-fixation pathways are also active even during low fertilizer conditions so that the crop has a constant metabolic rhythm. Epigenetic changes including methylation of DNA and acetylation of histones are also known to enhance a stressful resistance through the regulation of gene action in reaction to environmental signals. In the long-term, these genomic modifications will lead to a higher resilience and a stable operation of low-input farming systems (Kashyap et al., 2025) (Fig. 15.8).

15.6 Morphological and Anatomical Adaptations for Low-Input Systems

15.6.1 Root System Architecture

Kodo millet grows deep, dense and highly branched roots that facilitates effective penetration of the soil and exploration even with shallow or nutrient deficient soils. This root structure increases the ability of the soil to extract water at deeper levels, which is important in case of drought or unpredictable rains. Such long root system also enhances nutrient acquisition, and the sparse soil minerals are available in the soil, which can be captured by means of minimum fertilization. The large root-shoot ratio is in favor of below ground biomass, which minimizes the water loss through transpiration and maximizes the uptake of moisture and nutrients. Kodo millet thrives and produces grain on low-input systems with no need to irrigate the fields and ply on the soil with fertilizers due to these root characters (Dalsaniya et al., 2024).

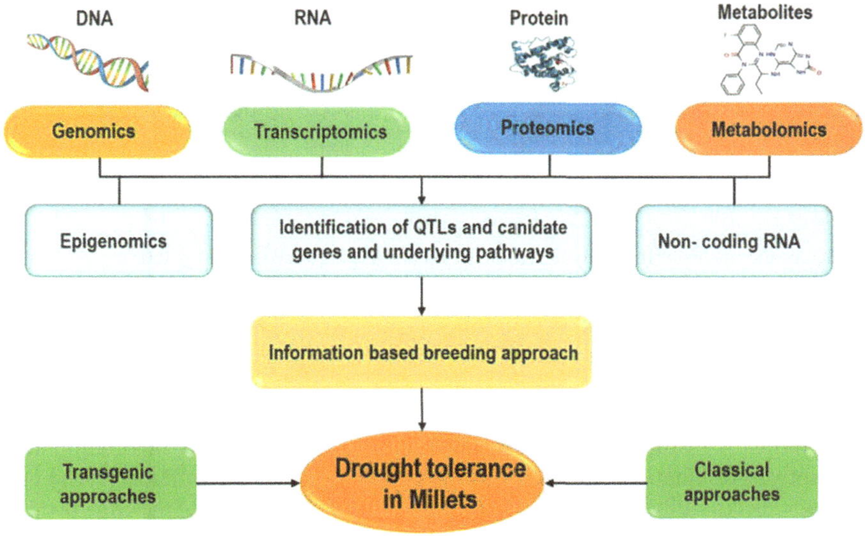

Fig. 15.8 Schematic representation of the molecular breeding approach for drought tolerance in millets crop (Tiwari et al., 2022)

15.6.2 Leaf and Stem Structural Responses

The adaptations of the kodo millet to heat and water loss are through the adaptations of its leaves. Its leaf cuticle is usually thick and waxy which restricts the transpiration in dry conditions. The plant can exhibit the middle stomatal density and relatively low stomatal pores, which should regulate both CO_2 uptake and the controlled amount of water loss. In certain dry zone adapted grasses, the density of trichome (leaf hair) or the amount of available wax on the leaf surface may further reduce water evaporation but species-specific measurements of kodo millet remain scanty—this is an indication that further anatomical work is required. The internal adaptations may include strong epidermal ridge, compact mesophyll and well developed vascular bundles that can efficiently transport water and preserve the turgor under the stress. A thick, fibrous stem that has well-formed sclerenchyma and vascular tissues helps support it on the poor soils and facilitates effective movement of water and nutrients through the roots to the grains. All these structural characteristics help kodo millet to be photosynthetic, to avoid wilting, and to spend resources on reproductive organs even in non-optimal soil moisture and fertility (Prabha et al., 2019).

Table 15.5 Estimated world and regional production of different millets (1981–1985) (Arendt & Dal Bello, 2008)

Region	Total	Pearl	Foxtail	Proso	Finger	Teff	Fonio	Others
Africa	9557	7330	–	–	855	1063[a]	309	–
Asia	17,048	6013	5462	2279	2905	–	–	386
World	29,295	13,351	5489	4931	3763	1063	309	387
World (%)	100	45.6	18.7	16.8	12.8	3.6	1.1	1.3

[a] Indicates the results of the statistical comparison test (e.g., Tukey test at $p \leq 0.05$)

15.7 Agronomic and Yield Responses Under Low-Input Management

Kodo millet is resilient to low-input agricultural system; this is the combination of adaptive germination, biomass allocation, and reproductive efficiency. The crop develops successfully even when there is minimal nutrient supply of the crop, because it depends more on the natural seed vitality and access to water in the environment. Sowing depth, moisture, and temperature of soil may also have an effect on germination rates, although the species remains early vigorous even in marginal soils. Canopy development, suppression of weeds, and root establishment are all-dependent on the early development of seedlings and all have an impact on the ultimate yield results. The biomass allocation of the Kodo millet is geared towards the capture of resources and reduction of effects of stress. During the early vegetative development, much of the assimilates is invested in roots, which boost exploration of the soil, and uptake of water. Overground biomass such as leaves and stems grows in relation to root capacity, which ensures that a balance is achieved that prevents excessive transpiration in conditions of limited moisture. Tillering behavior is a major determinant of yield; the crop has several productive tillers per plant under conditions of favorable conditions however, the survival of the tillers could be limited by the lack of nutrients or by drought. An effective tiller control, spacing, or crop density may be directly increased by reproductive output of low-input environments (Saleem et al., 2022) (Table 15.5).

15.8 Nutrient Management Strategies in Kodo Millet

15.8.1 Soil Fertility Challenges in Millet-Growing Regions

Kodo millet grows in most cases in marginal and rainfed areas where the soils are naturally low in fertility. Low organic matter content, inadequate nutrient availability, and successive erosion, nutrient mining, and low external contribution are common problems that are associated with it. These restrictions inhibit plant development, decrease the biomass retention, and inhibit the potential of grain yield. These soil fertility issues are important to the millet production in low-input systems that can be used sustainably (Rangaiah et al., 2024).

15.8.2 Conventional and Improved Nutrient Approaches

Kodo millet reacts favorably to balanced nutrient use that comprises of nitrogen, phosphorus, potassium and essential micronutrient use. Reduced nutrient availability suppresses the tillering, the growth of the leaf area, and the grain filling, which confirms the necessity of specific management by low-input practices. Organic manures include compost, farmyard manure and green manure, which improve the physical fitness of soils, water retention, and slow-release nutrients. These organic sources are used in low-input systems to supplement the low inorganic fertilizers and sustain crop production as well as ensuring environmental sustainability (Shikha et al., 2024).

15.8.3 Integrated Nutrient Management (INM)

INM is a blend of organic, inorganic, and microbial inputs to increase nutrient supply, increase the efficiency of plant uptakes, and sustain the soil health. This method is used in combination with seed priming, bio fertilizers, and soil-conditioning agents to enhance the efficiency of nutrient use, promote root development and enhance early seedling development. INM established appropriately lowers the reliance on chemical fertilizers and stabilizes production output in marginal conditions, as well as enhancing long-run soil fertility. It also fits the approach of low-inputs by combining the use of site-specific nutrient applications, organic amendments, and beneficial microbes to develop a resilient production system of Kodo millet (Joshi et al., 2025).

15.9 Ecosystem Services Provided by Kodo Millet

15.9.1 Soil-Related Ecosystem Services

Kodo millet has many benefits in improving soils, especially in soils that are marginal and degraded. Its growth improves soil structure by increasing the organic residues and root biomass, which improves aggregation and porosity. Due to the vast nature of the root system of the crop, organic matter builds up and enhances the action of microbes, which subsequently contribute to nutrient cycling and sequestration of carbon. Kodo millet also helps in eliminating soil erosion by the virtue of its ground-covering growth habit and thick canopy area, which cushions the surface soil against the impact of rainfall and surface runoff. Better infiltration and water retention within the soil profile is also an added advantage which helps conserve soil moisture which in turn positively affects crop productivity in low-input systems (Babele et al., 2022).

15.9.2 Biodiversity and Agroecosystem Functions

The development of Kodo millet promotes more agroecosystem biodiversity. Its roots and residues provide an optimal environment of advantageous soil microorganisms, such as nitrogen-fixing bacteria, phosphate solubilizing bacteria and decomposers that improve nutrient availability. The flowering season of the crop acts as a source of nectar and pollen to attraction of pollinators including bees and butterflies, as well as natural enemies of pests including predatory insect and parasitoid. Included in the agro ecological stability of Kodo millet mix are the ecosystem properties, better pest management, and excellent ecosystem stability through the support of diverse biological community (Das et al., 2025).

15.9.3 Climate Resilience and Environmental Benefits

The Kodo millet is pre-adapted to low-input rainfed systems, thereby minimizing the need to use irrigation and foreign nutrient application. It has a low water footprint, and it can withstand water stress, thus it is a climate-tolerant cereal that can withstand yields in semi-arid and drought-prone areas. The crop further supports climate mitigation by the root biomass and organic matter deposition in the soil, which helps in land stewardship to ensure that land is not abandoned but instead used to support sustainable landscapes and allow the growing crop to be produced in degraded, or marginal territories. Kodo millet would offer sustainable food systems and climate-smart agriculture with its potential to meet the ecological needs and low-input agronomic needs of smallholder farmers, enabling a sustainable solution to the food issues of millions of people in developing countries (Kumar et al., 2022). AS Shown in (Fig. 15.5), The phosphorus in Kodo millet is of relatively lower content as depicted in figure, but it is a nutrient powerhouse with high contents of fiber, protein, and vitamins as well as antioxidants. Kodo millet has various nutritional benefits—its high content of dietary fiber improves digestion, its good source of iron prevents anemia and its balanced protein-fibre content promotes good metabolism. The number also indicates the impressive fiber content of Kodo millet, which goes on to support the usefulness of Kodo millet as a functional and health-enhancing cereal (Figs. 15.9 and 15.10).

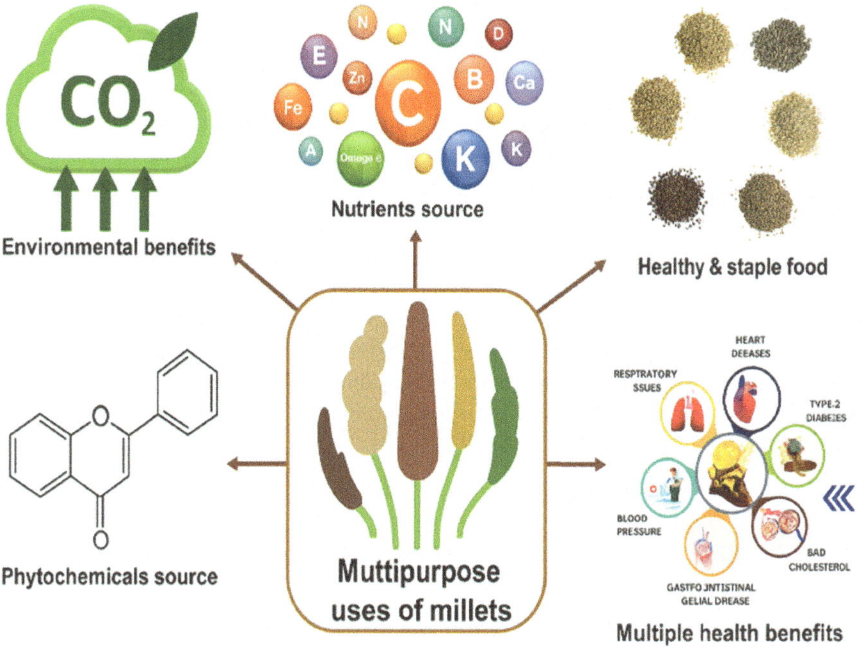

Fig. 15.9 Multipurpose use of millets

Fig. 15.10 Kodo millets benefits

15.10 Remote Sensing Applications in Monitoring Low-Input Performance

15.10.1 Remote Sensing Principles for Millet Assessment

Remote sensing offers an efficient and non-destructive technique of monitoring the Kodo millet performance under low inputs. Spectral reflectance data has the ability to detect changes in crop health due to either nutrient deficiency, water stress or pest pressure. The alteration of spectral signature through changes in chlorophyll content of leaves, canopy structure, and pigment composition under nutrient stress can be recorded with multispectral and hyperspectral sensors. Thermal sensing also enables the determination of canopy water condition as anomalies in temperature are related to the lack of moisture and dynamics of transpiration, which serve to give early signs of drought stress (Karimli & Selbesoğlu, 2023).

15.10.2 Relevant Vegetation and Stress Indices

The remote sensing data is necessary to calculate vegetation indices, which would help in assessing the performance of low-input. Normalized difference vegetation index (NDVI) indicates the total plant vigor and biomass accumulation whereas the Enhanced vegetation index (EVI) is more sensitive when the canopy is dense. Water stress and photosynthetic efficiency can be detected with the help of the Normalized Difference Water Index (ndwi) and Photochemical Reflectance Index (PRI). These indices are useful in the measurement of the level or extent of nutrient or moisture restriction in Kodo millet fields and also in the areas where some intervention might be necessary (Septem Riza et al., 2022).

15.10.3 Remote Sensing Platforms

Many platforms allow monitoring Kodo millet at spatial and temporal scales. Satellite images of Sentinel and Landsat also allow the extensive coverage of the region in terms of productivity and stress measurement, including the identification of temporal patterns and spatial variations in crop performance. Phu and Thermal sufficient sensors enable at-field level multispectral or thermal phenotyping of the expenses of UAVs to quantify how much stress patches and non-homogeneous functions. The Smart IoT sensor networks (ground-based) are introduced to complement remote platforms, offering data on the areas of moisture and temperature of the soil, as well as, the microclimatic conditions that can help to make the decision-making process more effective during low-input management (Solgi et al., 2023).

15.10.4 Integration with Crop Models

The combination of remote sensing data and models of crop growth and stress will help to gain a better insight into the Kodo millet performance under low-input conditions. Data assimilation can be used to map stress intensity, spatial variability of yield potential and temporal variations of biomass accumulation. These combined strategies help in accuracy management, water and nutrient consumption efficiency as well as enhancing the predictability of harvests in rainfed or marginal systems. Remote sensing together with modeling allows smallholder farmers and other researchers to use evidence-based measures to improve productivity without compromising on resource efficiency.

15.11 Case Study 1: Low-Input Kodo Millet Cultivation for Productivity and Soil Health Improvement in Madhya Pradesh, India

To determine the impact of the low-input management of Kodo millet in question of its productivity and soil health, Meena et al. organized a field investigation of the rainfed tribal areas in Jhabua, Madhya Pradesh, during a period of two years. The experiment involved comparison between the traditional broadcasting, line sowing using farmyard manure (FYM), and line sowing using FYM and seed priming on 60 fields of farmers. Measurements were of grain yield, soil organic carbon, bulk density, microbial biomass carbon as well as water-use efficiency (WUE). The findings have shown that the FYM + seed priming treatment was the one that had the highest grain yield of 1.42 t ha^{-1} by 41.7% over the traditional broadcasting, and the increase of soil organic carbon was 0.48% to 0.62 p. The loss in bulk density became 5.6, the rise in microbial biomass carbon was 32.9 and the increase in WUE was 3.8–5.1 kg ha^{-1} mm^{-1}. The research concluded that productivity and soil resilience are increased when FYM and seed priming is combined in low-input situations and suggested the implementation of these methods in tribal rainfed regions to increase long-term sustainability.

15.12 Case Study 2: Ecosystem Services and Economic Benefits of Traditional Kodo Millet Cultivation in Odisha, India

The article by Sahu and Patel explores the provision of the ecosystem services and the economic performance of traditional Kodo millet-based agricultural systems in the Eastern Ghats of the Kora put district of Odisha. Through participatory rural appraisal, household surveys, 120 farmers, and the ecological field assessment, the research measured biodiversity indicators, biological activity of soils and provisioning benefits. The results found that Kodo millet plots had 26–32 weed species that increased pollinator diversity and the soil respiration was 18.6% greater than upland paddy fields nearby. The cost of production was 48% less than the cost of growing maize, and the ecosystem service value was estimated to be USD 312 ha^{-1} $year^{-1}$, which was 1.7 times more than the maize system. Dietary diversity in households increased by 22% because of frequent intake of Kodo millet. The authors suggested including millet-based ecosystem services value into the spatial planning and encouraging the community seed banks to preserve the local landraces and their associated biodiversity.

15.13 Case Study 3: Productivity and Soil Fertility Enhancement Through Kodo Millet—Legume Intercropping in Andhra Pradesh, India

A field experiment study was completed by Reddy et al. at the semi-arid district of Anantapur in Andhra Pradesh to determine the agronomic and economic efficacies of low-input Kodo millet-legume intercropping systems. The experiment involved comparison of sole Kodo millet and Kodo + red gram (2:1) and Kodo + cowpea (2:1) inter cropping systems on a randomized block design. The major measurement parameters were grain yield, soil nutrient status, rainwater productivity and system profitability. The findings revealed that, the Kodo + cowpea recorded the highest yield of grains, whereby there was an increment of 34.5% as compared to monocropped Kodo millet. Under the intercropping system, soil nitrogen, phosphorus and potassium increased by 19.8, 14.6, and 11.2, respectively. The net economic returns were increased to [?]28,400 ha^{-1}, which is a 42% higher than that of solo cropping and the rainwater productivity rose to 4.3 kg ha^{-1} mm^{-1}. The research proposed a broader use of Kodo-legume intercropping to increase nutrient cycling, resistance to drought and profitability of the farm in arid lands.

15.14 Case Study 4: Climate Resilience and Carbon Footprint Reduction in Organic Kodo Millet Systems in Chhattisgarh, India

Sharma and Singh conducted a field and life cycle assessment study in the low-input organic management of Kodo millet that has been cultivated in Dhamtari district, Chhattisgarh in order to measure the climate resilience and carbon footprint of Kodo millet. In the experiment, organic Kodo millet was compared with chemically controlled Kodo millet and irrigated rice systems based on the indicators of carbon footprint (CO_2-eq ha^{-1}), stability index of yield, energy-use efficiency, soil organic carbon, and greenhouse gas emission intensity. The results indicated that organic Kodo millet would cut carbon footprint by 54.2% than chemical-input rice with an energy-use efficiency density of 3.8, which is almost twofold compared to the rice with an energy-use efficiency of 1.9. Kodo millet yield stability index was found to be 0.82 compared to 0.56 of rice in case of heat stress, and soil organic carbon gained 0.21% in two years. The intensity of emission became lower at 0.88 kg CO_2-equivalent, 1 kg of grain. The authors suggested organic Kodo millet cultivation to be promoted in climate-prone zones to minimize emissions and enhance the carbon stocks of the soil, in addition to promoting the policy of promoting climate-intelligent crop diversification (Fazli et al., 2025).

15.15 Future Directions

15.15.1 Enhancing Value Chains and Policy Support

Further improvement of breeding Kodo millet derived by genomic selection and detection of stress-resistance will expedite the production of cultivars in low-input and marginal environments. By focusing on drought and runaway heat and salinity and on the efficiency with which it employs nutrients the focus will facilitate more resilience in times of climate variability. A hybrid approach of combining traditional breeding and the use of molecular marker-assisted selection can reduce the breeding period and enhance stability in yield and nutritional quality (Choudhary et al., 2023).

15.15.2 CRISPR-Based Opportunities

Gene-editing tools like CRISPR/Cas9 have potential in accuracy of introduction of characters improving yield, efficiency in nutrient-use and tolerating to abiotic stresses. With this method, elite cultivars can be developed quickly without the need to subject the plant to a program of widespread hybridization, therefore resulting in a process that is not as lengthy as that of hybridization (Kudapa et al., 2023).

15.15.3 High-Throughput Phenotyping and Digital Agriculture

The high-throughput phenotyping of growth, stress response, and productivity observing of UAVs, multispectral imaging, remote sensing, and IoT-based sensors requires integration in solving field-born problems. The insights that these technologies provide through data can inform agronomics, realize maximization in resource-use, and inform breeding programs to suit low- input environments (Fluck & Jaffe, 1975).

15.15.4 Scaling Kodo Millet Cultivation

An increase in Kodo millet production to appropriate agroecological areas needs modification of best management practices, creation of awareness to the farmers, and integration into the available cropping systems. Maximizing productivity through promotion of intercropping, crop rotation and conservation agriculture methods can ensure that the soil is healthy and input dependency is minimal (Atia et al., 2024).

15.15.5 Enhancing Value Chains and Policy Support

The economic viability of Kodo millet can be boosted through deploying post-harvest infrastructural, processing and market nexuses to enhance its economic viability. Subsidies, crop insurance, and research incentives will be used as policy measures that will promote the adoption and boost food and nutritional security as well as position Kodo millet as a strategic cereal crop embedded in a sustainable and climate-resilient food system (Mishra et al., 2024a, 2024b).

15.16 Conclusion

Kodo millet has proven to have an impressive combination of physiological, biochemical and agronomic features making it work in low-input, marginal, and rainfed environments. The low water-use solution, its photosynthetic stress-tolerance, adjustment of osmotic pressure, its antioxidant defense mechanisms all help it stand up to nutrient-infertile soils and semi-arid abiotic conditions. Morphological and anatomical features such as deep and branched roots and leaf structural adaptation further support its survival and productivity under resource-limited conditions. Besides that, Kodo millet has significant impact on the ecological services as it enhances soil health, enhances carbon sequestration, sustains biodiversity, and keeps the agroecosystem stable despite having the lowest external requirements. The opportunities to increase yield, stress tolerance, and nutritional quality can be opened by technological innovations (remote sensing, high-throughput phenotyping, and digital agriculture tools), as well as genomic and molecular breeding methods. Having been coupled with farmer-based management approaches, streamlined nutrient plans, and sustainable crop tone, Kodo millet will present a robust and climate wary solution to the concerns of world food and nutrition security. Its capacity to survive in unfavorable environment, sustain ecosystem, as well as retains its productivity with a modest input level, places it in a potential crop in sustainable agricultural activities and future climatological adaptation programs.

References

Almaski, A., Thondre, S., Lightowler, H., & Coe, S. (2017). Determination of the polyphenol and antioxidant activity of different types and forms of millet. *Proceedings of the Nutrition Society, 76*, E5. https://doi.org/10.1017/S0029665117000052

Arendt, E. K., & Dal Bello, F. (2008). *Gluten-free cereal products and beverages* (1st ed). Academic.

Atia, M., Jiang, W., Sedeek, K., Butt, H., & Mahfouz, M. (2024). Crop bioengineering via gene editing: Reshaping the future of agriculture. *Plant Cell Reports, 43*(4), 98. https://doi.org/10.1007/s00299-024-03183-1

Azar-Kia, B., Palacios, E., & Churchill, R. (1975). Aqueductal stenosis and Parinaud's syndrome. *IMJ Illinois Medical Journal, 148*(5), 532–533.

Babele, P. K., Kudapa, H., Singh, Y., Varshney, R. K., & Kumar, A. (2022). Mainstreaming orphan millets for advancing climate smart agriculture to secure nutrition and health. *Frontiers in Plant Science, 13*, 902536. https://doi.org/10.3389/fpls.2022.902536

Bao, L., Liu, J., Mao, T., Zhao, L., Wang, D., & Zhai, Y. (2024). Nanobiotechnology-mediated regulation of reactive oxygen species homeostasis under heat and drought stress in plants. *Frontiers in Plant Science, 15*, 1418515. https://doi.org/10.3389/fpls.2024.1418515

Ceasar, S. A., & Ignacimuthu, S. (2010). Effects of cytokinins, carbohydrates and amino acids on induction and maturation of somatic embryos in kodo millet (*Paspalum scorbiculatum* Linn.). *Plant Cell, Tissue and Organ Culture (PCTOC), 102*(2), 153–162. https://doi.org/10.1007/s11240-010-9716-6

Choudhary, P., Shukla, P., & Muthamilarasan, M. (2023). Genetic enhancement of climate-resilient traits in small millets: A review. *Heliyon, 9*(4), e14502. https://doi.org/10.1016/j.heliyon.2023.e14502

Dalsaniya, D., Das, A., & Parmar, D. J. (2024). Variability Study and DUS Characterization in Kodo Millet (*P. scrobiculatum* L.). *International Journal of Plant and Soil Science, 36*(4), 408–417. https://doi.org/10.9734/ijpss/2024/v36i44494

Das, J., Sharma, U., Sankhyan, N., Sharma, S., Chauhan, V., Parida, S., et al. (2025). Integrating millets into agroforestry systems: A climate-smart strategy for sustainable land use and livelihood improvement with special emphasis on India. *Trees, Forests and People, 22*, 101087. https://doi.org/10.1016/j.tfp.2025.101087

Dey, S., Saxena, A., Kumar, Y., Maity, T., & Tarafdar, A. (2023). Optimizing the effect of ultrasonication and germination on antinutrients and antioxidants of kodo (*P. scrobiculatum*) and little (*P. sumatrense*) millets. *Journal of Food Science and Technology, 60*(12), 2990–3001. https://doi.org/10.1007/s13197-023-05837-6

Fazli, S., Maharjan, S., Li, W., Fisher, J. B., Thomas, R., Galvan, F. R., Shirkey, G., & El-Askary, H. (2025). Unraveling crop nitrogen-water dynamics with hyperspectral-thermal sensing in Northern Central Valley, California. *Earth Systems Environments, 25*, 41748. https://doi.org/10.1007/s41748-025-00794-w

Fluck, R. A., & Jaffe, M. J. (1975). Cholinesterases from plant tissues. VI. Preliminary characterization of enzymes from *Solanum melongena* L. and *Zea mays* L. *Biochimica Et Biophysica Acta, 410*(1), 130–134. https://doi.org/10.1016/0005-2744(75)90213-2

Ghatak, A., Pierides, I., Singh, R. K., Srivastava, R. K., Varshney, R. K., Prasad, M., Chaturvedi, P., & Weckwerth, W. (2025). Millets for a sustainable future. *Journal of Experimental Botany, 76*(6), 1534–1545. https://doi.org/10.1093/jxb/erae507

Hariprasanna, K. (2017). Kodo millet, *P. scrobiculatum* L. In J. V. Patil (Ed.), *Millets and sorghum* (1st ed., pp. 199–225). Wiley. https://doi.org/10.1002/9781119130765.ch8

Jeeva, G., Suhasini, B., Pramitha, L., Jency, J. P., Joshi, P., Ravikesavan, R., & Elango, D. (2024). Unlocking the potential of Kodo millet: Reviving an indigenous super grain for tomorrow's nutrition. *Planta, 259*(6), 140. https://doi.org/10.1007/s00425-024-04414-1

Joshi, S., Joshi, N., Tariq, A., Khursheed, M., Jan, R., Singh, Kh. C., Sultan, A., Kashyap, M., Rajgonda, K. P., & Tutlani, A. (2025). Integrating biofertilizers in nutrient management of millets: A review of advances and prospects. *Journal of Experimental Agriculture International, 47*(11), 638–653. https://doi.org/10.9734/jeai/2025/v47i113903

Kannababu, N., Deepika, C., Venkateswarlu, R., Hariprasanna, K., & Das, I. K. (2023). Physiological maturity of seeds in kodo millet (*P. scrobiculatum* L.) cultivars across seasons. *Plant Physiology Reports, 28*(2), 209–220. https://doi.org/10.1007/s40502-023-00725-9

Karimli, N., & Selbesoğlu, M. O. (2023). Remote Sensing-based yield estimation of winter wheat using vegetation and soil indices in Jalilabad, Azerbaijan. *ISPRS International Journal of Geo-Information, 12*(3), 124. https://doi.org/10.3390/ijgi12030124

Kashyap, Y., Salam, P., Thakur, A., Chandrakar, T., Singh, D. P., Toppo, N. R., & Lakshmi, P. M. (2025). Response of kodo millet (*P. scrobiculatum* L.) to nutrient levels and planting geometry for growth and yield attributes. *International Journal of Research in Agronomy, 8*(7), 1693–1696.

Kudapa, H., Barmukh, R., Vemuri, H., Gorthy, S., Pinnamaneni, R., Vetriventhan, M., Srivastava, R. K., Joshi, P., Habyarimana, E., Gupta, S. K., & Govindaraj, M. (2023). Genetic and genomic interventions in crop biofortification: Examples in millets. *Frontiers in Plant Science, 14*, 1123655. https://doi.org/10.3389/fpls.2023.1123655

Kumar, L., Naresh, R. K., Tiwari, H., Kataria, S. K., Saharan, S., Reddy, B. R., Singh, O., Qidwai, S., & Singh, R. P. (2022). Millets for food and nutritional security in the context of climate resilient agriculture: A review. *International Journal of Plant and Soil Science*. https://doi.org/10.9734/ijpss/2022/v34i232504

Luo, D., Usman, M., Pang, F., Zhang, W., Qin, Y., Li, Q., Li, Y., Xing, Y., & Dong, D. (2024). Comparative transcriptomic and physiological analyses unravel wheat source root adaptation to phosphorous deficiency. *Scientific Reports, 14*(1), 11050. https://doi.org/10.1038/s41598-024-61767-z

Mathur, S., Khanduri, S., Gosai, N., Sharma, P. S., & Singh, M. (2025). Nutritional and nutraceutical properties of kodo millets (*P. scrobiculatum*): A comprehensive review. *Asian Journal of Dairy and Food Research, 6*, 2340.

Mehta, D., Vyas, S., Dudhagara, D., Patel, A., & Parmar, V. (2024). Significance of Indian millets in enhancing global food security: A comprehensive review. *Trends in Food Science and Technology, 149*, 104527. https://doi.org/10.1016/j.tifs.2024.104527

Mishra, G., Tiwari, S., Paliwal, S., & Sharma, S. (2024a). Assessment of kodo millet (*P. scrobiculatum*) genotypes for salt stress tolerance at the seedling stage using germination tray. *International Journal of Plant and Soil Science, 36*(8), 276–288.

Mishra, S., Kumar, S., & Srivastava, R. C. (Eds.). (2024b). *Genetic improvement of small millets*. Springer, Singapore. https://doi.org/10.1007/978-981-99-7232-6

Mishra, P., Nanda, S. R., Barpanda, T., Dash, M., Dash, S., Choudhury, S., Roul, S., & Mishra, A. (2025). The complexity of kodo millet: Genomic analysis and implications in crop improvement. *Planta, 261*(1), 15. https://doi.org/10.1007/s00425-024-04588-8

Mondal, S., & Jespersen, D. (2025). Understanding salinity tolerance mechanisms in finger millet through metabolomics. *Plant Physiology and Biochemistry, 222*, 109742. https://doi.org/10.1016/j.plaphy.2025.109742

Mukherjee, B., Jha, R. K., Sattar, A., Dutta, S., Bhattacharya, U., Samanta, S., Huirem, B., Singh, S. K., Das, S., & Bal, S. K. (2025). Harnessing the potential of millets for climate-resilient and sustainable agriculture. *Frontiers in Plant Science, 16*, 1574699. https://doi.org/10.3389/fpls.2025.1574699

Nirubana, V., Ravikesavan, R., & Ganesamurthy, K. (2020). Evaluation of underutilized kodo millet (*P. scrobiculatum* L.) accessions using morphological and quality traits. *Indian Journal of Agricultural Research, 5*, 5462.

Prabha, R., Singh, D. P., Gupta, S., Gupta, V. K., El-Enshasy, H. A., & Verma, M. K. (2019). Rhizosphere metagenomics of *P. scrobiculatum* L. (kodo millet) reveals rhizobiome multifunctionalities. *Microorganisms, 7*(12), 608. https://doi.org/10.3390/microorganisms7120608

Prashanth, K. S., Reddy, M. N., Rajasekaran, R., & Muthamilarasan, M. (2025). Kodo millet. In S. A. Ceasar, S. Penna, C. W. P. Carvalho, & S. M. Jain (Eds.), *Millets: Crops for climate resilience and for food and nutritional security* (pp. 93–125). Springer. https://doi.org/10.1007/978-981-95-1256-0_4

Rangaiah, K. M., Nagaraju, B., Kasturappa, G., Kadappa, B. P., Narayanaswamy, U. K. S., Sab, M. S. H., Veerabadraiah, G. G., Srivastava, S., & Dey, P. (2024). Optimizing nutrient management for Kodo millet (*P. scrobiculatum* L) in the *Alfisols* of Southern India through modeling soil, plant, and fertilizer interactions. *Scientific Reports, 14*(1), 31852. https://doi.org/10.1038/s41598-024-83265-y

Ru, C., Hu, X., Chen, D., & Wang, W. (2023). Droughts and thermo-priming enhance acclimation to later drought and heat stress in maize seedlings by improving leaf physiological activity. *Agronomy, 13*(4), 1124. https://doi.org/10.3390/agronomy13041124

Saleem, M. F., Ghaffar, A., Rahman, M. H. U., Imran, M., Iqbal, R., Soufan, W., Danish, S., Datta, R., Rajendran, K., & El Sabagh, A. (2022). Effect of short-term zero tillage and legume intercrops on soil quality, agronomic and physiological aspects of cotton under arid climate. *Land, 11*(2), 289. https://doi.org/10.3390/land11020289

Septem-Riza, L., Yudianita, A. H., Nugraha, E., Somantri, L., Sitanggang, I. S., Abu-Samah, K. A. F., & Nazir, S. (2022). Remote sensing and machine learning for yield prediction of lowland paddy crops. *F1000Research, 11*, 682.

Shikha, D., Kumar, Y., Sharanagat, V. S., & Saxena, D. C. (2024). Kodo millet: Technological impact and nutritional benefits for value-addition in food products. *Journal of Food Process Engineering, 47*(6), e14655. https://doi.org/10.1111/jfpe.14655

Solgi, S., Ahmadi, S. H., & Seidel, S. J. (2023). Remote sensing of canopy water status of the irrigated winter wheat fields and the paired anomaly analyses on the spectral vegetation indices and grain yields. *Agricultural Water Management, 280*, 108226. https://doi.org/10.1016/j.agwat.2023.108226

Tiwari, A., Kesarwani, K., Sharma, A., Ghosh, T., Bisht, N., & Punetha, S. (2022). Drought Stress in Millets and Its Response Mechanism. In J. Ngui Kimatu (Ed.), *Advances in plant defense mechanisms*. IntechOpen. https://doi.org/10.5772/intechopen.105942

Vetriventhan, M., Azevedo, V. C. R., Upadhyaya, H. D., Nirmalakumari, A., Kane-Potaka, J., Anitha, S., Ceasar, S. A., Muthamilarasan, M., Bhat, B. V., Hariprasanna, K., Bellundagi, A., Cheruku, D., Backiyalakshmi, C., Santra, D., Vanniarajan, C., & Tonapi, V. A. (2020). Genetic and genomic resources, and breeding for accelerating improvement of small millets: Current status and future interventions. *The Nucleus, 63*(3), 217–239. https://doi.org/10.1007/s13237-020-00322-3

Vetriventhan, M., & Upadhyaya, H. D. (2019). Variability for productivity and nutritional traits in germplasm of kodo millet, an underutilized nutrient-rich climate smart crop. *Crop Science, 59*(3), 1095–1106. https://doi.org/10.2135/cropsci2018.07.0450

Wang, Z., Yung, W.-S., Gao, Y., Huang, C., Zhao, X., Chen, Y., Li, M.-W., & Lam, H.-M. (2024). From phenotyping to genetic mapping: Identifying water-stress adaptations in legume root traits. *BMC Plant Biology, 24*(1), 749. https://doi.org/10.1186/s12870-024-05477-8

Zulfiqar, B., Raza, M. A. S., Saleem, M. F., Ali, B., Aslam, M. U., Al-Ghamdi, A. A., Elshikh, M. S., Hassan, M. U., Toleikienė, M., Ahmed, J., Rizwan, M., & Iqbal, R. (2024). Abscisic acid improves drought resilience, growth, physio-biochemical and quality attributes in wheat (*Triticum aestivum* L.) at critical growth stages. *Scientific Reports, 14*(1), 20411. https://doi.org/10.1038/s41598-024-71404-4

Chapter 16
Foxtail Millet (*Setaria italica*): Linking Nutrient Use Efficiency and Climate Resilience

Muqaddas Noor, Abdullah Jan, Umar Farooq, Amjad Malik, and Shakeel Ahmad

Abstract Foxtail millet (*Setaria italica L.*) is a climate-resilient and nutrient-efficient cereal suitable for low-input agricultural systems. Climate change, soil nutrient depletion, and rising input costs have increased interest in such crops for sustainable production. The objective of this chapter is to synthesize physiological, agronomic, and molecular mechanisms governing nutrient use efficiency (NUE) and climate resilience in foxtail millet. The chapter also evaluates its role in strengthening productivity under rainfed and marginal environments. An integrative review approach was adopted using evidence from crop physiology, genomics, agronomy, remote sensing, modeling tools, and field-based case studies. Comparative analyses with major cereals were used to highlight efficiency and resilience advantages. Quantitative results from case studies show that optimized NPK application (30:20:20 kg ha^{-1}) improves biomass production and grain yield while maintaining high NUE. Moderate irrigation regimes (120–150 mm) enhance water productivity and root development under stress conditions. DSSAT-based simulations predict yield responses across nitrogen and water gradients with errors generally below 10–15%. Field evaluations further reveal substantial genetic variability in phenology, biomass partitioning, and stress tolerance. Future research should integrate multi-omics-assisted breeding with precision nutrient and water management. Policy and value-chain support are essential to scale foxtail millet adoption in climate-vulnerable regions.

Keywords Foxtail millet · Nutrient use efficiency · Climate resilience · Sustainable agriculture · Crop modeling

---

M. Noor (✉) · A. Jan · U. Farooq
Department of Agronomy, Pir Mehr Ali Shah Arid Agriculture University, Rawalpindi, Pakistan
e-mail: noormuqaddas954@gmail.com

A. Malik
Kenanow Bachelor of Education Program Faculty of Education, University College of the North (UCN), Thompson, MB, Canada

S. Ahmad
Institute of Agronomy Faculty of Agricultural Sciences and Technology, Bahauddin Zakariya University, Multan, Pakistan

M. Ahmed (ed.), *Climate Change and Neglected Underutilized Crops*,
https://doi.org/10.1007/978-3-032-21005-0_16

16.1 Introduction

Global agricultural systems are growing under the pressure of climate change, reduced soil fertility, water scarcity and increased costs of production. The persistent dependence on a set of high-input cereal crops has diminished resilience of agroecosystems, enhanced environmental degradation as well as exposure to climatic extremes. In this regard, the neglected and underused crops have become a new focus of scientific and policy interest owing to their capacity to yield consistent yields under the conditions of poor environment with lesser external inputs. These crops provide viable solutions that can be used to improve agroecosystems in weak ones by improving climate resilience, efficient resource use, and food security (Bandyopadhyay et al., 2017a, 2017b; Farooq et al., 2023). Among these crops, millets have become climate-smart crops because of their natural resistance to drought, heat, and low soil fertility. One of the earliest domesticated millets is foxtail millet (*Setaria italica* (L.) P. Beauv.) due to the wide agroecological range, short development cycle, and efficient C4 photosynthetic cycle. The crop is a good culture in semi-arid and arid regions and is productive when production is low-input. New physiological and molecular evidence also has foxtail millet as a model to understand stress tolerance, nutrient efficiency and adaptive metabolism in cereals (Aggarwal et al., 2022).

Foxtail millet is still very useful in rainfed and marginal agricultural systems in South Asia and some parts of Africa, where it has been customarily grown on degraded soil with low irrigation and fertilizer levels. It is also resistant to unpredictable rainfall and low soil fertility, which makes it a consistent crop that the farmers of smallholder can use as climatic unpredictability, rises. In addition to agronomic resilience, foxtail millet also leads to nutritional security due to the elevated level of dietary fiber, balanced amino acids, and dense levels of a range of micronutrients, including iron and zinc. Its low level of glycemic index makes it even more useful as a functional food crop that could help in curbing malnutrition and metabolic disorders contracted due to dietary causes (Gouvêa et al., 2018; Ma et al., 2024). One of the traits that are important in associating productivity and climate adaptation in foxtail millet is nutrient use efficiency (NUE). However, during abiotic stresses, nutrient uptake and assimilation in cereals is usually disrupted yet under stress, foxtail millet continues to achieve an efficient acquisition of nutrients, internal remobilization and coordinated carbon–nitrogen metabolism. These characteristics have been known to facilitate long-term photosynthesis, enhanced antioxidant protection and effective osmotic regulation hence minimizing yield losses in low input conditions. New findings also show that NUE in foxtail millet is closely combined with physiological/molecular stress-adaptive responses as opposed to acting as an independent agronomic trait (Nadeem et al., 2018, 2020). This chapter provides a thorough evaluation of foxtail millet as an efficient nutritious and resistant to climate cereal. It is a combination of physiology, genomics, agronomy, and nutrition to study the origin of the crop, botanical and phenological characteristics, and the processes involved in nutrient efficiency and abiotic tolerance to stress. Recent developments in breeding and management practices to enhance resilience to changing climatic conditions are also discussed in the chapter with a strategic role of foxtail millet in sustainable, low input, and climate smart agricultural systems.

16.2 Botanical and Physiological Profile of *S. italica*

16.2.1 Taxonomy and Botanical Classification

Foxtail millet morphologically is an annual, tufted grass, which has erect and slender culms, linear-lanceolate leaves, and a thick panicle as shown in (Fig. 16.1), resembling a foxtail, which is the origin of the name. The small inflorescence design makes it possible to pollinate and set fruits in changing environmental conditions, which leads to a stable yield in the marginal agroecosystem (Farooq et al., 2023; Ma et al., 2024). *Setaria italica* has close evolutionary relatedness with *Setaria viridis*, its wild ancestor, phylogenetically. The Setaria genus has become an effective comparative system to study domestication, stress tolerance and nutrient-use characteristics in C4 grasses through this relationship. It is also necessary for them to be able to manage different situations that could lead to accidents (Bandyopadhyay et al., 2017a, 2017b).

16.2.2 Domestication History and Relationship with S. viridis

The history of domestication of *S. viridis* is very similar to that of rice, as neither has a domestication history. One of the oldest domesticated cereals is foxtail millet (*S. italica*), the archaeological and genetic evidence show that it originated in northern China around 8000–10,000 years ago, and in rainfed, low-input farming systems, which means it well adapted to drought, nutrient-deficient soils, and climatic variations (Diao & Jia, 2017; Farooq et al., 2023). Domestication selection mechanisms were chosen in support of reduced seed shattering, compact non-brittle panicles, larger grains and synchronized flowering, which enhanced harvestability and retained stress tolerance. The *S. italica* genome comparative genomics of cultivated and wild progenitor *S. viridis* have shown that domestication bottlenecks were relatively mild in these cases, and therefore both could retain a lot of adaptive plasticity (Nadeem et al., 2020; Singh, 2023). *Setaria viridis*, which carries more genetic variation and is better adapted to nutrient-limited and stressful conditions, is an excellent genetic source of enhancing nutrient use efficiency, abiotic stress tolerance, and yield stability in foxtail millet, as breeding methods with genomics aid in extracting and implementing a more effective allele of the GSK gene. Figure 16.2 illustrates the morphological comparison between foxtail millet and *S. viridis* (Aggarwal et al., 2022; Fukunaga & Kawase, 2024).

Fig. 16.1 Botanical description of foxtail millet (Kumari et al., 2024)

Fig. 16.2 Morphological comparison of domesticated foxtail millet and its wild ancestor green foxtail. Whole and longitudinal section of foxtail millet (**a**, **b**) and green foxtail (**c**, **d**) panicles. Whole plant of foxtail millet (**e**) and green foxtail (**f**) (Hu et al., 2018)

16.2.3 Morphological Characteristics: Roots, Shoots, and Reproductive Structures

Foxtail millet is morphologically structured to suit low-input and stressful environments. Due to the fibrous and very plastic root system, the crop will be able to explore soil volume effectively in search of water and nutrients in drought and nutrient limited environment. Root plasticity also helps the crop to keep growing and absorbing nutrients when the rainfall is erratic (Ardie et al., 2025; Ma et al., 2024). The plant bears erect and slender shoots, which are normally 60–150 cm high aboveground. The reduced transpiration surface area of the narrow leaves ensures a good photosynthetic capacity especially in the high temperature and radiations. The reproductive structure is a dense and bristly panicle with tiny caryopses which make it easy to fill and complete the life cycle in short growing seasons (Farooq et al., 2023). Early flowering and maturity are the two important characteristics of foxtail millet, which enables it to avoid terminal drought and heat stress to increase yield consistency in marginal agroecosystems (Ma et al., 2024).

16.2.4 Physiological Features: Water-Use Efficiency and Photosynthetic Adaptation

The foxtail millet has high physiological efficiencies in environments with water limitation and high temperature, which is mostly attributed to the C4 photosynthetic system. C4 mechanism concentrates CO_2 at the site of Rubisco, which significantly decreases photorespiration and enhances carbon assimilation during high light intensity and temperature. This gives it a high level of water-use efficiency over the C3-type cereals like rice and wheat (Gao et al., 2025; Nadeem et al., 2020). Foxtail millet also exhibits C4 photosynthesis, effective stomatal control, morphological adaptation of leaves, and high osmotic adjustment ability that reduce water loss and reduce the expense in maintaining photosynthesis. Another role played by improved antioxidant defense mechanisms is to safeguard the cellular structures against oxidative damages caused by drought, heat, and salinity stress (Fukunaga & Kawase, 2024; Venkatesh Babu et al., 2025). These physiological characteristics are closely connected with the efficient nutrient uptake and internal remobilization which supports the crop resilience to the nutrient-poor conditions and underlines the strong interconnection between nutrient-use efficiency and the ability to endure the abiotic stressors (Ardie et al., 2025; Farooq et al., 2023).

16.2.5 Comparison with Other Millets and Major Cereals

The foxtail millet is characterized by a reduced duration of growth, increased water use efficiency, and drought and heat stress tolerance than other small millets. These characteristics render it especially appropriate in the semi-arid and rainfed conditions where the productivity of the crops with more extended maturity periods is restricted by the existence of the climatic uncertainty (Bandyopadhyay et al., 2017a, 2017b; Ma et al., 2024). As compared with other major cereals including rice, wheat, and maize, foxtail millet does not need large quantities of inputs like water and nitrogen, and its yield is constant even in conditions of nutrient deficiency and stress. In spite of the fact that maize also has a C4 photosynthetic system, foxtail millet outperforms it due to its low soil fertility and water stress tolerance (Bandyopadhyay et al., 2017a, 2017b; Farooq et al., 2023). Also, foxtail millet has a high micronutrient level and dietary fiber, which compared to most of the staple cereals supports the idea of nutrition-sensitive and climate-resilient food systems (Ardie et al., 2025). All these features make foxtail millet an excellent stepping stone between minor and major cereals in the process of technological change to sustainable and climate-smart farming.

16.3 Global Production Landscape and Agronomic Importance

16.3.1 Global Distribution and Environmental Suitability

The foxtail millet (*S. italica* (L.) P. Beauv.) is a common crop grown in Asia and some parts of Africa, especially in the semi-arid and arid agro-ecological conditions with low and unreliable levels of rainfall, high temperatures and marginal soils. Archaeologists and modern evidence in northern China and South Asia have established its major centers of cultivation with enormous production happening in India, Nepal, Mongolia and adjacent areas and in Africa, it is cultivated under rainfed conditions in Ethiopia, Niger and Mali as shown in (Fig. 16.3). This crop has a large ecological range because of being able to grow and achieve its full life cycle in well-drained sandy to loamy low-inherent fertility soils with high inherent fertility. The fact that it is moderately tolerant to soil acidity and salinity only adds to its flexibility in that it can be used in an environment where the large cereals like rice and wheat are often characterized by yield variability or even failure (Aggarwal et al., 2022; Farooq et al., 2023). Foxtail millet is also environmentally favorable, and has a short growth period, early establishment, and a high water-use efficiency, which enable the crop to dodge terminal drought and take advantage of short rainfalls. The physiological plasticity and its C4 photosynthetic metabolism provide it with efficiency in the production of biomass at high temperature and moisture-limited conditions, which gives it yield stability in resource-poor agroecosystems. All the previously mentioned features make foxtail millet a trustworthy drought-resistant cereal in climate-sensitive areas where it directly leads towards food security amidst climate variability (Deevi et al., 2024; Farooq et al., 2023).

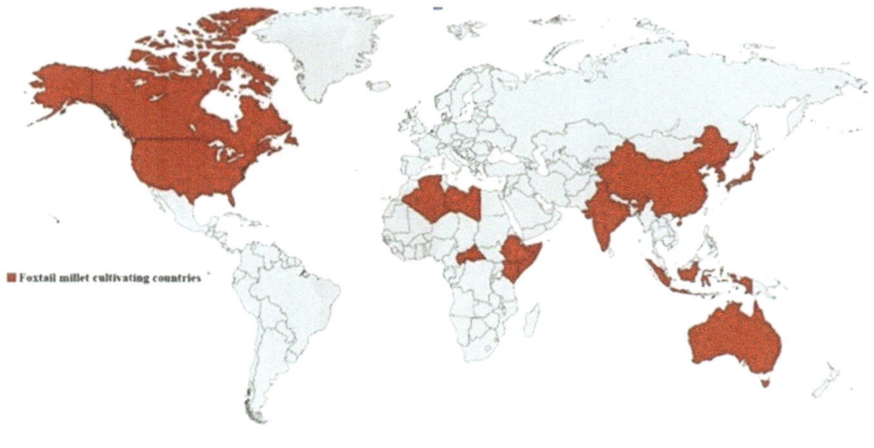

Fig. 16.3 Geographical distribution of foxtail millet cultivation (red color) (Kumari et al., 2024)

16.3.2 Production Trends and Yield Potential

Foxtail millet production is also still concentrated in Asia, especially in China and India as shown in (Fig. 16.4) where it is customarily grown under low-input and rainfed cultivation. In spite of the small area of total cultivation in comparison with the main cereals, foxtail millet has recently regained scientific and policy interest due to the importance attached to climate and sustainable food systems. The increased stability in the yield has been a result of improved agronomic management, easier and better knowledge of the stress physiology especially when drought and heat stresses are involved (Deevi et al., 2024).

The yield potential of foxtail millet is affected by genotype, environmental factors and intensity of management significantly. In good conditions when there is sufficient moisture and the nutrient levels are not very low, grains have been shown to be producing up to 3–4 t ha^{-1}, and in marginal conditions with low inputs; the yield is normally between 1–2 t ha^{-1}. Although foxtail millet yields are lower in absolute terms than those of high-input cereals, the crop has a higher yield stability, less production cost, and lower risk of crop failure, which can be regarded as the key to its inclusion in resilient agro-climatic systems (Ma et al., 2025; Sachdev et al., 2021) (Table 16.1).

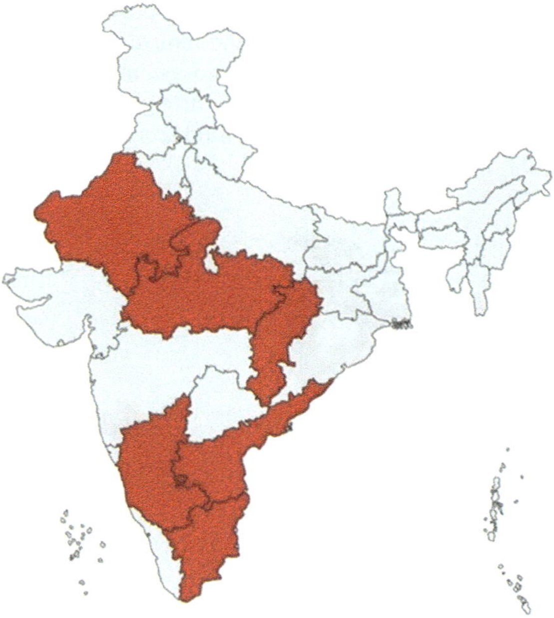

Fig. 16.4 Geographical distribution of foxtail millet cultivation in India (Kumari et al., 2024)

Table 16.1 Recent production status of millets and foxtail millet (Liu et al., 2025; Muskan et al., 2025)

Parameter	Recent value	Reference
Global millet production	~30–31 Mt (2023)	(Muskan et al., 2025)
Leading producing countries	India, China, Niger	(Muskan et al., 2025)
India's share	~40% of global production	(Muskan et al., 2025)
Foxtail millet major region	Asia (China, India)	(Liu et al., 2025)
Foxtail millet production potential	~9–10 Mt (China model)	(Liu et al., 2025)

16.3.3 Nutritional Attributes and Utilization in Food and Feed Systems

Foxtail millet has become one of the most admired wheat varieties due to its high-quality nutritional value, which goes beyond the caloric value. It has 11–15% protein, high dietary fiber, a balanced amino acid profile, and relatively high concentrations of essential micronutrients including iron, zinc and calcium. Foxtail millet has low glycemic index compared with the refined cereals making it especially appropriate to diabetics diet as well as in treating diet related metabolic disorders (Arora et al., 2023; Kalsi & Bhasin, 2023). In addition to macronutrients, foxtail millet is also rich in bioactive compounds (phenolics, flavonoids, antioxidant molecules) that make them function as a functional food. These compounds have been linked to less oxidative stress, better cardiovascular health and better metabolism. Historically, foxtail millet has been used as flat bread, porridges, and fermented foods but the residues of the straw and grain have been used as livestock food. The increased interest in the world demand in gluten-free, nutritionally balanced, and functional foods has also added to the topicality of foxtail millet in contemporary food and feed systems (Bandyopadhyay et al., 2017a, 2017b; Sharma & Niranjan, 2018) (Table 16.2).

Table 16.2 Nutrition composition of foxtail millet (Kalsi & Bhasin, 2023)

Parameter	Subtype	Composition
Moisture	–	12%
Protein	Total proteins	12.3/100 g
Fats	Crude fat	4.3/100 g
Starch/carbohydrates	Total carbohydrates	55–69/100 g
Ash	Mineral matter	3.3/100 g
Energy	–	351 kcal
Minerals	Calcium	2.4 mg/100 g
	Iron	81 mg/100 g
Crude fiber	Dietary fiber	6–8/100 g

16.3.4 Importance for Climate-Smart and Low-Input Farming Systems

The foxtail millet is also naturally adapted to climate-smart and low-input agricultural techniques because it is drought-, heat-, and nutrient-tolerant. Its C4 photosynthetic pathway also gives it high photosynthetic performance and water-use performance in high temperature and very low moisture conditions, thus it is able to build stable biomass using minimal water. These characteristics can greatly decrease irrigation and synthetic fertilizer reliance, and improve the sustainability of the environment and economic sustainability of smallholder farmers when paired with an effective nutrient uptake and remobilization within the body (Gao et al., 2025; Kumar et al., 2024).

Besides, foxtail millet is well-suited to diversified cropping systems, such as intercropping and crop rotation, in which foxtail millet helps to enhance resource-use efficiency, soil health and system stability. Its resistance to marginal land, and resilience in climatic uncertainty make it an option crop to build food and nutritional security as well as to maintain sustainable and resilient agroecosystems in case of climate change (Farooq et al., 2023; Kasanaboina et al., 2025).

16.4 Phenology and Developmental Dynamics of Foxtail Millet (*S. italica*)

Foxtail millet (*S. italica* (L.) P. Beauv.) has a highly plastic and rapid phenological development that is key to its stress-responsive adaptation. The crop has a normal period of 70–100 days with the life cycle depending on the genotype, temperature, moisture content and sensitivity to photoperiod. Such a short growth period enables foxtail millet to match the most important developmental stages with positive environmental periods and avoid terminal drought, heat stress, and nutrient constraints, which has been found to characterize foxtail millet compared to longer-duration cereals that grow this crop using rainfed systems (Aggarwal et al., 2022; Sachdev et al., 2021). The germination and early seedling development is fast and in most cases it takes 3–5 days when the soil is properly moist and of the correct temperature. The process of early vigor is facilitated by the effective mobilization of seed reserves and quick root growth to allow the crops to be planted in coarse-textured and nutrient-deplete soils. In early vegetative development, active tillering and leaf growth is supported by effective nitrogen absorption and the C4-mediated carbon assimilation to enable the stable canopy development during moderate drought and nutrient stress conditions (Aggarwal et al., 2022). The vegetative to reproductive development is a highly significant adaptive checkpoint. The foxtail millet panicles are usually initiated 30–40 days following sowing then flowered with a brief and synchronized flowering duration. In many genotypes early flowering or stress-induced flowering also occurs and allows the genotype to undergo a reproductive completion process before the severe heat or moisture stress kicks in. This shortened reproductive period mitigates this risk of floral sterility, which is a significant yield problem in crops that adjust to conditions of high temperature and water shortage. The grain filling process is effective and relatively short, and is usually 20–30 days in duration, and it depends on the capacity to remobilize carbon and nitrogen in vegetative tissues. Foxtail millet is capable of sustaining photosynthesis, effective nitrogen remobilization, and antioxidant activity, which increases the grain filling of foxtail millet during terminal drought and heat stress. Physiological and molecular-based evidence suggests that major metabolic activities are active during grain filling in the presence of stress and reduce yield losses, common to less tolerant cereals (Ma et al., 2024; Nadeem et al., 2020) Fig. 16.5 depicts the different growth stages of Foxtail millet. It matures very fast, and the physiological maturity occurs before the soil moisture becomes too low to allow seed viability and yield stability in low-input conditions. In general, the phenological approach of the foxtail millet is an indication of a highly synchronized interrelation of quick growth, nutrient economy and resistance to stress. This integration explains why it is appropriate to climate-smart agriculture and points to phenological traits as potential objectives of breeding programs to improve nutrient use efficiency and adaptability to changing climatic conditions (Farooq et al., 2023; Nadeem et al., 2020).

16.5 Nutrient Use Efficiency (NUE) in Foxtail Millet

16.5.1 Concepts of Uptake and Utilization Efficiency

The nutrient use efficiency (NUE) of crops is the combined ability to obtain nutrients in the soil (uptake efficiency) and transform nutrients obtained to biomass and economic yield (utilization efficiency). In foxtail millet (*S. italica*), the characteristic feature of adaptive trait is high NUE, which allows maintaining stable productivity in conditions of low inputs and marginal environments. Uptake efficiency is mainly controlled by prolific root development, extensive nutrient transport systems of high affinity and the nutrient mobilization by the rhizosphere, and utilization efficiency is determined by efficient internal nutrient distribution, remobilization and co-ordination of carbon and nitrogen metabolism during vegetative growth and grain filling. NUE-related phenotypic variance has been also reported to be of significant genotypic variation in foxtail millet, and this suggests that it offers a solid foundation of breeding cultivars to meet the requirements of agroecosystems that are resource-limited and vulnerable to climatic conditions (Sharma & Niranjan, 2018).

Fig. 16.5 Different growth stages of foxtail millet (*S. italica* L.) (Farooq et al., 2023)

16.5.2 Root Architecture and Rhizosphere Interactions

The main factor of nutrient uptake in the foxtail millet is root system architecture. The crop acquires dense fibrous and highly branched root system that is known to explore the soils optimally and support efficient uptake of nitrogen and phosphorus by the surface and subsoil layers. This structural plasticity is in favor of long-term development under poor-in-nutrient and water-scarce conditions. Moreover, foxtail millet roots do not only interact positively with beneficial rhizosphere microorganisms, such as plant growth-promoting rhizobacteria and arbuscular mycorrhizal fungi, but also promote phosphorus solubilization, availability of nitrogen, and micronutrient uptakes. Such root-microbe interactions greatly enhance the ability of the crop to obtain nutrients and lead to the better performance of the crop when used in low-input systems (Ahmad et al., 2018; Jin et al., 2017).

16.5.3 Physiological and Biochemical Basis of NUE

There is a cooperation of physiological and biochemical processes in the regulation of NUE in foxtail millet. C4 photosynthetic pathway guarantees high photosynthetic efficiency and better carbon assimilation of each unit of nitrogen and hence nutrient productivity in stressful situations. On the biochemical level, the growth under low nutrient status is supported by efficient nitrogen assimilation through the synchronized action of nitrate reductase and glutamine synthetase, effective phosphorus transportation and intracellular recycling. Moreover, improved antioxidant defense mechanisms defend metabolic activities against oxidative stress during drought and heat stress and enables the use of nutrients and formation of yield. All these combined mechanisms allow foxtail millet to be twice as productive with significantly reduced fertilizer application in contrast with the major cereals (Chen et al., 2023; Deng et al., 2024).

16.5.4 Role of Major Nutrients (N, P, K)

Nitrogen, phosphorus and potassium are key factors in revealing growth, yield, and NUE of foxtail millet. The nitrogen aids in the synthesis of chlorophyll, enzymes, and protein formation; phosphorus helps in the transfer of energy, root formation, and reproductive achievement; where potassium helps in controlling the osmotic potential, enzyme activation, and functions of stomata during periods of stress. The foxtail millet has been shown to have an efficient uptake and internal remobilization of these macronutrients and some genotypes of foxtail millet are able to maintain a grain yield in a lower fertilizer regime. This effective intake of macronutrients is especially beneficial to rain fed and low input systems, which strengthen the crop against nutrient and climatic stress (Xing et al., 2023; Yin et al., 2023).

16.5.5 Micronutrient Dynamics

Indispensable roles of micronutrients in foxtail millet are iron, zinc, manganese and copper, which are important in photosynthesis, redox regulation, and activities of enzymes. Although they are of low numerical necessity, the acquisition and translocation of micronutrients are essential in sustaining grain production and nutrient value. Even in low inherently fertile soils, foxtail millet has been shown to have an excellent capacity of micronutrient uptake with adaptive root characteristics and rhizosphere-mediated solubilization. These micronutrients are also cofactors of antioxidant enzymes and they increase tolerance of drought and heat stress as well as promote efficient metabolism of nutrients. These characteristics make foxtail millet a potential bio fortification and nutrition-sensitive crop to grow (Ceasar, 2023; Gouvêa et al., 2018).

16.5.6 Soil Health and Microbial Associations

One of the major stimulating factors of NUE in foxtail millet is soil health. Balanced organic matter soils, favorable PH and good microbial communities increase nutrient availability, root development and efficiency of nutrient uptake. Plant growth-promoting rhizobacteria and mycorrhizal fungi are associated with enhanced nutrient cycling, solubilization of phosphorus, biological fixing of nitrogen and uptake of water. These microbial activities supplement the nutrient efficiency of crop and minimize dependence on exogenous inputs, which promotes sustainable crop production on marginalized and degraded soils (Jin et al., 2017; Yu et al., 2025).

16.5.7 NUE Comparison with Rice, Wheat, and Maize

Foxtail millet has been found to have a significantly higher NUE compared to major cereals like rice, wheat and maize in low-input and stressful situations. Normal levels of fertilizer and water are usually needed to get high yields of rice and wheat but foxtail millet is self-sustained due to good nutrient uptake, low water use and because of good internal nutrient cycling. Foxtail millet has high efficiency in the remobilization of nutrients when under stress of drought and nutrients, even compared to maize, another C4 crop. The features render foxtail millet a valuable factor of sustainable crop rotation to decrease reliance on fertilizers, the costs of production, and ecological impacts during the climate change (Deng et al., 2024; Lu et al., 2024).

16.6 Climate Resilience Attributes of Foxtail Millet

16.6.1 Spectrum of Abiotic Stress Tolerance

Foxtail millet (*S. italica* L.) has a broad-spectrum tolerance to several of the largest abiotic stresses such as drought, heat, salinity, nutrient limitation, and moderate levels of chilling stress making it one of the climate-resilient cereal crops that is currently grown. This plasticity is essentially connected to its C4 photosynthetic pathway, brief growth cycle, high water- and nutrient-use efficiency and strong physiological plasticity. All of them allow the crop to endure both environmental conditions in which major cereals like rice and wheat suffer serious yield penalties, especially in semi-arid and marginal agroecosystems which have become more vulnerable to climate variability (Farooq et al., 2023; Nadeem et al., 2020). In addition to being tolerant to stresses, foxtail millet is highly moderately saline, nutrient poor, and degraded soils. The ability to retain metabolic activity and reproductive growth even in the event of resource inadequacy reflects its appropriateness in the low-input and climate-wise agricultural systems (Ma et al., 2024).

16.6.2 Drought and Heat Resilience Mechanisms

The most outstanding elements of climatic resilience in foxtail millet include drought and heat tolerance. This is as a result of high water-use efficiency that is realized in tight stomatal control, minimized transpiration losses and efficient osmotic adjustment by the buildup of compatible solutes like proline and dissolved sugars. Having a thick and fibrous root system increases the availability of remaining soil moisture, and high early vigor and speedy phenology lead to the life cycle completion before severe terminal drought and heat stress occur (Mundhe et al., 2025; Terfa et al., 2025). Anatomical and biochemical changes also support heat tolerance and these include cuticular thickening, membrane stability, and expression of heat-shock proteins. Simultaneously, a well-developed antioxidant defense system, including superoxide dismutase, catalase, and peroxidases helps to prevent oxidative stress during heat and drought stress, which guarantees continuation of photosynthesis and grain filling (Ardie et al., 2025; Mundhe et al., 2025).

16.6.3 Salinity and Chilling Stress Adaptation

Foxtail millet is moderately tolerant to stress, as it retains homeostasis of Na+/K+, and efficiently regulates osmotic stress, in response to salinity stress. The limitation of sodium translocation to aerial tissues and chronic retention of potassium retain cell integrity and maintain both cellular and enzymatic activities in saline conditions.

Salinity-induced oxidative stress is further alleviated by accumulation of compatible solutes and activation of antioxidant enzymes, which allows further growth in soils of moderate salinity (Ma et al., 2024; Nadeem et al., 2020). Though foxtail millet is not a cold-hardy cereal, it is tolerant enough to chilling stress to allow it to be established early in the season and have short-lived exposure to low temperatures. Chilling adaptation consists of stabilization of the lipids in the membrane, preservation of photosynthetic functionality, and soluble sugar accretion which preserves the cell structures in temporary cold events. These reactions enhance its agroecological flexibility, especially in temperate and high altitude areas that have changing temperatures (Nahar et al., 2018; Sreenivasulu et al., 2000).

16.6.4 Multi-stress Physiological Plasticity and Field-Level Resilience

The tolerance to various abiotic stresses simultaneously is a characteristic feature of foxtail millet, and it becomes even more relevant in the conditions of changing climate. This is a multi-stress resilience which is regulated in a coordinated manner through regulation of the root architecture, stomatal behavior, stress signaling networks, and antioxidant metabolism. This type of integration enables the crop to have the capability of sustaining carbon uptake, nutrient uptake and reproductive success in areas of drought, heat, salinity, and nutrient stress (Aggarwal et al., 2022; Li & Brutnell, 2011). South Asian and sub-Saharan African field evidence always show the high performance of foxtail millet in low rainfall, high temperatures, and unfavorable soil conditions in comparison with major cereals. Its predictable yields, high level of external inputs, and low chances of crop failure present it as a viable crop in marginal farming systems as a climate-smart cereal. Such observations support physiological and molecular evidence and strength the part of foxtail millet to increase food security in the changing climatic conditions (Farooq et al., 2023).

16.7 Integrating Nutrient Use Efficiency and Climate Resilience in Foxtail Millet

16.7.1 Functional Linkage Between Nutrient Use Efficiency and Stress Adaptation

The topic of nutrient use efficiency (NUE) and abiotic stress tolerance in foxtail millet (*S. italica* L.) is interrelated i.e. it is associated with physiological, biochemical and developmental processes. Effective intake and uptake of nitrogen, phosphorus and potassium promotes fundamental physiological functions, such as photosynthesis, osmotic regulation, and root development, which have a direct effect on drought, heat, and salinity tolerance. On the other hand, stress-adaptive characteristics inclusive of deep and plastic root systems, osmotic adjustment and flexible phenology enhance nutrient capture, and internal remobilization during nutrient-poor and water-limited environments. The two-way communication creates a positive cycle where stress resilience increases with the improvement of NUE, and tolerance to stress increases with the efficiency of nutrients, especially confirmed in the cases of low-input and marginal conditions (Deng et al., 2024; Panchal et al., 2023). The overlapping signaling networks that mediate growth, metabolism, and defense are elicited by the nutrient limitation and abiotic stress at the regulatory level. The availability of nutrients determines the antioxidant capacity and the expression of stress-reactive genes, whereas environmental stress alters the activity of nutrient transporters and internal recycling of nutrients. Such common regulatory processes imply the centrality of NUE as a climatic resilience factor as opposed to a yield-related quality (Nadeem et al., 2020; Panchal et al., 2023).

16.7.2 Shared Molecular, Hormonal, and Root-Mediated Mechanisms

NUE and stress adaptation are intertwined and mediate each other, in which foxtail millet integrates via a complex of interconnected molecular, hormonal, and root-level processes. The transcriptional networks controlled by stress and nutrients maintain genes that deal with nutrient uptake, carbon–nitrogen metabolism, and abiotic stress tolerance, which guarantee integration of growth under joint limitations. The nutrient sensing is connected to the root system architecture, stomatal regulation and assimilate partitioning via hormonal signaling, specifically the use of abscisic acid (ABA), auxin, and cytokinins. As an illustration, ABA accretion due to drought favors deeper root penetration and water-phobic water use, which indirectly boosts the nitrogen uptake, and the balance between the levels of auxin and cytokinins controls the lateral expansion of the roots and nutrient foraging in heterogeneous soils (Muthamilarasan et al., 2016; Yuan et al., 2025). Root architecture is a unifying characteristic since it enhances water and nutrient accessibility in one way. Its root system is fibrous and highly plastic, which allows adapting to nutrient gradients and environmental stress, a fact that can be found together in the literature and breeding and management interventions have a single physiological basis to improve the NUE and climate resilience (Farooq et al., 2023).

16.7.3 Metabolic Coordination Under Combined Nutrient and Abiotic Stress

Foxtail millet has been found to possess a high degree of metabolic plasticity following a combination of nutrient restriction and abiotic stress. Carbon and nitrogen metabolism are re-packaged to put more emphasis on synthesizing compatible solutes, such as proline and soluble sugars, which serve as osmoprotectants, redox buffers and cell structures stabilizers. These metabolic changes only decrease oxidative injury, preserve membrane integrity and sustain photosynthetic efficiency during combined drought, salinity or heat stress and low nutrient conditions (Ardie et al., 2025; Li & Brutnell, 2011). Coordinated action of the enzymes involved in carbon metabolism, nitrogen assimilation and antioxidant systems at the enzyme level is known to control efficient resource utilization and continued growth under stress. This close metabolic integration promotes the wind-break millet to remain yield constant with less external treatments and thus its application in the environmentally sensible and a resource-efficient agricultural regimes (Ma et al., 2024; Nadeem et al., 2020).

16.7.4 Systems Biology Insights, Challenges, and Opportunities

Systems biology methods can be used to gain greater insight into the interactions between NUE and climate resilience in foxtail millet by combining genomic, transcriptomic, and proteomic and metabolomics with physiological and agronomic phenotypes. These integrative studies make it easier to determine regulatory centers and major genes that regulate nutrient efficiency and stress resistance simultaneously to make breeding decisions more focused and specific (Ardie et al., 2025; Panchal et al., 2023). Although these have been improved upon, parallel enhancement of NUE and stress resilience has been difficult because of the high genotype x environment interactions and low throughput phenotyping platforms using combined nutrient stress phenotypes. However, the current developments in the field of molecular breeding, gene editing, and combination with multi-omics, as well as precision nutrient and water management, have a significant potential to address these limitations. Capacitance-based genetic enhancement could be used to develop nutrient- and climate-resistant foxtail millet varieties using strategic integration of adaptation agronomic management to enable the crop to become the backbone of a sustainable food system in the era of climate change (Ardie et al., 2025; Farooq et al., 2023).

16.8 Agronomic Interventions to Enhance NUE and Climate Resilience

16.8.1 Integrated Fertilizer and Nutrient Management Practices

The key to enhanced nutrient use efficiency (NUE) and stress tolerance in foxtail millet (*S. italica* L.) grow in rainfed and low-input production systems is efficient management of fertilizers. Nitrogen, phosphorus and potassium have been applied in a balanced and site-specific manner and improve nutrient availability, as well as minimizing volatilization, leaching and immobilization losses. Recent reports with foxtail millet and related millets show that the carefully timed nutrient supply and, particularly, split nitrogen application, in accordance with the phenological requirement, leads to a decrease in nutrient salvage and maintenance of yield in the water-prone and heat-prone environments (Ardie et al., 2025; Ma et al., 2024) Control of nutrient availability increases physiological responses related to coping with stress as well. Proper nutrition in nitrogen and phosphorus improves the growth of roots, photosynthetic efficiency, and antioxidant activity, which improves the performance of plants during water and temperature stress. All these answers designate the mutual connectivity between NUE and climate resilience; along with integrated nutrient management as the main approach to stabilizing the foxtail millet production in the conditions of climate fluctuations (Aggarwal et al., 2022; Fukunaga & Kawase, 2024).

16.8.2 Organic Amendments, Biofertilizers, and Soil Health Enhancement

It has been established that organic amendments and biofertilizers help to enhance the NUE and adaptive capacity of foxtail millet through the positive effects on the physical structure of soil, its ability to retain nutrients, and its microbial activity. The use of farmyard manure, compost and vermicompost enhances the soil organic carbon, aggregation properties and the soil moisture holding property that in turn promote the effective absorption of nutrients in the marginal soil situations. Facts on the millet-based systems show that organic sources of nutrients enhance better nutrient synchronization and elimination of chemical dependence on fertilizers without affecting the stability of yields (El-Beltagi et al., 2025). Fertilizers that contain nitrogen-fixing bacteria, phosphate-solubilizing bacteria and mycorrhizal fungi also increase the nutrient availability and assimilation efficiency. The nutrient mobilization that occurs through rhizosphere enhances root development, acquisition of nutrients and stress buffering in nutrient-limiting and drought-stressful conditions. Those biological inputs can thus be used as bi-primer to boost NUE and climate resilience in agroecosystems (Aggarwal et al., 2022; Gobikashri et al., 2025).

16.8.3 Soil Amendments and Water-Efficient Agronomy

The addition of water-efficient agronomic practices like silicon and biochar to soil, together with the strategy of improving NUE and the biochemical stress tolerance of foxtail millet can be used effectively. Silicon usage has demonstrated to enhance structural stability of the plant tissues, decrease transpirational vents and increase anti-oxidant defense and thus drought heat salt stress contributes to its tolerance to such stresses in millets. These physiological advantages are tangentially related to enhanced nutrient use in the case of stress (Fukunaga & Kawase, 2024; Nadeem et al., 2018). Biochar amendment enhances soil porosity, cation exchange capacity and moisture retention resulting into enhanced nutrient accessibility and root development with low-input systems. Biochar, when used with water saving technologies like mulching and deficit irrigation also increase soil moisture-nutrient interactions and to elevate the efficiency of fertilizer recovery. These combined soil–water-nutrient techniques are relevant especially when foxtail millet is grown in semi-arid and climate vulnerable environments (Ardie et al., 2025; Ma et al., 2024).

16.8.4 Cropping System Diversification and Precision Agriculture Approaches

Intercropping/rotation with legumes, diversifying the cropping system, in particular foxtail millet makes NUE and climate resilient, because of better nutrient cycling, soil fertility and system stability. Legume integration also brings in biologically fixed nitrogen and higher levels of microbial activity lowering the externality of nitrogen inputs whilst also increasing the general resource-use efficiency. It has been demonstrated, using millet-based diversified systems, that they are more stable and resilient to climatic stress than mono-cropping systems (Nadeem et al., 2020). Precision based agriculture methods also enable agriculture management of foxtail millet in terms of nutrients and water. Timely diagnosis of the deficit of nutrients and stress conditions, as well as precise application of inputs, is possible with the help of remote sensing, decision-support tools and site-specific nutrient management. These technologies have a high potential to increase NUE and lower production costs as well as enable better climate resilience when they are combined with naturally resilient crops like foxtail millet, even though they are still in the early stages of development in millet-based systems (Ardie et al., 2025; Ma et al., 2024).

16.9 Case Studies on Nutrient Use Efficiency (NUE) and Climate Resilience in Foxtail Millet (*S. italica*)

16.9.1 Case Study 1: Effects of NPK Fertilization on Growth, Yield, and Physiological Perfrormance of Foxtail Millet

In a controlled greenhouse study conducted by Shahzadi et al. (2025) in the University of Agriculture, Faisalabad, Pakistan they assessed the impact of graded fertilization of foxtail millet (*S. italica*) using NPK to determine the influence of these graded applications on growth, physiological characteristics, and yield of eight foxtail millet lines. They were treated by five nutrient regimes, namely, 0:0:0 (control), 20:15:15, 30:20:20, 40:25:25 and 50:30:30 kg ha^{-1}, at one and four weeks of emergence respectively. Some of the parameters measured were emergence traits, crop growth rate (CGR), net assimilation rate (NAR), chlorophyll content, biomass, and the components of grain yield. Genotype-uniform emergence was fast and consistent, with 50% emergence of between 4.33 and 5.92 days and endogenous emergence of over 89% in most of the lines. The growth performance was higher with increase in the level of NPK: CGR increased when fertilized (3.13 g plant^{-1} day^{-1}) was highest compared to the control (1.46 g plant^{-1} day^{-1}) and NAR was the highest (3.13 g plant^{-1} day^{-1}) which showed an improvement in photosynthetic efficiency. The greatest grain per panicle (2.12 g) and optimum plant height (up to 84.2 cm) were observed at moderate fertilization (30:20:20 kg ha^{-1}) and the highest value at maximum accumulation of biomass (12.59 g plant^{-1}) was at the highest NPK level. The contents of chlorophyll a and b were increased by a maximum of 23 and 15.8, respectively when high nutrient supply was achieved that would facilitate enhanced carbon assimilation. The researchers concluded that in foxtail millet, balanced NPK application and especially 30:20:20 kg ha^{-1} can help optimize growth, physiological efficiency, and yield. These findings demonstrate the high nutrient responsiveness of the crop and its ability to attain high nutrient use efficiency at moderate levels of input validating its suitability in sustainable and climate-resilient production systems (Shahzadi et al., 2025).

16.9.2 Case Study 2: Simulation of Nutrient–Water Interactions and Grain Quality in Foxtail Millet

Zhou et al. (2025) designed and tested a more efficient DSSAT-CERES-Millet model to simulate the effects of irrigation (rainfed to full irrigation) and nitrogen (0–270 kg N ha^{-1}) on foxtail millet production of grain, protein, and starch accumulation through field experiments that were carried out in northern China (2021–2023). The simulation of model performance was strong whereby the NRMSE values are 8.27–18.35%

of the leaf area index, 1.65–15.07% of the biomass, and approximately 9.7% of the grain yield. The integrated quality modules well modeled both grain protein and starch dynamics with NRMSE of 3.06–26.22 based on protein accumulation, and 4.06–26.88 based on starch accumulation, and amylase and amylopectin at maturity were simulated with <14% error. It was found that there was a high water-nutrient synergy and that optimized nitrogen supply and moderate irrigation enhanced yield stability and nutritional quality of the grain. The research showed that process-based modeling can be useful to promote nutrient-efficient and climatic robust management approaches to foxtail millet in fluctuating environmental settings (Zhou et al., 2025).

16.9.3 Case Study 3: Irrigation Management, Yield Response, and Water Productivity of Foxtail Millet in the North China Plain

Zhang et al. used a two-year field experiment (2020–2021) in the city of Hengshui located at the North China Plain to conclude the optimum irrigation requirement of summer foxtail millet (*S. italica* L.) and to estimate its water productivity and root distribution. Six irrigation regimes were tested in the study namely I0 (0 mm), I1 (60 mm), I2 (120 mm), I3 (150 mm), I4 (180 mm), and I5 (210 mm), where irrigation was applied during the jointing and grain-filling stages. Measures made were grain yield (GY), biomass, thousand grain weight (TGW), water productivity (WP), and root characteristics. In 2020, GY rose to irrigation levels to I3–I5 with maximum yield of 7.44 t ha^{-1} at I5 and minimum yield under the control (I0). The same trends are evident in 2021, and the maximum GY of 6.42 t ha^{-1} under I5, then 6.22 t ha^{-1} under I3 were reported, and the higher irrigation levels did not significantly differ. When irrigated, biomass, TGW and panicle dry weight improved, whereas abortive grain rate reduced at intermediate irrigation levels. The water productivity (WP) based on grain yield/Total water consumed revealed two large peaks under I0 (30.10 $kgmm^{-1}$ hm^{-2} in 2020 and 27.44 $kgmm^{-1}$ hm^{-2} in 2021) and I3 (27.84 $kgmm^{-1}$ hm^{-2} in 2020) and I2 (28.80 $kgmm^{-1}$ hm^{-2} in 2021). The highest water levels exhibited the lowest values of WP (I5 in 2020 having 23.70 kg mm^{-1} hm^{-2}, and I4 in 2021 with 21.48 kg mm^{-1} hm^{-2}). These findings suggest that the optimal irrigation (120–150 mm) is the one that maximizes the water productivity as well as maintains high yields. Root distribution study revealed that the intermediate level of irrigation (I2/I3) did improve the total root length density, plane area, root volume and root dry weight over no-irrigation and over irrigation activity. The analysis of grey relationship revealed that the effect of total water content at the shallow soil layer (0–40 cm) was the strongest in the two years resulting in the significance of water content in upper soil layer in facilitating water and nutrient absorption of millet roots. The authors concluded that foxtail millet prefers to irrigate with the amount of about 150 mm mostly because it has a good balance between yield, water productivity, root establishment, and economic returns. This paper highlights the significance of streamlined water management in enhancing efficiency in resource utilization and resilience of crops to water scarcity (Zhang et al., 2025).

16.9.4 Case Study 4: Phenotypic Diversity and Trait Correlations in Foxtail Millet Germplasm

Bimochana et al. (2023) carried out a field assessment of phenotypic diversity and relationship between traits in foxtail millet (*S. italica* L.) by using 120 accessions where accessions were trailed at the conditions of rainfed in Nepal based on a randomized complete block design with trio repetitions. Qualitative (panicle shape, grain colour) and quantitative traits (plant height, days to 50% flowering, days to maturity, number of tills per plant, and weight of grain per plant) were measured to determine the adaptive variability as it applies to low-input systems. The research found a broad range of numbers between the accessions and the plant height was 74.6–129.3 cm, days to flowering was 61–78 days, days to maturity was 100–125 days and the weight of the grains per plant was 18.2–32.7 g and these numbers illustrated high genetic plasticity. Correlation analysis demonstrated that yield and major characteristics like productive tillers and panicle characteristics were significantly positively associated whereas days to 50% flowering had a strong positive correlation with days to maturity ($r = 0.71$–0.83; $p < 0.05$). Plant height was also positively correlated to panicle length and grain weight indicating productivity may have been indirectly selected through indirect action. On the whole, the results reveal that the foxtail millet current phenotypic variation can be strategically utilised to produce cultivars that are the renunciation of equal parts of early maturity, stable output, and resource-usage efficiency, which enhances climate-pentacity and performance of the marginal rainfed agroecosystems. The findings suggest the strategic utilisation of current phenotypic variance in breeding cultivars that have unbiased maturity rates, consistent yield, and efficiency in resource usage (Bimochana et al., 2023).

16.10 Conclusion

Foxtail millet (*S. italica* L.) is a high potential climate-resistant and nutrient-efficient crop with huge potential in sustainable farming in the light of growing environmental limitations. The short life cycle, C4 photosynthetic pathway, adaptive canopy and root features allow efficient use of water and nutrients allowing it to remain stable in the face of drought, heat and low fertility soils. The mentioned features ensure that foxtail millet is a good fit in rainfed and marginal agroecosystems wherein resources are scarce. Combination of nutrient use efficiency (NUE) and stress tolerance in foxtail millet is regulated by the coordination of physiological, biochemical and molecular processes. Effective uptake of nutrients, internal recycling and harmonious metabolism of carbon-nutrient also promote continuous photosynthesis and grain completion in unfavorable environments. Recent directions in genomics, phenotyping, and crop modeling have only intensified the model as a C4 crop model of nutrient-stress interactions, as well as, in recognizing important traits of resilience and productivity. In addition to yield stability, foxtail millet is also a source of nutritional security because of its high dietary fiber, mineral compositions and low glycemic index, fitting in the nutrition-sensitive and climate-smart food systems. The fact that it is compatible with low-input management practices and diversified systems of cropping also contributes to soil health as well as curtailing environmental footprint. The foxtail millet is an overall promising direction to improve the agricultural sustainability through its resource-use efficiency and climate adaptability, as well as its nutritional value. Such integration must be well-chosen by strategic genetic improvement and adaptive agronomic management, and thus hold the potential of genetic improvement in the future food systems.

References

Aggarwal, P. R., Pramitha, L., Choudhary, P., Singh, R. K., Shukla, P., Prasad, M., & Muthamilarasan, M. (2022). Multi-omics intervention in Setaria to dissect climate-resilient traits: Progress and prospects. *Frontiers in Plant Science, 13*, 892736. https://doi.org/10.3389/fpls.2022.892736

Ahmad, Z., Nadeem, F., Wang, R., Diao, X., Han, Y., Wang, X., & Li, X. (2018). A larger root system is coupled with contrasting expression patterns of phosphate and nitrate transporters in foxtail millet [*S. italica* (L.) Beauv.] under phosphate limitation. *Frontiers in Plant Science, 9*, 1367. https://doi.org/10.3389/fpls.2018.01367

Ardie, S. W., Nugroho, R. B., Dirpan, A., & Anshori, M. F. (2025). Foxtail millet research in supporting climate change resilience efforts: Bibliometric analysis and focused literature review. *Heliyon, 11*(3), e42348. https://doi.org/10.1016/j.heliyon.2025.e42348

Arora, L., Aggarwal, R., Dhaliwal, I., Gupta, O. P., & Kaushik, P. (2023). Assessment of sensory and nutritional attributes of foxtail millet-based food products. *Frontiers in Nutrition, 10*, 1146545. https://doi.org/10.3389/fnut.2023.1146545

Bandyopadhyay, T., Jaiswal, V., & Prasad, M. (2017a). Nutrition potential of foxtail millet in comparison to other millets and major cereals. In M. Prasad (Ed.), *The foxtail millet genome* (pp. 123–135). Springer. https://doi.org/10.1007/978-3-319-65617-5_10

Bandyopadhyay, T., Muthamilarasan, M., & Prasad, M. (2017b). Millets for next generation climate-smart agriculture. *Frontiers in Plant Science, 8*, 1266. https://doi.org/10.3389/fpls.2017.01266

Bimochana, G. C., Upadhyay, A., Dhakal, P., Bashyal, S., Adhikari, S., Poudel, P. K., & Ayer, D. K. (2023). Characterization of qualitative and quantitative traits for diversity assessment and correlation studies in foxtail millet (*S. italica* (L.) BEAUV.). *Malaysian Journal of Sustainable Agriculture, 7*(2), 88–97.

Ceasar, S. A. (2023). Foxtail millet (*S. italica*) as a model system to study and improve the nutrient transport in cereals. *Plant Growth Regulation, 99*(1), 3–10. https://doi.org/10.1007/s10725-022-00878-x

Chen, E., Qin, L., Li, F., Yang, Y., Liu, Z., Wang, R., Yu, X., Niu, J., Zhang, H., Wang, H., Liu, B., & Guan, Y. (2023). Physiological and transcriptomic analysis provides insights into low nitrogen stress in foxtail millet (*S. italica* L.). *International Journal of Molecular Sciences, 24*(22), 16321. https://doi.org/10.3390/ijms242216321

Deevi, K. C., Swamikannu, N., Pingali, P. R., & Gumma, M. K. (2024). Current trends and future prospects in global production, utilization, and trade of pearl millet. In V. A. Tonapi,

N. Thirunavukkarasu, S. Gupta, P. I. Gangashetty, & O. Yadav (Eds.), *Pearl millet in the 21st century* (pp. 1–33). Springer. https://doi.org/10.1007/978-981-99-5890-0_1

Deng, Y., Chen, Y., Kou, T., Bo, Y., Zhao, M., & Zhu, F. (2024). Domestication affects nitrogen use efficiency in foxtail millet. *Euphytica, 220*(6), 93. https://doi.org/10.1007/s10681-024-03351-0

Diao, X., & Jia, G. (2017). Origin and domestication of foxtail millet. In A. Doust, & X. Diao (Eds.), *Genetics and genomics of setaria* (Vol. 19, pp. 61–72). Springer. https://doi.org/10.1007/978-3-319-45105-3_4

El-Beltagi, H. S., Abdel-Haleem, M., Khedr, N., Shalaby, T. A., El-Mogy, M. M., & Khedr, E. H. (2025). Millets as climate-resilient crops: Emerging technologies for sustainable postharvest management. *Journal of Food Composition and Analysis, 147*, 107999. https://doi.org/10.1016/j.jfca.2025.107999

Farooq, M., & Siddique, K. H. M. (Eds.). (2023). *Neglected and underutilized crops: Future smart food*. Academic Press.

Fukunaga, K., & Kawase, M. (2024). Crop evolution of foxtail millet. *Plants, 13*(2), 218. https://doi.org/10.3390/plants13020218

Gao, S., Wang, C., Yang, X., Ji, T., Shang, S., Li, S., Wen, Y., Ren, J., Li, X., Zhao, J., Hu, C., Yuan, X., & Dong, S. (2025). Adaptability of foxtail millet varieties based on photosynthetic performance and agronomic traits. *Plants, 14*(16), 2502. https://doi.org/10.3390/plants14162502

Gobikashri, N., Kumar, K. U., Nivetha, C., Punithkumar, S., Kousalya, A., Flora, G. J., Dhivyalakshimi, T., & Jeyasingh, R. A. I. (2025). A brief review on world of foxtail millet: From botanical characteristics to market trends. *Journal of Applied Biology and Biotechnology*. https://doi.org/10.7324/JABB.2025.243269

Gouvêa, L. F. C., Moreira, A. J., Freschi, C. D., & Freschi, G. P. G. (2018). Speciation of nitrite, nitrate and p-nitrophenol by photochemical vapor generation of NO using high-resolution continuum source molecular absorption spectrometry. *Journal of Food Composition and Analysis, 70*, 28–34. https://doi.org/10.1016/j.jfca.2018.04.003

Hu, H., Mauro-Herrera, M., & Doust, A. N. (2018). Domestication and Improvement in the model C4 grass, Setaria. *Frontiers in Plant Science, 9*, 719. https://doi.org/10.3389/fpls.2018.00719

Jin, T., Wang, Y., Huang, Y., Xu, J., Zhang, P., Wang, N., Liu, X., Chu, H., Liu, G., Jiang, H., Li, Y., Xu, J., Kristiansen, K., Xiao, L., Zhang, Y., Zhang, G., Du, G., Zhang, H., Zou, H., et al. (2017). Taxonomic structure and functional association of foxtail millet root microbiome. *GigaScience, 6*(10), 1–12. https://doi.org/10.1093/gigascience/gix089

Kalsi, R., & Bhasin, J. K. (2023). Nutritional exploration of foxtail millet (*S. italica*) in addressing food security and its utilization trends in food system. *eFood, 4*(5), e111.

Kasanaboina, K., Ishwarya Lakshmi, V., Nandigam, S. R., Jagadeesh, K., & Ceasar, S. A. (2025). The role of millets in climate-resilient agriculture for conserving global food security amidst climate change. *The Nucleus*. https://doi.org/10.1007/s13237-025-00599-2

Kumar, S., Kumar, A., Sen, H., Janeja, H. S., Maity, S., Banerjee, S., Singh, P., & Channapur, A. M. (2024). Small millets: A multifunctional crop for achieving sustainable food security under climate change. *Plant Science Today, 19*, 4113.

Kumari, N., Prakash, A., Tiwari, P., Kumar, A., Ranjan, S., Ray, P., Taku, M., Rajendran, A., & Ramlal, A. (2024). Origin, diversity, floral biology, pollination, and genetics in foxtail millet. In S. Mishra, S. Kumar, & R. C. Srivastava (Eds.), *Genetic improvement of small millets* (pp. 351–364). Springer. https://doi.org/10.1007/978-981-99-7232-6_15

Li, P., & Brutnell, T. P. (2011). *Setaria viridis* and *S. italica*, model genetic systems for the Panicoid grasses. *Journal of Experimental Botany, 62*(9), 3031–3037. https://doi.org/10.1093/jxb/err096

Liu, L., Zhang, J., Sun, Z., & Zhang, Q. (2025). Unlocking the potential of foxtail millet: Spatial strategy and multiple benefits of expanding cultivation on marginal lands in China. *Agricultural Systems, 229*, 104420. https://doi.org/10.1016/j.agsy.2025.104420

Lu, Y., Zhao, Y., Xia, X., Liu, M., Wei, Z., Wang, J., Liu, J., Cui, J., & Li, S. (2024). Evaluation of nitrogen fertilizer supply and soil nitrate thresholds for high yields of foxtail millet. *Agriculture, 14*(10), 1711. https://doi.org/10.3390/agriculture14101711

Ma, K., Jia, Z., Wen, X., & Chen, F. (2025). Genetic gain of yield and agricultural phenotype in foxtail millet from 1970s to 2020s in China. *European Journal of Agronomy, 170*, 127780. https://doi.org/10.1016/j.eja.2025.127780

Ma, K., Zhao, X., Lu, B., Wang, Y., Yue, Z., Zhang, L., Diao, X., & Yuan, X. (2024). Effect of ecological factors on nutritional quality of foxtail millet (*S. italica* L.). *Agronomy, 14*(2), 387. https://doi.org/10.3390/agronomy14020387

Mundhe, B. S., Sakhare, V. V., Andhale, M. D., Moharil, M. P., Gawali, S. R., & Penna, S. (2025). Foxtail millet. In S. A. Ceasar, S. Penna, C. W. P. Carvalho, & S. M. Jain (Eds.), *Millets: Crops for climate resilience and for food and nutritional security* (pp. 33–65). Springer. https://doi.org/10.1007/978-981-95-1256-0_2

Muskan, F., Redhu, M., Redhu, S., & Rahimi, M. (2025). Millets in the global market: A critical review of challenges and opportunities. *Food Production, Processing and Nutrition, 7*(1), 37. https://doi.org/10.1186/s43014-025-00316-1

Muthamilarasan, M., Dhaka, A., Yadav, R., & Prasad, M. (2016). Exploration of millet models for developing nutrient rich graminaceous crops. *Plant Science, 242*, 89–97. https://doi.org/10.1016/j.plantsci.2015.08.023

Nadeem, F., Ahmad, Z., Ul Hassan, M., Wang, R., Diao, X., & Li, X. (2020). Adaptation of foxtail millet (*S. italica* L.) to abiotic stresses: A special perspective of responses to nitrogen and phosphate limitations. *Frontiers in Plant Science, 11*, 187. https://doi.org/10.3389/fpls.2020.00187

Nadeem, F., Ahmad, Z., Wang, R., Han, J., Shen, Q., Chang, F., Diao, X., Zhang, F., & Li, X. (2018). Foxtail millet [*S. italica* (L.) Beauv.] grown under low nitrogen shows a smaller root system, enhanced biomass accumulation, and nitrate transporter expression. *Frontiers in Plant Science, 9*, 205. https://doi.org/10.3389/fpls.2018.00205

Nahar, A., Mannan, M., Mamun, M., & Ghosh, T. (2018). Growth and yield performance of foxtail millets under salinity. *Bangladesh Agronomy Journal, 21*(1), 51–59. https://doi.org/10.3329/baj.v21i1.39360

Panchal, A., Singh, R. K., & Prasad, M. (2023). Recent advancements and future perspectives of foxtail millet genomics. *Plant Growth Regulation, 99*(1), 11–23. https://doi.org/10.1007/s10725-022-00858-1

Sachdev, N., Goomer, S., & Singh, L. R. (2021). Foxtail millet: A potential crop to meet future demand scenario for alternative sustainable protein. *Journal of the Science of Food and Agriculture, 101*(3), 831–842. https://doi.org/10.1002/jsfa.10716

Shahzadi, H., Khan, S., Al-Hashimi, A., Basra, S. M. A., Mehmood, K., Wahid, M. A., Nawaz, M., Irshad, S., Haseeb, A., Rais, A., Gul, S., Ibrar, D., & Dwiningsih, Y. (2025). Preliminary study on unlocking growth and yield potential of USDA foxtail millet (*S. italica* L.) lines with NPK fertilization. *BMC Plant Biology, 25*(1), 22. https://doi.org/10.1186/s12870-024-05960-2

Sharma, N., & Niranjan, K. (2018). Foxtail millet: Properties, processing, health benefits, and uses. *Food Reviews International, 34*(4), 329–363. https://doi.org/10.1080/87559129.2017.1290103

Singh, A. K. (2023). Early presence/introduction of African and East Asian millets in India: Integral to traditional agriculture. *The Nucleus, 66*(3), 261–271. https://doi.org/10.1007/s13237-023-00435-5

Sreenivasulu, N., Grimm, B., Wobus, U., & Weschke, W. (2000). Differential response of antioxidant compounds to salinity stress in salt-tolerant and salt-sensitive seedlings of foxtail millet (*S. italica*). *Physiologia Plantarum, 109*(4), 435–442. https://doi.org/10.1034/j.1399-3054.2000.100410.x

Terfa, G. N., Pan, W., Hu, L., Hao, J., Zhao, Q., Jia, Y., & Nie, X. (2025). Mechanisms of salt and drought stress responses in foxtail millet. *Plants, 14*(8), 1215. https://doi.org/10.3390/plants14081215

Venkatesh-Babu, D., Raghavendra, T., & Satish-Kumar, Y. S. (2025). *Screening of foxtail millet (S. italica L.) varieties for water use efficiency traits and yield potential under rainfed situation* [Pdf]. 1.8 MB. https://doi.org/10.22004/AG.ECON.371476

Xing, G., Ma, J., Liu, X., Lei, B., Wang, G., Hou, S., & Han, Y. (2023). Influence of different nitrogen, phosphorus, and potassium fertilizer ratios on the agronomic and quality traits of foxtail millet. *Agronomy, 13*(8), 2005. https://doi.org/10.3390/agronomy13082005

Yin, M., Li, Y., Hu, Q., Yu, X., Huang, M., Zhao, J., Dong, S., Yuan, X., & Wen, Y. (2023). Potassium increases nitrogen and potassium utilization efficiency and yield in foxtail millet. *Agronomy, 13*(9), 2200. https://doi.org/10.3390/agronomy13092200

Yu, G., Han, Y., Liu, P., Hao, H., & Li, M. (2025). Response of foxtail millet yield, soil chemical property and bacterial community to different green manure-foxtail millet rotation models in North China. *Frontiers in Microbiology, 16*, 1558354. https://doi.org/10.3389/fmicb.2025.1558354

Yuan, X., Bai, X., Yu, J., Jia, Z., & Wang, C. (2025). Genome-wide identification of the BREVIS RADIX gene family in foxtail millet: Function, evolution, and expression. *Genes, 16*(4), 374. https://doi.org/10.3390/genes16040374

Zhang, W., Wang, B., Liu, B., Chen, Z., Lu, G., Bai, C., & Ge, Y. (2025). Response of summer foxtail millet yield and water productivity to water supply in the North China plain. *Agronomy, 15*(11), 2468. https://doi.org/10.3390/agronomy15112468

Zhou, S., Liu, Z., & Chen, F. (2025). Developing the DSSAT-CERES-millet model for dynamic simulation of grain protein and starch accumulation in foxtail millet (*S. italica*) under varying irrigation and nitrogen regimes. *Plants, 14*(6), 910. https://doi.org/10.3390/plants14060910

Chapter 17
Triticale: A Climate-Resilient Pseudocereal with Benefits Food and Energy Nexus

Zohaa Fatima, Muqaddas Noor, Abdullah Jan, Umar Farooq, Amjad Malik, Mukhtar Ahmed, and Shakeel Ahmad

Abstract A hybrid of wheat (***Triticum*** spp.) and rye (***Secale cereale***) is a pseudo-cereal known as triticale (× ***Triticosecale***) with enormous potential and can survive the climate in integrated food and energy systems. Its ability to adapt to marginal, low-fertility and rainfed agroecosystems coupled with moderate heat and water stress tolerance has enabled it to be grown in areas where other staple cereals cannot be grown. Triticale is a multi-purpose type of crop that can be used as nutritious grain food. It is also the source of straw that enables the production of bioenergy such as biogas, bioethanol, solid fuel and so on. Research shows that some genotypes of triticale produces grain yields ranging from 4.83 to 6.92 t ha^{-1} and fresh forage biomass between 20.06 and 29.78 t ha^{-1}, hence is viable in the production of renewable energy. Its moderate level of nitrogen and phosphorus demand coupled with ease of adaptation in the soil makes it quite effective especially in rotation or intercropping with legumes. It has been indicated that the inclusion of triticale into resilient cropping systems is both beneficial in land-use efficiency and soil health as well as climate adaptability. It also provides contribution to the renewable energy sector. The key aim of this chapter is to have a detailed overview of the agronomic characteristics of triticale, nutritional content and bioenergy capabilities in attainment of sustainable food and energy nexus. Combination of empirical data, crop modelling, use of remote sensing and socio-economic policies ensures that triticale produces the most benefits to climate-resilient and sustainable agricultural systems.

Z. Fatima (✉) · M. Noor · A. Jan · U. Farooq · M. Ahmed
Department of Agronomy, Pir Mehr Ali Shah Arid Agriculture University, Rawalpindi, Pakistan
e-mail: zohaafatimaniazi@gmail.com

A. Malik
Kenanow Bachelor of Education Program Faculty of Education, University College of the North (UCN), Thompson, MB, Canada

S. Ahmad
Institute of Agronomy Faculty of Agricultural Sciences and Technology, Bahauddin Zakariya University Multan, Multan, Pakistan

M. Ahmed (ed.), *Climate Change and Neglected Underutilized Crops*,
https://doi.org/10.1007/978-3-032-21005-0_17

Keywords Triticale · Pseudocereal · Food security · Climate resilience · Bioenergy

17.1 Introduction

Triticale (× *Triticosecale*) is a man-made hybrid of wheat (*Triticum* spp.) and rye (*Secale cereale*) combining the desirable traits of both parents. It is noticeable in cereal crops for its agronomic advantages such as high drought adaptability, short to medium growth cycle of 90–120 days, relatively low input requirements and tolerance to low-fertility soils. Even though it has dual-purpose potential and climate-resilience, triticale is underutilized due to limited awareness, insufficient breeding programs and underdeveloped value chains. These limitations causes decrease in its contribution to food and energy security (Gaviley et al., 2024; Malherbe et al., 2023). There is an increasing global need of food and renewable energy in parallel with the increasing population growth, urbanization and climate change issues. The existing agricultural systems have been exposed to various kinds of pressures such as water scarcity, soil degradation, nutrient depletion and extreme weather patterns. In this scenario, food crops that are concurrently used in food security production and renewable energy generation have been the subject of a new scientific attention. Triticale is an upcoming breed that could be used in such dual-purpose approaches because it is a hybrid crop with the yields of wheat and the versatility and stress-resistance of rye. This makes triticale suitable especially to rainfed, semi-arid and marginal lands where other cereals usually fail (Gaviley et al., 2024).

Triticale can yield grain that is rich in proteins, fibre and the essential micronutrients which can be used in the human and animals' diets. Its straw which has always been thought of as its by product is being recognized as a worthy source of biomass for energy generation. The use of anaerobic digestion to make biogas, fermentation to make bioethanol and direct combustion to make solid biofuel allow farmers to make good use of the residues. These processes help to increase the energy security of the farm and reduce waste (Malherbe et al., 2023). Because of food and energy production, triticale is part of the sustainable agricultural systems. It encourages the principles of the circular economy and provides potential income diversification to the smallholder farmers. This chapters covers characteristic of triticale such as botanical, agronomical and nutritional. In conclusion, triticale is a dual-purpose solution to the food security and renewable energy production intersecting problems. Its potential is supported by its agronomic resilience, nutritional importance, marginal lands adaptation and co-existence with other cropping systems used as a resilient crop in the sustainable food energy nexus solutions.

17.2 Botanical Identity, Taxonomy, and Distribution

17.2.1 Taxonomy and Botanical Description

Triticale (× ***Triticosecale***) is an amphidiploid hybrid crossing from a Poaceae family between wheat (***Triticum*** spp.) and rye (***Secale cereale***). The hybrid combines the high yielding nature of the wheat and the environmental stability of the rye. This makes the hybrid a very adaptable cereal that can be used in various agroecological regions. Triticale is mainly self-pollinating and there is a low-frequency rate of cross-pollination enabling stability of the traits as well as providing the opportunity to breed hybrids (Gaviley et al., 2024). The plant is an annual grass, which has a varying height between 80–120 cm depending on the cultivar and environmental conditions. The linear leaves are glaborus or has slightly hairy lamina and range between 25 and 40 cm. It has spike shaped inflorescence like a wheat but a little denser with sessile or a little pedicelled spikelet (Fig. 17.1). The grains are tiny, oval and compact containing a hard endosperm which makes them suitable for milling and animal feed. The straw is long, fibrous and sturdy with high lignocellulosic content and can be used for the production of solid biofuel, bioethanol and biogas (Faccini et al., 2023).

The triticale hybrid quality has the benefit of having a wider genetic base than either of the corresponding components of wheat and rye. This quality enhances its resistance to abiotic stresses like drought, salinity and moderate heat. Its quality of

Fig. 17.1 Triticale (× **Triticosecale**): **a** panicle, **b** inflorescence, **c** plant morphology and **d** grains (Faccini et al., 2023)

wheat like grains and environmental stability as rye enables it to be grown in the marginal and low-input environment thus generating food as well as biomass energy.

17.2.2 Phenology

Triticale has a rather short growth period of between 80–110 days, which varies with the cultivar and environmental factors. The period of early-maturing varieties is about 80–90 days to complete their life cycle and that of late-maturing cultivars can be up to 120 days. This variability gives great flexibility to the use of double cropping as well as relay cropping as well as rotation with the legumes or other cereals. Phenological flexibility guarantees that triticale is able to avoid end of season droughts or heat stress, which is a major benefit of semi-arid and rainfed farming systems (Alagoz et al., 2023). The growth stages in the development of triticale are unique and are usually referred using the BBCH scale (a cereal phenological scale). These stages are important to comprehend agronomic management, nutrient application time, irrigation schedule and harvest planning (Table 17.1 and Fig. 17.2).

Both graphs in Fig. 17.3 shows a positive relationship of average cumulative growing degree days (CGDD) with DAS and BBCH. As the average CGDD increases with the increase of DAS and BBCH scale, this shifts phenological stage of triticale from vegetative stage to reproductive stage.

Triticale is a vigorous species in its early stages. The competitive ability of rapid seedling establishment and tillering is good both in the grain yield and straw production. The flowering and spike development depend on temperature and photoperiod which is very important when addressing the adaptation to different climates. Grain filling is determined by the soil moisture and the temperature whereas maturation is accompanied by the build-up of residue to be utilized as bioenergy.

Table 17.1 Phenological stages of triticale (Gaviley et al., 2024)

BBCH stage No.	Stage name
00–09	Germination
10–19	Leaf development
20–29	Tillering
30–39	Stem elongation
40–49	Booting & heading
50–59	Flowering
60–69	Early grain development
70–79	Milk to dough stages
80–89	Ripening

Fig. 17.2 Phenological stages with BBCH codes for triticale (Bryan & Morrill, 2023)

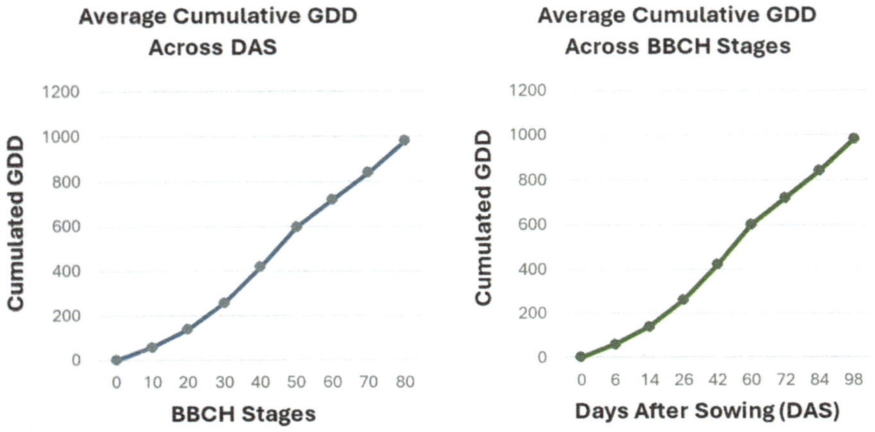

Fig. 17.3 The average CGDD across DAS necessary for triticale to reach the same BBCH stage (Coblentz et al., 2018)

17.2.3 Distribution and Agroecological Spread

Triticale is grown in temperate, sub-temperate and semi-arid parts of the world such as Europe, North America, China and India (Figs. 17.4 and 17.5). It is strongly associated with the areas where the other cereals cannot thrive like marginal and low-fertile lands. The major producers in Europe are Poland, Germany and France and the major areas of cultivation in Asia are China, India and some regions of Iran (Malherbe et al., 2023). The flexibility of triticale enables it to grow in 400–600 mm per year rain, medium soil with moderate fertility, low input and low water irrigation. This strength makes it perfect to smallholder farmers who desire to gain two benefits of grain and straw. It is also versatile because of its capability of growing in moderately acidic to neutral soils (pH 5.5 to 7.5). Triticale is primarily farmed in the northern plains and uphill slopes in India where it is planted as a dual crop with either pulses or oilseeds. This kind of integration offers benefits to the soil fertility and decreases the occurrence of any pathogen and increases the overall productivity of the farm and also helps in bioenergy production using the residues of the straw (Cui et al., 2025).

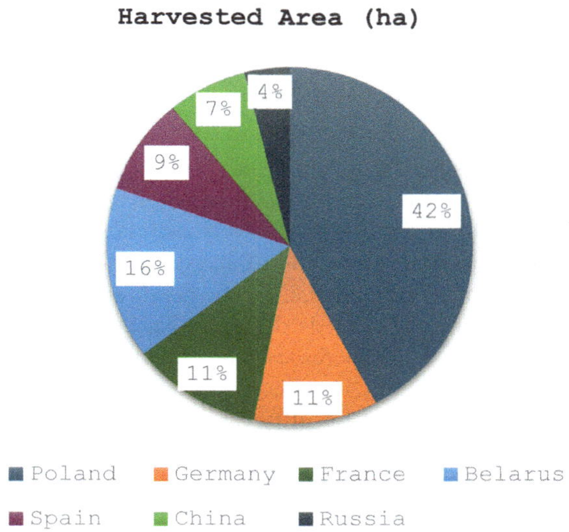

Fig. 17.4 Harvested area of triticale (ha) all over the globe in year 2023–2024 (Bonea, 2024)

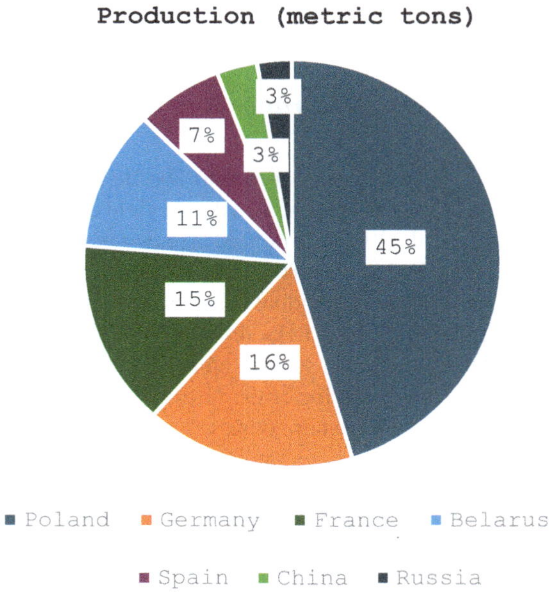

Fig. 17.5 Production of triticale (metric tons) all over the globe in year 2023–2024 (Bonea, 2024)

17.3 Agronomic Traits and Climatic Resilience

17.3.1 Agronomic Traits

Research shows that some genotypes of triticale grain yields ranging from 4.83 to 6.92 t ha^{-1} and fresh forage biomass gives yields between 20.06 and 29.78 t ha^{-1} (Cui et al., 2025). This yield is very useful in systems where dual purpose is required. With a growth cycle of 80–110 days, depending on cultivar, climatic conditions and management procedures, it has the flexibility of undergoing relay crop or alternating with legume and oilseeds. The plant has good initial vigour, fast tillering and strong root system which increases competitiveness with weeds and increases the biomass accumulation to be used as bioenergy (Faccini et al., 2023). The structure of the triticale is different for different cultivars. The semi dwarf plants are favoured for grain production because they have less lodging risk in contrast to the tall cultivars that are favoured in the production of straw as source of energy. The straw lignocellulosic contains proportions of cellulose, hemicellulose and lignin ranging between 35 to 40%, 25% to 30%, and 15% to 20% respectively. This makes triticale the best source of biogas and solid biofuel (Cui et al., 2025).

17.3.2 Stress Tolerance and Adaptability

Triticale has average tolerance to abiotic stresses, especially drought and low soil fertility. In contrast to wheat, triticale yields can be obtained with a modest irrigation (Mosaad et al., 2025). In rainfed area, it can grow on yearly precipitation amounts as low as 400–500 mm. Its root system has the capability of penetrating the deeper soils hence accessing the residual soil moisture, which facilitates filling of grains in cases of terminal drought (El Haddoury et al., 2024). Tolerance of triticale to heat stress is moderate and it can survive up to 35–38 °C (Table 17.2). Extreme heat during the anthesis stage could lower the fertility and grain weight. It can grow in upland and low-input systems, where other cereals fail as it is resistant to marginal soils such as the sandy-loam and moderately acidic soils (pH 5.5 to 7.5) (Camerlengo & Kiszonas, 2023).

Table 17.2 The optimal environmental and soil conditions for triticale (Camerlengo & Kiszonas, 2023; Cui et al., 2025)

Parameter	Range/threshold
Drought tolerance	Moderate, 400–600 mm rainfall
Heat tolerance	Up to 38 °C
Soil pH tolerance	5.5–7.5
Nutrient requirement	Low to moderate

Table 17.3 Water and temperature requirements of triticale at different growth stages (Cui et al., 2025; Malherbe et al., 2023)

Growth stage	Optimum temperature (°C)	Water requirement (mm)	WUE (kg grain/mm)
Germination	15–20	50–60	2.8
Vegetative growth	18–25	150–200	3.1
Reproductive stage	20–28	100–150	3.5–4.0
Grain filling	18–25	50–70	3.5

17.3.3 Water-Use Efficiency (WUE)

Triticale being a C_3 crop has moderate water-use efficiency with an average yield 2.5–4 kg per mm water depending on the soil type, rainfall and management procedures. The conservation agriculture practices such as low tillage, mulching and retention of the residue can also increase water-use efficiency and promote biomass growth to generate bioenergy. It has been found that triticale has consistent yields in unpredictable rainy climates because of its adjustable growth phases as well as rapid tillering (Pieniążek & Kaszuba, 2024) (Table 17.3).

17.3.4 Nutrient-Use Efficiency

Triticale has low to moderate nutrient needs, especially nitrogen (50–80 kg/ha) and phosphorus (20–30 kg/ha) and is thus applicable in low-input farming systems. Its effectiveness in the uptake of nutrients is increased when it is planted in rotation with legumes which fixes the nitrogen in atmosphere and increases soil fertility (Salama & Badry, 2021). Pulse crop rotation like chickpea or lentils will help to eliminate dependence on artificial fertilizers and enhance the productivity of the whole system (Malherbe et al., 2023). The straw that remains in the fields does not only serve to increase the accumulation of organic matter in the soil, but it also enables the energy conversion of biomass simultaneously. One can say it is a repeated loop of a food and energy soil nexus. Research has shown that integrated management of triticale residues can add 0.15–0.25 t C/ha/year of carbon to soil and they can be used to generate bioethanol or biogas (Camerlengo & Kiszonas, 2023).

Table 17.4 Macronutrient composition of triticale grain (per 100 g dry weight) (Gaviley et al., 2024)

Component	Content
Energy	~ 335 kcal
Carbohydrates	65–70 g
Protein	12–15 g
Fat	1.5–2.5 g
Dietary fiber	8–12 g

17.4 Nutritional Value

Triticale is a healthy cereal with protein content like that of wheat and the high fibre content of rye. It provides a good supply of macronutrients, essential minerals and dietary fibre in its grain making it suitable for human consumption and is included in the livestock diets.

17.4.1 Macronutrient Composition

The protein levels of triticale (12–15 g) are better than those of wheat and rye and provide essential amino acids to human diets especially lysine which is inadequate in wheat. The dietary fibre is high and enhances the gastrointestinal, helps regulate blood sugar and promotes cardiovascular health. These properties render triticale grain an option for functional foods, multigrain products and health based dietary interventions (Alagoz et al., 2023) (Table 17.4 and Fig. 17.6).

17.4.2 Micronutrients and Bioactive Compounds

The micronutrients that support metabolism, bone health and immunity are found in triticale include iron, zinc, magnesium, phosphorus and potassium. There are also high concentrations of antioxidants in the grain such as phenols, flavonoids and lignans. They play a role in reducing oxidative stress and may help prevent lifestyle-related diseases (Santana et al., 2025). Triticale has bioactive components that provide anti-inflammatory, antimicrobial and antioxidant properties making it suitable in the development of functional foods. Its bioactive form also increases the nutritional value of animal feeds indirectly promoting food security (Alagoz et al., 2023) (Table 17.5 and Fig. 17.7).

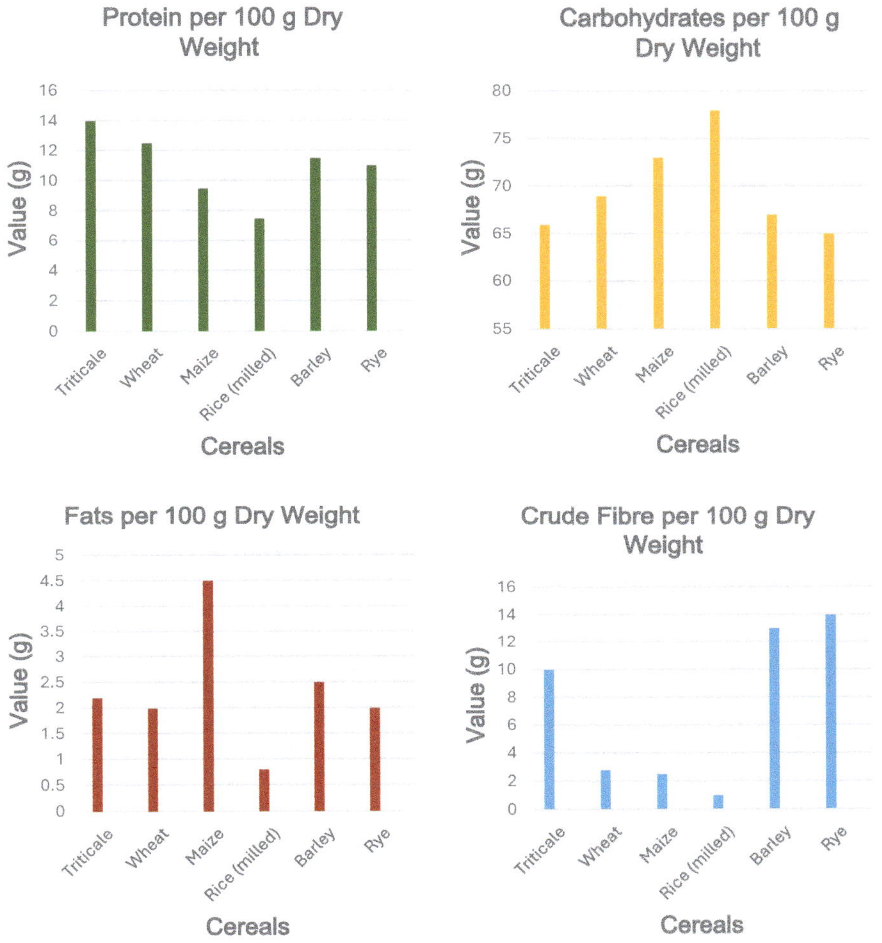

Fig. 17.6 Macronutrient (protein, fats, carbohydrates and crude fibre) composition of triticale versus other cereals (Camerlengo & Kiszonas, 2023)

Table 17.5 Micronutrient composition of triticale grain (per 100 g dry weight) (Gaviley et al., 2024)

Component	Content (mg)
Iron (Fe)	4–6
Zinc (Zn)	2–3
Magnesium (Mg)	120–140
Phosphorus (P)	250–300
Potassium (K)	350–400

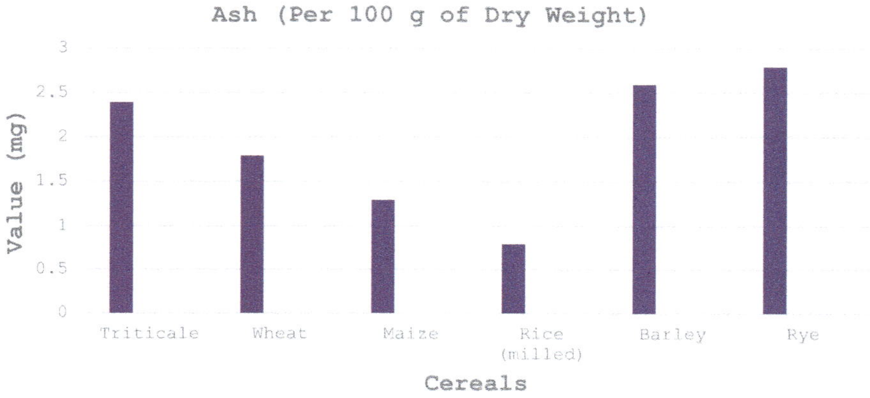

Fig. 17.7 Micronutrient ash composition mg/100 g dry weight of triticale versus other cereals (Kamanova et al., 2023)

17.5 Food and Energy Nexus Benefits of Climate Resilient Triticale

Triticale is progressively considered an eco-friendly climate-insensitive cereal which stabilizes food production in marginal environments, drought-prone and heat-stressful conditions. It is a promising high-yield and stress-resistant crop thus a perfect candidate in a robust food system.

17.5.1 Stability of Grain Yield Under Climate Stress

Triticale is highly stable in grain yield against numerous climatic stresses such as drought, heat and low fertility. It is also resistant because of its intensive root system that is deep fibrous and is efficient in searching the soil profile of moisture and nutrients, enabling the crop to grow even when there is a shortage of water (Pieniążek & Kaszuba, 2024). Moreover, triticale is tolerant to heat up to 38–40 °C, as well as low sensitivity to early-season droughts, which is essential in ensuring crop establishment during unpredictable rainfalls. Its ability to flexibly adjust its phenology with both early and late-maturing cultivars enables farmers to adjust the time of sowing based on the local rainfall patterns and therefore reduce the chances of crop failure (Ozkurt, 2025). Triticale yield has the potential to sustain food security in marginal lands avoiding the occurrence of complete crop failure in case of extreme weather conditions. This stability is especially important to the areas that are highly susceptible to unpredictable rainfall or frequent droughts (Table 17.6).

Table 17.6 Grain yield of triticale under diverse climatic conditions (Faccini et al., 2023; Gaviley et al., 2024)

Condition/environment	Grain yield (kg/ha)
Rainfed semi-arid	3500–3800
Moderate drought stress	3200–3500
Low fertility soils	3000–3400
Heat stress (up to 40 °C)	3200–3600

17.5.2 Nutritional Security

In addition to stability in yields, triticale has a major role to play in nutritional security. It contains high level of protein (12–15%) so that it forms a key source of dietary protein in those areas where people rely on cereals. In addition, triticale is rich in fibre and other important micronutrients that are essential in preventing micronutrient deficiencies and hunger. The bioactive compounds in triticale grains promotes metabolic, cardiovascular and gastrointestinal well-being. These nutritional qualities not only make triticale a robust crop for its yields but also a very important source of human nutrition and health (Gaviley et al., 2024).

17.5.3 Contribution to Smallholder and Marginal Farming

Triticale is particularly profitable for smallholder and marginal production systems. It is also resistant to unpredictable rain fall which enables small scale farmers to produce a consistent cereal crop in harsh conditions. Certain cultivars can be used in a short growth cycle and can be used in double cropping or relay cropping systems to maximize land use efficiency and farm productivity. Moreover, triticale is able to grow under low-input conditions, meaning that it needed very little fertilizer and irrigation which led to low production costs and has been economically viable to farmers with limited resources (Malherbe et al., 2023). These three attributes of yield stability, nutritional quality and low-input compatibility make triticale a fundamental crop of climate-resilient food systems especially for smallholder farmers who work under marginal agroecosystems.

17.5.4 Bioenergy Potential in Climate-Resilient Systems

The dual-purpose characteristic of triticale enables it to provide grain for food and straw biomass for renewable energy. This dual-purpose nature is the key to the food and energy nexus especially in areas with scarce land and water supplies where a high output per unit area is critical to ensuring sustainability (Ghimire et al., 2025).

Table 17.7 Bioenergy potential of triticale straw (Faccini et al., 2023)

Straw yield (t/ha)	Energy potential (GJ/ha)	Conversion pathway
6	102	Combustion
7	122.5	Bioethanol
8	144	Biogas

17.5.4.1 Straw Biomass and Energy Yields

Triticale germinates well, has a calorific value of 16–18 MJ/kg, and yields high straw (usually 6–8 t/ha) which can be used in various bioenergy applications that include combustion, bioethanol production and biogas production. The straw contains a high level of cellulose, hemicellulose and lignin which are ideal for lignocellulosic biofuels production. These properties render triticale a very practical energy crop that compliments it as a source of food (Malherbe et al., 2023). With the use of straw in the farming systems, the small holders will produce renewable energy that can be used in the households or local agro-industrial activities without relying on fossil fuels as agro-industrial production will not decrease (Xue et al., 2025) (Table 17.7).

17.5.4.2 Climate Resilient Bioenergy Systems

The ability of Triticale to grow on low-fertile soils and water scarce ecologies means that production of biomass is consistent even where the environment is not ideal. Household level biogas digesters or small-scale thermal energy systems can help smallholder farmers to capitalize on this potential. Notably, the duality of triticale has been found to be useful in creating energy without compromising food security as it is an important factor in planning climate resilient agriculture (Prášil et al., 2025).

17.5.5 Energy Indices Calculations

Calculations for energy efficiency of triticale could be calculated by using following equations Furthermore, food and energy outputs of triticale under climate stress have been given in Table 17.8.

Human Power (Labour Energy)

Human Power (MJ/ha) = Total worked hours (hr/ha) × $E_{\text{eq of labour}}$(MJ/hr)

$E_{\text{eq of labour}}$ is the energy equivalent of human labour.

- Accounts for all manual activities in triticale cultivation (sowing, weeding, harvesting).

Table 17.8 Food and energy outputs of triticale under climate stress (Gaviley et al., 2024)

Output	Yield/ha	Use	Advantage
Grain	3800–4200 kg	Human/animal food	Stable under drought/heat
Straw	6–8 t	Bioenergy/biogas	Reliable under low water
Residues returned	2–3 t	Soil fertility/carbon	Improves soil resilience

Machine Energy

$$\text{Machine Energy (MJ/ha)} = \frac{W_{\text{m}}}{E_{\text{life}}} \times \text{ME}_{\text{eq}} \times M_{\text{t}}$$

where

- W_{m} = weight of the machine (kg)
- E_{life} = economic life of the machine (hr)
- ME_{eq} = energy coefficient of machine (MJ/kg)
- M_{t} = working time of machine in the field (hr/ha).

Includes ploughing, sowing, harvesting etc., for triticale.

Fuel Energy

$$\text{Fuel Energy (MJ/ha)} = \text{Total fuel used (1/ha)} \times E_{\text{eq of diesel}}(\text{MJ/1})$$

- Considers diesel consumption of tractors, harvesters, irrigation pumps.
- For triticale, calculate per hectare for all operations.

Fertilizer Energy

$$\text{Fertilizers Energy (MJ/ha)} = \frac{\%\ \text{Nutrient per bag}}{100} \times E_{\text{eq of nutrient}} \times \text{No. of bags/ha}$$

- Converts the chemical energy embedded in fertilizers (N, P, K) into MJ/ha.
- Triticale has low-to-moderate nutrient requirements, so this input is lower compared to wheat.

Total Input Energy (TIE)

$$\begin{aligned}\text{TIE (MJ/ha)} = {} & \text{Human Power} + \text{Machine Energy} \\ & + \text{Fuel Energy} + \text{Fertilizers Energy} \\ & + \text{Other Inputs (Seeds, Pesticides, Irrigation, etc.)}\end{aligned}$$

Total Output Energy (TOE)

$$\text{TOE (MJ/ha)} = \text{Grain Energy} + \text{Residue Energy}$$

- Grain energy: Grain yield (kg/ha) × Energy content (MJ/kg)
- Residue (straw) energy: Straw yield (kg/ha) × Energy content (MJ/kg).

Energy Efficiency (EE)

$$\text{EE} = \frac{\text{TOE (MJ/ha)}}{\text{TIE (MJ/ha)}}$$

EE > 1 → net positive energy, crop contributes significantly to food–energy nexus (Ditta et al., 2025).

Food Output Per Area

$$\text{Food Yield}_{\text{triticale}} \approx 4.8\text{–}6.9\ \text{t/ha}$$

Biomass Energy Potential

$$\text{Biomass}_{\text{triticale}} \approx 20\text{–}30\ \text{t/ha}$$

Bioethanol Energy Yield

Convert to energy terms via bioethanol calorific value (~ 21 MJ L^{-1}):

$$\text{Bioethanol Energy} = 1.6\ \text{to}\ 1.9\,\text{m}^3 \times 21\,\text{MJ/L} = 33.6\text{–}39.9\,\text{GJ/ha}$$

17.6 Genetic Resources, Breeding and Approaches for Food-Energy Nexus

17.6.1 Genetic Diversity and Germplasm Resources

Triticale is a hybrid cross between wheat and rye and offers a wide range of diverse germplasm. This genetic variation is important in breeding programs focused on climate resistance and dual-purpose use. Triticale specie is very adaptable and can survive under various stress conditions that include droughts, saline and low-fertility soils. Besides stress resistance, the triticale germplasm exhibits a great biomass potential and thus is a great choice when it comes to the production of straw that can serve in generating bioenergy (Yang et al., 2024). Triticale grain has a great deal of variability in terms of its nutrient composition; protein (12–15%), essential amino acids and micronutrients. These characteristics facilitate human as well as

Table 17.9 Major gene banks conserving germplasm of triticale Worldwide (Ma et al., 2024)

Repository/gene bank	Country/institute	Approx. triticale accessions
CIMMYT Germplasm Bank	Mexico	17,394
NSGC (National Small Grains Collection) (USDA-ARS)	USA	2009
SCRDC-AAFC (Canadian Small Grains)	Canada	2000
VIR (Vavilov Institute of Plant Industry)	Russia	2030
IR (Institute/Gene bank)	Ukraine	1748
Lublin (Poland)	Poland	1748
IPK Gatersleben	Germany	1577
Other small collections	Various countries	8934

animal nutrition and improves the crop production towards food security. Due to this genetic diversity, breeders have been able to utilize it on the level of parent lines to achieve a balance between grain yield as a source of food and straw biomass as source of energy. This diversity also add stress tolerant traits to achieve climate resilience (Malherbe et al., 2023) (Table 17.9).

17.6.2 Conventional Breeding Approaches

Traditional breeding has continued to be a base tool in the development of dual-purpose triticale varieties. The hybridization and the mechanism of selective breeding encompass crossbreeding between high-grain and high-biomass to integrate the preferred traits in the same cultivar. Phenotypic selection is the selection based on the observable traits like early vigour, root structure and drought resistance. According to studies, multi-location trials will enable breeders to determine cultivars under semi-arid and heat prone conditions that can be used interchangeably in various stress conditions to achieve consistency in yield and biomass production. Although they are time-consuming, these conventional methods are important in the validation of trait expression in the field (Alagoz et al., 2023).

17.6.3 Molecular Breeding and Marker-Assisted Selection (MAS) Approaches

Molecular breeding methods have enhanced breeding to produce climate resilient and dual-purpose triticale within a short period. Genetic markers of grain yield, protein content and biomass traits allow selection of early-generation and reduces the number of cycles needed to produce better cultivars. Quantitative trait loci (QTL) mapping is used to determine the genomic regions involved in heat stress tolerance, drought resistance and disease resistance. These can be used to select breeding strategies. Through marker-assisted selection (MAS), breeders can combine these traits effectively to come up with lines that are good as a food crop as well as a bioenergy source (Pieniążek & Kaszuba, 2024).

17.6.4 Genomic Selection and CRISPR/Cas9 Approaches

Genomic selection has become an influential technique of enhancing various traits simultaneously and includes grain yield, straw biomass and stress tolerance. Genome prediction models enable breeders to be able to predict the performance of the breeding lines without having to undergo extensive trials in the field and save a lot of time and money. Moreover, the methods to increase the specific traits are under investigation with the use of gene editing technologies such as CRISPR/Cas9. Studies show that specific changes can be made to boost cellulose and hemicellulose straw to produce bioethanol and enhance the root structure to improve drought resistance. Through these strategies, it is possible to manipulate genetic pathways with precision to produce cultivars that can be used to both fulfil nutritional and energy requirements in climatic challenging areas (Yang et al., 2024) (Fig. 17.8).

17.6.5 Genetic Breeding in Climate Resilient Food and Energy Nexus

Triticale breeding programs are emphasizing traits that can guarantee food and energy production to survive in the climate stress scenario. Deep and fibrous roots increase uptake of water and nutrients during drought. High biomass allocation permits straw production without decreasing grain quality. Early maturing varieties are especially adapted to areas with unpredictable rainfall allowing repeated crop production within a season (Pieniążek & Kaszuba, 2024). The tolerance to heat and cold conditions stabilize the performance at extreme temperatures whereas protein content and micronutrient levels ensure that nutritional quality of the grain is not impaired. Combining these characteristics, triticale will be able to sustain food and energy reliably even in the marginal environments. This makes it a direct contributor to the climate resilient agricultural systems (Gołębiowska-Paluch et al., 2025).

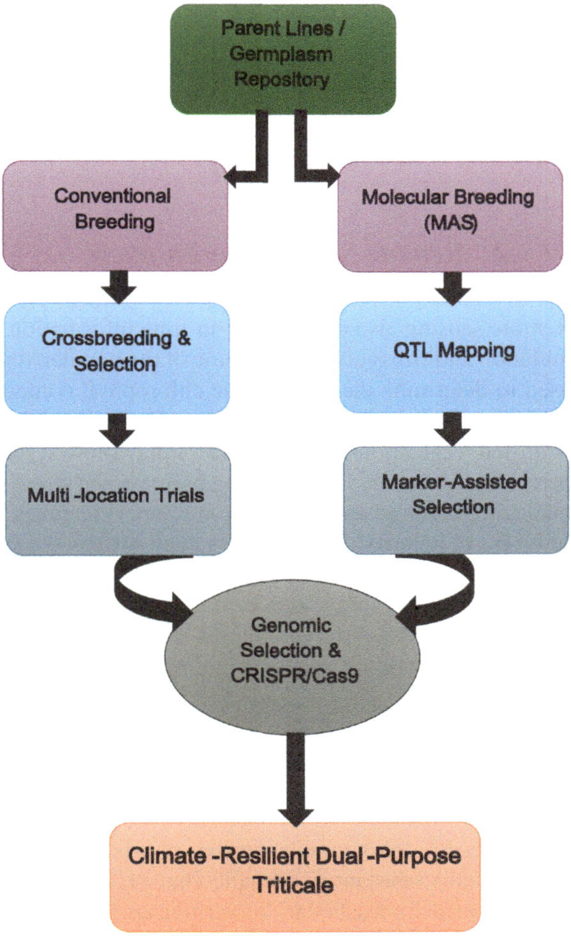

Fig. 17.8 Roadmap for genetic and genomic improvement in triticale

17.7 Crop Modeling and Remote Sensing

17.7.1 Crop Modeling Approaches

The crop modeling offers an imaginary set up of simulation of growth, yield and biomass production of triticale under various management and climatic conditions. Commonly employed models are DSSAT, APSIM and InfoCrop which combine genetic, environmental as well as management variables to project the outputs without encountering the hazards of experimentation in real-world (Januszkiewicz et al., 2024). Studies for simulation indicate that triticale grains under rainfed conditions give a yield of between 3500 to 4500 kg/ha whereas the straw biomass yields 6 to 8 t/ha depending on varieties, soil fertility and planting density. The models are useful when trying to determine the impacts of drought, nutrient limitation and temperature stress which are primary limitations of marginal agroecosystems. To illustrate this point, simulations using APSIM showed that at an early sowing time and moderate irrigation level, the yields of the grain could remain at 3800 kg/ha and still have sufficient straw to be used as bioenergy (Januszkiewicz et al., 2024).

17.7.2 Remote Sensing Techniques

Remote sensing also offers close-in-time information of crop health, biomass accumulation and stress condition. Some of the popular vegetation indices that have been used to determine the canopy, the chlorophyll content and water stress in the triticale fields include NDVI (Normalized Difference Vegetation Index) and SAVI (Soil Adjusted Vegetation Index). Research has showed that these indices can exceed 80 percentage accuracy in forecasting biomass and stress levels and are very suitable for managing the food and energy production (Pieniążek, 2023). An example is the use of NDVI to identify drought stress at an early stage so that the farmers can regulate their irrigation patterns leading to stabilization of the grain production and biomass of straw as a bioenergy source. It is important in rainfed systems where rainfall is unpredictable. An early warning of stress will allow avoiding losses in the form of yield and maintain the feedstock to produce renewable energy (Karaman, 2023).

17.7.3 Integration of Crop Modeling and Remote Sensing

Crop modeling plus remote sensing boosts the support of decision making in triticale systems resistant to climatic change. DSSAT, APSIM and INFOCrop generated seasonal forecasts can be verified and adjusted with satellite derived vegetation indices. This provides a dynamic feedback loop that enhances the accuracy of prediction and resource use efficiency (Januszkiewicz et al., 2024). This integration will enable the farmers and policymakers to predict food and energy availability in different climatic conditions and support nutrition and renewable energy planning. The stakeholders can be able to maximize land utilization by using predicted grain yields and straw biomass data to produce two products through the dual purposes (Pieniążek, 2023).

17.7.4 *Implications on Climate-Resilient Food-Energy Systems*

With support of both crop modeling and remote sensing, precision agriculture is possible. This reduces the input waste while keeping grain quality and biomass yield remaining high. Optimized fertilizer and irrigation control will boost the protein level of grains which would be important in nutritional security and aid in adequate biomass build up to bioenergy yield. Marginal agroecosystems also benefit by ensuring dual-purpose outputs because of these technologies supporting sustainable intensification. In the case of triticale, the synergistic relationship between grain and food and the energy source of straw is emphasized as an example of the possible food and energy nexus of resilient and resource-efficient agricultural systems, where food and energy requirements are sustainably fulfilled.

17.8 Case Studies

Case Study 1: Intercropping Triticale with Legumes Under Rainfed Conditions

Pieniążek and Kaszuba (2024) examined the triticale and black gram as well as other legumes integration within the semi-arid rainfed agroecosystems as a way of understanding the food and energy nexus. This was aimed at testing the aspect of grain yield as well as biomass production and the enhancement of the soil fertility by biological fixation of nitrogen. Results were the yield of grains as 3850 kg/ha and straw yield as 7.2 t/ha which was much more than that of triticale monoculture. The analysis of nutrient uptake indicated that the system increases availability of nitrogen and phosphorus in the intercropped plots. It was suggested that straw could be collected to make bioenergy without affecting the food yield because of the dual purpose of triticale.

Case Study 2: Conservation Agriculture Practices for Climate Resilience

Malherbe et al., (2023) conducted field experiment to assess conservation agriculture methods such as no-tillage, reduced tillage and mulching on performance of triticale. The objective of the study was to establish the impacts of these practices on the yield of grain, biomass, water-use efficiency and weed suppression against marginal soils. Results showed that mulching treatment gives grain yield of triticale 4 t/ha and the straw 7.5 t/ha which is 20 to 25% higher than the conventional tillage plots at both treatments. In addition to the yields, this system retained the carbon sequestration and the higher levels of soil organic matter. All this indicates that conservation agriculture can increase production of food as well as bioenergy biomass which can benefit smallholder and marginal land farmers.

Case Study 3: Optimizing Variety and Row Spacing

Cui et al. (2025) conducted experiment to investigate the impact of the variety of triticale, as well as the row spacing on the production of grain and straw in rainfed situations. Three better triticale varieties were cultivated using three rows spacing namely 22.5 × 10 cm, 30 × 10 cm and 45 × 10 cm. It was aimed at finding combinations that can maximize dual-purpose outputs and efficiency of resource use. The findings revealed that the type of variety which gives the closest gap between rows (22.5 × 10 cm) yielded the most grains (4150 kg/ha) and the straw (7.8 t/ha). The increased separation between rows lowered the production of grains and biomasses because of the inability to establish a canopy and increase soil evaporation. Improved nutrient uptake efficiency was also measured in plots that were spaced close together in the study supporting the food and energy outputs. The findings show the significance of maximizing agronomic management such as spacing and the type of variety used to achieve climate resilient productivity in a dual-purpose crop system.

Case Study 4: Genetic Variability and Yield-Trait Diversity

Alagoz et al. (2023) assessed seventeen triticale genotypes with phenotypic and genotypic variability under semi-arid conditions. It was aim for determining characteristics of climate resilience, dual-purpose productivity and stress tolerance. Some characteristics studied were days to flowering, plant height, biomass allocation, grain yield and straw yield. Results showed moderate to high genetic control over yield and biomass traits. Early maturing varieties kept stable yields under drought stress, while genotypes having deeper roots gave higher straw biomass for bioenergy. This study proves the significance of genetic diversity in developing climate resilient triticale varieties that can supply both food and energy sustainably.

17.9 Challenges and Constraints

Although triticale has an outstanding potential of climate resistant pseudocereal with food and energy output, there are various challenges that limit its mass implementation. These issues are in agronomic, socio-economic, genetic and policy areas. They collectively constrain the crop to be integrated into viable food and energy systems.

17.9.1 Limited Research Attention and Genetic Bottlenecks

The fact that there has been limited research on triticale in comparison to staple grains such as wheat, rice and maize is also one of the main limitations. Although triticale is resistant to drought, heat and low-fertilized soils, the breeding and genomics research has not been as invested as it has been in other species which explains why the development of high-yielding and nutritionally superior varieties have lagged

(Januszkiewicz et al., 2024). Conventional methods of breeding are sluggish unless there is a combination of modern genomic solutions like marker-assisted selection, genomic selection and transcriptomics. The possibility of swift enhancement of the stress tolerant and dual-purpose varieties is then attained (Pozo et al., 2023). In addition to this, there is insufficient attention to research studies both public and private, and this leads to insufficient elite lines that suit particular marginal agroecosystems.

17.9.2 Agronomic and Management Constraints

Triticale production generally faces agricultural drawbacks in crop production. There is limit in mechanization especially in sowing, harvesting and post-harvest processing. Most smallholder farmers in rainfed or marginal lands are still using manual practices which are very laborious and diminish economic conditions. The agroclimatic zones have not been fully optimized to the fertilizer recommendations and irrigation schedules leading to variability in the yield results. Farmers will not risk investing in triticale without agronomical packages and services thus they prefer other cereals that assure them of good yields or easy access to the market (Faccini et al., 2023).

17.9.3 Seeds Availability and Systems

There is a lack of certified and high quality of triticale seed especially to smallholder and marginal farmer. Lack of seed supply chains by organized means less production of better varieties and limited stability of yields. It is common practice of farmers to use stored seed from previous harvests thus resulting in low vigour, germination and prone to pests and diseases. This difficulty is found in areas with high climatic variability where the seed of the right phenology is important for climate resilience (Pieniążek, 2023).

17.9.4 Post-harvest and Value Chain Constraints

The dual purpose of Triticale being both grain and straw requires a favourable value chain to be economically feasible. Processing facilities like mills, pelletized bioenergy and combined grain and strand systems are not well developed. The low market availability of both grain and straw will lower the incentives of farmers to grow triticale at a larger scale. Moreover, poor storage systems may cause post-harvest losses because of the pests, moisture or fungi which can impact the food security as well as the energy production. Such after harvest restrictions make the crop less desirable than more commercially established grain (Malherbe et al., 2023).

17.9.5 Gaps in Institutional, Policy and Awareness

The support for triticale policy has not been as strong as other staple crops. Triticale is not a priority in national food security programs and subsidies or extension campaigns as compared to rice, wheat or maize. Farmers have little knowledge of its food and energy benefits and climate resilient characteristics and economic potential. The rate of adoption would remain low without the specific extension services, public awareness and incentives like energy production subsidies or including the nutritional program. Also, energy, agriculture and nutrition policies do not always go hand in hand which weakens the inclusion of triticale into the local food and energy platforms (Gaviley et al., 2024).

17.10 Future Potential Strategies

To address the current limitations and embrace the full service of triticale as dual-purpose and climate resistant crop, several future strategies are being introduced. These strategies aim at breeding innovation, integration of cropping systems, improved crop surveillance, market creation and policy intervention.

17.10.1 Advanced Breeding and Genomic Approaches

The modern genomic tools can greatly help in the creation of triticale varieties with better yield, nutritional value and stress resistance. The use of marker-assisted selection, genomic selection, and CRISPR-based gene editing techniques may make it possible to target specific traits of drought tolerant, heat resistant and straw quality for bioenergy applications (Pozo et al., 2023). Micronutrient enrichment can also be prioritized in future breeding programs to overcome hidden hunger at the expense of energy producing biomass (Januszkiewicz et al., 2024).

17.10.2 Integrated Food-Energy Systems

Attracting the attention of future directions, designing the integrated systems of crop cultivation in which triticale is a constituent of an expanded food-energy strategy is required. It involves dual-purpose cultivation in which grain is grown to be used as food to humans or livestock feed and straw is grown to be used as bioenergy to be either burned for bioethanol or biogas. Circular systems have the potential to be applied through the recycling of a specific part of straw residues to the soil to increase organic matter and soil fertility. The triticale and leguminous intercropping or rotations would also help to enhance the efficiency of nitrogen utilization and decrease the input needs contributing to the resource saving food and energy production (Cui et al., 2025).

17.10.3 Post-harvest Technologies and Value Chain Development

Post-harvest management and development of value chains should be done efficiently to enhance economic viability of triticale. Grain storage, milling and straw processing can be innovative, minimize losses and provide value addition to food and energy products. The decentralized bioenergy plants can be developed to empower the small holder farmers by availing energy solutions at the local level and at the same time offer market opportunities to the straw. By reinforcing market connections and collaborative models can deliver equitable prices and broader utilization of dual role triticale varieties (Cui et al., 2025).

17.10.4 Policy, Institutional and Awareness Interventions

The use of triticale should be scaled up with the help of policy support. Subsidies on dual-purpose types, energy incentive programs to produce bioenergy and nutrition-based programs should be part of the future intervention to incorporate triticale into the food security strategy. It has been found that institutional support obtained by the extension services, farmer training and awareness can enhance better adoption by ensuring that not only the nutritional benefits but also the energy benefits are highlighted (Gaviley et al., 2024).

17.10.5 Climate-Resilient and Digital Agriculture Integration

New digital technologies in agriculture including AI-based predicting models, IoT-based soil moisture sensors and precision nutrient management systems can be used to improve climate resiliency of triticale. The predictive analytics can predict the stress events and optimize the allocation of resources so that the grain and straw yield remain stable even during extreme weather conditions. The climate-sensitive activities that include conservation agriculture, holding back of residues and water-efficient irrigation strategies will also make the triticale food and energy systems more resilient.

17.11 Conclusion

Triticale is a unique type of climate resistant crops that have great potential in providing food security as well as renewable energy, at the same time. In different agroecological zones, semi-arid marginal lands to temperate Mediterranean climate, triticale has shown impressive adaptability. It has not shown a decline in yields due to drought, heat, low fertility and salinity stress. The food and energy nexus represented by triticale highlights the importance of ensuring that the land use is optimized in landscapes with limited resources. Through the production of grain and biomass on the same piece of land, farmers will be able to produce more products without having to expand the size of their land. Furthermore, the recycling of the residues into the soil does not only aid the fertility and carbon-fixation of soils but also establishes a circularity which increases the long-term sustainability. Crop modeling and remote sensing technology enable accurate predictions of grain and biomass yields and enhances decision making in rainfed and marginal conditions. These tools help farmers, researchers and policymakers to prepare to meet climate variability, are efficient to use inputs and will be productive on a regular basis. Moreover, current breeding efforts using marker-assisted selection, genomic selection and transcriptomics are continuously improving the dual-purpose breeds of triticale which are bound to be in a better position to withstand the extreme conditions of climate. At the same time, they are optimizing the food and energy yields of future breeds. To conclude, triticale represents an integrated solution to the interrelated issues of food security, energy production and climate resilience. Its dual potential of use as well as its adaptability and efficiency of resource use makes it a central crop in terms of sustainable intensification. Its contribution to the world needs can be maximized with strategic intervention in the breeding, agronomy, value chains and policy frameworks. The results of experimental trials, case studies and modeling strategies clearly show that application of triticale in food-energy nexus is very useful to sustainable agroecosystem in the future.

References

Alagoz, S. M., Hadi, H., Toorchi, M., Pawłowski, T. A., Asgari Lajayer, B., Price, G. W., Farooq, M., & Astatkie, T. (2023). Morpho-physiological responses and growth indices of triticale to drought and salt stresses. *Scientific Reports, 13*(1), 8896. https://doi.org/10.1038/s41598-023-36119-y

Bonea, D. (2024). Triticale—An alternative cereal for food industry in the world, EU-27 and Romania, *24*(3).

Bryan, K. A., Morrill, K. (2023). Cover crop or winter forage? AgProud. https://www.agproud.com/articles/56979-cover-crop-or-winter-forage?utm

Camerlengo, F., & Kiszonas, A. M. (2023). Genetic factors influencing triticale quality for food. *Journal of Cereal Science, 113*, Article 103744. https://doi.org/10.1016/j.jcs.2023.103744

Coblentz, W. K., Akins, M. S., Kalscheur, K. F., Brink, G. E., & Cavadini, J. S. (2018). Effects of growth stage and growing degree day accumulations on triticale forages: 1. Dry matter yield, nutritive value, and in vitro dry matter disappearance. *Journal of Dairy Science, 101*(10), 8965–8985. https://doi.org/10.3168/jds.2018-14868

Cui, L., Xu, L., Wang, H., Fan, X., Yan, C., Zhang, Y., Jiang, C., Zhou, T., Guo, Q., Sun, Y., Yang, F., & Li, H. (2025). Evaluation of dual-purpose Triticale: Grain and forage productivity and quality under semi-arid conditions. *Agronomy, 15*(4), 881. https://doi.org/10.3390/agronomy15040881

Ditta, S. A., Ullah, A., Mahar, R. B., Ansari, K., Suhail, T., & Ahmed, W. (2025). Analyzing the food-energy nexus of irrigated crops in Pakistan: Implications for food security and agricultural sustainability. *Agricultural Research.* https://doi.org/10.1007/s40003-025-00878-z

El Haddoury, J., Rajae, K., & Fatima, B. (2024). Triticale, céréale résiliente face aux changements climatiques. *African and Mediterranean Agricultural Journal—Al Awamia, 143*, 295–304. https://doi.org/10.34874/IMIST.PRSM/AFRIMED-I143.49548

Faccini, N., Morcia, C., Terzi, V., Rizza, F., & Badeck, F.-W. (2023). Triticale in Italy. *Biology, 12*(10), 1308. https://doi.org/10.3390/biology12101308

Gaviley, O. V., Katerynych, O. O., Ionov, I. A., Dekhtiarova, O. O., Griffin, D. K., & Romanov, M. N. (2024). Triticale: A general overview of its use in poultry production. *Encyclopedia, 4*(1), 395–414. https://doi.org/10.3390/encyclopedia4010027

Ghimire, R. P., Russell, A., Pariyar, D., Hampton, J., Armstrong, K., Shrestha, K. K., Rolston, M. P., Stevens, E. J., & Amgain, R. (2025). Fodder potential of triticale (× *Tritisecale wittmack*) genotypes in Nepal: Evaluation of biomass and seed productivity. *Nepal Agriculture Research Journal, 16*(1), 1–11. https://doi.org/10.3126/narj.v16i1.80421

Gołębiowska-Paluch, G., Stawoska, I., Mucha, Z., Wesełucha-Birczyńska, A., & Kornaś, A. (2025). Physiological effects in genotype-dependent plants of winter triticale growing in previously salinized soil. *Plant and Soil.* https://doi.org/10.1007/s11104-025-07934-1

Januszkiewicz, R., Kulczycki, G., & Sacała, E. (2024). The effect of multi-component foliar fertilisers on yield and quality of grain and straw of triticale. *Agronomy, 14*(12), 2846. https://doi.org/10.3390/agronomy14122846

Kamanova, S., Yermekov, Y., Shah, K., Mulati, A., Liu, X., Bulashev, B., Toimbayeva, D., & Ospankulova, G. (2023). Review on nutritional benefits of triticale. *Czech Journal of Food Sciences, 41*(4), 248–262. https://doi.org/10.17221/67/2023-CJFS

Karaman, M. (2023). Relationship between physiological characteristics and grain yield of triticale genotypes. *Bangladesh Journal of Botany, 52*(1), 187–196. https://doi.org/10.3329/bjb.v52i1.65250

Ma, Z., Liu, Z., & Wang, W. (2024). Taxonomy and genetic resources of Triticeae: Exploring wild and cultivated species. *Triticeae Genomics and Genetics.* https://doi.org/10.5376/tgg.2024.15.0013

Malherbe, S. J. M., Cripwell, R. A., Favaro, L., Van Zyl, W. H., & Viljoen-Bloom, M. (2023). Triticale and sorghum as feedstock for bioethanol production via consolidated bioprocessing. *Renewable Energy, 206*, 498–505. https://doi.org/10.1016/j.renene.2023.02.047

Mosaad, I. S. M., Selim, E.-M.M., Gaafar, D. E. M., & Al-Anoos, M. A. T. (2025). Effects of humic and fulvic acids on forage production and grain quality of triticale under various soil salinity levels. *Cereal Research Communications, 53*(3), 1811–1829. https://doi.org/10.1007/s42976-024-00609-0

Ozkurt, M. (2025). The effect of different harvest stages on yield and quality of triticale grown under rainfed conditions. *Journal of King Saud University—Science, 37*, 2422024. https://doi.org/10.25259/JKSUS_242_2024

Pieniążek, M. J. (2023). The influence of cropping systems on photosynthesis, yield, and grain quality of selected winter triticale cultivars. *Sustainability, 15*(14), 11075. https://doi.org/10.3390/su151411075

Pieniążek, M. J., & Kaszuba, J. (2024). The influence of agrotechnical factors on the yield and quality parameters of winter triticale grain. *Agriculture, 14*(12), 2219. https://doi.org/10.3390/agriculture14122219

Pozo, A. D., Méndez-Espinoza, A. M., & Castillo, D. (2023). Triticale. In *Neglected and underutilized crops* (pp. 325–362). Elsevier. https://doi.org/10.1016/B978-0-323-90537-4.00029-6

Prášil, I. T., Musilová, J., Prášilová, P., Janáček, J., Coufová, M., Kosová, K., Klíma, M., Hermuth, J., Holubec, V., & Vítámvás, P. (2025). Effect of geographical origin, regional adaptation, genotype, and release year on winter hardiness of wheat and triticale accessions evaluated for six decades in trials. *Scientific Reports, 15*(1), 5961. https://doi.org/10.1038/s41598-025-89291-8

Salama, H. S. A., & Badry, H. H. (2021). Forage and grain yields of dual-purpose triticale as influenced by the integrated use of Azotobacter chroococcum and mineral nitrogen fertilizer. *Italian Journal of Agronomy, 16*(2), 1719. https://doi.org/10.4081/ija.2020.1719

Santana, D. C., Ratke, R. F., Gava, R., Ceron, A. C. C., Martins, E. V., De Oliveira, I. C., Oliveira, G. S., Campos, C. N. S., Teodoro, L. P. R., Coradi, P. C., & Teodoro, P. E. (2025). Interplay between irrigation and genotype determines biofortification and flavonoid enrichment in wheat and triticale leaves. *Scientific Reports, 15*(1), 37895. https://doi.org/10.1038/s41598-025-21675-2

Xue, F., Xu, C., Zhao, Y., Pu, X., Wang, J., Wei, X., & Wang, W. (2025). The impact of varieties and growth stages on the production performance and nutritional quality of forage triticale in the Qaidam Basin. *Plants, 14*(19), 2942. https://doi.org/10.3390/plants14192942

Yang, Y., Zhang, Q., Weng, Y., Jiao, A., & Jin, Z. (2024). Effects of hydrocolloids on the structure and physicochemical properties of triticale starch during fermentation. *International Journal of Biological Macromolecules, 282*, Article 137327. https://doi.org/10.1016/j.ijbiomac.2024.137327

Part III
Food Legume Crops: Nutrition, Water Efficiency, and Environmental Protection

Chapter 18
Ground Bean (*Kerstingiella Geocarpa*): Enhancing Soil Fertility and Food Security

Umar Farooq, Murtaza Ali, Muqaddas Noor, Abdullah Jan, Umar Farooq, and Amjad Malik

Abstract Ground bean (*Macrotyloma geocarpa*) is an underutilized legume with strong potential to address soil fertility decline and food insecurity in smallholder farming systems worldwide. Climate variability, low soil nitrogen, and reliance on a few staple crops limit sustainable food production in rainfed and marginal environments. This highlights the need for climate-resilient, low-input legumes that enhance both soil health and household nutrition. This chapter aims to synthesize agronomic, nutritional, and socio-ecological evidence on the role of ground bean in improving soil fertility and food security. It integrates botanical and agronomic characterization, climate adaptation strategies, nutritional analyses, crop modelling, and documented case studies from major production regions. Case studies showed grain yields ranging from 0.6–1.2 t ha^{-1} under rainfed conditions. Soil nitrogen increased by 15–20% through biological nitrogen fixation. Household nutrition studies reported daily protein contributions of 18–22 g per capita from ground bean consumption. Processing interventions increased seed protein content by 12% and reduced phytic acid by 35%, improving mineral bioavailability. These findings confirm the dual role of ground bean in soil fertility restoration and nutrition security. Future efforts should focus on breeding stress-tolerant genotypes. They should also improve rhizobial inoculation, strengthen value chains, and integrate ground bean into climate-smart agricultural policies.

Keywords *Macrotyloma geocarpa* · Underutilized legumes · Soil fertility · Climate resilience · Food and nutrition security

U. Farooq (✉) · M. Ali · M. Noor · A. Jan · U. Farooq
Department of Agronomy, Pir Mehr Ali Shah Arid Agriculture University, Rawalpindi, Pakistan
e-mail: umerfaroooq42@gmail.com

A. Malik
Kenanow Bachelor of Education Program Faculty of Education, University College of the North (UCN), Thompson, MB, Canada

M. Ahmed (ed.), *Climate Change and Neglected Underutilized Crops*,
https://doi.org/10.1007/978-3-032-21005-0_18

18.1 Introduction

There are long-term problems associated with smallholder farming systems in sub-Saharan and other third-world countries, including the problem of declining soils, unpredictable rainfalls, increase in temperature, and the frequent occurrence of droughts. These agro environmental strains compromise the agricultural productivity, lower crop yields, and increase susceptibility to hunger and malnutrition of the rural households. Meanwhile, the challenge of food insecurity is a burning problem because of the increase in population, poor access to farming resources, and destruction of natural resources. In these circumstances, they cannot fully rely on a few types of staple crops to acquire stable and healthy food (Vilakazi et al., 2025).

Legumes crops have two functions in such fragile systems. Legumes are a biological source of nitrogen supplement and enhance the soil fertility through symbiotic fixation of nitrogen hence limiting reliance on expensive synthetic fertilizers as well as enhancing soil health to support future crops. Moreover, legumes are also source of necessary nutrients such as high quality of proteins and essential amino acids, as well as micronutrients that are lacking in cereal-based diets. These attributes of ecological functions and nutritional values pre-empt the position of legumes as significant elements of sustainable intensification plans that could be used to enhance food and nutrition security during climate change (Sibabi, Palanga et al., 2025a).

The ground bean (*Kerstingiella geocarpa*), Kersting's groundnut, geocarpa groundnut or Hausa groundnut are an indigenous and underutilized legume that deserves special consideration. The legume is acclimatized to various marginal environments and low soils fertility, which is characteristic of rainfed smallholder systems. The fact that it can fix nitrogen into the atmosphere, survive water-limited environment and provide nutrient rich grain highlights its potential as a soil fertility enhancer as well as a food security crop. Regardless of these benefits, ground bean has been relatively overlooked in official research, extension and agricultural policy structures and consequently underutilized and under exploited to achieve its potential (Cheung et al., 2025).

The production of ground bean is small all over the world, and the FAO has registered approximately 1.2 million tonnes each year. Nigeria, Ghana, Cameroon and Benin are the most productive countries that produce the majority of the crop and contribute most of the global output (Coulibaly, Bodjrenou et al., 2022a).

Introduction of ground bean into the areas of crop integration is in line with greater aspirations of resilience and sustainability. Ground bean can be used to support more sustainable food systems and enhance household food security by improving the fertility of soil with nitrogen, increasing the diversity of crops, and providing food that is rich in nutrients and thus more resilient to their food system. Hence, its agronomic properties, ecological flexibility, and socio economic importance must be known and its use promoted to achieve its possible opportunities (Ayilara et al., 2022).

This chapter explores the aspect of ground bean in the framework of sustainable farming, rebuilding soil fertility, and food security. It describes botanical profile of the crop, agro ecological fit and climate variability interaction. The chapter also discusses yield performance, nutritional value, processing and usage, modelling and digital evaluation tools, the case studies in major production regions, socioeconomic limitations, future research and policy focus. This is aimed at giving a holistic overview of existing information and map the way forward on how to improve the part played by ground bean in ensuring the food system is sustainable and resilient (Oyetayo & Ajayi, 2011).

18.2 Botanical, Agronomic, and Ecological Profile of Ground Bean

18.2.1 Taxonomy and Botanical Identity

Ground bean (*Kerstingiella geocarpa*) which is also called Kersting groundnut, geocarpa groundnut, Hausa groundnut is a family in the Fabaceae. Correct taxonomic nomenclature and characterization of germplasm: Germplasm characterization, conservation, and breeding programs should have accurate taxonomic nomenclature. It is native to West and Central Africa and commonly known by many different vernacular naming, which is indicative of its ancient cultural, agronomic and nutritional significance in the conventional agricultural practices (Maphosa et al., 2022) (Fig. 18.1).

18.2.2 Floral Morphology and Reproductive Biology

An axillary pseudo raceme has four sessile or sub sessile flowers on each node (Fig. 18.2a). Depending on the genotype, flower color varies between white-greenish and yellowish to purple or pink. Flowers are zygomorphic having one bract and two persistent bracteoles; both calyx and corolla are pentamerous and the petals are arranged vexillarily, a normal petal (9.5–19 mm × 6–17 mm), two wing ones (10–12 mm × 2–2.4 mm) and two keel ones (10–13 mm × 2.5–3.5 mm) wrapping the reproductive organ (Fig. 18.2b, c).

The androecium is gamostemonous, diadelphous and has ten stamens (7–9 mm long). Others are dorsifixed, longitudinally dehisced bilobed, ellipsoid (Fig. 18.2e, f). The grains of pollen are minute (7.9–38.9 um in diameter). The gynoecium is sessile or sub sessile (8–17 mm long) having two or three ovules with erect dorsal placentation in a superior ovary (1.2–5 mm long). The stigma is capitate and short thick hairs enclose it.

Flowers are bisexual, cleistogamous before the process of anthesis and incompletely protogynous. The normal petal has two nectary glands (Fig. 18.2d) though

Fig. 18.1 Kersting's groundnut plant showing **a** upright and **b** open growth habits, **c** pubescent, and **d** glabrous main stems, and **e** a trifoliate leaf, and **f** pods. Arrows indicate main stem (Agoyi et al., 2023a)

insects have not been reported to visit them. It has been reported to have a specialized forced self-pollination mechanism, whereby early style bending is involved. The period of Anthesis takes place around three days following the bud appearance and takes 2–4 days. Stigma receptivity is also observed before anther dehiscence, after which it continues until flower wilting. The viability of pollen is great (88–98%), but in vitro germination is low (8–37%), and it depends on landrace and developmental stage (Akohoue et al., 2020).

Fig. 18.2 Flower structure of Kersting's groundnut. **a** Emergence of four flower buds at a node, **b** floral diagram, **c** opened flower showing the different parts of the corolla, **d** exudate glands on the inner face of the standard petal (white arrows), **e** dorsifixed immature anther, and f dehiscing anther. Scale bar ¼ 4 mm in (**c**) and (**d**) (Agoyi et al., 2023a)

18.2.3 Fruit, Seed, and Developmental Stages

The fruits are indehiscent, pods 0.5–1.0 cm × 0.6–2.0 cm, which are elliptical to elongate, and have one to three seeds separated by constrictions (Fig. 18.1f). The seeds are oblong, oblong-ovoid (kidney-shaped), between 5–10 mm long, 4–7 mm wide and deep, 3–5 mm thick, outward-colored (Fig. 18.3) with the color of the seed coats cream, black, brown, or mottled and the shape of the hilum irregularly distributed across different seeds.

Seed germination takes place in 3–5 days following sowing and is epigeal leading to appearance of opposite lanceolate primary leaves; the first trifoliate leaf is observed 5–10 days after germination. There are six flower and fruit developmental stages (S1–S6 and F1–F6) between flower initiation and development of seed to maturity (Figs. 18.4 and 18.5).

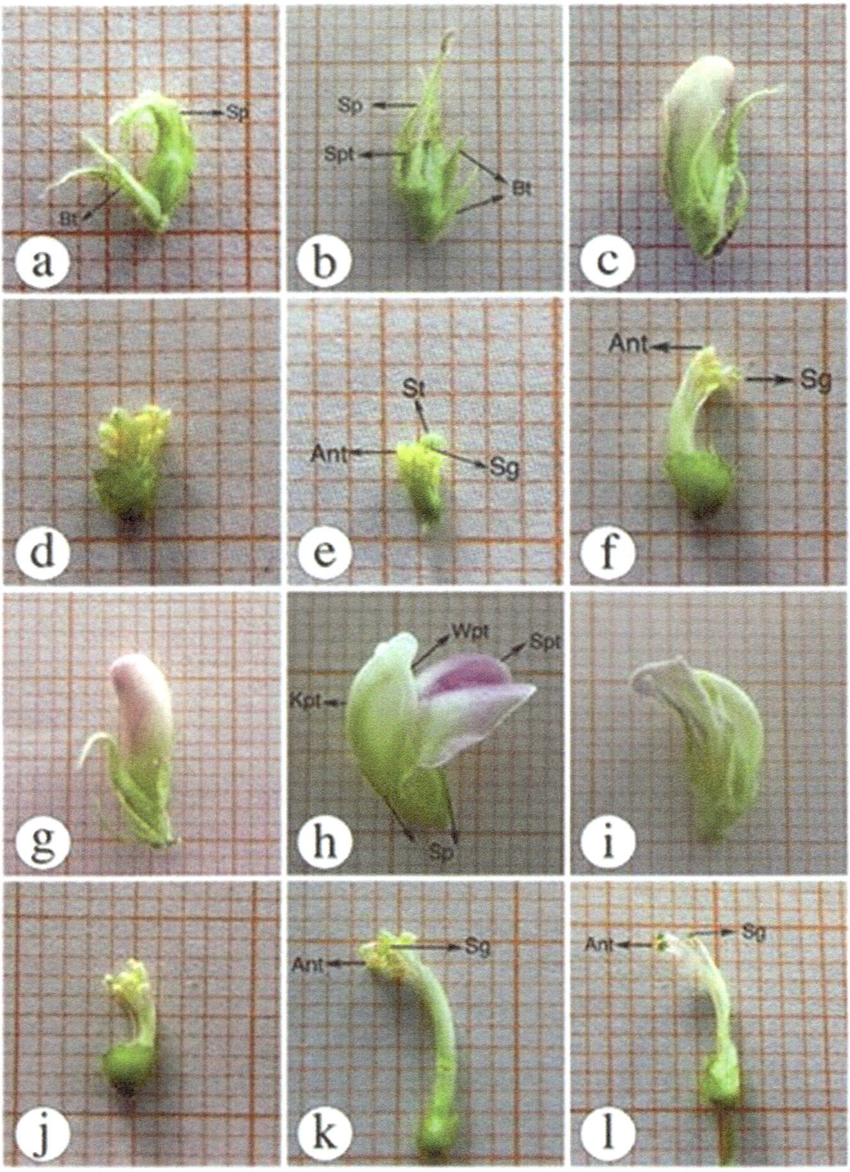

Fig. 18.3 Stages of floral development in Kersting's groundnut. **a**, **d** Initiated flower, **b**, **e** young bud, **c**, **f** developed bud, **g**, **j** mature bud, **h**, **k** open flower, **i**, **l** wilted flower. Ant—anther; Bt—bracteole; Sg—stigma; Sp—sepal; Spt—standard petal; St—style; Kpt—keel petal; Wpt—wing petal. Grid cells ¼ 1 mm (Agoyi et al., 2023a)

Fig. 18.4 Stages of fruit development in Kersting's groundnut. **a** Beginning peg, **b** beginning pod, **c**, **g** full pod, **d**, **h** beginning seed, **e**, **i** full seed, and **f**, **j** mature seed (Akohoue et al., 2020)

18.2.4 Growth Habit, and Geocarpic Pod Development

The ground bean has a short life cycle as it has a prostrate to semi erect growth habit and is highly suitable to the rainfed farming system that has a short season. It has such phenological stages as germination, vegetative growth, flowering, pod formation and maturity. One of the most significant peculiarities of ground bean is that the pods are geocarpic, which means that they develop underground and provide resistance to pests, diseases, and other adverse environmental factors. This is a special property that makes the crop to be resilient to the changing rain and temperatures that are characteristic of the small holder farming systems (Adu-Gyamfi et al., 2011).

Fig. 18.5 Kersting's groundnut landraces

18.2.5 Root System Architecture, Nodulation, and Biological Nitrogen Fixation

Ground bean root system is also deep and highly branched, which allows the ready access to the moisture and nutrients in the soil that is not very fertile. Close association with rhizobia promotes efficient nodulation and biological fixation of nitrogen, increases soil fertility, and limits the use of artificial fertilizers in the production of later crops. Both strong rooting structure and the ability to fix nitrogen makes ground bean an important element of the sustainable cropping systems (Chiurazzi et al., 2025) (Fig. 18.6).

18.2.6 Agronomic Traits Supporting Low-Input and Climate-Resilient Production

Ground bean possesses characteristics that render it a great production venture with low input and climate enduring characteristics. It has resistance to water-stressed environments, low-nutrient soils and moderate resistance to local pests and diseases. Its inclusion in the cropping systems by way of intercropping, rotation, or relay cropping maximizes the healthiness of soils, diversification of production, and grain of nutritious composition to boost home food security. All these ecological and agronomic properties indicate the promise of ground bean as a strategic crop in resilient

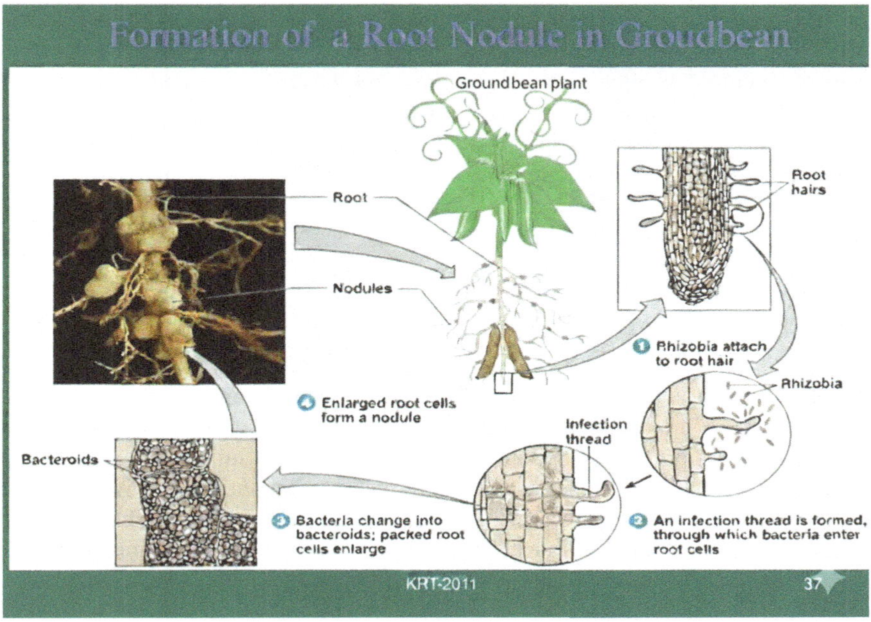

Fig. 18.6 Biological nitrogen fixation in ground bean roots (Shen & Feng, 2024)

and sustainable smallholder farming systems in an environment that is changing with climatic variations (Nhamo et al., 2022).

18.3 Phenology, Distribution, and Production Systems

18.3.1 Phenological Development

The phenology of ground bean (*Kerstingiella geocarpa*) has a characteristic that germinates seeds soon after the initiation of a rainy season then continues to vegetative growth, flowering and pod formation. The germination will usually be achieved in days when the right amount of moisture is present in the soil then the vegetative growth will be speedy and will be characterized by the appearance of robust stems and canopy branches. The onset of flowering is a result of plant biomass accumulation, and small self-pollinating flowers are placed along the stems. Another peculiarity of the ground bean phenology is that it forms geocarpic pods whereby the pollinated flowers above ground place the stolons into soil where the pods grow and mature. This is an adaptive feature that is characteristic of underground fruiting, which shields the plant against pest attacks and drought in changing moisture levels, thereby improving survival in rainfed areas. These developmental stages have different levels of sensitivity to moisture and temperature stress, germination and early vegetative development are highly dependent on soil moisture, and flowering and pod formation are prone to dry spells and heat stress respectively of savanna climates. These phenological features imply a developed resiliency in favor of performance of the crop in low-input and changing environmental conditions (Mohammed et al., 2018).

18.3.2 Geographical Distribution and Agro-ecological Adaptation

The ground bean is native to West Africa and its common agro ecological zone is the savanna region of this region. It grows in a very wide latitude zone, where it is found in the northern part of Sudan to the southern part of Guinea, and is adaptable to rainfall gradients and temperature changes that are characteristic of West African habitat. The crop farming area cuts across a number of countries such as Nigeria, Ghana, Togo, Benin, and Burkina Faso where it has been incorporated into the small-scale farming systems, which are under rain-fed condition. Its capacity to expand in regions that have moderate to low annual precipitation usually after the start and end of rainy season indicates its adaptation to semi-arid and sub humid weather. The ecological adaptation of ground bean goes downwards to the soil characteristics in which the bean thrives well on sandy loam and light textured soils with good drainage that is widespread all over the savanna regions. This ecological plasticity has helped it to persist in a variety of landscapes and to aid in making it a climate resilient species adapted to marginal environments (Sibabi, Palanga et al., 2025a) (Fig. 18.7).

18.3.3 Production Systems and Cropping Practices

Rainfed and low input conditions are widespread in the production systems of West Africa, where ground bean is used; this demonstrates the restricted availability to irrigation and modern agronomic inputs by the smallholder farmers. Farmers usually use ground bean as a single crop at the onset of rainy season, though, it is also intercropped or rotated with other cereals like maize and sorghum to maximize land use and enhance the soil moisture. The introduction of the crop in the mixed systems of the cropping also serves to increase the productivity and sustainability of the farms since it enriches the soil with nitrogen through the biological fixation of nitrogen and assists in the coverage of the soil during the changeable weather patterns of rain. Bean cultivation is not complicated, and it requires only small land preparation, planting in ridges or mounds, and little use of external fertilizers or pesticide. Seed systems are not formal, and therefore, most farmers use the seeds they have retained in their past harvests or share them with members of the community, which

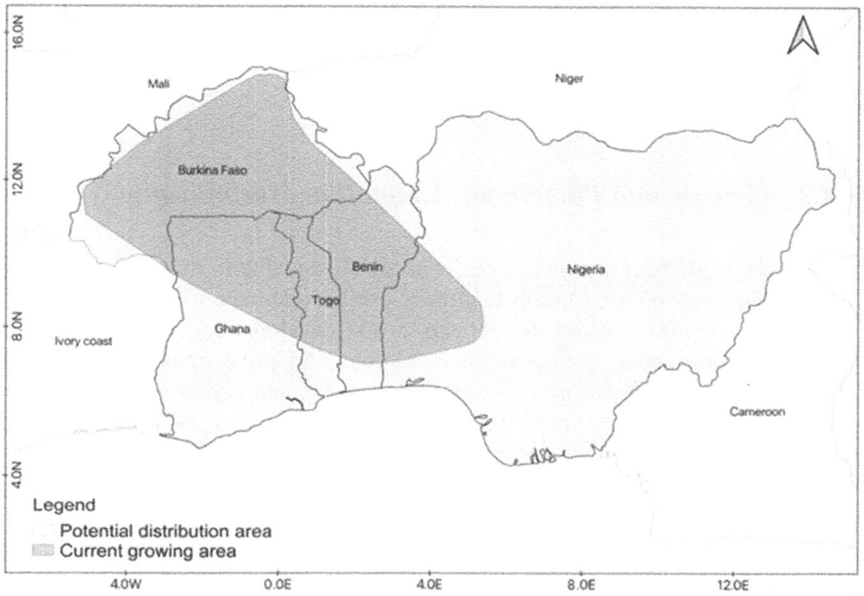

Fig. 18.7 Map of distribution and production areas of Kersting's groundnut (*Macrotyloma geocarpum* (Harms) Marechal & Baudet) (Coulibaly, Idohou et al., 2022b).

affects both the practices of cropping and the varietal diversity. These production models emphasize the importance of ground bean in sustainable agricultural systems that incorporate sustainability, reduced external input, and food insecurity in rainfed landscapes (Coulibaly et al., 2020).

18.4 Climate Change Impacts, Adaptation, and Soil Fertility Enhancement

18.4.1 Climate Risks Affecting Ground Bean Production

Ground bean (*Kerstingiella geocarpa*) cultivation is increasingly vulnerable to climate change-induced stresses such as erratic rainfall patterns, extended dry spells, elevated temperatures, and unexpected flooding events. These climatic fluctuations disrupt germination, vegetative growth, flowering, and geocarpic pod development, ultimately constraining yield stability. In rainfed systems, particularly in West and Central Africa, drought and heat stress pose the greatest threats, reducing soil moisture availability and accelerating evapotranspiration. Such climatic risks exacerbate the challenges faced by smallholder farmers who often have limited access to irrigation, adaptive technologies, and formal weather forecasting systems (Sibabi, Tchabi et al., 2025b).

18.4.2 Genetic and Phenotypic Adaptation Mechanisms

The inherent resilience of ground bean is partly attributable to its genetic diversity and phenotypic plasticity. Certain landraces exhibit tolerance to moisture deficits and high temperatures, while the geocarpic pod development ensures reproductive success even under surface stress conditions. Adaptive traits, such as rapid early growth, drought-resistant root systems, and efficient water-use mechanisms, help maintain productivity under fluctuating environmental conditions. Breeding programs targeting the enhancement of these traits, coupled with the selection of locally adapted genotypes, are essential for strengthening the crop's resilience to climate variability (Hummel et al., 2018).

18.4.3 Agronomic Adaptation Strategies

Several agronomic practices can mitigate climate risks and sustain ground bean productivity. These include optimized sowing dates aligned with rainfall onset, intercropping with cereals to reduce water stress, use of mulches to conserve soil moisture, and minimal tillage to protect soil structure. Crop rotation with cereals and other legumes not only breaks pest and disease cycles but also enhances nutrient cycling, thereby improving system resilience. Low-input approaches that emphasize soil and water conservation are particularly effective in marginal lands where resource availability is limited (Mohammed et al., 2021).

18.4.4 Role of Biological Nitrogen Fixation in Soil Fertility Restoration

Ground bean contributes significantly to soil fertility through symbiotic nitrogen fixation, enriching soils with bioavailable nitrogen for subsequent crops. The crop's nodulation and nitrogen-fixing capacity reduce dependence on synthetic fertilizers, lower production costs, and improve soil health in smallholder farming systems. Incorporating ground bean in legume-based rotations or intercropped systems can enhance nutrient cycling, increase organic matter, and improve the long-term productivity of degraded soils, aligning with the principles of sustainable intensification (Hu et al., 2024).

18.4.5 Implications for Climate-Smart and Sustainable Farming Systems

Integrating ground bean into climate-smart agriculture frameworks offers multiple benefits. Its resilience to water-limited conditions, combined with soil fertility enhancement and nutrient-rich grain production, positions the crop as a key contributor to sustainable and resilient food systems. Promoting adaptive management practices, cultivar selection, and system diversification ensures stable yields, improved household nutrition, and enhanced soil health. Consequently, ground bean represents a strategic crop for addressing the dual challenges of climate change and food insecurity in smallholder farming landscapes (Tabe-Ojong et al., 2023).

18.5 Yield Potential, Agronomic Performance, and Food Security Outcomes

18.5.1 Grain Yield Ranges Under Different Agro-ecological Conditions

Ground bean (*Kerstingiella geocarpa*) has been found to have inconsistent grain yield in relation to the agro-ecological environment, soil fertility, and the management practices. On its native West and Central African territories, the production would only be 0.5–2.0 tonnes per hectare under rainfed conditions, and better results could be obtained under properly managed plots with sufficient moisture and organic inputs. The environmental factors that lead to variability in yield are the type of soil, the distribution of rainfall, and the presence of pests or diseases, which illustrate the sensitivity of the crop to environmental factors and the need of site-specific management techniques (Ayenan & Ezin, 2016).

18.5.2 Yield Stability Under Rainfed and Low-Input Environments

The ability of ground bean to sustain decent yields even in the low-input and rainfed systems is one of the strengths associated with the product. Its extensive root system and adaptive characteristics to drought enable it to survive moisture stressful periods whereas its geocarpic pods shield the reproductive structures to surface-level alterations of the environment. These attributes render ground bean a good alternative to the smallholder farmers in marginal regions where the chances of crop failure are high and access to irrigation or fertilizers is either low. Seasonal stability in terms of

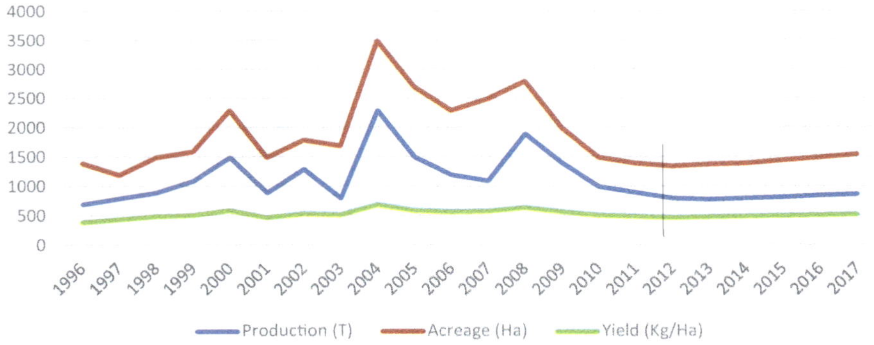

Fig. 18.8 Production history of Kersting's groundnut in Benin from 1996 to 2017. MAEP, Benin

yields improves the resilience of households because it offers stable food and income (Anderson et al., 2016) (Fig. 18.8).

18.5.3 Genotype × Environment Interactions

The existence of a performance difference between ground bean genotypes under different environmental conditions indicates that there are a substantial genotype × environment interactions. Some landraces have been found to work well in drier, low-fertile soils, whilst others respond well to better soil management and average rainfall. Knowledge of these interactions is critical to the selection of suitable cultivars to particular agro-ecological areas and to breeding schemes to help increase climate resilience and yield potential (Coulibaly et al., 2024).

18.5.4 Constraints to Productivity and Yield Improvement

Even though it is resilient, there are a number of constraints that constrain productivity of ground beans. They are vulnerability to pests and diseases, low quality of seeds, poor agronomic management, and ineffective research focus. The potential yield is also limited by nutrient depletion in the soil that has been caused by soil erosion and poor supply of better varieties. A combination of pest management, seed systems, and involvement of climate-sensitive agronomic methods is important to improve the productivity and role of the crop in the food security (Kafoutchoni et al., 2022).

18.5.5 Comparison with Cowpea and Bambara Groundnut

Ground bean has shown similar or better results in most low-stress prone, low-input environments as compared to other underutilized legumes like cowpea (*Vigna unguiculata*) and Bambara groundnut (*Vigna subterranea*). Its geocarpic pod structure provides a benefit in terms of preventing predation and environmental harm, and its adaptability to marginal soils provides a greater range of cultivation. These comparisons reveal the prospect of ground bean as an auxiliary crop in the legume based cropping systems (Table 18.1).

18.5.6 Contribution of Yield Stability to Food Availability and Household Food Security

The constant production of ground bean in adverse environments directly leads to food supply, nutrition, and earnings in the household. Ground bean prevents seasonal shortages and malnutrition as well because it offers a source of protein-rich grain even in low-input systems. Its adoption in smallholder agricultural systems is consistent with the larger objective of sustainable agriculture and climate resiliency and local food security, which supports its usefulness in terms of strategic culture in West and Central African agriculture (Melesse et al., 2023).

18.6 Nutritional Composition and Contribution to Food and Nutrition Security

18.6.1 Protein Content, Amino Acid Profile, and Energy Value

Ground bean (*Kerstingiella geocarpa*) is considered as a good source of seed protein with crude protein ranging between 20 and 28% depending on the genotype and soil fertility. It has a good amino acid composition with the necessary amino acids like lysine and leucine, which are otherwise lacking in cereal-based diets. Ground bean seeds have also a good energy content of 350,370 kcal per 100 g, thus a good source of dietary calories to the rural households. This high protein and energy density makes ground bean a strategic legume to enhance nutritional adequacy especially in the smallholder farming communities (Fassinou et al., 2023) (Table 18.2).

Table 18.1 Proximate composition of groundnut, Bambara groundnut, and cowpea by Kersting cowpea (Agoyi et al., 2023b)

Proximate	Kersting's groundnut		Bambara groundnut		Cowpea	
	Values	References	Values	References	Values	References
Protein (g/100 g)	19.91–22.36	Ajayi and Oyetayo, Chickwendu (2007)	16.6–30.37	Adebowale et al., Enwere and Hung, Halimi et al.	17.5–32.5	Boukar et al., Nielsen et al.
Crude fat (g/100 g)	1.00–2.42	Ajayi and Oyetayo, Chickwendu (2007)	5.9–6.51	Enwere and Hung, Halimi et al.	1.4–2.7	Akissoé et al., Halimi et al., Nielsen et al.
Carbohydrate (g/100 g)	68.82–70.52	Ajayi and Oyetayo, Chickwendu (2007)	44.94–64.43	Adebowale et al., Enwere and Hung, Halimi et al.	58–71.6	Akissoé et al., Halimi et al., Nielsen et al.
Dietary fiber (g/100 g)	3.99–4.9	Ajayi and Oyetayo, Chickwendu (2007)	1.9–5.49	Adebowale et al., Enwere and Hung, Halimi et al., Onimawo et al.	0.43–3.65	Ajayi et al., Halimi et al.

Table 18.2 Proximate composition, nitrogenous constituents, sugar and energy values of four varieties of groundbean (dry matter basis) (Chickwendu, 2007)

Nutrients	RGBG	BGB	LGBG	DCGB
Crude protein (%)	21.58 ± 0.53	20.91 ± 0.44	19.91 ± 0.65	22.36 ± 0.06
Total nitrogen (%)	3.45 ± 0.15	3.35 ± 0.10	3.16 ± 0.06	3.61 ± 0.21
Non-protein nitrogen (%)	0.70 ± 0.20	0.42 ± 0.02	0.70 ± 0.20	0.70 ± 0.10
True protein nitrogen (%)	2.75 ± 0.17	2.93 ± 0.46	2.46 ± 0.20	2.91 ± 0.10
True protein (%)	17.19 ± 0.11	18.31 ± 0.30	15.50 ± 0.50	18.19 ± 0.09
Crude fat (%)	2.42 ± 0.07	2.18 ± 0.03	2.21 ± 0.95	2.15 ± 0.57
Crude fibre (%)	4.17 ± 0.40	3.99 ± 0.21	4.09 ± 0.91	4.10 ± 1.00
Ash (%)	2.91 ± 0.08	2.57 ± 0.05	2.46 ± 0.12	2.57 ± 0.07
Carbohydrate (%)	68.92 ± 0.35	70.35 ± 0.03	70.52 ± 0.22	68.82 ± 0.22
Total sugar (g/100 g)	0.60 ± 0.30	0.80 ± 0.10	0.50 ± 0.02	0.30 ± 0.10
Energy (kJ)	1713	1705	1688	1697

Mean ± SD of 3 determinations
RGBG—Reddish-brown ground bean
BGB—Brown ground bean
LGBG—Light brown ground bean
DCGB—Deep-cream ground bean

18.6.2 Mineral and Micronutrient Composition

Besides the macronutrients, ground bean is an important source of the essential minerals and micronutrients. Seeds are rich sources of calcium, iron, magnesium, and zinc, which are essential to bone, cognitive and immune processes. The presence of these micronutrients is very vital in regions where deficiencies are so common because of the monotonous cereal diets. Ground bean may be a possible solution to the hidden hunger and micronutrient malnutrition in resource-poor populations by taking frequent portions of the food (Abiola & Oyetayo, 2015).

18.6.3 Comparison with Cowpea and Bambara Groundnut

In comparison to the other under-exploited legumes like cowpea (*Vigna unguiculata*) and Bambara groundnut (*Vigna subterranea*) ground bean is shown to have competitive or better nutritional value. Its lysine content is also higher than that of cereals and moderate fat content is lower than the Bambara groundnut thus making it very useful in the complementary feeding programs. These comparative advantages give emphasis on the potential of ground bean as an important ingredient in smallholder systems as an ingredient of nutritionally balanced diets (Aremu et al., 2007) (Fig. 18.9).

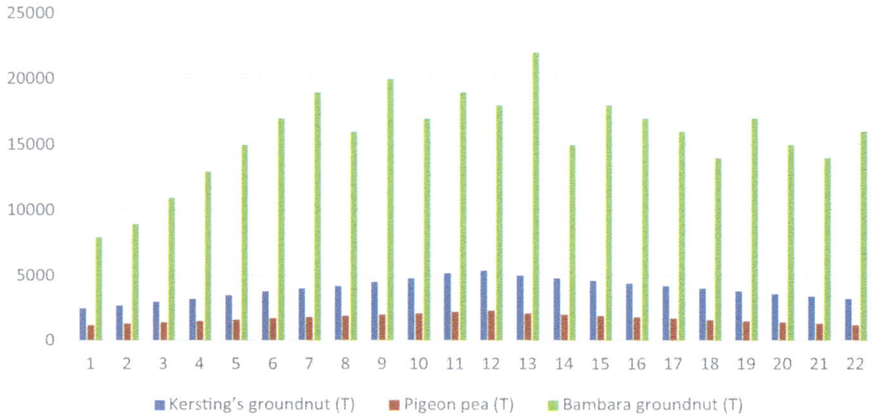

Fig. 18.9 Evolutionary trend of Kerstings groundnut production in comparison with other orphan legume crops commonly grown in Benin. MAEP, Benin (Agoyi et al., 2023b)

18.6.4 Importance in Protein-Deficient and Resource-Poor Diets

Ground bean will also help in reducing protein shortage among the resource-deficient communities. Ground bean is a cheap source of plant-based protein in areas where plant sources of animal protein are scarce or unaffordable. Its incorporation into everyday meals contributes to lowering the risk of malnutrition, child growth, and development as well as enhancing the household nutritional security in general (Fassinou et al., 2023).

18.6.5 Role in Dietary Diversity and Nutrition Security

In addition to the separate nutrient intake, ground bean improves dietary variation by incorporating it in the traditional stews, soups and complement food. Legume diversity enhances the overall quality of the diet, resolves the resistance to food insecurity and fosters sustainable nutrition outcomes. Connecting agronomic performance and nutritional value, ground bean has a dual role in food and nutrition security, which is strengthened due to emerging conditions of changed climatic and socioeconomic conditions in smallholder farming systems (Ng'endo et al., 2018).

18.7 Processing, Utilization, Value Addition, and Food Security Enhancement

18.7.1 Traditional Processing Methods and Food Uses

The main traditional way of processing ground bean (*Kerstingiella geocarpa*) through dehulling, soaking, boiling, roasting, and milling methods are instrumental in lowering the consumption barriers and making it palatable. Ground bean is used to make soups, stews, porridges and snacks in most indigenous communities. Such traditional ways of carrying out such practices contribute to not only the preservation of the nutritional value of the crop but also make it a part of culturally acceptable diets to ensure its continued use in smallholder farming systems (Muhammad et al., 2020).

18.7.2 Effects of Fermentation and Germination on Nutritional Quality

Fermentation and germination have become efficient methods of post-harvest to improve nutritional value of ground bean. The germination enhances the action of endogenous enzymes that promote protein digestibility and bioavailability of essential amino acids. On the same note, the nutritional value of the legume could be improved through controlled fermentation process, enriched with useful microorganisms and increasing the levels of vitamins, decreasing the levels of anti-nutritional factors and enhancing the digestibility. These procedures, therefore, provide low-cost, easy to implement interventions to enhance the nutrition value of ground bean in home meals (Abiola & Oyetayo, 2015) (Table 18.3).

18.7.3 Reduction of Anti-nutritional Factors and Improved Bioavailability

Similar to other legumes, ground bean has naturally occurring anti-nutritional components, such as tannins, phytates and trypsin inhibitors, which may reduce the absorption of nutrients. These compounds are significantly reduced due to traditional processing, germination as well as fermentation which enhances mineral and protein bioavailability. These approaches are essential and should be optimized to achieve maximum nutritional value but maintain the necessary macro- and micronutrients especially in communities that lack enough resources and where protein deficiency occurs (Okaiyeto et al., 2025).

Table 18.3 Essential amino acids in raw and fermented Kersting's groundnut (g/100 g protein) compared with FAO standard (Abiola & Oyetayo, 2015)

Amino acid (g/100 g protein)	0	FAO ref value	24L	24S	48L	48S	72L	72S
Lysine	6.71 ± 0.01	4.20	7.22 ± 0.01	6.49 ± 0.01	6.62 ± 0.02	6.90 ± 0.01	5.62 ± 0.01	6.52 ± 0.01
Threonine	3.43 ± 0.01	2.60	4.14 ± 0.01	3.53 ± 0.01	3.28 ± 0.01	3.81 ± 0.01	3.03 ± 0.01	3.13 ± 0.01
Histidine	2.88 ± 0.01	2.40	3.02 ± 0.01	2.93 ± 0.01	2.71 ± 0.01	2.91 ± 0.01	2.21 ± 0.01	2.40 ± 0.01
Arginine	6.45 ± 0.01	2.00	1.65 ± 0.01	1.37 ± 0.01	1.24 ± 0.00	1.45 ± 0.00	0.76 ± 0.00	1.18 ± 0.00
Valine	4.38 ± 0.00	4.20	5.20 ± 0.00	4.56 ± 0.00	4.22 ± 0.00	4.91 ± 0.00	3.10 ± 0.00	4.02 ± 0.00
Methionine	1.21 ± 0.01	2.20	1.55 ± 0.00	1.26 ± 0.01	0.98 ± 0.01	1.42 ± 0.00	0.62 ± 0.00	0.91 ± 0.00
Isoleucine	3.68 ± 0.01	4.20	4.12 ± 0.00	3.62 ± 0.00	3.56 ± 0.01	3.97 ± 0.01	2.91 ± 0.00	3.36 ± 0.00
Leucine	7.19 ± 0.00	4.20	8.10 ± 0.00	7.39 ± 0.00	6.97 ± 0.01	7.49 ± 0.01	5.50 ± 0.01	6.78 ± 0.00
Phenylalanine	5.55 ± 0.01	2.80	5.81 ± 0.00	5.11 ± 0.01	5.38 ± 0.01	5.62 ± 0.01	3.71 ± 0.01	4.83 ± 0.01

Key S—solid fermentation, L—liquid fermentation

18.7.4 Development of Value-Added Food Products

Ground bean has significant prospects of value-added food products. It can be made into flour, protein concentrates, extruded snacks, energy bars, and bakery products that have an extension of shelf life, providing consumers with a wide range of choices, and it can make new market prospects. Value addition can help in food security as well as create income especially among women and smallholder farmers who are engaged in post-harvest processing (Awolu et al., 2020).

18.7.5 Potential for Infant, Weaning, and Functional Foods

Ground bean has a good potential in infant and weaning foods because of its high protein content, digestibility and high micronutrient profile. Moreover, the legume bioactive compounds have functional capabilities which include antioxidant and immunomodulatory potential, which can be harnessed in functional foods formation in addressing certain nutritional requirements. These applications highlight the further applications of the crop to the nutrition of the general population (Coulibaly, Bodjrenou et al., 2022a).

18.7.6 Contribution to Food Access, Utilization, and Income Security

Processing and value addition of ground bean in foods will improve access and consumption of foods. Ground bean can raise nutrition security through its direct impact of providing more nutrients and diversity in the diet. At the same time, the sale of processed products generates income-generating prospects among rural households, so the production of agriculture is connected to economic sustainability. All these make ground bean a multi-purpose crop that can serve the purpose of nutrition and livelihood in the smallholder systems (Ikujenlola et al., 2022).

18.8 Crop Modelling, Remote Sensing, and Digital Approaches for Productivity and Soil Fertility Assessment

18.8.1 Role of Crop and Soil Models in Simulating Yield and Nitrogen Fixation

Crop and soil models give important resources in the interpretation of the growth, yield potential, and nutrient dynamics of ground bean (*Kerstingiella geocarpa*) in various agro-ecological conditions. These models enable researchers and farmers to increase their assumption and decrease negative results by simulating phenology, biomass, and biological nitrogen fixation and making forecasts of productivity, industry with regards to planting and estimating the contribution of it to the soil fronts by rotations of legumes. These predictive features are especially useful in less input and rainfed systems when trial and error in the field can be either minimal or resource-intensive (Verburg et al., 2025) (Fig. 18.10).

18.8.2 Modelling Ground Bean Performance Under Climate Variability

Climate changes such as changes in the amount of rainfall and temperatures are also a major problem to the cultivation of ground beans. Crop models that are process based can be used to model the impact of these climatic factors on the growth stages, formations of yields and the efficiency of fixation of nitrogen. The models make it possible to test the scenario when using various sowing dates, cultivar selection, and management approaches and provide evidence-based recommendations to reduce risks created by climate. This means that modelling makes agronomic decisions climate-resilient, better, and more efficient in managing soil fertility (Nykytiuk & Kravchenko, 2025).

18.8.3 Application of Remote Sensing for Monitoring Crop Growth and Stress

In monitoring the growth of ground beans, canopy development and even stress responses, remote sensing device has been utilized such as multispectral and hyper-spectral imaging which offer non-destructive approaches. Vegetation indices based on satellite or drone imagery, e.g. NDVI or LAI, are capable of measuring biomass accretion, water or nutrient stress, and general crop health. Remote sensing and then combined with ground truth information will facilitate accuracy agriculture

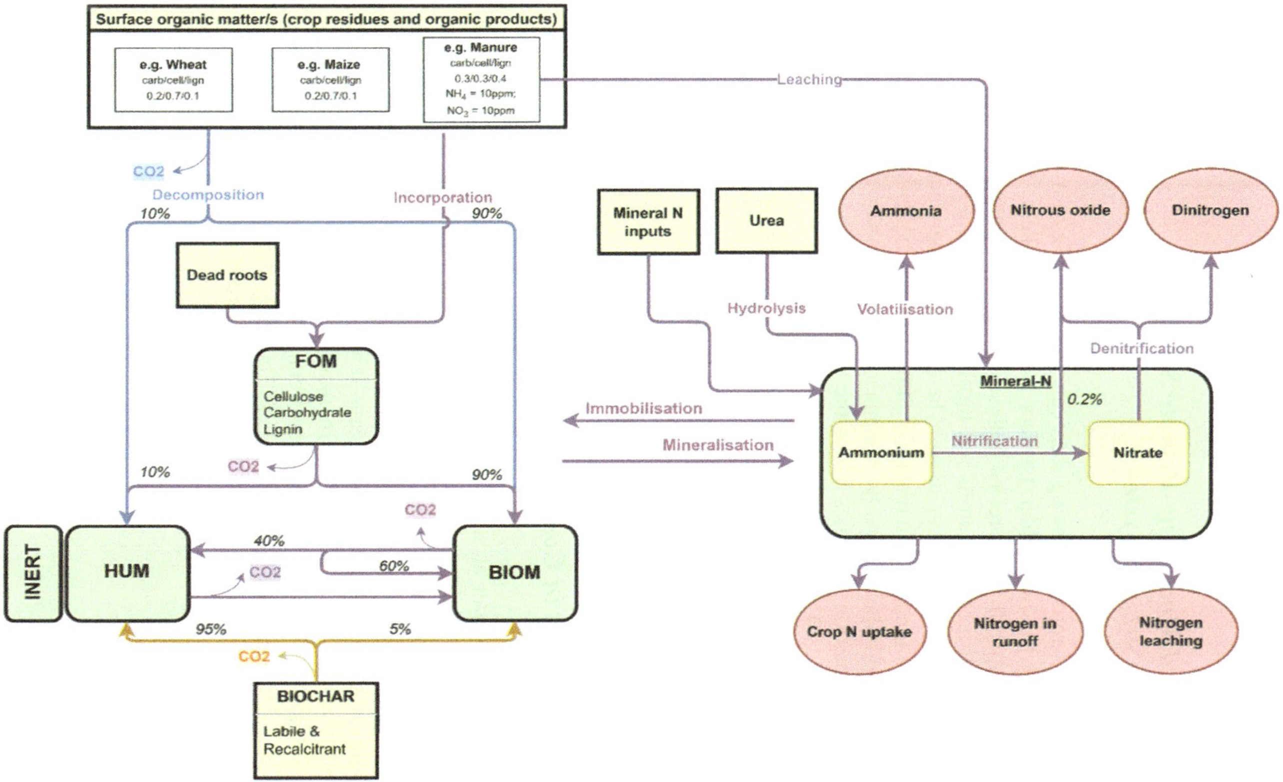

Fig. 18.10 Schematic representation of APSIM's conceptual organic matter pools and processes affecting the carbon and nitrogen (N) flows and transformations; FOM = fresh organic matter pool, BIOM = more labile organic matter pool notionally representing soil microbial biomass and microbial products; HUM = more stable, humus-like organic matter pool; Inert = most stable, assumed to not decompose; green boxes represent soil pools, yellow boxes the inputs to the system and pink boxes the losses from the system (Verburg et al., 2025)

methods to provide interventions like supplemental irrigation or nutrients application to optimize output and available resources.

18.8.4 Integrated Modelling–Remote Sensing Decision-Support Systems

Both crop models and remote sensing data can be combined to make a powerful decision-support tool to smallholder farmers and policymakers. Such combined structures have the capability to forecast both site-specific productivity, to estimate the contribution of nitrogen fixation to soil fertility, and to give early information on the occurrence of drought or heat stress. The connection between the outputs of modelling and observational data in real time allows the stakeholders to have actionable information to optimize ground bean management, enhance yield stability, and augment climate adaptation strategies within the legume-based networks of farming systems (Mehedi et al., 2024).

18.9 Case Studies Demonstrating the Role of Ground Bean in Soil Fertility and Food Security

18.9.1 Case Study I: Nodulation, Yield, and Soil Fertility Improvement in Coastal Savannah of Ghana

The case study by Asiedu et al. aims at the Coastal Savannah of Ghana to determine the effect of nodulation, yield, and soil fertility of ground bean (*Kerstingiella geocarpa*) using low-input farming systems. The research took the form of a randomized complete block design where there were three types of ground bean planted on marginal soils and no use of synthetic fertilizers. It was found that 12–18 nodules per plant were observed which led to a 15–20% increase in the soil nitrogen levels during the cropping season. The grain yield was between 0.8–1.2 t ha^{-1}, which revealed average productivity due to the lack of nutrients. The paper suggested future research on whether ground bean should be combined with cereal rotations and the analysis of inoculant strains to maximize the fixation of nitrogen in agro-ecologies in the same manner.

18.9.2 Case Study II: Ground Bean and Household Nutrition Security in West Africa

Kouadio et al. evaluated the role played by ground bean in the household nutrition security of rural Cote d'Ivoire communities. The researchers identified protein and micronutrient intake based on the ground bean consumption through dietary surveys and through household consumption measurements during a 12-month period. The results showed that habitual inclusion of ground bean in the family diets served 18–22 g of protein per capita of protein/day and greatly enhanced intake of iron, zinc and essential amino acids. The paper has emphasized the promise of ground bean in alleviating protein-energy malnutrition and suggested strategies aimed at promoting ground bean to the community and enhancing the use and consumption.

18.9.3 Case Study III: Processing, Fermentation, and Nutritional Enhancement of Ground Bean Products

The study conducted by Bello et al. was aimed at determining the impacts of germination and fermentation on the nutritional quality of ground bean in Northern Nigeria. The process involved a germination period of 72 h and fermentation of 48 h after which the process was subject to proximate and mineral analysis. The outcome revealed that protein content had risen by 12% whereas anti-nutrition factors like phytic acid had reduced by 35%, improving the mineral bioavailability. These advancements indicate the appropriateness of processed ground bean in infant, weaning and functional food items. The authors suggested that investigations should be done on scalable processing technologies that can be used in commercialization and broad application in nutrition-sensitive services.

18.9.4 Case Study IV: Ground Bean in Climate-Resilient and Sustainable Food Systems

The study conducted by Ouattara et al. aimed to investigate the behaviour of ground bean in the Sahelian region of Burkina Faso in the context of climate variability. Field experiments were done in rainfed condition with different sowing dates and with intercropping with millet. The experiment recorded grain production (between 0.6 and 1.1 t ha^{-1}) and found that ground bean could be sustained with consistent growth and nodulation even during times of low rainfall (250–400 mm seasonal rainfall). Ground bean also added to both the nitrogen of the soil (+ 18%) and the overall system production. The study highlighted the ground bean promise of increasing resilience of marginal environments and proposed new research on drought-tolerant genotypes and approaches to integrating the systems to smallholders.

18.10 Future Research Needs and Development Pathways

18.10.1 Breeding for Yield, Nodulation Efficiency, and Stress Tolerance

The future research must emphasize the production of high yielding ground bean varieties, more efficient nodulated varieties and tolerant to biotic and abiotic stress. It is possible to use breeding programs to select genotypes with better drought tolerance, resistance to heat stress and early maturity to help the rainfed and low-input farming systems. Besides, the focus on genetic diversity of native landraces can assist with the identification of features that maximize productivity and preserve adaptability during the different agro-ecological areas (Mohammed et al., 2018).

18.10.2 Rhizobial Inoculation and Soil–Plant–Microbe Interactions

The symbiotic relationship between ground bean and rhizobia is still of concern and understanding of the relationship and its improvement is critical in order to maximize the biological fixation of nitrogen. Studies must be on the identification of the most effective rhizobial strains, best inoculation methods, and the contribution of soil microbiomes to the improvement of plant health. Microbial interventions in production can enhance the productivity of the soil, decrease the use of synthetic fertilizers, and promote the sustainable cyclic processes of nutrients in smallholder systems (Desmae et al., 2019).

18.10.3 Mechanization, Processing Technologies, and Value Chains

In order to maximize the potential of ground bean, research and development must aim at mechanization of the cultivation, harvesting, and after harvest operations. Fermentation, germination, and other product value-adding innovations have the potential to increase nutritional quality, minimize anti-nutritional properties, and develop market-ready products that could meet the functional, weaning, and infant food needs. Value chains will be enhanced by storage, packaging, and market linkages to boost the farmer income, spur local economies, and enhance household access to nutrient rich foods (Woomer et al., 2023).

18.10.4 Policy and Institutional Support

The introduction and the expansion of the cultivation of ground bean need to be facilitated through institutional structures and the policies. Favorable policies are to be encouraged in terms of research capital, extension services and incentives on the incorporation of ground bean in the cropping systems. It should be encouraged through capacity building, programs of training farmers, and awareness campaigns to promote adoption and the cooperation between governments, research institutions and private sector actors can be used to promote sustainable production, market development and climate-resilient food security strategies (Breuer et al., 2023).

18.11 Conclusion

Ground bean (*Kerstingiella geocarpa*) is a potentially effective legume crop that can both improve soil fertility and help to provide food and nutrition security, especially in smallholder, rainfed-farming systems in West and Central Africa. Its natural ability to adapt to marginal soils, resistance to moisture stress and the ability to fix nitrogen biologically, predisposes it as a climate-resilient crop that can be used to maintain productivity when subjected to fluctuating and challenging agro-ecological conditions. Agronomic research shows that ground bean enhances nitrogen levels in soils, crop diversification, as well as stabilization of yield in low-input systems, and the fact that its grain is nutrient-dense provides answers to protein and micronutrient deficiencies that are common in low resource-diets.

Nutritional studies, processing experiments, and value-addition studies indicate that it can be used to increase the nutritional variety in the diet, lessen the anti-nutritional properties of the diet, and create functional, infant, and weaning foods. Its role in household nutrition, sustainable cropping systems and restoring soil fertility under natural conditions has been highlighted in case studies in Ghana and Nigeria among other parts of the West Africa.

In order to maximize the potential of ground bean, breeding against stress tolerance, enhance the efficacy of nodulation, enhance rhizobial inoculation, and mechanize the processing and value chains should be the priorities of research and development. Policy assistance, service extension and integration into climate-smart agriculture systems are key in ensuring that adoption is at scale and resilience is attained both within the farm and community. All these measures make ground bean an important part of sustainable resilient and nutrition sensitive agricultural systems under climate change.

References

Abiola, C., & Oyetayo, V. (2015). Effect of liquid and solid fermentation on mineral and amino acid contents of Kersting's groundnut (*Macrotyloma geocarpum*). *Journal of Scientific Research and Reports, 8*(7), 1–12. https://doi.org/10.9734/JSRR/2015/19980

Adu-Gyamfi, R., Fearon, J., Bayorbor, T. B., Dzomeku, I. K., & Avornyo, V. K. (2011). The status of Kersting's groundnut (*Macrotyloma Geocarpum* [Harms] Marechal and Baudet): An underexploited legume in Northern Ghana. *Outlook on Agriculture, 40*(3), 259–262. https://doi.org/10.5367/oa.2011.0050

Agoyi, E. E., Kafoutchoni, K. M., & Tele Ayenan, M. A. (2023). Kersting's groundnut (*Macrotyloma geocarpum* (Harms) Maréchal & Baudet). In *Neglected and underutilized crops* (pp. 365–389). Elsevier. https://doi.org/10.1016/B978-0-323-90537-4.00004-1

Akohoue, F., Achigan-Dako, E. G., Sneller, C., Van Deynze, A., & Sibiya, J. (2020). Genetic diversity, SNP-trait associations and genomic selection accuracy in a west African collection of Kersting's groundnut [*Macrotyloma geocarpum* (Harms) Maréchal & Baudet]. *PLoS ONE, 15*(6), Article e0234769. https://doi.org/10.1371/journal.pone.0234769

Anderson, W., Johansen, C., & Siddique, K. H. M. (2016). Addressing the yield gap in rainfed crops: A review. *Agronomy for Sustainable Development, 36*(1), 18. https://doi.org/10.1007/s13593-015-0341-y

Aremu, M. O., Olaofe, O., Akintayo, E. T., & Adeyeye, E. I. (2007). Foaming, water absorption, emulsification and gelation properties of Kersting's groundnut (*Kerstingiella geocarpa*) and Bambara groundnut (*Vigna subterranean*) flours as influenced by neutral salts and their concentrations. *Pakistan Journal of Nutrition, 7*(1), 194–201. https://doi.org/10.3923/pjn.2008.194.201

Awolu, O. O., Magoh, A. O., & Ojewumi, M. E. (2020). Development and evaluation of extruded ready-to-eat snack from optimized rice, Kersting's groundnut and lemon pomace composite flours. *Journal of Food Science and Technology, 57*(1), 86–95. https://doi.org/10.1007/s13197-019-04033-9

Ayenan, M. A. T., & Ezin, V. A. (2016). Potential of Kersting's groundnut [*Macrotyloma geocarpum* (Harms) Maréchal & Baudet] and prospects for its promotion. *Agriculture & Food Security, 5*(1), 10. https://doi.org/10.1186/s40066-016-0058-4

Ayilara, M. S., Abberton, M., Oyatomi, O. A., Odeyemi, O., & Babalola, O. O. (2022). Potentials of underutilized legumes in food security. *Frontiers in Soil Science, 2*, 1020193. https://doi.org/10.3389/fsoil.2022.1020193

Breuer, A., Leininger, J., Malerba, D., & Tosun, J. (2023). Integrated policymaking: Institutional designs for implementing the sustainable development goals (SDGs). *World Development, 170*, Article 106317. https://doi.org/10.1016/j.worlddev.2023.106317

Cheung, T.-Y., Kafoutchoni, K. M., Agoyi, E. E., Chan, T.-F., & Chapman, M. A. (2025). Exceptionally low genomic diversity in the underutilised legume Kersting's groundnut. *Nature Communications, 16*(1), 5183. https://doi.org/10.1038/s41467-025-60494-x

Chickwendu, N. J. (2007). Chemical composition of four varieties of groundbean (*Kerstingiella geocarpa*). *Agro-Science, 6*(2), 79–84. https://doi.org/10.4314/as.v6i2.1575

Chiurazzi, M., Frugis, G., & Navazio, L. (2025). Symbiotic nitrogen fixation: A launchpad for investigating old and new challenges. *Journal of Experimental Botany, 76*(6), 1473–1477. https://doi.org/10.1093/jxb/erae510

Coulibaly, M., Agossou, C. O. A., Akohoué, F., Sawadogo, M., & Achigan-Dako, E. G. (2020). Farmers' preferences for genetic resources of Kersting's groundnut [*Macrotyloma geocarpum* (Harms) Maréchal and Baudet] in the production systems of Burkina Faso and Ghana. *Agronomy, 10*(3), 371. https://doi.org/10.3390/agronomy10030371

Coulibaly, M., Bodjrenou, G., Akohoue, F., Agoyi, E. E., Merinosy Francisco, F. M., Agossou, C. O. A., Sawadogo, M., & Achigan-Dako, E. G. (2022a). Profiling cultivars development in Kersting's groundnut [*Macrotyloma geocarpum* (Harms) Maréchal and Baudet] for improved yield, higher nutrient content, and adaptation to current and future climates. *Frontiers in Sustainable Food Systems, 5*, Article 759575. https://doi.org/10.3389/fsufs.2021.759575

Coulibaly, M., Idohou, R., Akohoue, F., Peterson, A. T., Sawadogo, M., & Achigan-Dako, E. G. (2022b). Coupling genetic structure analysis and ecological-niche modeling in Kersting's groundnut in West Africa. *Scientific Reports, 12*(1), 5590. https://doi.org/10.1038/s41598-022-09153-5

Coulibaly, M., Bodjrenou, G., Fassinou Hotègni, N. V., Akohoue, F., Agossou, C. A., Azon, C. F., Matro, X., Bello, S., Adjé, C. O. A., Sanou, J., Batieno, B. J., Sawadogo, M., & Achigan-Dako, E. G. (2024). Genotype by environment interaction and stability analysis of three agronomic traits in Kersting's groundnut (*Macrotyloma geocarpum*) using factor analytic modeling and environmental covariates. *Crop Science, 64*(4), 2095–2115. https://doi.org/10.1002/csc2.21249

Desmae, H., Janila, P., Okori, P., Pandey, M. K., Motagi, B. N., Monyo, E., Mponda, O., Okello, D., Sako, D., Echeckwu, C., Oteng-Frimpong, R., Miningou, A., Ojiewo, C., & Varshney, R. K. (2019). Genetics, genomics and breeding of groundnut (*Arachis hypogaea* L.). *Plant Breeding, 138*(4), 425–444. https://doi.org/10.1111/pbr.12645

Fassinou, F. T. K., Agoyi, E. E., Affonfere, M., Sacla Aide, E., Assogbadjo, A. E., & Chadare, F. J. (2023). Comparative analysis of proximate compositions, physical properties, and sensory attributes of Kersting's groundnut (*Macrotyloma geocarpum* (Harms) Maréchal & Baudet) accessions. *Journal of Food Quality, 2023*, 1–13. https://doi.org/10.1155/2023/6372248

Hu, W., Wang, X., Xu, Y., Wang, X., Guo, Z., Pan, X., Dai, S., Luo, Y., & Teng, Y. (2024). Biological nitrogen fixation and the role of soil diazotroph niche breadth in representative terrestrial ecosystems. *Soil Biology and Biochemistry, 189*, Article 109261. https://doi.org/10.1016/j.soilbio.2023.109261

Hummel, M., Hallahan, B. F., Brychkova, G., Ramirez-Villegas, J., Guwela, V., Chataika, B., Curley, E., McKeown, P. C., Morrison, L., Talsma, E. F., Beebe, S., Jarvis, A., Chirwa, R., & Spillane, C. (2018). Reduction in nutritional quality and growing area suitability of common bean under climate change induced drought stress in Africa. *Scientific Reports, 8*(1), 16187. https://doi.org/10.1038/s41598-018-33952-4

Ikujenlola, A. V., Osungbade, O. R., & Gbadamosi, S. O. (2022). Bioactive and chemical properties of Kersting's groundnut proteins. *Food Hydrocolloids for Health, 2*, Article 100043. https://doi.org/10.1016/j.fhfh.2021.100043

Kafoutchoni, K. M., Agoyi, E. E., Sossou, H. S., Agbahoungba, S., Agbangla, C., Assogbadjo, A. E., & Sinsin, B. (2022). Influence of socio-demographic factors on the production system of Kersting's groundnut in Benin. *African Crop Science Journal, 30*(s1), 21–36. https://doi.org/10.4314/acsj.v30is1.2S

Maphosa, Y., Jideani, V. A., & Maphosa, L. (2022). Bambara groundnut production, grain composition and nutritional value: Opportunities for improvements. *The Journal of Agricultural Science, 160*(6), 448–458. https://doi.org/10.1017/S0021859622000521

Mehedi, I. M., Hanif, M. S., Bilal, M., Vellingiri, M. T., & Palaniswamy, T. (2024). Remote sensing and decision support system applications in precision agriculture: Challenges and possibilities. *IEEE Access, 12*, 44786–44798. https://doi.org/10.1109/ACCESS.2024.3380830

Melesse, M. B., Miriti, P., Muricho, G., Ojiewo, C. O., & Afari-Sefa, V. (2023). Adoption and impact of improved groundnut varieties on household food security in Nigeria. *Journal of Agriculture and Food Research, 14*, Article 100817. https://doi.org/10.1016/j.jafr.2023.100817

Mohammed, M., Jaiswal, S. K., Sowley, E. N. K., Ahiabor, B. D. K., & Dakora, F. D. (2018). Symbiotic N_2 fixation and grain yield of endangered Kersting's groundnut landraces in response to soil and plant associated Bradyrhizobium inoculation to promote ecological resource-use efficiency. *Frontiers in Microbiology, 9*, 2105. https://doi.org/10.3389/fmicb.2018.02105

Mohammed, M., Mbah, G. C., Sowley, E. N. K., & Dakora, F. D. (2021). Bradyrhizobium inoculation of field-grown Kersting's groundnut [*Macrotyloma geocarpum* (Harms) Marechal & Baudet] increased grain yield and N_2 fixation, measured using the Ureide, and ^{15}N natural abundance techniques. *Frontiers in Sustainable Food Systems, 5*, Article 672247. https://doi.org/10.3389/fsufs.2021.672247

Muhammad, I., Rafii, M. Y., Ramlee, S. I., Nazli, M. H., Harun, A. R., Oladosu, Y., Musa, I., Arolu, F., Chukwu, S. C., Sani Haliru, B., Silas Akos, I., Halidu, J., & Arolu, I. W. (2020). Exploration of Bambara groundnut (*Vigna subterranea* (L.) Verdc.), an underutilized crop, to aid global food security: varietal improvement, genetic diversity and processing. *Agronomy, 10*(6), 766. https://doi.org/10.3390/agronomy10060766

Ng'endo, M., Bhagwat, S., & Keding, G. B. (2018). Contribution of nutrient diversity and food perceptions to food and nutrition security among smallholder farming households in Western Kenya: A case study. *Food and Nutrition Bulletin, 39*(1), 86–106. https://doi.org/10.1177/0379572117723135

Nhamo, L., Paterson, G., Van Der Walt, M., Moeletsi, M., Modi, A., Kunz, R., Chimonyo, V., Masupha, T., Mpandeli, S., Liphadzi, S., Molwantwa, J., & Mabhaudhi, T. (2022). Optimal production areas of underutilized indigenous crops and their role under climate change: Focus on Bambara groundnut. *Frontiers in Sustainable Food Systems, 6*, Article 990213. https://doi.org/10.3389/fsufs.2022.990213

Nykytiuk, Y., & Kravchenko, O. (2025). Spatial modelling of bean suitability in the context of climate change: Analysis for Polissia and Forest-Steppe of Ukraine. *Agrology, 8*(2), 103–110. https://doi.org/10.32819/202514

Okaiyeto, S. A., Liu, D., Zhang, C., Bai, J.-W., Chen, C., Sharma, P., Venugopal, A. P., Asiamah, E., Ketemepi, H. K., Imadegbor, F. A., Gabriel, O. T., Lv, W., & Xiao, H.-W. (2025). Anti-nutrients of plant-based food: Physicochemical properties, effects on health and degradation techniques—A comprehensive review. *Journal of Future Foods*, S2772566925003088. https://doi.org/10.1016/j.jfutfo.2025.06.022

Oyetayo, F. L., & Ajayi, O. B. (2011). Kersting's nut (*Kerstingiella geocarpa*): A source of food and medicine. In *Nuts and seeds in health and disease prevention* (pp. 693–698). Elsevier. https://doi.org/10.1016/B978-0-12-375688-6.10081-7

Shen, L., & Feng, J. (2024). NIN—At the heart of nitrogen-fixing nodule symbiosis. *Frontiers in Plant Science, 14*, 1284720. https://doi.org/10.3389/fpls.2023.1284720

Sibabi, M., Palanga, K. K., Agboka, K., Tchabi, A., Banla, E. M., Kamde, K., & Houngo, A. M. E. (2025a). Geographical distribution, utilization and farmer's knowledge of Kersting's groundnut [*Macrotyloma geocarpum* (Harms) Maréchal & Baudet] in Togo. *Plant Genetic Resources: Characterization and Utilization*, 1–12. https://doi.org/10.1017/S1479262125100348

Sibabi, M., Tchabi, A., Agboka, K., Palanga, K. K., Banla, E. M., Kamde, K., & Houngo, A. M. E. (2025b). Production system and constraints of the Kersting's groundnut [*Macrotyloma geocarpum* (Harms) Maréchal & Baudet] in Togo. *International Journal of Current Microbiology and Applied Sciences, 14*(6), 181–195. https://doi.org/10.20546/ijcmas.2025.1406.017

Tabe-Ojong, M. P. J., Lokossou, J. C., Gebrekidan, B., & Affognon, H. D. (2023). Adoption of climate-resilient groundnut varieties increases agricultural production, consumption, and smallholder commercialization in West Africa. *Nature Communications, 14*(1), 5175. https://doi.org/10.1038/s41467-023-40781-1

Verburg, K., Pasley, H. R., Biggs, J. S., Vogeler, I., Wang, E., Mielenz, H., Snow, V. O., Smith, C. J., Pasut, C., Basche, A. D., He, D., Archontoulis, S. V., Gaydon, D. S., Huth, N. I., Holzworth, D. P., Sharp, J. M., Cichota, R., Khaembah, E. N., Teixeira, E. I., … Thorburn, P. J. (2025). Review of APSIM's soil nitrogen modelling capability for agricultural systems analyses. *Agricultural Systems, 224*, 104213. https://doi.org/10.1016/j.agsy.2024.104213

Vilakazi, B., Mafongoya, P. L., Odindo, A. O., & Phophi, M. M. (2025). The role of neglected grain legumes in food and nutrition security and human health. *Sustainability, 17*(1), 350. https://doi.org/10.3390/su17010350

Woomer, P. L., Zozo, R. M., Lewis, S., & Roobroeck, D. (2023). Technology promotion and scaling in support of commodity value chain development in Africa. In J. Stanton & R. Caiazza (Eds.), *Sustainable development* (Vol. 7). IntechOpen. https://doi.org/10.5772/intechopen.110397

Chapter 19
Lablab Bean (*Lablab Purpureus*): Agroforestry and Water-Use Efficiency

Umar Farooq, Muqaddas Noor, Abdullah Jan, Mukhtar Ahmed, Amjad Malik, and Shakeel Ahmad

Abstract *Lablab purpureus* is a multipurpose drought-tolerant leguminous crop that is central to the boost in resilience and sustainability of the small holder farming systems in the dry and marginal conditions. As the challenges of climate variability, soil erosion and water scarcity continue to increase, there is a rising demand to find crops that have high water-use efficiency (WUE), low-input adaptability, and a variety of ecosystem advantages. The objective of this chapter is to give a comprehensive insight of the physiological, biochemical, morphological and agronomic characteristics which allow Lablab to survive in a water-limited and stressful environment, and highlight the inclusion of Lablab in agroforestry systems to achieve sustainable production. Lablab has been found to enhance soil moisture capture by 15–20%, above ground biomass by 18–25%, and grain yield to 2.4 t ha^{-1} in conservation agriculture as compared to conventional methods. This is due to its deep rooting, osmotic adjustment, and antioxidant defenses, adaptive leaf-stem characteristics, all of which make it drought resilient and efficient in the use of water. Intercropping of cereals or incorporation in alley cropping systems also stabilize yield, increase the quality of fodder, and increase nutrient cycling. Residue retention, reduced tillage, and the diversification of legumes have been identified as mitigation strategies that are more effective in enhancing yield stability, decreasing soil erosion, and enhancing the overall agroecosystem services to be more climate resilient. Future research aims at enhancing genomic improvement, high-throughput phenotyping and scaling up agroforestry-based Lablab systems.

U. Farooq (✉) · M. Noor · A. Jan · M. Ahmed
Department of Agronomy, Pir Mehr Ali Shah Arid Agriculture University, Rawalpindi, Pakistan
e-mail: umerfaroooq42@gmail.com

A. Malik
Kenanow Bachelor of Education Program Faculty of Education, University College of the North (UCN), Thompson, MB, Canada

S. Ahmad
Institute of Agronomy Faculty of Agricultural Sciences and Technology, Bahauddin Zakariya University, Multan, Pakistan

M. Ahmed (ed.), *Climate Change and Neglected Underutilized Crops*,
https://doi.org/10.1007/978-3-032-21005-0_19

Keywords *Lablab purpureus* · Drought tolerance · Water-use efficiency · Agroforestry · Conservation agriculture

19.1 Introduction

Lablab bean (*Lablab purpureus* L. Sweet) is a versatile leguminous crop, which is appreciated due to its flexibility and drought-resistance and capacity to tolerate marginal weather conditions. The rising level of droughts, land degradation, and water scarcity are weakening agriculture and jeopardizing food security all over the world. There are growing international interests in crops such as lablab that have resilience and nutritional and fodder value in such harsh environments as strategic tools to achieve climate-smart agriculture (MiSsanga et al., 2022).

Lablab also has high adaptability to varied agro ecological settings such as arid and semi-arid areas, sloppy terrains, and low fertility soil. Its extensive root system, drought resistance and capacity to fix nitrogen in the atmosphere make it the best one to incorporate in agro forestry systems, alley cropping and intercropping with other cereals like maize and sorghum. Such properties help not only in increasing the fertility of the soil and stabilizing production, but also in terms of fodder, green manure, and food that will benefit not only humans but also the economy in general (Tasnia et al., 2024).

In international level, lablab has been helping the smallholder farmers in sub-Saharan Africa, South Asia and some parts of Australia, to earn their livelihoods. It is a good source of protein, essential amino acids, minerals, and bioactive substances and can be used to remedy nutritional deficiency among rural populations. Lablab is water-scarce, temperature-tolerant, and low-input-farmed, which makes it resistant to all forms of erratic rainfall and marginal soils in which major cereals and pulses tend to die (Gs et al., 2024).

In spite of these benefits, the use of lablab on the global level is underused because of the lack of research, the absence of integration into the market, and the implementation of the modern agronomic methods. Its growth in agroforestry and systems that are conservation oriented should be encouraged to strengthen the soil structure, lessen soil erosion, boost moisture storage, and stabilize growth across climate variability. It has been revealed that lablab combined with trees or cereal crops enhance the water-use efficacy, soil fertility, and resilience of the whole system, highlighting its ability to make the farming in drylands sustainable (Njaci et al., 2023).

The chapter discusses agro forestry cultivation of lab using water-use efficiency. It gives a detailed description of its botanical and agronomic description, phenology, physiological and molecular drought adaptations, agroforestry performance, and growth and water-use monitoring using modeling and remote sensing techniques. Through synthesis of existing knowledge, this chapter has given insights that lablab can help in building climate-resilient, sustainable, and productive farming systems around the world (Pandey et al., 2023) (Fig. 19.1).

Fig. 19.1 Varietal diversity of *L. purpureus* (L.) sweet in Eastern Uttar Pradesh, India. Plate no (1) to (14) represents different varieties i.e. AS-PCA-Lp (1); AS-PCA-Lp (2); AS-PCA-Lp (3); AS-PCA-Lp (4); AS-PCA-Lp (5); AS-PCA-Lp (6); AS-PCA-Lp (7); AS-PCA-Lp (8); AS-PCA-Lp (9); AS-PCA-Lp (10); AS-PCA-Lp (11); AS-PCA-Lp (12); AS-PCA-Lp (13); and AS-PCA-Lp (14). These varieties were collected from diverse habitat such as kitchen garden/backyard garden, roadside, pond side, disturbed side and other geographical areas of Eastern Uttar Pradesh and field gene bank of these varieties is maintained at Rajgarh block of Mirzapur District of Eastern Uttar Pradesh, India (Singh & Abhilash, 2019)

19.2 Botanical and Agronomic Profile of *L. purpureus*

19.2.1 Taxonomy, Botanical Description, and Domestication Pathway

Lablab purpureus belongs to family Fabaceae and genus Lablab. It was domesticated in Africa, used as food and livestock nourishment, and then spread over to Asia and other tropical areas. The species exhibits a wide morphological diversity, which encompasses the presence of trifoliate leaves, axillary racemes, and the presence of different flower colors such as purple, white or blue. The size, shape, and number of seeds in pods are genetically varied, and this has been used to breed new plants (MiSsanga et al., 2022). Lablab has a broad morphological variation in most of its parameters, including the color of the leaves, color of the leaf veins, pigmentation of stems, flower, pods and seeds color and growth habit. Figure 19.2. Below is some of the images of the varied seed and flower morphology.

Fig. 19.2 Diversity in flower and seed morphology of lablab **a** white flowers color, **b** purple flowers, **c** white seed color, **d** white seed with black hilum edges, **e** cream seed color, **f** brown seed color, **g** red seed color, **h** black seeds (Letting et al., 2021)

19.2.2 Growth Characteristics, Branching Pattern, and Perennial/Annual Behavior

Its growth habit is trailing or climbing but this is based on the availability of support. It may act as an annual when subjected to under intensive cultivation or as a short-lived perennial in less managed. The branching is extensive, augmenting the canopy cover and adding to the generation of biomass. Photoperiod has an effect on flowering and the maturation of the pod depends on both the genotype and environmental factors. These characteristics enable lablab to fit in various types of farming systems such as monocrop, intercropping, and agroforestry systems (Gs et al., 2024) (Table 19.1).

19.2.3 Ecological Adaptation to Drylands and Marginal Environments

A highly adapted plant is *L. purpureus*, adapted to low fertility soils (sandy loams to heavy clays) with a pH bearing of 4.5–7.5. It has a good root system, which enables it to get the subsoil moisture, hence survival during dry seasons. The crop has a broad tolerance to rainfall regimes and can tolerate high temperatures hence it can be used in rainfed and marginal conditions. Such adaptability of the ecosystem makes it especially useful in the smallholder and agroforestry systems in arid and semi-arid regions (Gs et al., 2024).

Table 19.1 Estimates of variance for the traits studied in lablab bean genotypes (Shubha et al., 2024)

Trait	Min	Max	Mean	Skewness	Kurtosis	Std error	Std Dev	Coeff Var
Days to 50% flowering	40.67	57.33	50.93	−0.72	−0.89	1.98	5.94	11.66
Plant height (cm)	60	154	110.11	−0.28	−2.08	12.84	38.53	34.99
No. of pod per plant	58	145	91.56	0.67	−0.91	10.66	31.97	34.92
No. of seed per pod	4.33	6.67	5.48	0.18	−0.38	0.24	0.73	13.29
Avg pod weight (g)	5.33	9.23	6.19	2.44	6.3	0.4	1.21	19.55
Pod length (cm)	8	13.67	11.19	−0.5	−0.93	0.64	1.92	17.16
Days to 1st picking	57	75.67	68.19	−0.71	−0.89	2.16	6.49	9.51
Pod yield per plant (g)	373	997	593.22	−0.28	−1.56	73.94	221.81	37.39
Protein (g)	19.01	25.47	21.8	0.74	0.93	0.63	1.89	8.68
Vit C (mg)	4.61	9.45	6.12	1.62	3.77	0.48	1.44	23.55
Vit A (mcg)	6.27	30.86	17.23	0.26	−0.71	2.7	8.11	47.08
Zinc (mg)	1.7	2.59	2.28	−0.84	−0.29	0.1	0.31	13.53
Iron (mg)	5.97	10.5	7.92	0.93	−0.01	0.5	1.49	18.84
Calcium (mg)	321.6	459.1	395.3	−0.27	−1.13	16.86	50.58	12.79
Phosphorous (mg)	317.89	460.85	363.89	1.24	1.19	15.55	46.66	12.82

19.2.4 Nutritional, Fodder, Socio-economic, and Medicinal Significance

Lablab is a multipurpose crop. Leaves, seeds, and pods are sources of food to humans, and biomass is good fodder to animal livestock. Its growth improves the fertility of soils by fixation of nitrogen and adds organic material to the soil. Lablab contributes to the ecosystem services by ensuring that the soil is not eroded, it is kept moist and it supports the organisms that are beneficial. It serves in traditional medicine too (considered a number of purposes), which is why it makes it a socio-economically and ecologically useful species in smallholder agricultural systems (Pandey et al., 2023) (Table 19.2).

Table 19.2 A summary of the comparison of the nutritional composition of lablab with other legumes (Letting et al., 2021)

Component	Lablab	Cowpea	Bambara groundnut	Common bean	Soybeans	Chickpea	Lentils
Carbohydrates (g)	62	61	61	60	8.36	27.42	20.13
Fiber (g)	8.6	5.4	4.8	6	7.6	7.9	–
Fat (g)	1	1.4	6.2	1.5	8.97	2.59	0.38
Protein (g)	22.8	22.5	18.8	21.7	18.21	16.4	9.02
Calcium (mg)	90	104	62	120	102	49	19
Iron (mg)	9	N/A	12.2	8.2	5.14	4.39	3
Phosphorus (mg)	328	416	276	323	245	168	180
Ascorbic acid (mg)	Trace	Trace	Trace	1	0	0.52	1.65
β-carotene (mg)	N/A	70	10	1	9	–	–
Niacin (mg)	0.2	1.9	0.1	0.7	0.399	0.526	1.05
Riboflavin (mg)	0.1	0.9	0.1	0.2	0.285	0.063	0.077
Thiamin (mg)	0.5	0.1	0.5	0.4	0.155	0.116	0.169

19.2.5 Agronomic Traits Contributing to Resilience in Low-Input Agroforestry

The species exhibits traits favorable for sustainable, low-input agriculture. Its drought tolerance, ability to grow in nutrient-poor soils, and compatibility with other crops in intercropping and agroforestry systems make it highly resilient. Lablab's canopy provides ground cover to reduce soil erosion, and its residues improve soil structure and nutrient cycling. These characteristics make it a suitable legume for climate-smart agroforestry and mixed cropping systems (Teshome et al., 2024).

19.3 Phenology, Distribution, and Production Zones

19.3.1 Phenology

The phenological pattern of *L. purpureus* is flexible, thus allowing the species to live in a great diversity of environmental conditions. Germination is successfully carried out in uneven soil moisture, but best emergence is seen in moist but not waterlogged soil. The crop exhibits robust vegetative growth where it has a vining or twining habit, which is favorable to heavy canopy development, which is useful in an intercropping system or agro forestry. The growth of leaves does not stop and trifoliate leaves grow to reach the optimum photosynthetic potential.

The photoperiod and genotype influence flowering and the majority of cultivars are short-day neutral. It is after flowering and depends on the environmental stress, especially water. The development of the seed proceeds continuously in the sufficient amount of moisture up to the physiological maturity of the seed, which takes 90–120 days, depending on the cultivar and the conditions of growth. The seasonal flexibility of Lablab enables its successful implementation as a companion crop in agroforestry or intercropping projects especially as a companion crop during periods of the year when it thrives (Hanapiah et al., 2022) (Fig. 19.3).

19.3.2 Geographical Distribution

Lablab bean has its origin in Africa and since then it has been extensively spread in the tropical and subtropical areas of the world. It is widely grown in Africa, Asia, and Australia and some parts of the Americas and grows under rainfed and irrigated conditions. Its indigenous distribution covers eastern and southeastern Africa, and Asia, especially India and southeastern Asia have a wide variety of landraces which are adapted to local climatic and soil conditions. Agro ecological areas that sustain lablab are semi-arid tropics, subtropical lowlands and hilly areas where moderate rainfall and warm weather are experienced. This broad spread is indicative of its ecological adaptability and the ability to adapt to different systems of agriculture (Das & Fakir, 2014) (Fig. 19.4, Table 19.3).

19.3.3 Production Zones and Practices

19.3.3.1 Major Producer Regions and Farming Communities

Lablab is mainly grown in East and South Africa, South and Southeast Asia as well as some parts of Australia. Lablab is widely available as a multipurpose food, fodder, and soil-enriching legume that is grown by smallholder farmers in India, Kenya, Uganda and Tanzania. The crop is part of the traditional farming system and contributes to the nutrition of the household, the food of livestock, and the enhancement of soil fertility (Mehetre et al., 2025).

19.3.3.2 Role in Smallholder Mixed Farming and Agroforestry

Lablab gives a lot to smallholder systems in the form of mixed cropping, silvopastoral set-ups, and alley cropping. This is because of its climbing growth habit as it can be

Fig. 19.3 Phenological development of the Lablab Bean (*L. purpureus*) according to the BBCH (Biologische Bundesanstalt, Bundessortenamt and Chemische Industrie) scale. The sequence illustrates the key principal stages from germination to senescence, including: BBCH 09 (emergence), BBCH 13 (3rd true leaf unfolded), BBCH 21 (first side shoot visible), BBCH 65 (full flowering), BBCH 75 (main pod development/filling), BBCH 89 (fully ripe), and BBCH 97 (plant dead and dry) (Muchow, 1985)

integrated with support crops or trees to cover the ground, enhancing the soil organic matter, and reducing soil erosion. Being a legume, it improves the nitrogen supply of the companion crops and helps to maintain low-input agriculture.

19.3.3.3 Yield Trends

Yield is highly diverse based on system of cultivation, genotype, and environment. Monocropped lablab can usually attain high yields of seeds when well-managed, yet intercropping or agroforestry systems offer more ecosystem services and biomass of

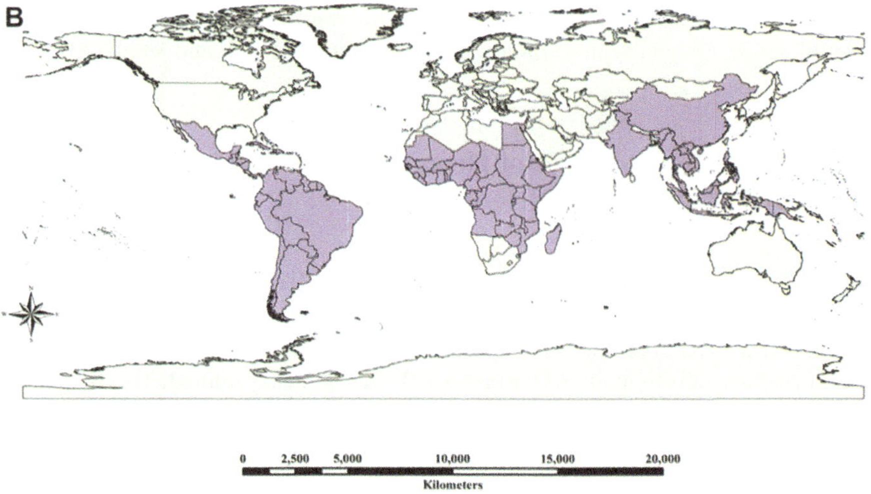

Fig. 19.4 Global distribution map of *L. purpureus* (L.) sweet was prepared using ArcGIS platform based on information collected from literature as well as from the databases of international agencies such as FAO, wild crop relatives, diversity international and international legume database and information service (ILDIS). The light purple color shows the global distribution range (Singh & Abhilash, 2019)

Table 19.3 Global distribution of Lablab bean in the world (Ceasar et al., 2023)

Type of Lablab bean	Countries	References
Indigenous	Indian subcontinent; African countries like Egypt, Sudan, East, West, South Africa	Singh and Abhilash (2019)
Naturalized	Asia: Malaysia, Indonesia, Philippines, China, Papua New Guinea and CaribeanEurope: ItalyAmerica: Central and South AmericaAustralia: Northern New South Wales and Southern Queensland	Singh and Abhilash (2019)

fodder, with only minor decreases in seed yield. Nutrient supplementation, irrigation and pest management are sustainable management practices that enhance total productivity and stabilize yields in changing conditions (Moniruzzaman et al., 2023).

19.3.3.4 Traditional and Modern Agronomic Practices

Traditional farming methods are being used by the farmers like staking, rotation and mixed planting, whereas the modern methods of agronomy intervention are optimum spacing, better choice of cultivars, use of mechanized planting, and use of water-efficient irrigation methods. The incorporation of better agronomic techniques will improve the yield of seeds and biomass, tolerance to abiotic stress, and water use efficiency, especially in semi-arid and marginal lands (Revanth et al., 2024).

19.4 Physiological Basis of Drought Tolerance and Water-Use Efficiency

19.4.1 Water Relations and Stomatal Behavior

Lablab purpureus supports water potential in the leaves by osmotic adjustment, which enables cells to retain water in the soil when water is limited. This adaptation helps to maintain metabolic process even when there is a long dry season. This is tightly regulated through stomatal conductance that minimizes transpiration but maintains a low level of photosynthesis. The conductance of the cuticles also helps to reduce water loss by limiting the non-stomatal evaporation to increase the total water conservation. Under conditions of agro forestry or semi shade, lablab roots can draw the moisture of deeper layers of soil and this makes the most of water uptake during season of limited rainfall. All these mechanisms make the legume resilient and efficient in the use of water in marginal environments (Kolesnikova et al., 2025).

19.4.2 Photosynthesis and Gas Exchange Under Water Stress

Photosynthesis in lablab is impacted by water deficits mainly by holing up stomata that restricts the availability of CO_2. There are mesophyll-level limitations that also affect carbon fixation especially with intense stress. The PSII photochemistry regulates through non-photochemical quenching and thus the photosynthetic apparatus is not overheated by excess light but rather by oxidative damage. Carbon assimilation conditions and shading in agroforestry systems are both regulated by electron transport, which balances energy, and environmental influence conditions. The intrinsic water-use efficiency increases with moderate stress that indicates the ability of the plant to balance between a gain of carbon and water loss. The mechanism of adaptation to changing microclimates guarantees a consistent performance under unstable agro ecological conditions (Tasnia et al., 2024).

19.5 Biochemical and Molecular Mechanisms

19.5.1 Oxidative Stress and ROS Dynamics

The reactive oxygen species (ROS) within *L. purpureus* are caused by drought and interspecific competition in the agroforestry. High ROS such as superoxide, hydrogen peroxide and hydroxyl radicals may cause damages to lipids, proteins, pigments, and cell membranes. Oxidative stress causes long-term effects of cell homeostasis and leads to disruption of photosynthesis and metabolic efficiency. Lablab triggers defense signaling to reduce oxidative stress and preserve cellular integrity when there is a water scarcity (Kongjaimun et al., 2022).

19.5.2 Antioxidant Defense Systems

Lablab uses enzymatic and non-enzymatic antioxidants to overcome the effects of ROS. Enzymes like Superoxide dismutase (SOD), catalase (CAT), ascorbate peroxidase (APX), glutathione reductase (GR) and guaiacol peroxidase (GPX) are effective in mitigating oxidative stress. Molecules that are not enzymatic such as glutathione, ascorbate, phenolic, carotenoids, flavonoids, among others, also stabilize cell structures and help with stress tolerance. The synergistic effect of these antioxidants promotes the long-term growth in drought or shaded agro forestry environments (Adekunle et al., 2023).

19.5.3 Compatible Solutes and Osmolyte Accumulation

In water deficit, lablab stores compatible solutes like proline, soluble sugars and glycine betaine. These osmolytes preserve the turgor of the cell, stabilize the proteins, and ensure that the membranes are not damaged by dehydration. Osmolyte also increases the water retention in tissues, boosts up the metabolism, and helps in intrinsic water-use efficiency during intermittent or extended drought stress (Wood et al., 2001).

19.5.4 Molecular Regulation

The expression of drought responsive genes in lablab is coordinated by abscisic acid (ABA) signaling. DREB, NAC and MYB transcription factors are involved in the regulation of downstream stress-responsive genes, such as controlling the synthesis of osmolytes, antioxidant defense and stomatal behavior. The level of genetic variation in domesticated and wild accessions of the lablab has an impact on the strength of the molecular reactions. There is a high potential of breeding and genomic improvement as wild germplasm has better drought resilience (Missanga et al., 2021).

19.6 Morphological and Anatomical Adaptations

19.6.1 Root System Architecture

Lablab purpureus forms a massive and deep root system, which allows it to access subsoil moisture to aid in avoiding droughts. Lateral root development improves soil exploration to increase water and nutrient uptake of heterogeneous soil profiles. Root biomass grows in agro forestry and intercropping systems and thus lablab takes advantage of soil moisture effectively without over competing with its companion species. Root-soil interactions enhance the infiltration, retention, and hydraulic conductivity, which helps plants to survive during intermittent or long-lasting water shortage (Khan et al., 2020).

19.6.2 Leaf and Stem Structural Responses

Lablab leaves exhibit adaptive characteristics to sustain water conditions during times of stress, such as the modified stomatal density and leaf thickness and improved cuticle development. The presence of trichomes and bulliform cells helps to decrease the transpiration and ensure turgor, whereas leaf rolling helps to decrease the water loss during drought events. During periods of limited water supply, Stem vascular architecture changes to maximize water movement to maintain photosynthetic functions and growth in agro forestry shade or marginal environments. All these anatomical adaptations are in favor of high intrinsic water-use efficiency and extended lifecycle under changing climatic conditions (Naeem et al., 2023).

19.7 Water-Use Efficiency and Drought Resilience

19.7.1 Germination, Early Vigor, and Growth Under Water Stress

Lablab purpureus exhibits excellent germination ability in fluctuating moisture conditions and is therefore very suitable to semi-arid and drought prone areas. Seedlings grow very fast and they experience intensive early development during

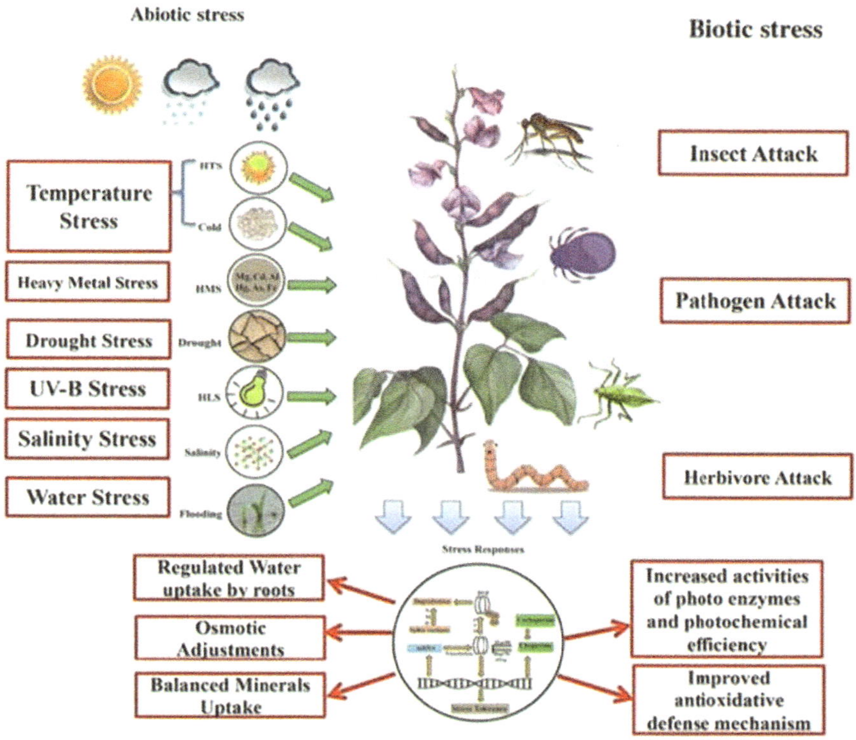

Fig. 19.5 Lablab bean tolerates various abiotic and biotic stresses (Naeem et al., 2023)

periods of intermittent water shortage. The crop has a vining and twining growth habit, which enables it to maximize the light interception and canopy as well as save soil moisture. The development of leaves is highly controlled in water-stressed environments where there is a balance between the processes of transpiration and photosynthesis. These characteristics are a guarantee of successful planting and a competitive edge in low-input systems at which the availability of moisture is unpredictable (Kongjaimun et al., 2022) (Fig. 19.5).

19.7.2 Yield Components and Biomass Allocation Under Drought

Lablab can partition biomass effectively under water deficit into roots, stems, leaves and reproductive organs. Deep rooting increases the ability to obtain water at the lower levels of the soil, whereas low vegetation growth decreases the amount of water that is lost. Moderate stress does not affect flowering, pod set, seed development as

Table 19.4 Aboveground dry matter production of six *L. purpureus* accessions evaluated under two water treatments at the time of flowering and physiological maturity at KARI, Katumani; no significant differences ($p \leq 0.05$) according to Tukey test were found

Plant stage	Accession ID	Fully irrigated	CI lower (FI)	CI upper (FI)	Rainfed	CI lower (R)	CI upper (R)
90% flowering	Q6880B	0.824^{c}	0.626	1.053	0.943^{c}	0.729	1.190
90% flowering	CPI 60795	1.019^{bc}	0.794	1.277	1.271^{ab}	0.827	1.230
90% flowering	CPI 52508	$1.322^{a\text{–}c}$	1.058	1.620	1.582^{a}	1.090	2.178
90% flowering	CPI 52513	1.041^{bc}	0.813	1.332	$1.194^{a\text{–}c}$	0.942	1.515
90% flowering	CPI 52535	0.935^{c}	0.722	1.181	1.025^{bc}	0.809	1.195
90% flowering	CPI 81364	0.974^{c}	0.756	1.226	0.935^{c}	0.722	1.181
Physiological maturity	Q6880B	0.392^{ac}	0.268	0.545	0.351^{ac}	0.236	0.495
Physiological maturity	CPI 60795	$0.162^{b\text{–}d}$	0.092	0.257	$0.159^{b\text{–}d}$	0.090	0.254
Physiological maturity	CPI 52508	$0.319^{a\text{–}d}$	0.210	0.455	$0.156^{b\text{–}d}$	0.087	0.265
Physiological maturity	CPI 52513	$0.296^{a\text{–}d}$	0.192	0.447	0.120^{d}	0.062	0.204
Physiological maturity	CPI 52535	$0.284^{a\text{–}d}$	0.170	0.432	0.147^{cd}	0.077	0.229
Physiological maturity	CPI 81364	0.659^{a}	0.430	0.946	0.499^{ab}	0.355	0.673

the crop exhibits a constant harvest index. Even in cases where water is limited, fodder quality does not reduce significantly in terms of protein and digestibility. This adaptive allotment facilitates food and fodder production, which offers constant productivity in a fluctuating moisture condition (Gachuiri et al., 2021) (Table 19.4).

19.7.3 Drought Tolerance Indices and Water-Use Efficiency

The crop has high intrinsic water-use efficiency that is manifested by stomatal control, osmotic adjustment, and efficient transpiration. The metrics of Drought Tolerance Index (DTI) and yield resiliency bring to the limelight the ability of lablab to stay productive even when water is limited. Lablab can endure dry agroecosystems by means of physiological adjustments, such as improved leaf water potential, limited transpiration and favorable photosynthetic carbon assimilation. All these characteristics contribute to the fact that lablab is a strong crop that can be used to promote climate resilience in low-input and semi-arid farming systems (Sharma et al., 2023).

19.8 Agroforestry Performance of *L. purpureus*

19.8.1 Interactions with Shade and Tree–Crop Competition

Lablab can be grown in partial shade although it is not able to thrive in the shade as well. Excessive transpiration is minimized by light competition with the canopy trees and leaf turgor is preserved. Root competition is avoided by deep rooting and further an extension laterally, which enable effective water and nutrient absorption. Microclimates are also buffered by Agroforestry conditions that reduce heat stress and improve physiological efficiency of lablab (Rolo et al., 2023).

19.8.2 Intercropping with Maize, Sorghum, and Other Species

When lablab is intercropped with cereals like maize or sorghum, this enhances productivity in terms of land-use efficiency. Lablab helps in fixation of nitrogen thus making soils fertile to other companion crops. The vining habit permits spatial complementarity, which reduces the losses to shade of cereals. The intercropped systems have always been found to give higher fodder and grain yield than the monocultures. The calculations of Land Equivalent Ratio (LER) show that theproductivity of the system is improved in the case of lablab-based intercropping (Castle et al., 2021).

19.8.3 Soil Fertility, Nitrogen Fixation, and Ecosystem Service Enhancements

Lablab is a leguminous crop that absorbs nitrogen in the atmosphere, thus developing the soil organic content and minimizing the use of artificial fertilizers. Its thick cover prevents erosion of soil, controls temperature and enhances the retention of moisture. The crop also favors Biodiversity since it offers shelter to useful insects and microbial communities. The role of integration can provide agroforestry with benefits that are multifunctional, with food, fodder, soil conservation, and climate-resilient (Akchaya et al., 2025) (Figs. 19.6, 19.7, and 19.8).

Fig. 19.6 Lablab as intercrop with maize, Bergville, KwaZulu-Natal (Singh & Abhilash, 2019)

Fig. 19.7 Lablab as part of a mixture with Sunhennop and fodder sorghum, Ottosdal, North West Province (Singh & Abhilash, 2019)

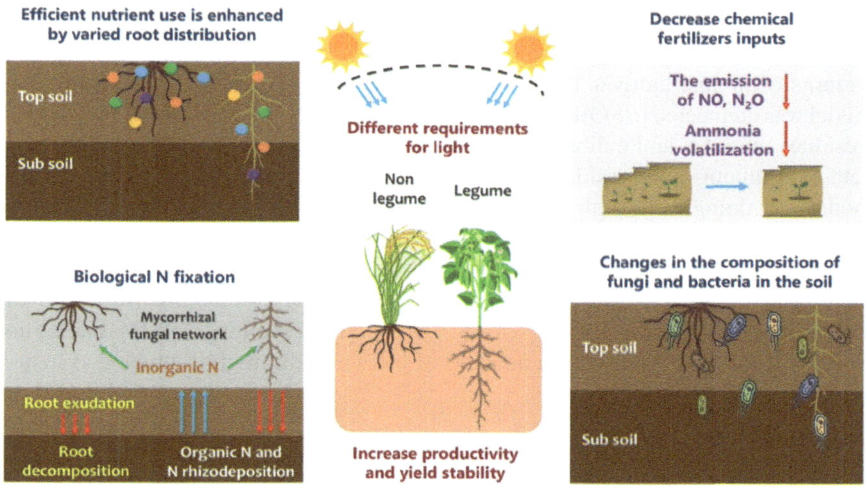

Fig. 19.8 Legume intercropping boosts resource use efficiency by enhancing interspecific facilitation and niche differentiation

19.9 Crop Modelling for Agroforestry and Water-Use Efficiency

19.9.1 Simulation of Growth and Biomass Allocation

The growth of *L. purpureus* under agroforestry and environmental conditions can be properly simulated by use of crop modelling. The models are based on soil moisture, nutrient availability, radiation interception, and canopy dynamics, which are used to predict the biomass allocation between roots, stems, leaves and pods. The simulation studies indicate the adaptive nature of Lablab to the variable availability of water, which serves as an insight into how the crop will respond to drought in terms of yield stability. The predictions are useful in scheduling intercropping and ensuring maximum productivity of the systems (Wang et al., 2024).

*19.9.2 Simulation Modelling of Lablab (*L. purpureus*) Pastures in Northern Australia (APSIM)*

This paper is an article conducted by Hill et al., in northern Australia titled Simulation modelling of lablab (*L. purpureus*) pastures in northern Australia, where the study was done using the Agricultural Production Systems Simulator (APSIM) model to determine whether or not it would have the ability to represent lablab pastures correctly. This was done by parameterizing the APSIM-Lablab module using field data of Gatton, Gayndah and Downfall Creek to specify phenology, leaf area development, radiation interception, biomass accumulation, fixation of nitrogen and regrowth patterns of annual cultivar High worth and the perennial cultivar Endurance. The model was connected to SOILWAT2 (soil water), SOILN2 (nitrogen) and RESIDUE2 (residue) modules and calibrated using field data and verified against independent data of Kununurra, Emerald, and Samford of irrigated and dryland conditions. The model was doing well, with $R^2 = 0.96$ and RMSD = 304 kg DM/ha in calibration and $R^2 = 0.92$ and RMSD = 670 kg DM/ha in validation and showed good prediction of biomass ranges of 63–13,055 kg DM/ha, cultivar variations, and responses of soil moisture stress and grazing episode. Future suggestions involve improving simulations of high-biomass late-season canopies and combining APSIM-Lablab and livestock models to improve the analysis of the whole-farm grain-grazing system.

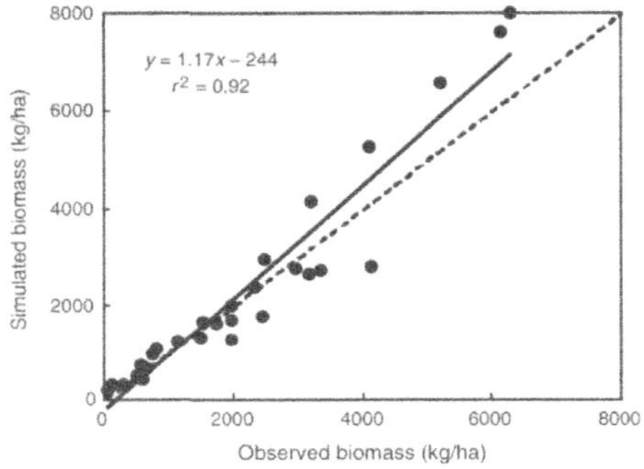

Relationship between observed and predicted biomass for the calibration datasets. RMSD=304kgDM/ha, n=37. The 1:1 line (dashed) of perfect

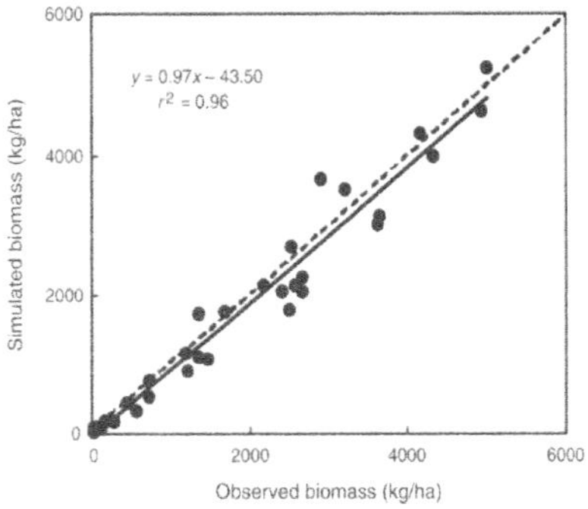

Relationship between observed and predicted biomass for the testing datasets. RMSD=670kgDM/ha,n=29.The1:1line(dashed) of perfect agreement is shown.

19.9.3 *Modelling Water-Use Efficiency and Drought Responses*

Crops models also measure water-use efficiency by showing relationships of transpiration, photosynthesis, and stomatal conductance in changing soil moisture. The stomatal control, root water uptake, and osmotic adjustment of lablab-specific modules are simulated, and thus, intrinsic WUE and drought resilience indices could be assessed. Yield losses in the case of intermittent drought can be predicted by using scenario-based simulations and managing the situation through the best irrigation plan and residue retention plan (Burgess et al., 2019).

19.9.4 Decision Support for Agroforestry Management

Combination of remote sensing and modelling results helps site-based management in agro forest systems. Models are used to help in the choice of appropriate species mixtures, optimal planting densities, and soil conservation methods. Smallholder farmers and researchers can use decision support tools based on these models to improve the productivity of Lablab without losing the soil moisture, fertility, and the overall ecosystem resilience. The modelling techniques can also be used to determine the long-term effects of climatic variations on crop performance (Deepana et al., 2025).

19.10 Remote Sensing Applications for Monitoring Lablab in Agroforestry Systems

19.10.1 Principles of Remote Sensing for Stress Assessment

Remote sensing would be used to monitor *L. purpureus* in non-destructive conditions of different agro forestry. Canopy reflectance patterns give information about water status of plants, deficiency of nutrients and the impact of light stress. Thermal imaging also enables capture of canopy temperatures variation, which is related to transpiration rate and water stress. These types of assessments assist in prompt warning of drought or stress-related to shade and in the optimization of irrigation or management response to predicted stress (Thapa et al., 2023).

19.10.2 Tools and Indicators

Numerous indices based on vegetation are popular to measure biomass, greenness, and water content, such as NDVI, EVI, NDWI, and PRI. Hyperspectral sensor has the capability to detect pigment composition, potential fixation of nitrogen, and other biochemical characteristics and provide an accurate monitoring of plant health and physiology. They can be used to monitor growth and stress continuously without phasing out even in mixed-species agroforestry systems or shaded agroforestry systems (Hoover et al., 2023).

19.10.3 Remote Sensing Platforms

The large scale monitoring of Lablab over different landscapes through satellite based platforms are able to capture the seasonal changes as well as agro ecological trends. UAVs can be used to conduct high-resolution phenotyping of water-use efficiency and canopy architecture of experimental plots or smallholder fields. In addition, sensors on soils and canopy with an IoT can be used to provide real-time information about moisture, temperature, and microclimatic conditions to help make timely management decisions (Singh et al., 2024).

19.10.4 Integration with Crop Models

Crop models can be used to assimilate remote sensing data to improve yield, water-use efficacy, and stress reactions forecasting. The spatial mapping of drought, shade gradient and inter-species competition are used to calibrate and validate the models. Integration enhances decision-support in agroforestry management whereby the researchers and the farmers can ensure that Lablab performs optimally under fluctuating environmental conditions (Ahmad & Sohel, 2025).

19.11 Case Study 1: Enhancing Productivity of Lablab-Maize Intercropping in Rajasthan, India

Singh et al. (2024) designed a study in the semi-arid areas in Rajasthan, India, to determine the productivity, water-use efficiency and the agronomic compatibility of the lablab-maize intercropping systems. The experiment involved comparison of sole maize, sole lablab and intercropping pattern with different row ratios (1:1, 1:2 and 2:1) in two successive growing seasons. Parameters of growth, the yield of the pod, accumulating biomass, the content of crude proteins and the water-use efficiency (WUE) were measured. Findings indicated that, 1:2 maize-lablab intercropping system had the highest lablab pod yield of 2.1 t ha^{-1}, maize grain yield of 4.5 t ha^{-1} and a 20% increase in WUE than the sole cropping systems. The distribution of biomass was equal and the intercrop plots showed an enhanced resistance to moisture stress. It was advised that optimized ratio of intercropping would be adopted to achieve optimal productivity, enhancement of soil health, and water saving of semi-arid farming regime to improve food security and sustainability of smallholder farmers in the study.

19.12 Case Study 2: Evaluating Drought-Tolerant Lablab Accessions in Machakos, Kenya

Okoth et al. carried out research in Machakos County, Kenya, aiming to identify lablab accessions with superior drought tolerance suitable for rainfed agriculture. Six accessions were tested under controlled drought stress at 30, 60, and 90 days after sowing, measuring leaf area index (LAI), relative water content (RWC), stomatal conductance, and final pod yield. Among the tested accessions, LP-3 maintained 78% RWC under drought conditions, produced 3.0 t ha^{-1} pods, and exhibited a Drought Tolerance Index (DTI) of 1.12, outperforming other varieties in both physiological stability and yield. The study highlighted that selecting drought-resilient accessions can enhance productivity in water-limited environments and reduce the vulnerability of smallholder farmers to climate variability. It also emphasized the importance of integrating physiological screening with traditional agronomic evaluations for sustainable crop improvement.

19.13 Case Study 3: Lablab Performance in Alley Cropping Systems with Grevillea in Tanzania

To determine whether lablab serves as an effective component of alley cropping systems that can be used to increase the soil fertility and yield stability in Grevillea robust in the Eastern Highlands of Tanzania, Msuya et al. evaluated the effectiveness of using lablab in such systems. Plants were planted in rows with trees and optimal growth period comparing shade and open-field environment across two seasons. The measurements were done in terms of biomass accumulation, pod yield, soil moisture content, and the potential of fixation of nitrogen. The shaded lablab plots had an increase in pod yield of 15% (2.6 t ha^{-1}) and an increase in the rate of nitrogen fixation of 12% than the open-field plots. The experiment also established that, moderate tree shading could enhance the microclimatic conditions, reduce the evapotranspiration and enhance nutrient cycling. It proposed that the alley cropping systems have the capacity to promote soil conservation, maintain the productivity of the lablabs and facilitate diversified smallholder strategies of farming in the tropical highlands.

19.14 Case Study 4: Enhancing Fodder Quality and Nutritional Value of Lablab in Central Uganda

The study by Namutebi et al. was held in Central Uganda to evaluate the fodder yield and nutritive quality of three lablab varieties and their agronomic performance on the smallholder farms. Measurement parameters comprised of dry matter yield, crude protein content, digestibility and regrowth capacity after two harvest cycles. The variety of LP-U2 was found to yield the most desirable dry matter (4.1 t ha^{-1}), crude protein (22%), and digestibility (78), which made it ideal in production of livestock fodder. The paper also emphasized the flexibility of the crop to low-input farming systems, and the benefits of the crop under mixed crop-livestock farming. It was recommended to increase the use of high-protein lablab varieties to enhance the nutrition of livestock and incorporate lablab into smallholder agroforestry or mixed farming systems to enhance the livelihood of the rural population as well as help mitigate seasonal feed deficiencies.

19.15 Future Directions

19.15.1 Genomic Improvement and Molecular Breeding

The use of molecular breeding has provided possibilities to develop water-use efficiency and drought resistance in *L. purpureus*. High-throughput phenotyping can be used with marker-assisted selection to develop better cultivars that are adapted to low-input and marginal environments. Root architecture, stomatal conductance, and osmolyte accumulation breeding programs can be used to enhance plant performance under changing moisture conditions (Sserumaga et al., 2021).

19.15.2 CRISPR, GWAS, and Pan Genomics Approaches

Through modern technologies in genome editing, like CRISPR-Cas technology, it is possible to create at will all the drought-responsive and shade-responsive genes. Several studies, which include genome-wide association studies (GWAS) and pangenomic analyses, help to uncover alleles, which are linked to stress tolerance, biomass, and nutrient-use efficiency. These methods offer knowledge on genetic diversity and allow coming up with Lablab lines that are resistant to climate variation (Njaci et al., 2023).

19.15.3 Digital Agriculture and High-Throughput Phenotyping

The digital technologies in agriculture, such as the UAV-based imaging, IoT-based sensors, and automated phenotype imaging platforms, offer real-time observation of the plant development and water conditions. The tools aid accuracy in the management of Lablab in agro forestry systems to enable timely interventions and better utilization of the resources. When combined with AI-based analytics, the prediction of yield, responses to droughts, and the productivity of the system are improved (Visakh et al., 2024).

19.15.4 Scaling Agroforestry-Based Lablab Systems

There is the opportunity to increase food, fodder, and nutritional security by increasing the Lablab production in the smallholder agro forestry landscapes, which facilitates the ecosystem sustainability. Strategies of scaling consist of involving farmers in participatory efforts, dissemination of knowledge, and inter-cereal/legumes integration. Agricultural communities should be resilient and broadly adopted, which can only be achieved by policy support, market linkages, and capacity building under climate stress (Seghieri et al., 2021).

19.15.5 Phenotypic Correlations for Pod Yield Components and Nutritional Quality Traits

Phenotypic correlation is a method of estimating the connection amid two categories of phenotypic values or variables of a populace but does not denote the strength of the association. The correlation between traits was visualized using a heat map (as in Fig. 19.9). D50F was significantly and positively correlated with DFP (D50F versus DFP, $p = 0.001$, rp = 0.964) and NPP with pod yield (D50F versus pod yield, $p = 0.001$, rp = 0.991). The correlation of D50F with PH (rp = 0.806), APW with NPP (rp = 0.825), protein with Zn (rp = 0.855), and APW (rp = 0.863) with the pod yield was found to be significant ($p < 0.01$). PL with NPP (rp = 0.755), APW with Vit A (rp = 0.729), protein with Ca (rp = 0.671), protein with P (rp = 0.28) and Vit C with Zn (rp = 0.669) had significant ($p < 0.05$) positive correlations. PH had significant negative ($p < 0.01$) correlations with NSP (rp = [?]0.839) and DFP with Zn (rp = [?]0.680) had significant negative correlations ($p < 0.05$). Figure 19.9 indicated positive, and strong, relationships between pod yield and pods per plant (NPP), average pod weight and pod length.

19.16 Conclusion

Lablab purpureus is a very multifaceted and climatic-adaptive legume with extraordinary physiological, biochemical, and structural alterations that allow it to survive in an agro ecological environment with limited water resources. The system of deep-rooting, osmotic adjustment, the control of stomata, and the increased antioxidant

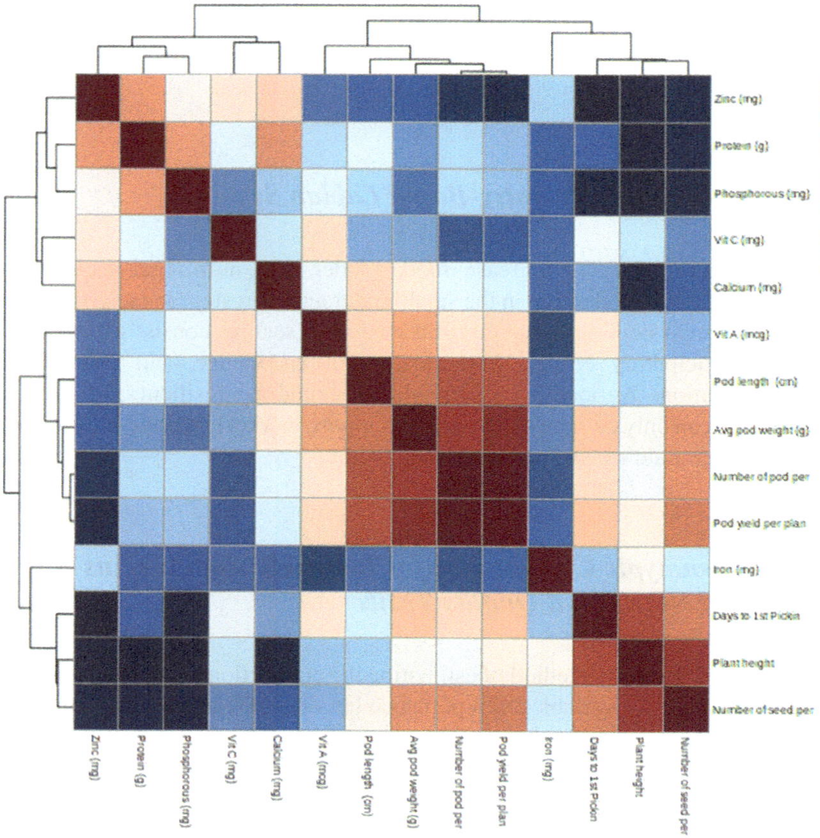

Fig. 19.9 Heat map representing phenotypic correlation between lablab bean pod yield components and nutritional quality traits (Shubha et al., 2024)

defenses are all factors that lead to the high drought resistance and water-use efficiency. This plasticity in the morphology of leaves and stems of the crop and the capacity to accumulate compatible solutes make this crop more resilient to changing environmental stresses. Biochemically, the antioxidant systems and osmolyte buildup in Lablab is key in alleviating the effects of oxidative stress and stabilizing cellular activity during moisture deficit thus maintaining a steady productivity even in adverse conditions. These natural benefits are exalted by agro forestry based modes of cultivation that offers moderated microclimates, ameliorated soil structure, enhanced nutrient recycling and better moisture retention. Incorporation of Lablab into intercropping, alley cropping, and silvopastoral products is not only useful in stabilizing yields, but also in improving fodder, food and nutritional insecurity among smallholder farmers. The latest developments in remote sensing, high-throughput phenotyping and modeling methods provide the accurate monitoring of water-use efficiency, stress responses, and productivity dynamics to make data-driven decisions and management. Overall, *L. purpureus* is a climate-aware, multipurpose legume usable to support livelihood, enhance ecosystem services and shield the smallholder-farming systems against changes in weather conditions. Its practical incorporation, coupled with conservation farming and new breeding strategies, can help to heavily empower the resilience, production, and sustainability of dryland and rainfed farming landscapes throughout the world.

References

Adekunle, A., Isiaka, M. B., Temitayo, U. I., Augustina, S., & Olubunmi, A. (2023). Phytochemicals, antioxidants and glycemic index assessment of *L. purpureus* (Lablab Bean) and *Phaseolus lunatus* (Lima Bean) seeds. *Asian Journal of Applied Chemistry Research, 13*(1), 54–63. https://doi.org/10.9734/ajacr/2023/v13i1237

Ahmad, U., & Sohel, F. (2025). Evaluating decision support systems for precision irrigation and water use efficiency. *Digital Engineering, 4*, 100038. https://doi.org/10.1016/j.dte.2025.100038

Akchaya, K., Parasuraman, P., Pandian, K., Vijayakumar, S., Thirukumaran, K., Mustaffa, M. R. A. F., Rajpoot, S. K., & Choudhary, A. K. (2025). Boosting resource use efficiency, soil fertility, food security, ecosystem services, and climate resilience with legume intercropping: A review. *Frontiers in Sustainable Food Systems, 9*, 1527256. https://doi.org/10.3389/fsufs.2025.1527256

Burgess, P., Graves, A., De Jalón, S. G., Palma, J., Dupraz, C., & Van Noordwijk, M. (2019). Modelling agroforestry systems. In M. R. Mosquera-Losada, & R. Prabhu (Eds.), *Burleigh dodds series in agricultural science* (pp. 209–238). Burleigh Dodds Science Publishing. https://doi.org/10.19103/AS.2018.0041.13

Castle, S. E., Miller, D. C., Ordonez, P. J., Baylis, K., & Hughes, K. (2021). The impacts of agroforestry interventions on agricultural productivity, ecosystem services, and human well-being in low- and middle-income countries: A systematic review. *Campbell Systematic Reviews, 17*(2), e1167. https://doi.org/10.1002/cl2.1167

Ceasar, S. A., Maharajan, T., Krishna, T. P. A., & Ignacimuthu, S. (2023). Finger millet (*Eleusine coracana* (L.) Gaertn). In *Neglected and underutilized crops* (pp. 137–149). Elsevier. https://doi.org/10.1016/B978-0-323-90537-4.00031-4

Das, S. S., & Fakir, M. S. A. (2014). Pod growth and seed composition in two genotypes of *L. purpureus*. *Legume Research an International Journal, 37*(3), 306. https://doi.org/10.5958/j.0976-0571.37.3.046

Deepana, K., Geetha, K., Jeyaprakash, P., Sangeetha, M., & Govindan, K. (2025). The Dolichos bean: A versatile legume with potential to address food security challenges and strategies for improvement. *Genetic Resources and Crop Evolution, 72*(2), 1313–1331. https://doi.org/10.1007/s10722-024-02085-1

Gachuiri, C. K., Muthomi, J. W., Ojiem, J. O., & Wangila, A. J. (2021). Biomass yield and quality of fodder from selected varieties of lablab (*L. purpureus* L.) in Nandi South Sub-County of Kenya. *Online Journal of Animal and Feed Research, 11*(1), 28–35.

Gs, D., Bt, S., Hanchinamani, C. N., et al. (2024). Studies on drought tolerance in dolichos bean (*L. purpureus* L. Sweet). *International Journal of Advanced Biochemistry Research, 8*(2), 635–639.

Hanapiah, N. F. H., Sinniah, U. R., & Yusoff, M. M. (2022). Seed quality of Lablab bean (*L. purpureus*) as influenced by seed maturity and drying methods. *Agronomy, 12*(2), 363.

Hoover, D. L., Abendroth, L. J., Browning, D. M., Saha, A., Snyder, K., Wagle, P., Witthaus, L., Baffaut, C., et al. (2023). Indicators of water use efficiency across diverse agroecosystems and spatiotemporal scales. *Science of the Total Environment, 864*, 160992. https://doi.org/10.1016/j.scitotenv.2022.160992

Khan, A. U., Choudhury, M. A. R., Talucder, M. S. A., Hossain, M. S., Ali, S., Akter, T., & Ehsanullah, M. (2020). Constraints and solutions of country bean (*L. purpureus* L.) production: A review. *Acta Entomology and Zoology, 1*(2), 37–45.

Kolesnikova, A., Hammond, J., & Chapman, M. A. (2025). Drought response in the transcriptome and ionome of wild and domesticated *L. purpureus* L. sweet, an underutilized legume. *Plant-Environment Interactions, 6*(1), e70027.

Kongjaimun, A., Takahashi, Y., Yoshioka, Y., Tomooka, N., Mongkol, R., & Somta, P. (2022). Molecular analysis of genetic diversity and structure of the Lablab (*L. purpureus* (L.) sweet) gene pool reveals two independent routes of domestication. *Plants, 12*(1), 57.

Letting, F. K., Venkataramana, P. B., & Ndakidemi, P. A. (2021). Breeding potential of lablab [*L. purpureus* (L.) Sweet]: A review on characterization and bruchid studies towards improved production and utilization in Africa. *Genetic Resources and Crop Evolution, 68*(8), 3081–3101.

Mehetre, P., Chavan, S., Mane, A., Kunkerkar, R., Pethe, U., Mahadik, S., Potadar, S., & Dhone, K. (2025). Effect of plant growth regulators on yield of Wal (*L. purpureus* (L.) sweet). *International Journal of Advanced Biochemistry Research, 9*(10S), 14–17.

MiSsanga, J., NdakiDemi, P., & Venkataramana, P. (2022). *Lablab purpureus*: Evaluation and selection of drought-tolerant—high-yielding accessions in dry farming systems based on drought tolerance indices and multi-environmental yield trials. *Tarım Bilimleri Dergisi*. https://doi.org/10.15832/ankutbd.1169256

Missanga, J. S., Venkataramana, P. B., & Ndakidemi, P. A. (2021). Recent developments in *L. purpureus* genomics: A focus on drought stress tolerance and use of genomic resources to develop stress-resilient varieties. *Legume Science, 3*(3), e99. https://doi.org/10.1002/leg3.99

Moniruzzaman, M., Khatoon, R., Moniruzzaman, M., & Hossain, M. (2023). Growth, yield and profitability of summer country bean (*L. purpureus* L.) as influenced by exogenous application of plant growth regulators. *Bangladesh Journal of Agricultural Research, 47*(1), 13–22. https://doi.org/10.3329/bjar.v47i1.64840

Muchow, R. C. (1985). Phenology, seed yield and water use of grain legumes grown under different soil water regimes in a semi-arid tropical environment. *Field Crops Research, 11*, 81–97. https://doi.org/10.1016/0378-4290(85)90093-0

Naeem, M., Shabbir, A., Aftab, T., & Khan, M. M. A. (2023). Lablab bean (*L. purpureus* L.)—An untapped resilient protein reservoir. In *Neglected and underutilized crops* (pp. 391–411). Elsevier. https://doi.org/10.1016/B978-0-323-90537-4.00018-1

Njaci, I., Waweru, B., Kamal, N., Muktar, M. S., Fisher, D., Gundlach, H., Muli, C., Muthui, L., Maranga, M., Kiambi, D., Maass, B. L., Emmrich, P. M. F., Domelevo-Entfellner, J.-B., Spannagl, M., Chapman, M. A., Shorinola, O., & Jones, C. S. (2023). Chromosome-level genome assembly and population genomic resource to accelerate orphan crop lablab breeding. *Nature Communications, 14*(1), 1915. https://doi.org/10.1038/s41467-023-37489-7

Pandey, D. K., Singh, S., Dubey, S. K., Mehra, T. S., Dixit, S., & Sawargaonkar, G. (2023). Nutrient profiling of lablab bean (*L. purpureus*) from north-eastern India: A potential legume for plant-based meat alternatives. *Journal of Food Composition and Analysis, 119*, 105252. https://doi.org/10.1016/j.jfca.2023.105252

Revanth, P. B., Lakshmana, D., Manjunatha, B., & Rajashekharappa, K. (2024). Genetic variability studies in dolichos bean (*L. purpureus* L.) for yield and yield traits. *Journal of Advances in Biology and Biotechnology, 27*(11), 926–931. https://doi.org/10.9734/jabb/2024/v27i111676

Rolo, V., Rivest, D., Maillard, É., & Moreno, G. (2023). Agroforestry potential for adaptation to climate change: A soil-based perspective. *Soil Use and Management, 39*(3), 1006–1032. https://doi.org/10.1111/sum.12932

Seghieri, J., Droy, I., Hadgu, K., & Place, F. (2021). Introduction to the special issue "scaling up of agroforestry innovations: Enhancing food, nutrition and income security." *Agroforestry Systems, 95*(7), 1245–1249. https://doi.org/10.1007/s10457-021-00689-5

Sharma, N., Raman, H., Wheeler, D., Kalenahalli, Y., & Sharma, R. (2023). Data-driven approaches to improve water-use efficiency and drought resistance in crop plants. *Plant Science, 336*, 111852. https://doi.org/10.1016/j.plantsci.2023.111852

Shubha, K., Choudhary, A. K., Dubey, A. K., Tripathi, K., Kumar, R., Kumar, S., Mukherjee, A., Tamta, M., Kumar, U., Kumar, S., Layek, J., & Das, A. (2024). Evaluation of lablab bean [*L. purpureus* (L.) sweet] genotypes: Unveiling superior pod yield, nutritional quality, and collar rot resistance. *Frontiers in Nutrition, 10*, 1243923. https://doi.org/10.3389/fnut.2023.1243923

Singh, A., & Abhilash, P. C. (2019). Varietal dataset of nutritionally important *L. purpureus* (L.) sweet from Eastern Uttar Pradesh. *India. Data in Brief, 24*, 103935. https://doi.org/10.1016/j.dib.2019.103935

Singh, N., Kaur, S., Mithraa, T., Verma, V. K., Kumar, A., Choudhary, V., & Bhardwaj, R. (2024). ProTformer: Transformer-based model for superior prediction of protein content in lablab bean (*L. purpureus* L.) using near-infrared reflectance spectroscopy. *Food Research International, 197*, 115161. https://doi.org/10.1016/j.foodres.2024.115161

Sserumaga, J. P., Kayondo, S. I., Kigozi, A., Kiggundu, M., Namazzi, C., Walusimbi, K., Bugeza, J., Molly, A., & Mugerwa, S. (2021). Genome-wide diversity and structure variation among lablab [*L. purpureus* (L.) Sweet] accessions and their implication in a Forage breeding program. *Genetic Resources and Crop Evolution, 68*(7), 2997–3010. https://doi.org/10.1007/s10722-021-01171-y

Tasnia, N., Haque, R. B., Islam, M. M., & Haque, Md. A. (2024). Morphological and molecular characterizations of country bean (*L. purpureus* L.) genotypes for drought tolerance. *American Journal of Plant Sciences, 15*(11), 1069–1090. https://doi.org/10.4236/ajps.2024.1511068

Teshome, A., Habte, E., Cheema, J., Mekasha, A., Lire, H., Muktar, M. S., Quiroz-Chavez, J., Domoney, C., & Jones, C. S. (2024). A population genomics approach to unlock the genetic potential of lablab (*L. purpureus* (L.) Sweet), an underutilized tropical forage crop. *BMC Genomics, 25*(1), 1241. https://doi.org/10.1186/s12864-024-11104-5

Thapa, B., Lovell, S., & Wilson, J. (2023). Remote sensing and machine learning applications for aboveground biomass estimation in agroforestry systems: A review. *Agroforestry Systems, 97*(6), 1097–1111. https://doi.org/10.1007/s10457-023-00850-2

Visakh, R. L., Anand, S., Reddy, S. B., Jha, U. C., Sah, R. P., & Beena, R. (2024). Precision phenotyping in crop science: From plant traits to gene discovery for climate-smart agriculture. *Plant Breeding, 6*, 13228. https://doi.org/10.1111/pbr.13228

Wang, T., Wang, B., Xiao, A., & Lan, J. (2024). Optimizing seeding ratio for legume forage to maximize system productivity and resource use efficiency in mixed cropping systems. *Agriculture, 14*(8), 1249. https://doi.org/10.3390/agriculture14081249

Wood, J. M., Bremer, E., Csonka, L. N., Kraemer, R., Poolman, B., Van Der Heide, T., & Smith, L. T. (2001). Osmosensing and osmoregulatory compatible solute accumulation by bacteria. *Comparative Biochemistry and Physiology Part a: Molecular and Integrative Physiology, 130*(3), 437–460. https://doi.org/10.1016/S1095-6433(01)00442-1

Chapter 20
Pearl Lupin (*Lupinus mutabilis*): Nitrogen Fixation and Climate Adaptation

Abdullah Jan, Umar Farooq, Amjad Malik, and Shakeel Ahmad

Abstract Pearl Lupin (*Lupinus mutabilis*), the nutrient-enriched and climate-resistant legume that was first introduced to the Andes have received a new focus due to its ability to reinforce food security and promote sustainable agriculture under the growing climatic loads. Its adaptation to the marginal environment, combined with the effective biological fixation of nitrogen by symbiosis with Brady rhizobium strains, has made it a valuable crop in terms of mitigating the reliance of the external fertilizer and improving the fertility of the soil. This chapter discusses its agronomic potential, phenological plasticity and adaptive interactions in various production systems with a focus on management approaches that can be implemented to reduce climate effects including selection of genotype-specific environment, organic fertility management, and resource-efficient crop production modes, among other strategies. Its promise is supported by evidence of past studies. In a case study, it was reported that grain yield would be about 18.1 g per plant when growing in the Mediterranean, versus 9.2 g per plant when growing in the Andes, and biomass accumulation would be as high as 2 kg per plant when growing in Northern Europe. In addition, multi-elevation experiments of a recent Andean experiment have shown that moderately warm temperatures did not impair yield or biomass and that organic manure performed just as well as synthetic fertilizers. The further realization of the potential of the crop is to advance breeding specifics; increase genotype/environment match; and improve modeling tools. Future research must increase the exploration of the responsive stress characteristics and further reinforce it into climate-resisting agricultural practice.

A. Jan (✉) · U. Farooq
Department of Agronomy, Pir Mehr Ali Shah Arid Agriculture University, Rawalpindi, Pakistan
e-mail: abdullahjanarid@gmail.com

A. Malik
Kenanow Bachelor of Education Program, Faculty of Education, University College of the North (UCN), Thompson, MB, Canada

S. Ahmad
Faculty of Agricultural Sciences and Technology, Institute of Agronomy, Bahauddin Zakariya University, Multan, Pakistan

M. Ahmed (ed.), *Climate Change and Neglected Underutilized Crops*,
https://doi.org/10.1007/978-3-032-21005-0_20

Keywords Pearl Lupin (*Lupinus mutabilis*) · Nitrogen fixation · Climate resilience · Genotype–environment interaction · Organic fertility management

20.1 Introduction

Pearl Lupin (*Lupinus mutabilis*) is becoming a noteworthy neglected and underutilized crop (NUC) to food and nutritional security in the world. These are marginalized crops with unique agronomic and nutritional characteristics highly useful in improving sustainable and climate-resistant food production. Application of NUCs is particularly applicable in regions where human beings are characterized by persistent food insecurity, erratic weather crises, and soil erosion. They have natural endurance and flexibility that allow farmers to cultivate food in areas where majority of the staple crops such as wheat, rice or maize do not grow/or they have high rates of input. Most NUCs use very minimum resources, including less water, fertilizers and pesticides, compared to conventional crops, which makes them a perfect choice in low-input and smallholder-based agricultural systems (Msaddak et al., 2023). According to the Food and Agriculture Organization (FAO), more than 1000 species of plants that have ever been used in human diet have been used by the modern-day farming industry, which uses approximately 150 species cultivated around the world (Keefe & Kharouba, 2024). This loss of crop diversity makes agroecosystems vulnerable to environmental variability, pests and market shocks, and the essential message of the necessity to reestablish and promote diverse, agroecosystem-adaptive, resilient plant species.

Pearl Lupin (*Lupinus mutabilis*) or Andean lupin is one of the potential NUCs due to its excellent agronomic properties, as well as its uses in the agro-systems. The Andean highlands of South America have Pearl Lupin, which is a unique food crop that has been consumed since early times by the inhabitants of the land as a high protein food crop, and it grows in very poor environmental conditions. The species is of the larger genus Lupinus; but it possesses some several positive features making it special among other kinds of lupins. It should be pointed that Pearl Lupin is well adapted to the low fertility, acidic and poor structure marginal soils. It is highly resistant to drought, cold temperatures and large daytime temperature changes—a scenario that is characteristic to mountainous dry or damaged ecosystems. These stress-tolerance traits indicate that Pearl Lupin will be good crop in future climatic conditions, which is more volatile and has a limited resources base.

The biological ability to meet atmospheric requirements of nitrogen in symbiotic associations with soil bacteria, and more specifically the species of Rhizobium and Brady rhizobium, also gives this crop its agricultural significance. Pearl Lupin can be employed to enhance the soil by adding nitrogen that will ensure farmers do not depend on synthetic fertilizers. The aspect is particularly important in low-input farming structures and regions where fertilizers are not only prohibitively costly, but also environmentally damaging. Nitrogen fertilizers may be regarded as renowned groundwater and soil contaminants, soil acidifiers, as well as greenhouse gas, specifically, of nitrous oxide (N_2O), a potent climate-heating gas. To promote the fertility of the soil, Pearl Lupin naturally enriches it by adding nitrogen to limit the effects of some of the negative externalities associated with industrial agriculture (Pueyo et al., 2021). Moreover, it has a deep root system, which increases the structure and mobility of the soil, increases organic matter and infiltration of water and is therefore a valuable crop in the regeneration of degraded soils.

As climate change keeps destabilizing the food production system with higher temperatures, greater variability of rains, greater weather extremes, and spread of crop pests, the first impulse is to diversify the agricultural landscapes with crops capable of enduring economic pressures. The integration of nutritional, ecological, and agronomic benefits of Pearl Lupin is in line with the global efforts in designing food systems of climate-resistance and environmental sustainability. Besides playing a role in fixation of nitrogen and soil enrichment, its seeds are rich in protein, dietary fibre, essential amino acids, and other beneficial phytochemical, which has led to the creation of its possible application as a protein substitute in the human diet as well as in the livestock diets. Despite the existence of these advantages, Pearl Lupin is not misused in areas of the world where it does not grow like in breeding since there is no understanding and the capability to tackle the problems relating to the level of alkaloid, which must be processed.

By considering its multipurpose value, Pearl Lupin can be viewed as a good chance to reach the sustainable agriculture, improved food security and ecosystem resilience. The rationale behind writing this chapter is to address the topic of Pearl Lupin use in agriculture at present, its ability to fix nitrogen, its ability to survive in changing climate, and its beneficial effect on the health of soils, and the potential application of Pearl Lupin to a diversified and robust crop production system. The chapter is a holistic treatise as to why Pearl Lupin deserves more attention in future agricultural research and development by highlighting the agronomic characteristics, environmental advantage, and the increased popularity of the crop in other regions of the world.

20.2 Taxonomy of Pearl Lupin (*Lupinus mutabilis*)

20.2.1 Botanical Classification

Pearl Lupin is a member of the fabaceae family that is popularly known as the legume family or pea family. Lupinus genus comprises of more than 200 species with the most common being *Lupinus mutabilis* which is extensively planted to be eaten by people and as food to feed animals (Table 20.1).

Lupinus mutabilis is also identified by the fact that its seeds are large and nutrient rich, and that it can tolerate a great variety of climates, including the cool high elevation Andes, and at lower altitudes. The species is also characterized by the

Table 20.1 Taxonomic hierarchy of pearl lupins

Kingdom	Plantae
Phylum	Angiosperms
Class	Eudicots
Order	Fabales
Family	Fabaceae
Genus	Lupinus
Species	*L.mutabilis*

possibility of fixing nitrogen and being an excellent source of protein, fiber and essential amino acids (Cheng et al., 2021).

20.2.2 Geographic Distribution and Domestication

Pearl Lupin, which is native to the Andean areas of South America, especially to such countries as Peru, Ecuador, and Bolivia, was cultivated centuries ago by the indigenous peoples. Conventionally, it has served as a source of food, both as a human food and as animal food. The domestication of Pearl Lupin can be dated over 3000 years ago when the ancient people have discovered the advantages of this plant in terms of soil fertility and crop rotation (Shen et al., 2010) (Fig. 20.1).

20.3 Phenology of Pearl Lupin

20.3.1 Growth Stages and Environmental Requirements

The phenology of Pearl Lupin as the part of the lifecycle between germination and harvest is dependent on a number of environmental factors including temperature, water, and soil conditions. The vegetative and reproductive growth pattern of the crop is usually as follows:

1. **Germination (0–10 days)**: Pearly Lupin seeds have a high rate of germination in a cool and moist environment. The highest germination is observed at a temperature of 10–20°C.
2. **Vegetative Growth (20–40 days)**: Once germinated the plant is in the vegetative growth. At this stage, leaves as well as stems are produced and nitrogen fixation will occur via root nodules that are developed by Rhizobium bacteria. Its ability to withstand moderate frost and its habit of thriving at an altitude of above 2000 meters make the plant very popular in altitude areas.

Fig. 20.1 Production yield and area harvested of pearl lupin since (1960–2020) (self drawn)

3. **Flowering (40–70 days)**: Pearl Lupin will start flowering 40–60 days after growing in the environment, based on the environmental conditions. The flowers are normally blue or purple and pollination is normally done by insects especially bees.
4. **Pod Formation and Seed Development (70–120 days)**: The plant produces pods with seeds, which take 30–50 days to mature. The seeds contain large quantities of protein and essential amino acids, although they need proper processing to lower the number of alkaloids, which are toxic to human beings.

20.3.2 Temperature and Water Requirements

Pearl Lupin is also adjusted to the cool climate, and it can withstand the temperatures down to 5°C. Nonetheless, it is also drought tolerant to some degree particularly in the developmental phase. It grows well in medium to high fertile soils that have a pH of 6.0–7.5. Though it tolerates bad soils, it enjoys enough moisture during flowering and seed formation to produce as high as possible yields (Chaves et al., 2022).

20.3.3 Visual Representation

See Fig. 20.2.

20.4 Zadok Scale, Growing Degree Days (GDD), and Crop Growth Stages

20.4.1 Zadok Scale Application to Pearl Lupin

The Zadok scale is an extensively used system to characterize the growth phases of cereals, although legumes such as Pearl Lupin can be developed using the scale. The stages go through 0 (germination) to 9 (maturity) with the in-between stages detailing important transitions, i.e. flowering and pod development. The scale proves effective in tracking the progress of crops and based on maximizing the field management activities (Fig. 20.3).

20.4.2 Growing Degree Days (GDD)

Growing Degree Days (GDD) is a very important parameter in determining the amount of heat needed to be accumulated in crop growth. The formula used in

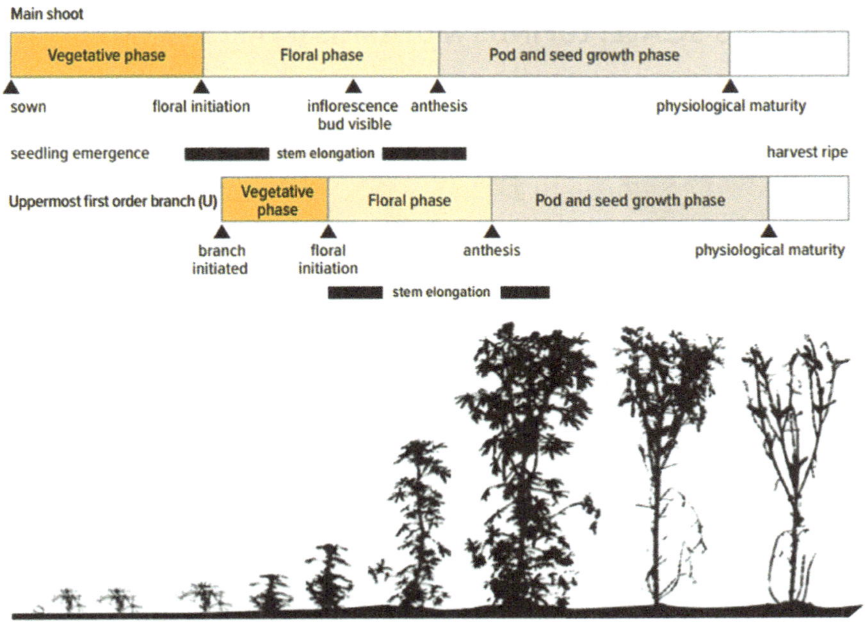

Fig. 20.2 Life cycle of lupin plant (Walker et al., 2010)

determining GDD is:

$$GDD = \frac{Tmax + Tmin}{2} - Tbase$$

Where

- (T_{max} = daily maximum temperature)
- (T_{min} = daily minimum temperature)
- (T_{base} = base temperature (usually around 5 °C for Pearl Lupin))

An optimal growth of Pearl Lupin can be attained through GDD of 1000–1500, whic is equivalent to the growing season of 90–120 days (Keefe & Kharouba, 2024) (Fig. 20.4).

20.5 Comparison with Other Leguminous Crops

In order to determine the promise of Pearl Lupin in the larger scope of legumes, it is necessary to make a comparison with other leguminous crops such as soybean, peas, and chickpeas, which are cultivated due to their ability to fix nitrogen and as source of protein (Table 20.2).

Fig. 20.3 Zadock scale growth stages of pearl lupin (Górski & Płaza, 2024)

20.5.1 Nitrogen Fixation

Pearl Lupin: Extremely efficient in the fixation of nitrogen up to 100 kg per hectare/year, and has a symbiotic relationship with Rhizobium. Soybeans are in the same category with nitrogen-fixing legumes but has a higher input in synthetic fertilizers' to grow well under non-optimal soils. Peas/Chickpeas fix a similar amount of nitrogen although Pearl Lupin has an advantage because it can be planted in poorer soils and in harsher conditions (Fig. 20.5).

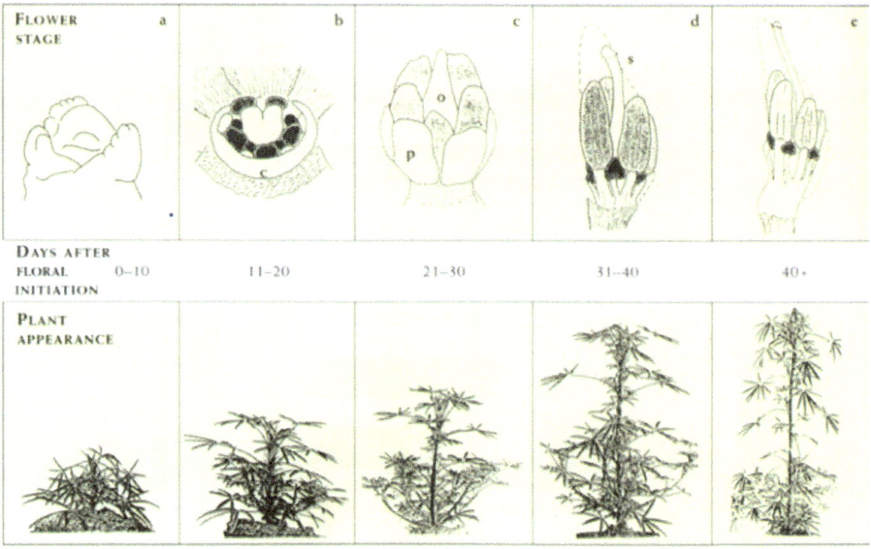

Fig. 20.4 Flower development of lupin plants (Dracup & Kirby, 1996)

Table 20.2 Nitrogen fixation (Kg/Acre) of pearl lupin and other crops

Crop (species)	Nitrogen fixation Kg/acre/season	reference
Pearl lupin/lupin (e.g. *Lupinus mutabilis*, *Lupinus* spp.)	110–300 kg N/acre	(Sulas et al., 2016)
Soybean (*Glycine max*)	40–220 kg N/acre	(Salvagiotti et al., 2008)
Faba bean (*Vicia faba*)	70–220 kg/acre	(Gidey et al., 2024)
Field pea / common pea (*Pisum sativum*)	110–165 kg N/acre	(Alitalo, 2025)
Chickpea (*Cicer arietinum*)	17–220 kg N/acre	(Mugerwa et al., 2022)
Alfalfa (*Medicago sativa*) (perennial forage)	120–220 kg N/acre	(Wang et al., 2023)
Red clover (*Trifolium pratense*)	150–220 kg N/acre	(Jing & Boelt, 2021)

20.5.2 Resilience to Climate Stress

Pearl Lupin is less prone to drought and frost as compared to soybeans and peas and thus it is suitable in high-altitude or arid areas. Soybeans need hotter weather and climate with regular rainfalls. Peas/Chickpeas are drought resistant but not as cold resistant as Pearl Lupin.

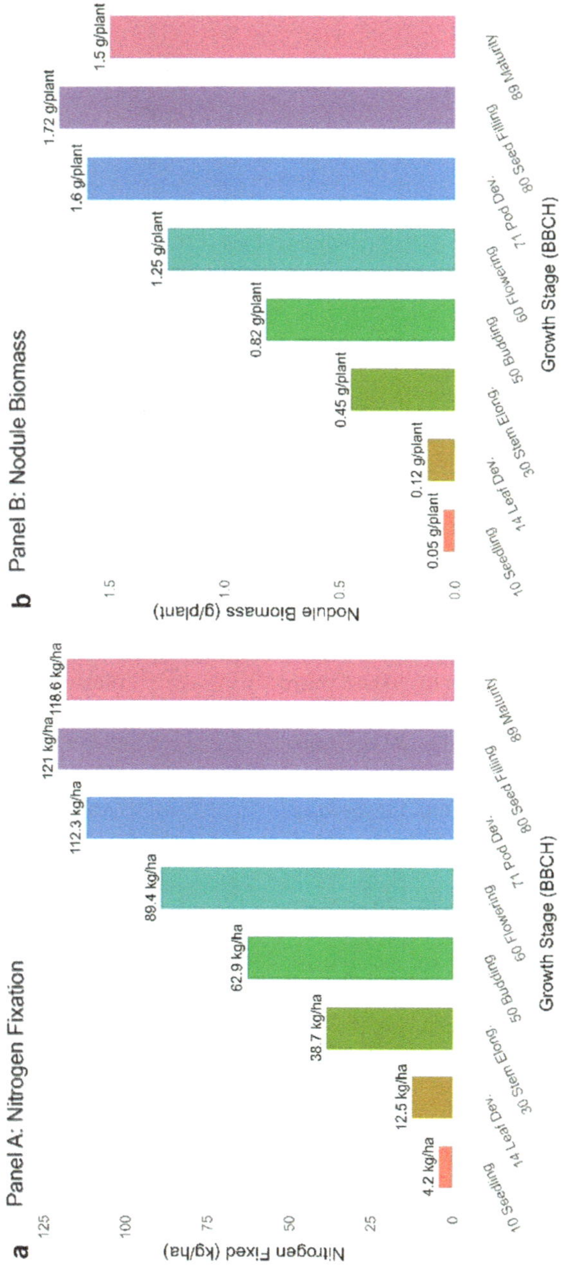

Fig. 20.5 Nitrogen fixation and nodule biomass production of pearl lupin according to (BBCH) growth stages

20.6 Edible form of Pearl Lupin and its Nutritional Value

Pearl Lupin (*Lupinus mutabilis*) is not only useful in terms of its nitrogen-fixing ability and resistance to adverse climatic conditions, but it can also be a valuable source of nutritious food crop. Pearl Lupin can eat its seeds, which are edible and have been eaten centuries in the Andean regions of South America. They are traditionally cooked in a variety of ways, such as boiled, roasted and even blended into flour. The raw seeds in most traditional dishes contain toxic alkaloids, and to make them edible, they are either soaked or fermented to inactivate the toxic alkaloids (such as quinolizidine alkaloids).

Pearl Lupin is regarded as high protein and high fiber legume with remarkable nutritional value. It is especially said to contain high levels of protein between 30 and 40%, by weight so it is an ideal substitute to soybean and other pulses. Furthermore, it contains many essential amino acids, one of them being lysine, which is commonly limited in other diets based on cereals. Minerals such as calcium, iron and magnesium are also present in substantial amounts in the crop and hence are useful in areas where micronutrient deficiencies are present.

Considering its nutritional properties, Pearl Lupin is becoming a stable foothold in terms of sustainable protein source in human consumption and animal feed. Its application is becoming popular in functional foods and plant-based protein products, which provide a possible solution to the world food security issues and the dietary inadequacies, particularly in those areas where animal-based proteins are insufficient (Table 20.3).

Table 20.3 Nutritional composition of pearl lupin seeds (per 100 g) (Zhao et al., 2024)

Nutrient	Amount per 100 g	RDA percentage (Adults)
Energy	350 kcal	–
Protein	30–40 g	60–80%
Fat	8–9 g	12–14%
Carbohydrates	45–50 g	–
Dietary fiber	25–30 g	80–100%
Calcium	200–250 mg	20–25%
Iron	5–7 mg	28–35%
Magnesium	160–180 mg	40–45%
Phosphorus	300–350 mg	40–45%
Lysine	4.5–5.5 g	–

20.6.1 Health Benefits of Pearl Lupin Consumption

1. **High-Quality Plant Protein**: Having a similar protein content with soybeans, Pearl Lupin is a great source of protein, especially to vegetarians, vegans, and those who live in the regions where animal proteins are scarce. It contains lysine as an essential amino acid, which increases its nutritional value and thus it is an appropriate substitute of animal proteins in food products (Gabur & Simioniuc, 2023).
2. **Dietary Fiber**: Pear Lupin contains a lot of dietary fiber that contributes to having a better digestive system, avoiding constipation, and a healthy gut microbiome. The diet with a high level of fiber is also linked to the lower risk of cardiovascular diseases and improved weight control (Silifonov, 2023).
3. **Micronutrient-Rich**: The seeds contain sufficient calcium, iron, and magnesium that are essential in the development of the bones as well as transportation of oxygen and general functioning of the cells. This is particularly significant in developing countries in which the micronutrient deficiencies (such as iron deficiency anemia) exist.
4. **Low Glycemic Index**: The glycemic index of pearl Lupin is quite low and therefore, it is a good food to a diabetic person because it assists in the stabilization of blood sugar levels.
5. **Sustainable Protein Source**: Nutrient-poor soils and low water demand require Pearl Lupin as a nutrient-depleted environment and low-water demand relative to conventional protein crops such as soy mean that it is an environmentally safe and sustainable solution to worldwide food security and protein shortage issues.

20.7 Nitrogen Fixation Mechanism

20.7.1 Symbiotic Nitrogen Fixation in Pearl Lupin

Similar to numerous other leguminous crops, pearl Lupin (*Lupinus mutabilis*) has a special characteristic of fixing atmospheric nitrogen as a form of use by plants via a symbiotic association with Rhizobium bacteria. The process is essential in the fact that it can grow well in soils with low nitrogen levels as well as decreasing the use of synthetic nitrogen fertilizers, which are not only expensive, but also harmful to the environment.

Nitrogen fixation starts with the infection of roots by Rhizobium bacteria (mainly Rhizobium leguminosarum (or Brady rhizobium-species)). This infection causes root nodules to develop which are special structures, in which the bacteria dwell and fix nitrogen. Within these nodules, the bacteria transform nitrogen gas (N_2) in the atmosphere into ammonia (NH_3) using an enzyme known as nitrogenase, which the plant utilizes to grow (Dawud, 2023) (Fig. 20.6).

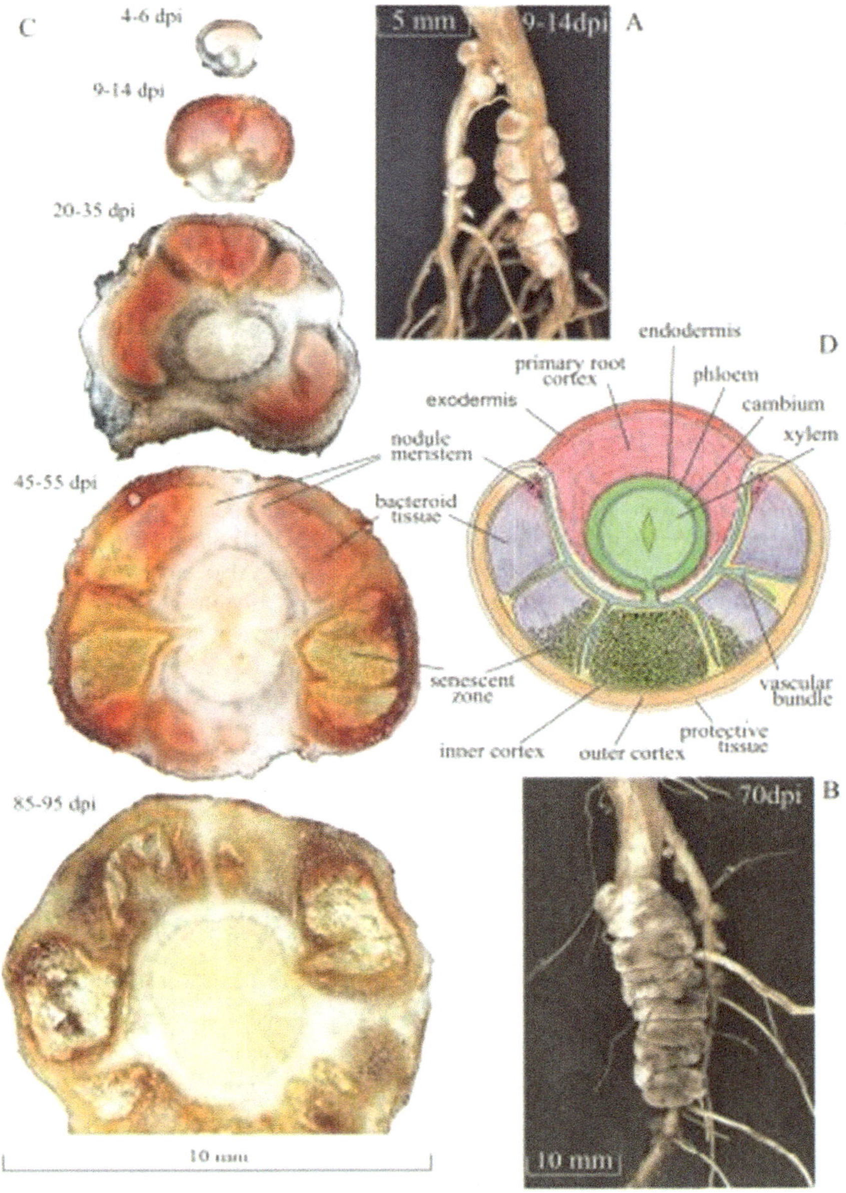

Fig. 20.6 *Lupinus luteus* root nodules. **a** Young nodules: because of very early division of laterally positioned nodule meristem and its growth in all directions, the lupin nodule escapes from its cylindrical shape. **b** As a result of the activity of lateral meristems, the initially spherical nodules grow laterally encircling the root (Golinowski et al., 1987). **c** Cross-sections through nodules at different stages of development. **d** Schematic diagram of the mature nodule (Strozycki et al., 2007)

20.7.2 Quantifying Nitrogen Fixation

Pearl Lupin has been demonstrated to be fixing a lot of nitrogen. Recent research shows that it is able to incorporate up to 100–120 kg of nitrogen per hectare/year in an ideal growing environment (Kebede, 2021). Recent research shows that it is able to incorporate up to 100–120 kg of nitrogen per hectare/year in an ideal growing environment. The fixation of nitrogen in Pearl lupin is particularly important to the sustainable agricultural system because it contributes to the maintenance of the soil health and fertility. In agroecosystems, the crop is normally employed in a legume-cereal rotation where the crop enriches the soil to allow other crops such as maize, wheat, or potatoes to grow, that need large quantities of nitrogen to grow well.

20.7.3 Impact on Soil Fertility and Ecosystems

Pearl Lupin improves soil microbial diversity and structure because of its capacity to fix nitrogen. It elevates the amount of organic matter (SOM) content in the soil, which makes the other plants and microorganisms more stable. Also, since the nitrogen is a limiting nutrient of plant growth, the rise in supply of nitrogen enhances crop productivity, especially in soils with low fertility (Akchaya et al., 2025).

20.7.4 Growing Degree Days (GDD) for Different Leguminous Crops

Crop name	Growing degree days GDD	References
Pearl Lupin (*Lupinus mutabilis*)	1000–1500 GDD	(Gulisano et al., 2022)
Soybean (*Glycine max*)	1200–1800 GDD	(Salvagiotti et al., 2008)
Chickpea (*Cicer arietinum*)	900–1200 GDD	(Suryavanshi et al., 2021)
Pea (*Pisum sativum*)	900–1300 GDD	(Alitalo, 2025)
Common bean (*Phaseolus vulgaris*)	1000–1400 GDD	(Pinzón-Sandoval et al., 2024)

Pearl Lupin improves soil microbial diversity and structure because of its capacity to fix nitrogen. It elevates the amount of organic matter (SOM) content in the soil, which makes the other plants and microorganisms more stable. Also, since the nitrogen is a limiting nutrient of plant growth, the rise in supply of nitrogen enhances crop productivity, especially in soils with low fertility (Keefe & Kharouba, 2024).

20.8 Adaptability and Resilience of Pearl Lupin Against Climate Change

20.8.1 Climate Resilience

a. Drought and Heat Tolerance

The climate resilience of Pearl Lupin is one of the outstanding characteristics of the crop especially its capacity to survive in drought and frost hence making it the best crop for areas with climate uncertainty. The pearl Lupin does well in cold, elevated settings (between 2000 and 4000 meters above sea level) yet can also endure medium dry conditions that are becoming more frequent in areas with water shortage.

Research established that the drought tolerance trait of Pearl Lupin is attributed to the fact that this plant regulates transpiration and is capable of sustaining root growth in water-limiting environments. Moreover, the deep root system of the plant enables the plant to tap into water distributed in lower layers of soil and is therefore less reliant on surface irrigation (Dutta et al., 2024). This renders it a very good rain-fed crop, which is common in most of the developing areas.

b. Frost Resistance

Pearl Lupin too is cold tolerant and only has a rare capacity to withstand light frosts. This characteristic has rendered it a significant crop in areas with short growing periods or chilly nights, which can be extremely destructive to other delicate crops such as maize or soybean. Pearl Lupin can also be cultivated in high-altitude regions using this cold resistance where other legumes would not develop or achieve a viable crop (Gabur & Simioniuc, 2023).

c. Adaptation to Marginal and Degraded Soils

Another key aspect of Pearl Lupin's adaptability is its ability to grow in marginal and degraded soils. Unlike many major crops that require high soil fertility, Pearl Lupin can thrive in soils with low organic matter and low levels of nitrogen. Its nitrogen-fixing ability, combined with its tolerance for poor soils, makes it a perfect candidate for soil rehabilitation projects. Pearl Lupin can contribute to improving soil structure and fertility, facilitating the growth of subsequent crops (Reinsch et al., 2024). In addition, it can grow in the harsh environment like in high salinity or acid soils, thus making it a good crop in climatically vulnerable areas where soil erosion is a major problem.

20.8.2 Climate Change Mitigation Potential

With the increased intensity of the climate change, the necessity of crops both resistant and able to enhance the health of the soil has become even more pressing than ever before. The solution decreased greenhouse gas emissions that are caused by the manufacture and use of fertilizers. It is also capable of fixing its nitrogen, which can mitigate the carbon footprint of agriculture making it one major element of a climate-smart agriculture plan (Teng et al., 2021). As shown in Fig. 20.3, Panel A depicts a steady increase in nitrogen fixation with crop advancement, attaining peak values during seed filling and maturity stages. Further, Panel B demonstrates high climate adaptability, particularly for drought tolerance, altitude adaptation, and low soil fertility, reflecting the crop's resilience under diverse environmental stresses (Fig. 20.7).

20.8.3 Nitrogen Fixation and Climate Adaptation Heatmap

Figure 20.3 indicates that the heatmap combines the dynamics of nitrogen fixation and climate adaptation properties of pearl lupin. The rate of fixation of nitrogen also shows a progressive rise with crop growth with maximum levels being recorded during the stages of pod growth, seed filling, and even at the stage of maturity meaning greater contribution of biological nitrogen at later phenophases. Further, the climate adaptation profile shows high resilience, together with high scores in drought tolerance, high altitude adaptation, low soil fertility tolerance, and the overall climate change suitability, and this indicates the potential of pearl lupin to grow in various and stress prone agro ecological conditions (Fig. 20.8) (Tables 20.4 and 20.5).

20.9 Application of Crop Models (DSSAT) for Pearl Lupin

Decision Support System on Agro technology Transfer (DSSAT) is a popular crop modeling application that assists in simulation of the growth and development of different crops in different environmental conditions to researchers, policymakers and farmers. DSSAT is a system of yield and performance prediction that applies against inputs such as temperature, precipitation, soil and farm management practices. The model can be particularly helpful in forecasting crop responses to the changing climatic conditions to develop the strategies of adapting to the impact of the climate change (Ahmed et al., 2024).

*20.9.1 Adapting the DSSAT-CROPGRO Model for Narrow-Leaved Lupin (*Lupinus Angustifolius*)*

Narrow-leaved lupin (*Lupinus angustifolius*) is one of the promising legumes to temperate regions of Germany, as it has a potential to boost the supply of protein,

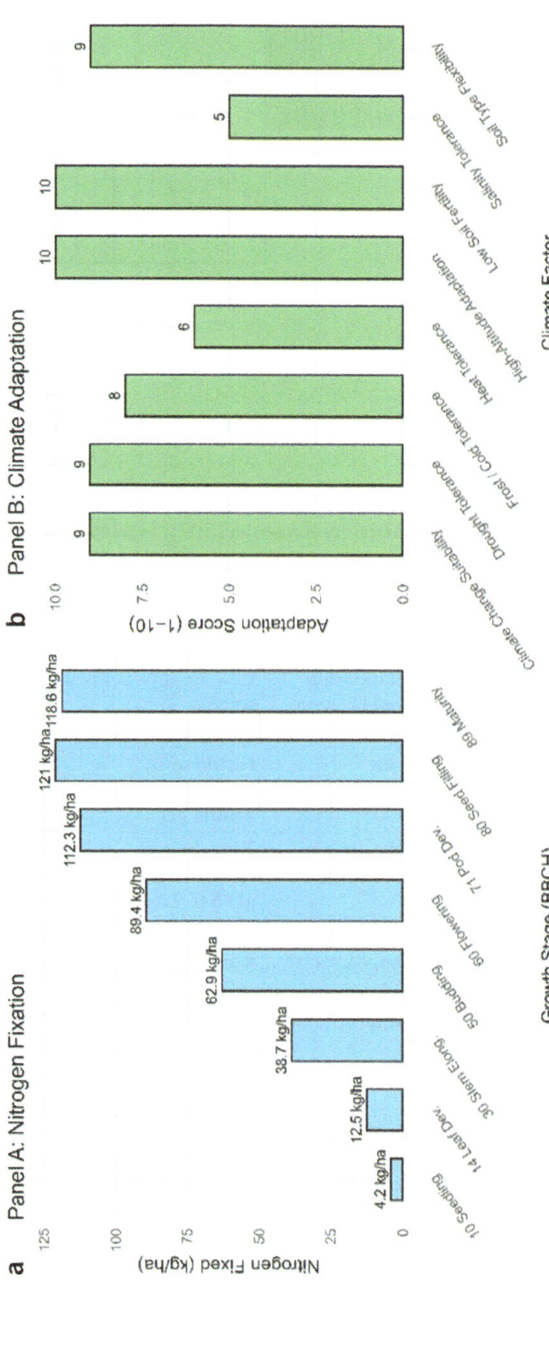

Fig. 20.7 Nitrogen fixation and climate adaptations of pearl lupin

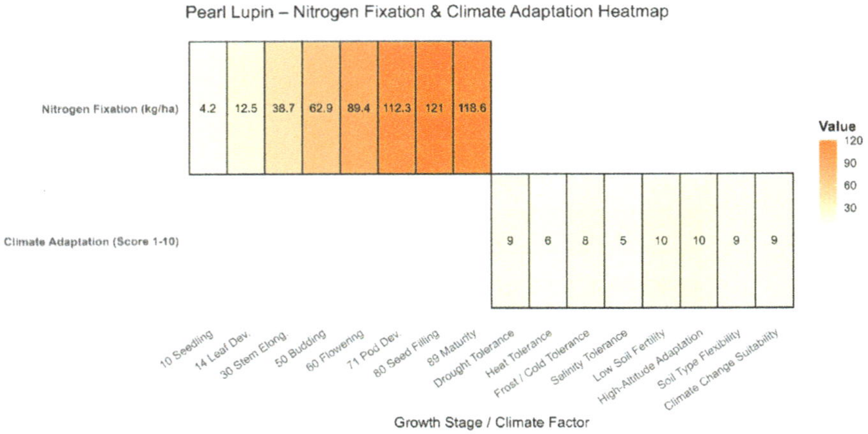

Fig. 20.8 Nitrogen and climate adaptation heat map of pearl lupin (self drawn)

Table 20.4 Nitrogen fixation rate, mechanism, soil requirements and climate tolerance of Perl Lupin in comparison with other crops: (Gabur & Simioniuc, 2023; Govindasamy et al., 2023)

Crop	Nitrogen fixation rate (kg/ha/year)	Mechanism of nitrogen fixation	Soil requirements	Climate tolerance
Pearl Lupin (*Lupinus mutabilis*)	100–120	Symbiotic relationship with *rhizobium* bacteria (high nitrogen fixation)	Low-fertility, acid soils, drought-resistant	Cold-tolerant, drought-tolerant, high-altitude
Soybean (*Glycine max*)	50–150	Symbiosis with *Bradyrhizobium* bacteria	Fertile soils, well-drained, neutral pH	Warm climate, moderate drought tolerance
Chickpea (*Cicer arietinum*)	60–100	Symbiotic relationship with *rhizobium* species	Well-drained, sandy soils, pH 6.0–7.5	Drought-tolerant, warm climate
Pea (*Pisum sativum*)	50–90	Symbiotic relationship with *rhizobium* bacteria	Fertile soils, neutral pH	Cold-tolerant, moderately drought-tolerant
Common bean (*Phaseolus vulgaris*)	30–70	Symbiosis with *rhizobium* species and *Bradyrhizobium* species	Fertile soils, slightly acidic to neutral pH	Warm climate, moderate drought tolerance

diversify crop rotations, and curtail the use of nitrogen fertilizer under climatic change. The DSSAT-CROPGRO model was fitted to measure its growth and yield under different environmental conditions using the faba bean module as a starting point with the data of almost 50 site-years, with the post-registration variety trials and field experiments on the JKI station in Berlin under four sowing dates and three

Table 20.5 Comparison of soil and climate requirements for leguminous crops (Ali et al., 2025; Amin et al., 2025)

Crop	Soil type	Soil pH range	Optimal temperature range	Water requirements	Altitude tolerance
Pearl Lupin	Well-drained, moderately fertile, acid soils	5.5–7.5	10–25 °C	Moderate (drought-resistant)	High altitude (up to 4000 m)
Soybean	Fertile, well-drained soils	6.0–7.0	20–30 °C	High (requires regular moisture)	Low to moderate altitude
Chickpea	Sandy, well-drained soils	6.0–7.5	18–30 °C	Low to moderate (drought-tolerant)	Low to moderate altitude
Pea	Fertile, well-drained soils	6.0–7.5	10–20 °C	Moderate (needs consistent moisture)	Low to moderate altitude
Common bean	Slightly acidic to neutral soils	5.5–7.0	20–30 °C	Moderate (prefers consistent moisture)	Low to moderate altitude

years. Calibration was conducted in which phenology, growth and yield parameters were adjusted to the parameters using the DSSAT time-series estimator whereby, a third of the dataset was used to calibrate and one third to validate. The adapted model was capable of simulating flowering, biomass accumulation and yielding reliably with early sowing and optimum plant density yielding increased productivity whereas simulation in a future climate scenario revealed no changes in yields with appropriate management of the soil and water. The model is a means to evaluate sowing plans, management plan, and nitrogen-use efficiency and the potential to reduce climate change mitigation of lupin growing, and future developments are scheduled in the form of indeterminate varieties and greater drought and heat stress tolerance.

20.9.2 Case Studies: DSSAT Simulations

20.9.2.1 Modeling Case Study: 1 South America

In their article, Diversity and Agronomic Performance of *Lupinus mutabilis* Germplasm in European and Andean Environments, Gulisano et al., (2022) assessed 225 Pearl Lupin accessions on their native Andean locations and compared them with an environment in Europe using multi-location field trials to determine phenology, morphology, grain yield, and biomass. They compared genotypic responses in Mediterranean winter-cropping, Northern European summer-cropping, and Andean highland conditions and found that, using their approach, the grain yields were found to be quantitative, e.g., at the Mediterranean, had a value of 18.1 g/plant and at the Andes, 9.2 g/plant and the biomass production was found to be 2 kg/ plant at the Northern Europe. They suggested just taking advantage of this genetic diversity to produce environment-specific high-yielding or early-flowering lines and apply genotype-environment congruence to scale up Pearl Lupin production beyond the introduction region. Likewise, the authors of the 2023 article "Traditional Soil Fertility Management Ameliorates Climate Change Impact on Traditional Andean Crops in Smallholder Farming Systems" visited the Andes and carried out multi-elevation field experiments with Pearl Lupin under organic (sheep manure) and synthetic fertilization, measuring the effect of climate change on this crop and monitoring the variables of temperature, precipitation, and soil nutrients. Their findings indicated that moderate warming did not decrease Pearl Lupin yield and biomass at any one elevation and that the performance of organic fertility management and biomass supported crop performance on a par with synthetic fertilizers. They suggested that traditional organic fertility methods should be incorporated with local crop production, including L. mutabilis, to be more resilient to climate change and sustain production in smallholder Andean systems in the future climate conditions.

20.9.2.2 Modeling Case Study 2: Sub-SaharanAfrica

In their article, DSSAT-Based Evaluation of Pearl Lupin (Lupinus mutabilis) as a Climate-Resilient Crop in Semi-Arid East Africa, Mwangi et al. (2024) ran a simulation in the LOCATION of the city of East Kenya and the nation of North Tanzania to analyze the potential of the crop in warmer and drier conditions with decreased rainfall. The researchers assessed the dynamics of soil moisture, nitrogen provision as well as the stability of the yield under stress (drought) through methodology that incorporated DSSAT crop with future climate scenarios (RCP 4.5 and RCP 8.5). Their Quantitative revealed that Pearl Lupin preserved 65–78% of their original yield under moderate drought, was a source of biological nitrogen fixation estimated to amount to 35–55 kg N ha-1 to soils and decreased simulated yield losses of maize when used in rotation systems by up to 22%. their recommendations based on these findings included encouraging Pearl Lupin as a drought-resilient legume in smallholder farming systems, interplanting with maize to enhance soil nitrogen and encouraging farmers to adopt the legume by developing seed systems and training in climate-smart legume production.

20.9.3 Case Studies of Cultivation

20.9.3.1 Case Study 1: Pearl Lupin in the Andes

In their article titled (Gabur & Simioniuc, 2023) integrating pearl lupin into andean cereal-based cropping systems to complement soil fertility and yield stability, conducted field tests in the area of the highlands of andean which covers much of the farmers in the area using maize-wheat rotations. The researchers quantified the effect of using the pearl lupin in conventional cereal systems using a multi-season field trial method expected to evaluate the effect of the soil nitrogen level and crop rotation performance. Their quantitative findings showed that pearl lupin had the ability to fix 40–60 kg n ha^{-1} through biological fixation of nitrogen and enhanced yields of maize and wheat that followed after by 18–32% without applying synthetic nitrogen fertilizers, thus saving farmers a lot of money in terms of inputs. On the basis of these results, their recommendations focused on advancing pearl lupin as a high-protein dual-purpose crop (food and forage) with the power to increase the soil fertility, the reduction of reliance on the external fertilizer, and improvement of the sustainability and output of the cereal-based farming systems in the andean highlands.

20.9.3.2 Case Study 2: Pearl Lupin for Drought-Prone Areas in Africa

Their work is called evaluating the drought resilience and nitrogen-fixing benefits of pearl lupin in semi-arid farming systems of east Africa and was conducted in the field of semi-arid Ethiopia and Kenya where the lack of water is a critical issue in the productivity of smallholder crops. The researchers studied the adaptability of pearl lupin to low-moisture conditions using a methodology that integrated experimental trials involving both field-testing of low-moisture conditions using water-stress and soil nitrogen measurements and crop performance indices of the crop. Their quantitative findings indicated that pearl lupin had 70–85% of optimum yield during severe drought stress and yielded 30–50 kg n ha^{-1} by bio fixation of nitrogen in the low nitrogen soils, which led to increase in soil fertility and performance of the succeeding cereal crops. Because of these findings, their suggestions gave significance to the incorporation of pearl lupin into semi-arid smallholder agricultural systems to minimize susceptibility to drought, improve soil condition without the use of synthetic fertilizers, and bolster long-term food security of resource poor farmers in sub-Saharan Africa.

20.10 Conclusion

The Pearl lupin (*Lupinus mutabilis*) is a neglected but untapped crop with a tremendous potential in ensuring world food security. This is because of its special capacity of fixing nitrogen, its hardiness to the weather extremes like droughts and fros, making it a valuable crop in sustainable food farming. As the problem of climate change keeps disrupting the very nature of traditional farming methods, Pearl Lupin is more than just an ecologically viable product because of its flexibility and the potential to combat food production and soil health issues.

More studies and field tests (especially with the help of modeling applications such as DSSAT) should also be conducted to maximize the possibilities of the crop and to inform its implementation in various agro-ecological regions. In the world where agriculture is becoming more resilient to climate change, Pearl Lupin has the potential to become instrumental in alleviating the effects of climate change and enhancing soil quality and minimize the use of synthetic inputs.

Pearl Lupin in the edible form is not only an important agricultural crop but also a superfood with an outstanding nutritional composition. High protein content, combined with high levels of fiber, calcium, iron, and magnesium, makes it an interesting substitute in the development of sustainable nutrition in the diet of humans and animals. With the increasing demand of plant-based proteins, Pearl Lupin may be valuable in providing protein solutions to the world, especially in areas where food insecurity and malnutrition are most significant issues.

References

Ahmed, M., Bilal, M., & Ahmad, S. (2024). Simulation of source sink partitioning in wheat under varying nitrogen regimes using DSSAT-CERES-wheat model. *Agricultural Water Management, 303*, 109028.

Akchaya, K., Parasuraman, P., Pandian, K., Vijayakumar, S., Thirukumaran, K., Mustaffa, M. R. A. F., ... Choudhary, A. K. (2025). Boosting resource use efficiency, soil fertility, food security, ecosystem services, and climate resilience with legume intercropping: A review. *Frontiers in Sustainable Food Systems, 9*, 1527256.

Ali, S. A., Krishna, K., Nithinkumar, K., Kumari, M., Jena, T., Ezing, U. M., et al. (2025). Soil-centric cropping systems building climate resilience. In *Climate-smart agricultural technologies: Approaches for field crops production systems* (pp. 247–283). Springer.

Alitalo, J. (2025). Precrop effects of faba bean and pea on oat yield and nitrogen uptake in Nordic cropping systems.

Amin, R., Jan, S. S., Butt, M., Manan, A., Khan, M., Khan, R. A., ... Ali, S. (2025). Soil carbon sequestration through cover crops and reduced tillage: A climate-SMART agricultural strategy for mitigating climate change and improving soil health. *Kashf Journal of Multidisciplinary Research, 2*(04), 20–37.

Awan, S. A., Khan, I., Rizwan, M., Brestic, M., Wang, X., Zhang, X., & Huang, L. (2024). Exogenous melatonin regulates the expression pattern of antioxidant-responsive genes, antioxidant enzyme activities, and physio-chemical traits in pearl millet under drought stress. *Journal of Plant Growth Regulation, 43*(4), 1061-1075.

Bebeli, P. J., Lazaridi, E., Chatzigeorgiou, T., Suso, M. J., Hein, W., Alexopoulos, A. A., ... & Van den Berg, M. (2020). State and progress of andean lupin cultivation in Europe: A review. *Agronomy, 10*(7), 1038.

Buccioni, F., Serio, A., Chaves-Lopez, C., Angelino, D., Rossi, C., Maggio, F., ... & Paparella, A. (2026). Nutrients and toxic compounds in lupin seeds: Potential bioactivity and health implications. *Critical Reviews in Food Science and Nutrition,* 1-17.

Cheng, M., Wang, H., Fan, J., Zhang, S., Wang, Y., Li, Y., … Zhang, F. (2021). Water productivity and seed cotton yield in response to deficit irrigation: A global meta-analysis. *Agricultural Water Management*, *255*, 107027.

Dawud, H. (2023). Exploring and screening potential Rhizobial isolates nodulating forage legumes white lupin (Lupinus Albus) and tree Lucerne (Chamaecytisus Palmensis). In *Growing in Ethiopia*. Arba Minch University.

Dracup, M., & Kirby, E. M. (1996). Lupin: Development Guide.

Dutta, D., Meena, A. L., Bhanu, C., Ghasal, P., Choudhary, J., Kumar, S., … Nutrition, P. (2024). Sustainable soil management for climate resilience: Long-term management effects on soil carbon sequestration and nitrogen dynamics in a semi-arid tropical Inceptisol of India. *24*(3), 4407–4426.

Gabur, I., & Simioniuc, D. P. (2023). Pearl lupin (*Lupinus mutabilis*): A neglected high protein and oil content crop. In *Neglected and underutilized crops* (pp. 413–436). Elsevier.

Gidey, T., Berhe, D. H., Birhane, E., Gufi, Y. & Haileslassie, B. J. S. (2024). Intercropping maize with faba bean improves yield, income, and soil fertility in semiarid environment. *Scientifica, 2024*(1), 2552695.

Górski, R. & Płaza, A. (2024). The effects of intercropping narrowleaf lupine with cereals under variable mineral nitrogen fertilization. *Agriculture*, *14*(7), 989.

Govindasamy, P., Muthusamy, S. K., Bagavathiannan, M., Mowrer, J., Jagannadham, P. T. K., Maity, A., … TK, D. (2023). Nitrogen use efficiency—a key to enhance crop productivity under a changing climate. *Frontiers in Plant Science*, *14*, 1121073.

Gulisano, A., Alves, S., Rodriguez, D., Murillo, A., Van Dinter, B.-J., Torres, A. F., … Paulo, M.-J. (2022). Diversity and agronomic performance of lupinus mutabilis germplasm in European and Andean environments. *Frontiers in Plant Science*, *13*, 903661.

Jing, S. & Boelt, B. (2021). Seed production of red clover (Trifolium pratense L.) under Danish field conditions. *Agriculture*, *11*(12), 1289.

Kebede, E. (2021). Contribution, utilization, and improvement of legumes-driven biological nitrogen fixation in agricultural systems. *Frontiers in Sustainable Food Systems*, *5*, 767998.

Keefe, H. E. & Kharouba, H. M. (2024). Growing degree-days do not explain moth species' distributions at broad scales. *Ecosphere*, *15*(7), e4885.

Kushwah, N., Billore, V., Sharma, O. P., Singh, D. & Chauhan, A. P. S. (2024). Integrated Nutrient management for optimal plant health and crop yield. *Plant Science Archives*, *8*(3), 10–12.

Leite Dias, S., Garibay-Hernández, A., Brendel, F. L., Gabriel Chavez, B., Brückner, E., Mock, H.-P., … D'Auria, J. C. (2023). A new fluorescence detection method for tryptophan-and tyrosine-derived allelopathic compounds in barley and lupin. *Plants*, *12*(10), 1930.

Medeot, D. B., Paulucci, N. S., Albornoz, A. I., Fumero, M. V., Bueno, M. A., Garcia, M. B., ... & Dardanelli, M. S. (2010). Plant growth promoting rhizobacteria improving the legume–rhizobia symbiosis. In *Microbes for legume improvement* (pp. 473-494). Vienna: Springer Vienna.

Msaddak, A., Mars, M., Quiñones, M. A., Lucas, M. M., & Pueyo, J. J. (2023). Lupin, a unique legume that is nodulated by multiple microsymbionts: The role of horizontal gene transfer. *International Journal of Molecular Sciences, 24*(7), 6496. https://doi.org/10.3390/ijms24076496

Mugerwa, T. M., Formann, P., & Shapland, R. (2022). *How important are arbuscular mycorrhizal fungi to chickpea yield? Tamworth 2021*. Department of Primary Industries.

Pinzón-Sandoval, E. H., Carvalho Silva, W. & Marchiori, P. E. (2024). Phenology, mass accumulation patterns and growing degree days in common bean. *Revista de Ciencias Agrícolas, 41*(3).

del Pozo, A., & Mera, M. (2021). Lupin. In *Crop physiology case histories for major crops* (pp. 430–450). Elsevier.

Pueyo, J. J., Quiñones, M. A., Coba De La Peña, T., Fedorova, E. E., & Lucas, M. M. (2021). Nitrogen and phosphorus interplay in Lupin root nodules and cluster roots. *Frontiers in Plant Science, 12*, 644218. https://doi.org/10.3389/fpls.2021.644218

Reinsch, S., Robinson, D. A., van Soest, M. A., Keith, A. M., Parry, S. & Tye, A. M. (2024). Temperate soils exposed to drought—Key processes, impacts, indicators, and unknowns. *Land, 13*(11), 1759.

Salvagiotti, F., Cassman, K. G., Specht, J. E., Walters, D. T., Weiss, A. & Dobermann, A. (2008). Nitrogen uptake, fixation and response to fertilizer N in soybeans: A review. *Field Crops Research, 108*(1), 1–13.

Shen, S., Basist, A. & Howard, A. (2010). Structure of a digital agriculture system and agricultural risks due to climate changes. *Agriculture and Agricultural Science Procedia, 1*, 42–51.

Silifonov, T. (2023). Grain yield and protein content in different ripening varieties of soft winter wheat using various types and doses of fertilizers. *Plant Varieties Studying and Protection, 19*(1), 44–51.

Strozycki, P. M., Szczurek, A., Lotocka, B., Figlerowicz, M., & Legocki, A. B. (2007). Ferritins and nodulation in Lupinus luteus: Iron management in indeterminate type nodules. *Journal of Experimental Botany, 58*(12), 3145–3153. https://doi.org/10.1093/jxb/erm152

Sulas, L., Canu, S., Ledda, L., Carroni, A. M. & Salis, M. (2016). Yield and nitrogen fixation potential from white lupine grown in rainfed mediterranean environments. *Scientia Agricola, 73*(4), 338–346.

Suryavanshi, A., Sahu, M., Bhan, M., Shrivastava, R. & Dubey, M. (2021). Response of varieties and sowing dates on water productivity of chickpea in kymore plateau and satpura hills agro-climatic zone of MP. *International Journal of Current Microbiology and Applied Sciences, 10*(01), 2283–2293.

Wang, Y., Li, M., Guo, J. & Yan, H. (2023). Alfalfa (Medicago sativa L.) nitrogen utilization, yield and quality respond to nitrogen application level with center pivot fertigation system. *Agronomy, 14*(1), 48.

Zhao, J., Yang, J., Xie, H., Qin, X. & Huang, R. (2024). Sustainable management strategies for balancing crop yield, water use efficiency and greenhouse gas emissions. *Agricultural Systems, 217*, 103944.

Concluding Chapter: Neglected and Underutilized Crops for Climate-Resilient Food Systems

Climate change is transforming the world agriculture. Increasing temperatures, unpredictable rain, droughts, floods and soil erosion are posing more threats to food production and sustenance. Smallholder farmers and vulnerable communities are some of the ones who are disproportionately impacted by these pressures. The existing food systems which are dominated by a small circle of crops that are highly inputted are becoming more fragile.

This book has shown that there are viable and viable ways to use neglected and underutilized crops (NUCs) to provide a route of adaptation and sustainability. These crops are highly adaptable to climatic stress in varying agroecological settings. They can grow in a restricted supply of water, low quality of soil and extreme temperatures. They are also high in nutritional value and have less externalities.

The chapters in totality demonstrate the fact that crops like millets, quinoa, amaranth, teff, fonio, legumes, and pseudo cereals are able to enhance resilience at the farm and landscape levels. Their incorporation in crop farming systems lowers climate variability risk. It improves biology of soil, biodiversity and ecosystem. It helps promote dieting and better nutrition as well.

Diversification comes out as one of the key strategies. Depending on few staple crops makes people prone. Conversely, shocks in diversified systems in NUCs buffer against shocks. They stabilize incomes. They enhance the use of water and nutrients. A few chapters show how NUCs of intercropping, agroforestry, and conservation agriculture improves the performance of the system in the climate stresses.

The mitigation capability of NUC-based systems is also featured in this volume. Reduced fertilizer and irrigation levels lead to the reduction of green house gases. Leguminous NUCs boost the biological nitrogen fixation. Millet and pseudocereals help in soil carbon sequestration when they are reared under low tillage and residue retention. Certain crops also have an additional connection between food production and renewable energy and circular bioeconomy pathways.

Agroecology offers a comprehensive set of principles. Agroecological approaches combine ecological processes, traditional knowledge and modern science when

M. Ahmed (ed.), *Climate Change and Neglected Underutilized Crops*,
https://doi.org/10.1007/978-3-032-21005-0

combined with NUCs. This integration reduces the environmental footprints without reducing the productivity. The indigenous knowledge systems recorded in various regions continue to be very important towards effective adaptation and localized acceptance.

In spite of their potential, NUCs are still crippled by structural constraints. Poor access to market, inadequate access to policy, poor research investment, and weak seed systems remain barriers to adoption. The issue of climate change increases these challenges. They need multi-stakeholder efforts in order to address them. The development of research, extension, digital advisory tools, value chain development and farmer led innovation should proceed in tandem.

Policy support is essential. These chapters continuously demand the incorporation of NUCs in national climate adaptation strategies, food and nutrition strategies, and agricultural investment systems. There is a need to have supportive policies, access to credit, insurance products and develop infrastructure to minimize the risk faced by farmers and to facilitate scaling. The issue of gender equity and community involvement should always be at the forefront since women and aboriginals are vital players in the NUC systems.

Food systems at a global scale are both sources of and victims of climate change. Their contribution to greenhouse gas emissions, loss of biodiversity and usage of freshwater is high. Meanwhile, food insecurity and malnutrition are still prevalent. In this respect, NUCs will provide the realistic possibility to diversify the production and decrease environmental pressure.

Finally, the book has placed neglected and underexploited crops as a strategic resource towards climate-resilient farming. They are not an artifact of the old. They are crops for the future. NUCs should be mainstreamed to favour adaptation, mitigation, nutrition, and sustainability at the same time. The facts provided herein support the point that it is time to go beyond individual solutions and to diversified, interconnected, and resilient food systems.

This book will inform research, policy and assist practice. It demands new consideration to crops that nourished communities long when conditions were unfriendly. Agriculture can address the challenge of a changing climate, and food security can be preserved by the future generation by removing them of their marginal position in the science, policy, and farming systems.

Editor
Mukhtar Ahmed

The manufacturer's authorised representative in the EU is Springer Nature Customer Service Centre GmbH, Europaplatz 3, 69115 Heidelberg, Germany. If you have any concerns regarding our products, please contact ProductSafety@springernature.com

Printed and bound by CPI Group (UK) Ltd, Croydon, CR0 4YY
05/10/2026
02228057-0015